# Functional Nanostructures

Processing, Characterization,
and Applications

T0185299

# Nanostructure Science and Technology

Series Editor: David J. Lockwood, FRSC
*National Research Council of Canada*
*Ottawa, Ontario, Canada*

*Current volumes in this series:*

Functional Nanostructures: Processing, Characterization, and Applications
*Edited by Sudipta Seal*

Nanotechnology in Catalysis, Volume 3
*Edited by Bing Zhou, Scott Han, Robert Raja, and Gabor A. Somorjai*

Controlled Synthesis of Nanoparticles in Microheterogeneous Systems
*Vincenzo Turco Liveri*

Nanoscale Assembly Techniques
*Edited by Wilhelm T.S. Huck*

Ordered Porous Nanostructures and Applications
*Edited by Ralf B. Wehrspohn*

Surface Effects in Magnetic Nanoparticles
*Dino Fiorani*

Alternative Lithography: Unleashing the Potentials of Nanotechnology
*Edited by Clivia M. Sotomayor Torres*

Interfacial Nanochemistry: Molecular Science and Engineering
at Liquid-Liquid Interfaces
*Edited by Hitoshi Watarai*

Nanoparticles: Building Blocks for Nanotechnology
*Edited by Vincent Rotello*

Nanostructured Catalysts
*Edited by Susannah L. Scott, Cathleen M. Crudden, and Christopher W. Jones*

Self-Assembled Nanostructures
*Jin Z. Zhang, Zhong-lin Wang, Jun Liu, Shaowei Chen, and Gang-yu Liu*

Semiconductor Nanocrystals: From Basic Principles to Applications
*Edited by Alexander L. Efros, David J. Lockwood, and Leonid Tsybeskov*

**Sudipta Seal**
**Editor**

# Functional Nanostructures

## Processing, Characterization, and Applications

 Springer

*Editor*
Sudipta Seal
Advanced Materials Processing and Analysis Center
University of Central Florida
4000 Central Florida Blvd.
Orlando, Florida 32816-2455

ISBN: 978-1-4419-2255-7        e-ISBN: 978-0-387-48805-9

Printed on acid-free paper.

9 8 7 6 5 4 3 2 1

springer.com

To my wife Shanta and daughter Anouska
My parents: father Prof. Bijoy and mother Mary
My brother and his wife: Pradipta and Gayatri
My in-laws: Subroto and Alokanda

# Contents

## 3. MEMS for Nanotechnology: Top-down Perspective ..........  107

*Ghanashyam Londe, Arum Han, Hyoung J. Cho*

## 6. Nanostructures: Sensor and Catalytic Properties ............ 305

*B. Roldan Cuenya, A. Kolmakov*

## 7. Nanostructured High-Anisotropy Materials for High-Density Magnetic Recording ........................ 345

*J. S. Chen, C. J. Sun, G. M. Chow*

8. **High-Resolution Transmission Electron Microscopy
for Nanocharacterization .............................. 414**

*Helge Heinrich*

## 9. Applications of Atomic Force Microscope (AFM) in the Field of Nanomaterials and Nanocomposites ......... 504

*S. Bandyopadhyay, S. K. Samudrala, A. K. Bhowmick, and S. K. Gupta*

# Preface

Functional nanostructures are materials considered to work at the molecular level comprised of nanometer scale components arranged in three-dimensions. These materials can be in the form of nanofilms, nanotubes, nanodots or nanowires, whose spatial arrangement and other attributes like size and shape can lead to various applications. While synthesis and fabrication of these nanostructures are always a challenge, however, tailoring these nanostructures can find applications biology, materials science, gas sensors, potential memory storage devices, surface plasmon waveguides, nanocatalysts, spatially ordered metal nanocatalyst seeds for high-efficiency flat-panel displays from carbon-nanotube arrays. Control fabrication to achieve unique spatial arrangement of the nanofeatures, control of its size, shape, composition, and crystallographic orientation is important and can be achieved by various synthesis routes. This book describes some of the aspects of functional nanostructures, from fabrication, characterization to novel applications for the 21$^{st}$ century nanotechnology boom.

The article by Ram et al., is primarily focused on the understanding of synthesis, fabrication, and control of the electronic structure and derived magnetic, GMR, and optical properties of a specific class of ferromagnetic ceramics and hybrid composites in forms of films, nanowires, and fine powders. Such materials, also called "half-metallic compounds" represent a new generation of spintronics, photonics, kinds of sensors, and nonlinear optical devices. In such systems, the electrical conduction is carried out exclusively by charge carriers of one spin direction, i.e., the conduction electrons are ~100% spin polarized. This distinctive property of spin polarization of charge carriers in such materials makes them the most appropriate to be utilized not only in magnetic tunnel junctions and spin polarizers for current injection into semiconductors, which are the constituent components of non-volatile magnetic random access memories and other spin-dependent devices, but also to increase the efficiency of optoelectronic devices and even in self-assembled quantum computers. Typical magnetic, GMR, optical, and other properties are reviewed and presented with identified mechanisms and models on case to case basis with selective examples. Finally, toxicities and hazards, which may be encountered during processing or handling such materials, are pointed out.

In the 2$^{nd}$ chapter, various processing techniques to create nanostructures are reviewed with an emphasis on self-assembly approaches. This section further discusses two important computer image analysis techniques that are important for characterization of nanoscale features like spatial order (Fourier analysis) and morphology (Minkowski analysis). The chapter concludes with examples of some important applications requiring ordered surface nanostructures.

The 3$^{rd}$ chapter deals with MEMS research, that has started from silicon microfabrication, emerged as novel methodology for the miniaturization and integration of sensors and actuators, and started providing top-down, cross-disciplinary solutions to nanoscale fabrication and characterization. The implementations and applications of those were found in the sensors and actuators using nanoparticles, nanobeams, nanoprobes, nanopores, nanogaps, nanochannels, nanotubes and nanowires are discussed.

A recent trend is shifting towards nanotechnology looking for improvement in the biological responses and overall life of medical implants, made of novel biomedical materials called 'biomaterials' is presented in Chapter 4. Designing of biomaterials at the nano scale level is critical for control of cell-biomaterials interactions, which can be used beneficially in developing implants with desired characteristic properties. Biological cells have dimensions within the range of 1–10 µm and contain many examples of extremely complex nano-assemblies; including molecular motors, which are complexes embedded within membranes that are powered by natural biochemical processes. Various examples of nano scale structures, nano motors and nano systems are present in abundance and *in vivo* in different biological systems. This chapter presents some of the current advances in the development of nanostructured metallic, ceramic, polymeric and composite biomaterials. In addition, the chapter also discusses various types of traditional biomaterials with their properties and applications to help new scholars in this field to gather solid background knowledge on biomaterials, without having to refer to other texts.

Chapter 5 describes self-assembly processes in polymeric systems. Self-assembly (SA) involves recognition or binding processes and the individual component contains enough information to recognize and interact with other appropriate unit to build a template for a structure composed of multiple units. The most widely studied aspect of self-assembly is molecular self-assembly and the molecular structure determines the structure of the assembly. Molecular self-assembly is also ubiquitous in chemistry and material science, the formation of molecular crystals, liquid crystals, semicrystalline polymers, colloids, phase separated polymers, and self-assembled monolayers are unique examples, which are reviewed in detail in this chapter.

Chapter 6 provides some examples of new experimental trends in the fields of catalysis and chemical sensing. In particular, sensors based on chemically-induced electronic phenomena on nanometer-scale metal films deposited on semiconductor junctions (MS and MOS devices), quasi-1D oxide chemiresistors and chemi-FET devices and chemically-tunable magnetic nanostructures have been discussed.

It has been shown, that nanometer-thick MS and MOS devices can be used as highly selective gas sensors. The main limitation of these devices is that they require low temperature (125–200 K) operation in order to minimize thermal noise. After less than a decade of development, chemiresistors and chemi-FET nanosensors made of quasi-1D oxides have demonstrated superior performance over conventional bulk-sensors and are currently considered the most promising sensing elements for the next generation of solid state gas sensors. There are still plenty of unexplored lines of fundamental research, particularly in the size range where quantum confinement dominates the electronic structure of nano-objects, which is reviewed herein. Furthermore, there exists a strong necessity of further developing new spectroscopic and nondestructive microscopic techniques to probe surface properties of individual nanostructures, in particular, the ones that can be operated under realistic sensing conditions. A brief overview is presented in this section.

Chapter 7 presents a brief review of the demand for the development of data storage technology to provide storage media with increased storage density. To achieve high density a reduction in the grain size is required so as to enable more grains per unit area on the magnetic media. In this chapter, the structural characterization and some intrinsic properties of $L1_0$FePt such as the origin of high anisotropy and finite size and interface effects were reviewed. However, practical applications of FePt films as recording media face many technical challenges including desirable reduction of ordering temperature, control of the FePt (001) texture and decrease of media noise. Therefore, more efforts were made to review the current status on solving above problems.

While processing and properties are important in nanostructures, detailed characterization is needed to understand the science behind nanstructures. Properties of nanoscaled materials are influenced by the size, distribution and orientation of objects, but they also depend on the internal structures and chemistries of these individual particles, precipitates or domains. Environmental effects like temperature, mechanical stresses, radiation damage and corrosion may reduce the long-term stability of devices or materials based on nanomaterials. Next two sections in the book deals with Transmission Electron Microscopy and Atomic Force Microscopy, one of the few essential analytical tools in nanocharacterization.

Chapter 8 reviews the role of Transmission electron microscopy for the study of internal structures requires special sample preparation, while scanning probe techniques and scanning electron microscopy for surface characterization are essentially non-destructive techniques. The most important transmission electron microscopy methods for nanomaterials are reviewed. Bright-field and dark-field imaging is employed for the analysis of the sizes and distribution of particles and for the identification of structural defects. High-resolution micrographs yield valuable data on interfaces and defects in crystallites. Diffraction techniques are used to identify local crystal structures. Analytical methods like energy dispersive X–ray analysis and electron energy-loss spectroscopy as well as high-angle annular dark-field scanning transmission electron microscopy and holography are important tools in qualitative and quantitative studies of the local chemistry. Transmission electron microscopy is employed for feedback on processing of nanomaterials and helps in the identification of methods to improve reliability and reproducibility of the nanomaterials properties. New developments in electron optics soon become available to an extended range of users are discussed. The prospects of these improved analytical methods for research on nanostructures are outlined.

It is well known based on the research in the field of nanotechnology that the full exploitation of fundamental characteristics of nano-reinforcements and development of nanostructures would facilitate the achievement of enhanced properties (for example, physical, mechanical, optical, magnetic, electric, and rheological properties) in these materials. So, an accurate characterization of these materials at an atomic level is needed that requires a precise understanding and measurement of the surface and interfacial phenomena along with the other factors like size of the nanoparticles, dispersion, crystallinity, and so forth. Despite the availability of different advanced characterization techniques for these purposes, AFM provides a better alternative due to the fact that it can be operated in most of the environments with excellent resolution, provides three-dimensional qualitative and quantitative information, and also used to characterize nano-mechanical and nanotribological properties of different materials. In this chapter 9, the authors review the recent progress and the fundamental aspects on the application of AFM to characterize nanomaterials and nanocomposites.

# Advanced Ceramics and Nanocomposites of Half-metallic Ferromagnetic CrO₂ for Magnetic, GMR and Optical Sensors

## S. Ram,[1,3] S. Biswas,[1] and H. J-Fecht[2]

[1]Materials Science Centre, Indian Institute of Technology, Kharagpur 721302, India.

[2]Division of Electronic Materials, University of Ulm, Albert Einstein Allee 47, Ulm, D-89061, Germany.

[3]To whom correspondence should be addressed. Prof. S. Ram, Materials Science Centre, Indian Institute of Technology, Kharagpur 721302, India. E-mail: sram@matsc.iitkgp.ernet.in

## 1. INTRODUCTION

### 1.1. Definition of Half-metals and Half-metallic Compounds

The physical properties of metal oxides are diverse, including semiconductors or insulators (ZnO, ZrO₂, TiO₂), good metals (RuO₂), and metal-insulator systems (VO₂) in terms of the electron band structure and variation of electrical resistivity ($\sigma$) as a function of temperature (T), i.e., thermal coefficient of $\sigma$, expressed as $\sigma^{-1}(\partial\sigma/\partial T)$. In terms of distributing the magnetic spins (of total value $S$) in the metal cations via $O^{2-}$ anions, a metal oxide behaves to be paramagnetic (Cr₂O₃), $S \neq 0$, or diamagnetic (YBa₂Cu₃O₇ and other ceramic superconductors[1-5]) with $S = 0$.[6-8] Depending on the relative strength of the spin-ordering over thermal effects, a paramagnetic oxide often behaves as a ferrimagnet ($\gamma$-Fe₂O₃ or Fe₃O₄) or antiferromagnet (FeO or MnO₂). Metallic CrO₂, with a Curie temperature $T_C = 390$ K, is the only ferromagnet in this class. Schwarz[9] used local-spin-density-approximation (LSDA) band theory to predict that the $S$-moment would be the full $2\mu_B$ required by Hund's rules for $Cr^{4+}$ (3d²) state in CrO₂. The Fermi level $E_F$ lies in a partly filled (metallic) band for the majority (up-spin) electrons, but for minority (down) spins lies in a semiconductor like energy bandgap

(1.5 eV) $E_g$ around the $E_F$-level, separating the filled $O^{2-}2p^4$-levels from the $Cr^{4+}$ $3d^2$-levels. Such a specific class of compound in which one spin species is metallic and the other is semiconducting is called a "half-metallic" compound. It includes Heusler alloys[10-12] or other dilute magnetic semiconductors.[13-17] All of them have ferromagnetic structure, with useful magnetic properties, very similar to rather conventional ferromagnets, e.g., Fe, Co, Ni, or similar metals,[18-23] and metal alloys,[24-30] or intermetallics.[31-40]

This specific class of half-metallic compounds has received renewed research interest nowadays due to promising potential for several applications, especially in the emerging fields of spintronics or photonics.[11,41-50] In such systems, the electrical conduction is carried out exclusively by charge carriers of one spin direction, i.e., the conduction electrons are 100% spin polarized (ideally).[47,51] This distinctive property of spin polarization of charge carriers in half-metals makes them the most apposite to be utilized not only in magnetic tunnel junctions and spin polarizers for current injection into semiconductors, which are the constituent components of nonvolatile magnetic random access memories and other spin-dependent devices, but also to increase the efficiency of optoelectronic devices and even in self-assembled quantum computers.[50,52-54]

In general, the half-metallic systems are referred as "half-metallic ferrromagnets (HMF)." Note the classical ferrromagnets, e.g., Fe, Co, or Ni metals, and derived intermetallics or alloys in selected compositions (i.e., the metal conductors) are not half-metals, although they have fully spin polarized $d$ bands with totally filled ↑ $3d$ band and only ↓ $d$ electrons at $E_F$.[11,48,55] The basic difference lies in the fact that the $E_F$ also crosses the unpolarized $4s$ band. As a result, there is plenty of both kinds of ↑ and ↓ electrons. Consequently, in order to obtain only ↑ or ↓ electrons at $E_F$, hybridization is necessary to reorder $3d$ and $4s$ bands, either by pushing the bottom of the $4s$ band up above the $E_F$ value or depressing the $E_F$ value in the $3d$ band below the bottom of the $4s$ band. This had been addressed in several examples in terms of experimental studies of the magnetic and electrical properties under selective conditions.[48,51,55]

All known HMF consist of more than one element. To be more specific, familiar examples include semi-Heusler alloys,[11,47,50,56] magnetite $Fe_3O_4$,[57-60] pyrite alloys,[43,47,60] $CrO_2$,[42,48] $LaMnO_3$ type of manganites and derivatives,[51,56,61] or in general perovskites of chemical formula $ABO_3^{49,53,62}$ (usually, A and B to be the divalent and tetravalent transition metal cations), full Heusler alloys,[46,52,54,63] semimetals,[64,65] pnictides and chalcogenides,[66] vacancy doped oxides,[45,67] diluted magnetic semiconductors (also called semimagnetic semiconductors[68]), etc. Most of them are stoichiometric compounds while others are solid solutions. Moreover, multiferroics, i.e., doped ferroelectric ceramics with a ferromagnetic ingredient, or simply a hybrid composite involving both ferroelectric and ferromagnetic components,[69-73] or ferromagnetic cermets of a robust shell[74-78] represent a kind of rather newer series of artificial HMF compounds. Table 1 summarizes representative examples of selective HMF materials along with the typical physical properties.

**TABLE 1.** Important series of half-metallic materials with selected examples and properties

| Half-metal types | Structure | Example | $T_C$ (K) | $\mu^S$-value ($\mu_B$/f.u.) | References |
|---|---|---|---|---|---|
| Semi-Heusler compounds | Ternary transition-metal intermetallic compounds (XYZ) | NiMnSb | 730 | 4 | 11, 47, 48, 50, 56 |
| | | PtMnSb | — | 4 | 11, 50 |
| | | NiCrP | — | 3 | 50 |
| | | NiCrSe, NiCrTe (Nearly HM: NiCrAs, NiCrSb, NiCrS) | — | 4 | 50 |
| Binary oxides | Rutile type | $CrO_2$ | 390-396 | 2 | 42, 48, 79 |
| Pyrite alloys | $Fe_{(1-x)}Co_x S_2$ $(0.25 \leq x \leq 0.9)$ | — | — | 1 per Co atom | 43, 47, 60 |
| Magnetite ($Fe_3O_4$) | Inverse spinel cubic (above Verwey transition temperature, $T_V = 122$ K; Monoclinic (below $T_V$) | — | 860 (Highest) | 4 or 2 (depending on the type of charge order on the octahedral sites). | 47, 48, 57-60 |
| Double perovskites | $A_2BB'O_6$ A = alkaline earth elements (Ca, Sr, Ba) B = 3d transition metal (Fe, Co) B' = 4d transition metal (Mo, W, Re) | $Sr_2FeMoO_6$ | 410-450 | 4 | 48, 49, 62 |
| | | $Sr_2FeReO_6$ | — | 3 | 49, 53 |
| | | $Ca_2FeMoO_6$ | — | 4 | 49 |
| | | $Ba_2FeMoO_6$ | — | 4 | 49 |
| | | $Ca_2FeReO_6$ | — | 3 | 49 |
| Mixed valence manganites | $La_{(1-x)}A_xMnO_3$ A = Ca, Ba, Sr | $La_{0.7}Sr_{0.3}MnO_3$ (transport half metal) | 350 | — | 48,51 |
| | | $La_{2/3}Sr_{1/3}MnO_3$ | 370 | — | 55, 62 |
| | | $La_{0.67}Ca_{0.33}MnO_3$ | — | — | 48 |
| | | $La_{0.3}Sr_{0.7}MnO_3$ | — | — | 61 |
| Heusler alloys | $X_2YZ$ X & Y = transition-metal elements (group 1B & VIIIB) Z = group III, IV or V element Both Heusler and Semi-Heusler crystal structure type are described by means of four interepenetrating fcc lattices. | $Co_2MnSi$ | 985 | 5 | 46, 52, 54 |
| | | $Co_2MnGe$ | 905 | 5 | 46, 52, 54 |
| | | $Co_2MnGa$ | — | 4 | 46 |
| | | $Co_2MnSn$ | 829 | 5 | 46, 52 |
| | | $Co_2MnAl$ | — | 4 | 46, 54 |
| | | $Co_2CrAl$ | — | 3 | 46, 54 |
| | | $Mn_2VGe$ | — | 1 | 46 |
| | | $Mn_2VAl$ | 760 | 2 | 46, 47, 48, 63 |
| | | $Fe_2MnSi$ | — | 3 | 46 |

**TABLE 1.** (*Continued*)

| Half-metal types | Structure | Example | $T_C$ (K) | $\mu^S$-value ($\mu_B$/f.u.) | References |
|---|---|---|---|---|---|
| Semi-metal pnictides and chalcogenides | — Compounds of transition metal elements of group V & VI atoms with zinc-blende structure. | $Tl_2Mn_2O_7$ VAs, VSb CrAs CrSb CrSe, CrTe | 120 — Over 400 Over 400 — | 6 — — — — | 48, 62, 64, 65 66 50, 66 50, 66 66 |
| Vacancy-doped transition metal monoxides | Point defect induced half-metallicity. | NiO (3% Ni vacancy) MnO (3% Mn vacancy) CaO (3% Ca vacancy) $Zn_xCo_{1-x}O_\delta, \delta \leq 1$ | — — — — | 0 3 — — | 45, 67 — 45 80-84 |
| Diluted magnetic semiconductors | — | (InMn)As (GaMn)As $Cd_xMn_{1-x}S$ | — — — | — — — | 68 68 85-88 |
| Ferromagnetic semiconductors | — | EuO & EuS with rare-earth $R^{3+}$ doping | 69 (EuO) 16 (EuS) | — — | 48 48 |
| Ferromagnetic cermets. | — | $Co:Al_2O_3$ | 1396 | — | 74–78 |
| Nitrides | $R_2M_{17}N_\delta, \delta \leq 3$ M = transition-metal elements. | $Sm_2Fe_{17}N_{2.6}$ $Li_3FeN_2$ | 743 10 | — 1 | 12, 89-92 47 |
| Half-metals with nonmagnetic atoms | — | $La_{0.005}Ca_{0.995}B_6$ Ferromagnetic carbon | — — | — — | 48 48 |
| Multiferroics (composites) | — | $BiFeO_3$, $BiMnO_3$, laminated composites of $Pb(Zr_{0.53} Ti_{0.47})O_3/$ $NiFe_2O_4$, $Pb(Fe_{0.5}Nb_{0.5})O_3$ | — | — | 69-73 |

In 1983, de Groot et al.[11] first introduced the concept of half-metallic ferro-magnetism by the electronic band structure calculations of Mn based semi-Heusler alloys. Coey et al.[48] proposed a classification scheme of HMF materials in terms of the electronic band structure together with localized and itinerant electron systems in addition to the semimetals and semiconductors. According to this classifica-tion, $CrO_2$ is a type $I_A$ half-metal having only ↑ electrons of $Cr^{4+}$ ($t_{2g}$) character at $E_F$ while the double perovskite $Sr_2FeMoO_6$ is a type $I_B$ half-metal with no ↑ electrons at $E_F$ but only ↓ conduction electrons of strongly hybridized $Mo^{3+}$ ($t_{2g}$), $Fe^{3+}$ ($t_{2g}$) and $O^{2-}$ ($2p$) character. On the other hand, $Fe_3O_4$ is a type $II_B$ half-metal, where ↓ electrons of $Fe^{3+}$ ($t_{2g}$) character form small polarons and hop among the B sites in the inverse spinel cubic structure. An interesting observation about type I or II half-metals is the spin magnetic moment $\mu^S$ (at 0 K) is precisely an integral $\mu_B$ multiple per formula unit (f.u.), e.g., $CrO_2$ ($2\mu_B$), $Sr_2FeMoO_6$ ($4\mu_B$) or $Mn_2VAl$ ($2\mu_B$).[48,49,62,63] The integer $\mu^S$-value criterion, or an extension of it to include other types of half-metals, is a necessary but not a sufficient condition for the half-metallicity. Spin-orbit coupling can disturb the half-metallicity.[48,93]

## 1.2. Spin Polarization

The spin polarization in HMF compounds is expressed as[48]

$$P_o = \frac{N^\uparrow - N^\downarrow}{N^\uparrow + N^\downarrow} \tag{1}$$

where $N^{\uparrow,\downarrow}$ are the densities of states at the $E_F$ level. In experiments involving ballistic or diffusive transport, the densities of states are generally weighted by the Fermi velocity $v$ (or its powers $v^n$), and the spin polarization is expressed as[48,51]

$$P_n = \left( \frac{< N^\uparrow v^{\uparrow n} > - < N^\downarrow v^{\downarrow n} >}{< N^\uparrow v^{\uparrow n} > + < N^\downarrow v^{\downarrow n} >} \right) \tag{2}$$

To date, several techniques have been developed to determine the value of $P_n$ precisely. This involves measuring the values of the spin-polarized photoemission, Andreev reflection and vacuum tunneling.[42,47,48,51] In an ideal case, if spin-orbit interactions are neglected, $P_n$ can assume a value to be as much as 100% at $T = 0$ K. The $P_n$ values observed in important half-metallic compounds are given in Table 2. As can be seen from the data in this table, the two series of the compounds $La_{0.7}Sr_{0.3}MnO_3$ and $CrO_2$ and derivatives show a maximum $P_n$ value, i.e., $\sim 100\%$, at low temperatures as $\sim 5$ K. In the following sections, we will limit ourselves to address mainly $CrO_2$ ceramics and hybrid composites in forms of thin films, powder compacts, or other useful shapes in rather detail in aspects of one of the futuristic materials for spintronics, photonics, different kinds of sensors, and several other possible applications. This is chosen due to technical reasons as follows.

A specific advantage with $CrO_2$ over other HMF compounds is that it offers excellent magnetic, GMR (giant magnetoresistance), as well as optical properties, which can be utilized at the same time with a single material in a specific device. It is an ideal material to develop the basic understanding of magnetic, GMR, optical, and other properties in such specific class of value added strategic materials for a new generation of spintronics, photonics or nonlinear optical devices. One can also develop magnetocaloric properties. The $CrO_2$ magnets have long been of use as a particulate magnetic recording medium and as a detector in image analysis in scanning electron microscope. As described in original research papers[9,41,94-100] the GMR, with excellent $P_n = 100\%$ values, and optical properties evolve in value added applications for spin-polarized electron injectors, spin valves, optical data storage systems and other magneto-electronic and magneto-optical devices and components.

**TABLE 2.** The values of spin polarization $P_n$ in some half-metallic compounds with selected experimental methods of the measurements

| Half-metals | $P_n$ values (%) | Methods | References |
|---|---|---|---|
| $CrO_2$ | $P_0 = P_1 = P_2 = 100$ | Spin polarized photoemission and vacuum tunneling. | 42, 48 |
| $Fe_3O_4$ | $P_0 = P_1 = P_2 = 80$ | Spin-polarized photoemission and vacuum tunneling. | 57-60 |
| NiMnSb | $P = 58$ | Andreev reflection | 47 |
|  | 28 | Spin-polarized tunneling |  |
|  | 100 | Normal incidence inverse photoemission |  |
|  | 40 | Normal photoemission. |  |
| $La_{0.7}Ca_{0.3}MnO_3$ | $P_0 = 36$ | — | 48, 51 |
|  | $P_1 = 76$ |  |  |
|  | $P_2 = 92$ |  |  |
| $La_{0.7}Sr_{0.3}MnO_3$ | $P = 78$ | Andreev reflection | 47, 51, 55 |
|  | 100 | Spin-resolved photo emission |  |
|  | 72 | Superconductor tunneling |  |
| $Tl_2Mn_2O_7$ | $P_0 = 66$ | — | 48, 64 |
|  | $P_1 = -5$ |  |  |
|  | $P_2 = -71$ |  |  |

## 2. CHROMIUM DIOXIDE CERAMICS AND NANOCOMPOSITES

### 2.1. Crystal Structure

The $CrO_2$ has a Rutile type of tetragonal crystal structure, of D $P4_2$ mnm space group, with $z = 2$ formula units per unit cell.[9,96] The distribution of the $Cr^{4+}$ cations in $2a$ sites 0,0,0; $^1/_2$, $^1/_2$, $^1/_2$ and $O^{2-}$ anions in $4f$ sites $\pm x, \pm x, 0; ^1/_2 \pm x, ^1/_2 \pm x, ^1/_2$, where $x = 0.302$, in the $CrO_2$ crystal unit cell is shown in Fig. 1. Local axes are defined with $x$ and $y$ towards the edge–sharing $O^{2-}$ anions, and $z$ towards the apical $O^{2-}$ anions. In this crystal structure, the $Cr^{4+}$ cations form a body-centered tetragonal lattice, with a slightly distorted octahedral array of $O^{2-}$ anions such that each of the $O^{2-}$ anions has three nearest $Cr^{4+}$ neighbors. Each $Cr^{4+}$ coordinates six $O^{2-}$ anions in an oxygen octahedron $CrO_6$, with $Cr^{4+}$ as the center in the octahedral site, with two short apical bonds (0.1890 nm) and four longer equatorial bonds (0.1910 nm). The space lattice of infinite size has the lattice parameters $a = 0.4421$ nm and $c = 0.2916$ nm, aspect ratio c $= a \cong 0.6596$,

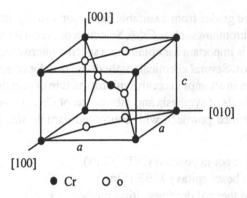

**FIGURE 1.** The Rutile crystal structure of ferromagnetic $CrO_2$.

with a lattice volume $V_0 \equiv a^2c = 0.05699$ nm$^3$ and density $\rho = 4.91$ g/cm$^3$.[101] The octahedra surrounding $Cr^{4+}$ at the body's center and corner positions differ by a 90° rotation about the $z$-axis, causing the nonsymmorphic space group.[9,96] The $CrO_6$ octahedra sharing a common edge form ribbons (planes) parallel to the crystallographic $c$-axis.

The lattice distortion increases further in $CrO_2$ crystallites of finite space lattice of a few nanometers.[79] X-ray powder diffraction has modified intensity distribution (as well as the average interplanar spacings $d_{hkl}$) in characteristic peaks in reflections form the different lattice planes (hkl). This occurs by modification in morphology and surface topology (which accompanies partial transformation $Cr^{4+} \rightarrow Cr^{3+}$ in a thin $Cr_2O_3$ surface layer of robust shell[79,102,103]) in such small $CrO_2$ particles. At this stage, we cannot ignore change in the $CrO_6$ polyhedral structure. For example, as small $CrO_2$ crystallites as 35 nm have $a = 0.4250$ nm and $c = 0.3190$ nm, with $c/a \cong 0.7506$, which involve a manifested $V_0 \equiv 0.0576$ nm$^3$ value, or a diminished $\rho = 4.86$ g/cm$^3$ value,[79] in comparison to the bulk equilibrium values mentioned above. This is according to the thermodynamics that small particles occur in high-energy states in support of a manifested value of the total surface energy (and in turn the internal energy and/or volume[76,104,105]). Technically, such kind of surface modification induces surface hardening with improved $CrO_2$ phase stability. Useful counterpart improvements occur in GMR, optical, magnetic, and other surface-sensitive physical parameters and in turn derived properties.

## 2.2. Methods of Synthesis

In ambient air atmosphere, virgin $CrO_2$ as such is not very stable. At atmospheric pressure, it degrades into $Cr_2O_3$ at the surface at an early temperature as 200°C.[94,97] As a result, very special methods need to be employed under controlled conditions in order to obtain substantially stable $CrO_2$ ceramics with stable chemical composition and properties. The degradation can be controlled under high oxygen pressures (>10 bar).[106,107] This pressure dependence $CrO_2$ stability could offer specific synthesis routes to grow good quality thin films, single crystals, or fine powders of selective grades from a suitable precursor, usually, involving a higher oxidation state of chromium such as $Cr^{6+}$. Selection of precursor with other experimental conditions is important for obtaining a specific microstructure of $CrO_2$ for possible applications. Several chemical methods are developed involving kinds of chemical precursors in attempt to control microstructure of functionalised properties. Identified methods of synthesis and fabrication of $CrO_2$ in form of thin films, bulk crystals, or refined powders, with controlled particle size to a nanometric scale, are as follows.

1. Chemical vapor deposition CVD (films),
2. Molecular beam epitaxy MBE (films),
3. Controlled thermal decomposition method,
4. Controlled hydrothermal decomposition method,
5. Controlled oxidation method,
6. Laser-induced transformation
7. Mechanical attrition in a closed reactor,
8. Chemical co-reduction method, and
9. Sol-gel type polymer precursor method.

Note the structural defects, which lead to deteriorate very sensitively the saturation magnetization, $M_s$ and other useful properties, are not easy to heal in a permanent basis by annealing in $O_2$ gas.[94,108] As analyzed with transmission electron microscopy and X-ray photoelectron spectrum, even a bulk $CrO_2$ crystal accompanies a surface layer of $Cr_2O_3$.[41] Such $Cr_2O_3$ surface layer, if not stabilized and controlled, propagates and destabilizes the $CrO_2$ phase. However, a dense and uniform stable $CrO_2$ surface layer of $Cr_2O_3$ (or other coating materials of ceramics, polymers, or metals such as silver) plays an ideal role of tunnel barrier (supports GMR and other useful properties) in small particles.[102,109]

### 2.2.1. CVD Method in Preparing $CrO_2$ Films

As early as 1978, Ishibashi et al.[110] applied a CVD method in processing epitaxially grown single crystal films of $CrO_2$ at atmospheric pressure in air. As described in the original article, the apparatus consists of a reaction tube and a two-zone furnace with a source zone and a growth zone. It allowed evaporation and deposition of $CrO_3$ over a lattice-matched substrate of Rutile crystals. Hot $CrO_3$ vapor splits and deposits as $CrO_2$. In a typical batch process, $CrO_3$ was kept at 260°C in a ceramic crucible in the source zone. While heating the substrate in the growth zone at 340-440°C, $O_2$ gas was introduced (at normal pressure) as carrier gas to transport the sublimed $CrO_3$ to the substrate, where it decomposes to $CrO_2$. As such, this is a very slow process. For example, growing a 200-nm-thick film applying such techniques, under such conditions, takes several hours.[97] According to high pressure (1.5 bar) $CrO_3$ decomposition at 400°C,[107] a filling factor of 50% of the reactor adds 800 bar $O_2$ in $CrO_3 \rightarrow CrO_2 + \frac{1}{2} O_2$ reaction. At higher temperatures, the $CrO_2$ decomposes to $Cr_2O_3$.

**FIGURE 2.** AFM images from granular $CrO_2$ film grown over Si (100) using $CrO_3$ as the precursor (Reprinted from the Journal of Magnetism & Magnetic Materials, Volume 239, J.J. Liang ot. al., Magnetotransport study of granular chromium, 213-216, 2002, with permission from Elsevier).

Because of the properties of $CrO_3$ as a precursor, a material that sublimes at ∼260°C and also partially decomposes, conventional CVD precursor method is not used. Specific additions include a precursor bubbler, automated valves, a pressure controller, mass flow controllers, and precursor flow sensors to deliver precise quantities of precursor into the reactor. Films of highly smooth $CrO_2$ surfaces were successfully achieved on a variety of substrates.[97,107,111] For example, Fig. 2 shows surface morphology of a typical granular $CrO_2$ film of average ∼1 µm grain size and rms value of surface roughness of 60 nm.[111] $O_2$ gas at 60 cc/min flow rate carried $CrO_3$ (heated at 280°C) to the substrate hold at 400°C.

All the above experiments were limited to a kind of a solid precursor. DeSisto et al.[112] and Anguelouch et al.[113] suggested using a liquid precursor such as chromyl chloride $CrO_2Cl_2$ in order to improve quality and control of $CrO_2$ films further. The $CrO_2Cl_2$, being a liquid, is compatible with conventional CVD precursor handling facilities. It deposits more efficiently of a given thickness. Atomically smooth $CrO_2$ films of rms value of surface roughness of less than 0.5 nm (at 100 nm thickness) were achieved, with as high $P_0$ value as 98%.[113] There are few reports of $CrO_2$ films grown by CVDs with other precursors such as $Cr_8O_{21}$,[114] photolytic and plasma assisted decomposition of $Cr(CO)_6$,[115] and RF sputtering of $CrO_3$ and annealing in a high-pressure cell.[103] A common observation is that $CrO_2$ films prepared by the various techniques do not have identical composition, microstructure, or properties. Surface smoothness, the kinds of the substrates (and crystallographic orientation) used and the substrate-induced lattice strain in $CrO_2$ play important role in controlling the microstructure and properties in such epitaxial $CrO_2$ films.

### 2.2.2. MBE Method in Preparing $CrO_2$ Films

Rabe *et al.*[97] applied MBE process to obtain highly textured $CrO_2$ films. A pure Cr instead of a chromium compound, often used in the CVD method, had been thermally evaporated and deposited on oriented $Al_2O_3$ (0001) and ($10\bar{1}0$) substrates (at 300-450 K) in ozone atmosphere. The ozone oxidizes Cr to $CrO_2$ during the deposition. Above 100°C, the $CrO_2$ surface degrades to lower oxidized states such as $Cr_2O_3$. Reducing the film surface roughness uses annealing in $O_2$ gas at an elevated temperature. The films have poor crystalline properties and adversely a high σ-value as $10^4 \Omega cm$, due to the restricted substrate temperature. Qualities of $CrO_2$ films of the CVD standard hardly appear.[97,111]

### 2.2.3. Controlled Thermal Decomposition Method

As mentioned above in Sec. 2.2, $CrO_2$ becomes stable in $O_2$ at pressures above 10 bar. Higher the temperature wider is the range of temperature of stability.[106,116] This offers synthesis of $CrO_2$ by decomposing a chromium (VI) compound such as $CrO_3$, $CrO_2Cl_2$ or $(NH_4)_2CrO_4$ in a closed vessel at an elevated temperature. The reaction is stopped at $CrO_2$ before reaching the stable oxide $Cr_2O_3$. This is demonstrated in Fig. 3 with thermopiezic analysis (TPA) by heating 16 mg of $CrO_3$ powder.[107] The TPA plot gives the $CrO_3$ decomposition temperature and the $O_2$ released during the heating at pressure under 1.5 kbar. $CrO_3$ starts decomposing at 250°C at atmospheric pressure. The plateau over 300-400°C (Fig. 3)

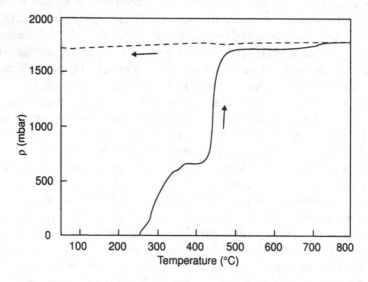

**FIGURE 3.** TPA curve measured during heating a $CrO_3$ powder at 2°C/min (Reprinted with permission from L. Ranno, A. Barry, and J. M. D. Coey, Production and magnetotransport properties of $CrO_2$ films, J. Appl. Phys. 81(8), 5774-5776. Copyright [1997], American Institute of Physics).

**FIGURE 4.** TEM images of acicular shaped magnetic $CrO_2$ particles from Micro Magnetics (Reprinted from J. Alloy. Compd., 379, Y. J. Chen, X. Y. Zhang, T. Y. Cai, and Z. Y. Li, Temperature dependence of ac response in diluted half-metallic $CrO_2$ powder compact, 240-246, 2004, with permission from Elsevier).

indicates occurrence of $CrO_2$ as a stable product in support of the byproduct gas $O_2$. Following the plateau, a further increasing value of the $O_2$ pressure results in due to $CrO_2 \rightarrow Cr_2O_3$ transformation, which almost completes before 500°C at pressure below 1.5 kbar.

Bajpai and Nigam[117] explored this novel idea in deriving a pure $CrO_2$ using a two-step heating process at 250°C followed by at 392°C in a sealed ampoule (at ambient pressure in air) for few hours. Needle-shaped $CrO_2$ occurs of an order of larger value of size reported for commercial powders from Dupont or Micromagnetics (USA).[102,118,119] In Fig. 4 are shown the single domain acicular particles (from Micromagnetics[119]) of an average 400 nm length and an aspect ratio 9:1. The powder of Dupont has more or less similar $CrO_2$ particles.[102] Such magnetic particles are suitable especially for ultra-high-density perpendicular magnetic recording and other applications.

## 2.2.4. Controlled Hydrothermal Decomposition Method

The hydrothermal process of forming $CrO_2$ involves a chemical decomposition of an aqueous solution of a chromium (VI) salt such as $CrO_3$ at elevated temperature and pressure in an autoclave or a similar high-pressure vessel or reactor.[120-124] Typically, at 480°C temperature and 2 kbar pressure, a pure polycrystalline $CrO_2$ powder had been obtained from aqueous $CrO_3$.[121] The powder so obtained had single crystallites of needle shapes (of few micrometers in size) with axial ratio ~20:1.[123] A combination of much higher temperature (900-1300°C) and pressure (60-65 kbar) is required in order to grow considerably bigger crystals.[121] The largest $CrO_2$ crystals obtained under optimal conditions were 1.5 mm × 0.33 mm. Chamberland[121] also performed the growth experiments in molten fluxes of PbO, $PbF_2$, $Cu_2O$, $K_2Cr_2O_7$, $KNO_3$, $K_2CrO_4$, $B_2O_3$, $Na_2CrO_4$, and $Na_2Cr_2O_7$, with no effective growth in the final size under optimal conditions of temperature, pressure, or other parameters. No impurity due to $Cr_2O_3$ or other possible chromium oxides was found at pressures above 30 kbar until the final temperatures lies below 900°C.

### 2.2.5. Controlled Oxidation Method

Oxidizing $Cr_2O_3$ with excess $CrO_3$ under pressure in presence of water produces commercial-grade magnetic $CrO_2$ particles for different kinds of magnetic tape storage media and detectors in image analysis in kinds of high-resolution microscopes. Cox describes the basic experimental facilities and conditions used in the original patent.[125] This in a simple chemical reaction,

$$Cr_2O_3 + CrO_3 \text{ (excess)} \rightarrow 3CrO_2 \qquad (3)$$

It is found that, during the reaction, adding a catalyst (e.g., $Sb_2O_3$, $Fe_2O_3$, etc.) to the starting $CrO_3$ solution controls final $CrO_2$ particle size. As a result, coercivity, $H_c$ can be tuned over an order of scale, with final value to a few hundreds Oe in reasonably small particles (but not too small to becoming superparmagntic of nature). In a typical example, the $H_c$ in particles of length $l = 3$-$10$ μm and diameter $d = 1$-$3$ μm, prepared without a catalyst, is only 5 $kAm^{-1}$ (or 63 Oe) whereas a sample processed with 2% $Sb_2O_3$ additive has smaller particles, $l = 0.1$-$2.0$ μm, $d = 0.04$-$0.08$ μm, with $H_c = 30$ $kAm^{-1}$ at room temperature.[123]

Such a catalytic additive often reduces the effective reaction temperature and pressure in reaction (3). Subletting part of $Cr^{4+}$ sublattices and/or a surface modification of $CrO_2$ particles can be found helpful in modifying $T_C$ value and other magnetic properties of interest. Part of diamagnetic coating (of the additives) of $CrO_2$ particles supports improved stability in ambient atmosphere as reported in several ferrites, $R_2Fe_{14}B$ intermetallics and other kinds of magnetic particles.[34,37,126] It is helpful in fabricating ideal single magnetic domain particles of maximum $H_c$ value depending on the morphology and surface topology (governs the shape or surface anisotropy) as well as the magnetocrystalline anisotropy. However, a kind of nonmagnetic substitution leads to decrease adversely the effective $M_s$ value in modified ferromagnetic $Cr^{4+}$ spin lattice structure.

### 2.2.6. Laser-induced Transformation

In addition to be using of a tool for preparing $CrO_2$ of thin films in CVD method,[103,115] a laser irradiation can be used to produce phase-controlled $CrO_2$

with a thin $Cr_2O_3$ surface layer. The surface $Cr_2O_3$ supports the core of virgin $CrO_2$ according to the oxygen partial pressure. Adjusting the duration and power of laser irradiation controls the relative fractions in the two phases.[127] The surface-modified (hardened) $CrO_2$ in this method becomes stable in air ambient. A self-confined $CrO_2 \rightarrow Cr_2O_3$ conversion on the surface leads to a threshold enhancement of the low field magnetoresistance, LFMR value.[102,127] The present results indicate that optical lithography is a potential method to control the magnetic, GMR, or optical properties, which are strong function of the surface insulator barrier layer. This opens a powerful approach for designing useful GMR properties for such kinds of materials, especially for materials with tunneling magnetoresistance (TMR).

### 2.2.7. Mechanical Attrition in a Closed Reactor

As described elsewhere,[21,128-134] mechanical attrition is one of the powerful tools in inducing (i) phase formation of metallic as well as nonmetallic compounds by reaction of the basic elements or derivatives, (ii) reconstructive phase transformation in small particles, and (iii) chemical dissociation or disintegration of metastable compounds. The selected examples include oxides such as $ZrO_2$, $TiO_2$, $Al_2O_3$, $Cr_2O_3$, and derivatives.[128,130,132-134] All this occurs according to nonequilibrium thermodynamics with redistributing of the ingredients in commensurate microstructure, exclusively with creation of new interfaces or phase boundaries, approaching the mechanical, chemical, or thermal equilibrium.

Our group had been successful in producing surface-stabilized $CrO_2$ by milling a series of chromium (VI) compounds, e.g., $CrO_3$, $(NH_4)_2CrO_4$, or $(NH_4)_2Cr_2O_7$, in a closed container.[135] It has been established that a small additive of a free carbon powder (2-5 wt%) promotes formation of stable $CrO_2$ particles with a surface interface layer due to carbon-stabilized $CrO_2$ or $CrO_2/Cr_2O_3$. While annealing, at 300-400°C in air, part of carbon burns along with the by-product $NH_3$ or $CO_2$ gases. The surplice partial precursor due to these gases, which relieve during the reaction, supports the $CrO_2$ phase control at these temperatures. This route to control the surface barrier layer of desired thickness is superior to the conventional ones using a thermal, or a laser, annealing $CrO_2$ in air or $O_2$.

### 2.2.8. Chemical Co-reduction Method

In this specific method, a chromium (VI) compound is co-reduced to $CrO_2$ by a chemical reaction with a suitable reducing agent at an elevated temperature. Ramesha and Gopalakrishanan[124] demonstrated the experiments with a simple solid-state reaction of $CrO_3$ with $NH_4X$ (X = Br or I) in vacuo at 120-150°C followed by annealing the product $CrO_2$ at temperatures (usually 195±5°C) below the $CrO_2$ decomposition point, i.e. $\geq 250$°C in atmospheric pressure.[107,124] The reaction is expressed as follows,

$$CrO_3 + 2NH_4X \rightarrow CrO_2 + 2NH_3 + X_2 + H_2O, \qquad (4)$$

which occurs exclusively over a narrow 120-150°C window.

In a typical reaction batch, stoichiometric quantities of $CrO_3$ (20 g) and $NH_4X$ (40-60g) were mixed under $CCl_4$. The mixture, taken in a Pyrex glass tube connected to a vacuum line, was heated slowly in vacuo ($10^{-5}$ Torr) under continuous pumping conditions. Evolution of $X_2$ occurs at $\sim 120°C$. The temperature was raised to 150°C and held at this value until the $X_2$ evolution ceased (2-3 h). Recovered blackish solid powder is nonmagnetic and X-ray amorphous. The Rutile $CrO_2$ structure occurs after annealing at $195 \pm 5°C$ (in an evacuated sealed tube) for one week, with $a = 0.4419$ nm and $c = 0.2915$ nm. The results demonstrate that capping $CrO_2$ in suitable films (possibly of polymer) and then auto-combusting in air at elevated temperature could lead to improve the $CrO_2$ quality of magnetic particles (of selective size and/or morphology), without involving a complex processing under a high pressure or a vacuum.

### 2.2.9. Sol-gel Type Polymer Precursor Method

This simple chemical route has been developed few years ago in our group.[79,135] It yields small $CrO_2$ particles of strictly controlled shape at a nanometric scale. The process has a hydrolysis reaction of a chromium (VI) compound in support of a polymer in aqueous solution followed by autocombustion of a kind of hydroxyl gel in ambient air pressure. The polymer serves as a capping agent to template the reaction in divided $Cr^{6+}$ groups capping in polymer micelles. In process to reaction, activated polymer molecules coordinate metal cations $Cr^{6+}$, forming a metal ion-polymer complex gel. At the reaction temperature (usually 60-70°C in air), a ligand charge transfer reaction $Cr^{6+} \rightarrow Cr^{4+}$ occurs via the polymer molecules in micelles, resulting in a phase stable $Cr^{4+}$-polymer complex in support of the polymer molecules. The process is briefed as follows:

In a typical reaction, a $Cr^{4+}$-polymer precursor is prepared by dispersing $CrO_3$ (dissolved in water in 2 M concentration) in activated polymer molecules of polyvinyl alcohol (PVA) and sucrose in solution in hot water (60-70°C in air), with 50 ml $CrO_3$ in 200 ml PVA-sucrose (44.0 g in 1:10 ratio). The PVA has weight average molecular weight $\sim$ 125,000, which implies $P_n \sim 2800$ repeat units of monomer. Assuming 0.2-0.3 nm lengths in the monomer, the value of $P_n$ represents a polymer molecular layer at a submicrometer scale. In a model reaction, as soon as adding $CrO_3$ (dropwise with magnetic stirring), active PVA molecules (with OH groups free from H-bonding) of planar surfaces enclose $CrO_3$ in form of micelles at early stage of reaction. As given in Fig. 5, the reaction proceeds by forming of intermediate polymer complexes in successive steps.

As given in Fig. 6, color of the sample changes from blackish red in the initial (intrinsic to $CrO_3$) to a pale yellowish to an orange yellowish to ultimately a dark blackish equilibrium one. After reaction, the sample is cooled and aged at room temperature for 15-25 h before drying at 70-80°C in air. A characteristic amorphous and nonmagnetic black colored fluffy voluminous mass lies of dried sample. The organic part decomposes and burns out if heating at 300-400°C in air, leaving behind recrystallized $CrO_2$ particles. Washing in benzene recovers $CrO_2$ from

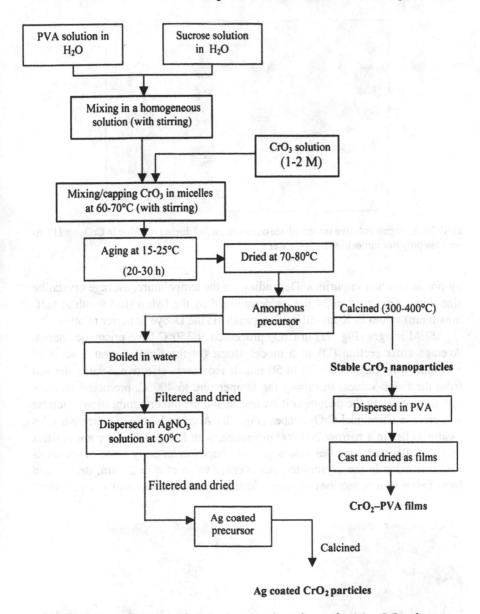

**FIGURE 5.** Schematic diagram showing the procedures for synthesizing $CrO_2$, phase stabilized $CrO_2$ particles with Ag-coating, and $CrO_2$ – PVA nanocomposite films (Reprinted from Chem. Phys. and Mater. Chem. Phys. (in press), 306, S. Biswas, and S. Ram, (a) Morphology and stability in a half-metallic ferromagnetic $CrO_2$ compound of nanoparticles synthesized via a polymer precursor, and (b) Synthesis of shape-controlled ferromagnetic $CrO_2$ nanoparticles by reaction in micelles of Cr6+−PVA polymer chelates, 163-169, 2004, with permission from Elsevier).

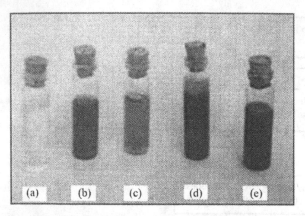

**FIGURE 6.** Representative intermediate colors recorded during reaction in $CrO_3$ and PVA-sucrose polymer molecules in hot water.

by-product carbon impurities. Depending on the temperature, average crystallite size $d$ varies from 18 to 24 nm, as determined by the fwhm (full width at half-maximum) values in X-ray diffraction peaks via the Debye Scherrer relation.[136]

SEM images (Fig. 7a) in $CrO_2$ processed at 350°C have prismatic shapes. Average cross section (D) in a model shape (in the right) lies at a scale of $\sim$18 nm, with length L = 20 to 50 nm. It represents effective $d$-value derived from the fwhm-values. In raising the temperature to 400°C, promoted reaction species, following the decomposition and in situ precursor combustion, nucleate and grow in ellipsoidal $CrO_2$ shapes (Fig. 7b). Average L-value in such particles (width $b$) lies in a narrow 200-300 nm range, with $L/b = 1$-3. They are clusters of $CrO_2$ allocated one after others in such shapes in an early stage of recombination reaction in the crystallites. An average value of $d = 24$ nm, determined from fwhm values, ascribes average width in the basic components (crystallites)

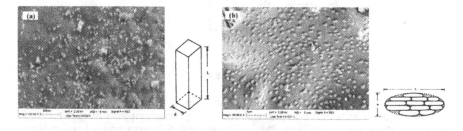

**FIGURE 7.** SEM images in $CrO_2$ derived from the polymer precursor at 350°C (a) and 400°C (b) for 1 h in air. The model shapes are given in the right (not to the scale) (Reprinted from Chem. Phys. and Mater. Chem. Phys. (in press), 306, S. Biswas, and S. Ram, (a) Morphology and stability in a half-metallic ferromagnetic $CrO_2$ compound of nanoparticles synthesized via a polymer precursor, and (b) Synthesis of shape-controlled ferromagnetic $CrO_2$ nanoparticles by reaction in micelles of Cr6+–PVA polymer chelates, 163-169, 2004, with permission from Elsevier).

in the clusters. In a simple case, it represents effective thickness in the clusters. An ellipsoidal shape of particles is not common in $CrO_2$. Commercial sample has acicular shapes (Fig. 4).

## 2.2.10. Phase Stabilized Particles of Surface Coating and Polymer Composites

### A. Diamagnetic Metal or Polymer Coating of $CrO_2$ Particles.

In order to support phase stability and to prevent undesirable reaction of $CrO_2$ with moisture, forming an artificial stabilization layer on the surface of the particles protects the particles. The coated $CrO_2$ particles have superior magnetic, GMR, optical, or other properties according to composition, microstructure, or other properties of the layer. The purpose of this procedure is the same as the stabilization of reactive metal particles. Of course, the chemistry is different. A desired $CrO_2$ surface layer can be achieved either by converting the top layer of the material into a $Cr^{3+}$-compound or putting a foreign layer (of a metal, metal oxide, or an organic compound) at expense of the top layer. It involves a topotactic $Cr^{4+} \rightarrow Cr^{3+}$ conversion reaction and that is not so easy to obtain under controlled conditions. In part of our work in this series,[79,135] we devised such a surface layer of Ag-metal (or gold) of $CrO_2$ particles with simple chemical reactions as follows.

In a typical reaction batch, part of the precursor powder, which consists of $CrO_2$ particles capping in modified PVA-sucrose polymer molecules (as described above in Sec. 2.2.9) is dispersed in hot water and boiled for 2 h, with ultrasonic stirring. In this process, excess polymer molecules separate and dispersed in the water. The recovered $CrO_2$ powder is dried at 50-70°C in air. If at all necessary, a gently milling in a ball mill is advised to break down soft agglomerates. This powder is allowed to react with a silver salt, such as nitrate, in aqueous solution (0.2-1.0 M typical concentration) at 50-60°C in dark, with magnetic stirring. A kind of ion-exchange reaction results in presumably Ag-coated $CrO_2$ particles. It occurs by displacing part of $Cr^{4+}$ cations by Ag atoms in a topotactic surface layer. Proposed chemical reaction can be expressed as follows:

A by-product $CrO(OH)$ as soon as appears, in this reaction, disperses and dissolves in the acidic water. The $Ag^+$ species, which are reacting with virgin $CrO_2$ surface, no longer allow $CrO(OH)$ to be depositing over the resulting $CrO_2$ surfaces. A topotactic $\beta$-$CrO(OH)$ layer does grow[126,137,138] in the case of the $CrO_2$ reaction with moisture. It is a very slow process. Under controlled conditions, it can be dried of a useful $Cr_2O_3$ layer. After 20-30 min of the reaction in hot conditions, the Ag-coated powder is recovered by washing in distilled water, and then dried in air ambient or a reduced pressure (10-100 mbar) at room temperature. Recrystallized Ag-coated $CrO_2$ occurs in 1 h of heating at 300-500°C in air. Similar experiments can be utilized in coating a thin layer of a polymer of such $CrO_2$ particles. Polymer coating is advised of $CrO_2$ crystallites. The applications are restricted exclusively to low temperatures. High $T_g$ (glass transition temperature) polymers are preferred

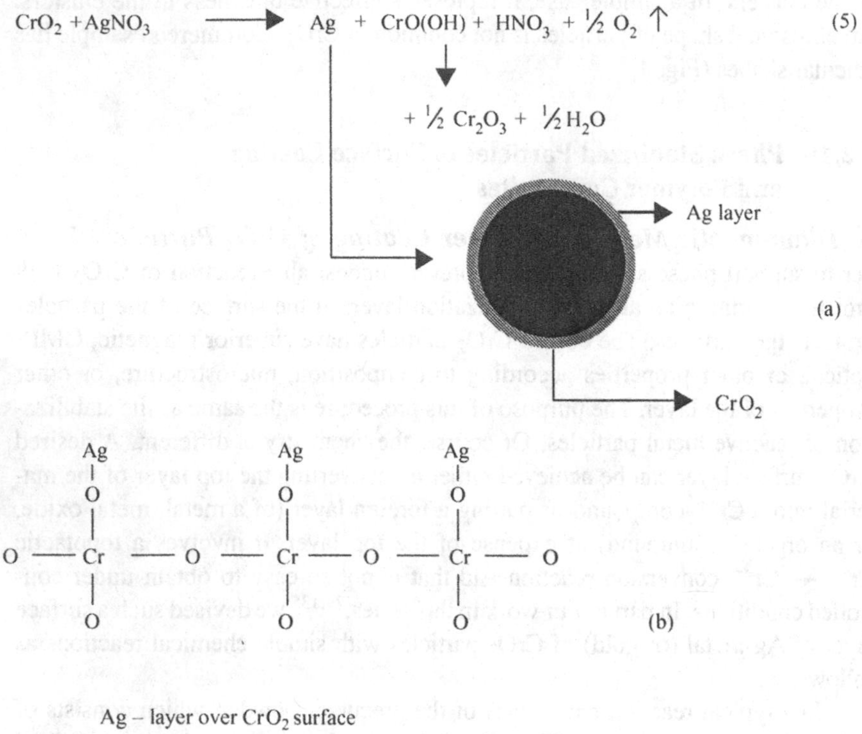
CrO2 + AgNO3 → Ag + CrO(OH) + HNO3 + ½ O2 ↑ (5)
+ ½ Cr2O3 + ½ H2O
Ag layer
(a)
CrO2
(b)
Ag – layer over CrO2 surface

**FIGURE 8.** A model reaction demonstrating (a) $Cr^{4+}$ of the top layer to be displacing by Ag atoms in a surface layer in $CrO_2$ particle of (b) a model $CrO_2$-Ag surface interfacing.

to extend the usable range of the temperature with stable properties. The $T_g$ affects thermal stability and other properties.

***B. Ceramic-polymer Composite Films.*** A ceramic-polymer composite of films in a simple way can be fabricated by dispersing ceramic particles of specific kinds in a polymer of solution at elevated temperature followed by casting the sample in such shapes. It involves a specific device (spin coating) to shape the liquid sample of film of control thickness and sooth surfaces. Dispersing ceramic particles in a requisite value in a polymer is a unique kind of problem in manufacturing useful products and components. In general, ceramics hardly mix in polymers.

For example, we fabricated $CrO_2$-PVA nanocomposite films ($t = 125$ μm thickness) of selected 1.0, 2.0, or 3.0 wt% $CrO_2$ contents. Finely divided loose powders for uncoated and Ag-coated $CrO_2$ were tested. They were dispersed over PVA polymer molecules in hot conditions at 50-60°C in aqueous solution with mechanical stirring. After 1 h of stirring, the temperature was raised to 80°C to obtain a viscous sample, by evaporating part of the water, which was cast as a film using a "spin coating." Thick films or laminates ($t = 1$-$5$ mm) were obtained

**FIGURE 9.** SEM images in (a) 2.0 wt% and (b) 3.0 wt% CrO$_2$-PVA composite films.

by sandwiching the sample between two plates in a mold. Free-standing films or laminates were achieved after drying in reduced pressure (10-20 mbar) at room temperature or a bit higher. They were used for the proposed studies in this chapter.

Figure 9 compares SEM micrographs in (a) 2.0 wt% and (b) 3.0 wt% CrO$_2$-PVA composite films. Both the samples have uniformly dispersed CrO$_2$ particles embeded in the PVA polymer molecules. Particles embeded in polymer are not easy to be imaged in the virginal features of the CrO$_2$ powder (Fig. 7b). Moreover, as high CrO$_2$ loading as 3.0 wt% used in these examples disrupts the film quality in terms of the surface topology with a rather inhomogeneous microstructure (Fig. 9b). The PVA polymer matrix, which exists in amorphous state, no longer imparts a fine structure at this scale. It is interesting to note that, in the PVA polymer; Ag-coated CrO$_2$ particles are aligned and grown in shape of fibrils.

A typical SEM micrograph, as given in Fig.10, presents 10-20 μm long fibrils of 300 nm average diameter and 30-50 aspect ratio. This presents a promising

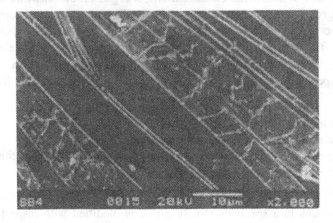

**FIGURE 10.** SEM images in Ag-coated CrO$_2$-PVA composite film (3.0 wt% CrO$_2$).

anisotropic texture of locally aligned $CrO_2$ crystallites along a specific crystallographic direction [110], analyzed with the X-ray diffraction pattern. Under hydrothermal conditions at 50-60°C, anisotropic $CrO_2$ crystallites align and grow in support over the polymer molecules. The Ag-coating (behaving as a thermal conductor) seems to be supporting the directional growth of the $CrO_2$ crystallites by driving the heat flow during the process primarily along the common interface between the particles and the PVA molecules, i.e., the film surface.

### 2.2.11. Porous Ceramics and Hybrid Composites

Porous ceramics and derived hybrid ceramic or polymer composites are a new class of applied materials. They are classified in three groups: (i) microporous, (ii) mesoporous, and (iii) macroporous according to the pore dimensions, i.e., in the ranges of 0.2-2 nm, 2-50 nm, and above 50 nm, respectively.[139-143] All these three primary classes of porous solids have their own values in properties and applications. The specific applications include microelectronic circuits, new phosphors, solid fuel cells, hydrogen (also others) gas storage systems, reaction catalysts, gas sensors, hot filters or separation membranes, drug delivery, thermal/electrical shockproof materials, etc.[144-153] Since the particles in such specific solids are distributed through pores, or vice versa, there exist both the interface between pore wall and particles and a rather free-standing surface of the particles within such pores. As a result, they are significantly different in atomic as well as electronic band structure and derived properties in comparison to the nonporous solids.

Several kinds of porous ceramics are available such as $SiO_2$, $Al_2O_3$, aluminosilicates, zeolites, $AlPO_4$, $ZrO_2$, ZnO, etc. Pores, especially of mesoporous or microporous kinds of solids, as such, have a reduced effective internal pressure $P_r$. As a result, they chemisorb much interstitial gas, usually 10-60 wt% depending on their volume and chemistry.[145,146,151] At ambient pressure, the interstitial gas desorbs depending on $P_r$ at specific temperatures. A mesoporous solid thus can be explored to dope desired metals, metal cations, or organic molecules into the pores in divided groups under $P_r$ at low temperatures. The dopant nucleates and grows in a specific structure led by the dynamics of the pores and other experimental conditions. This is a new method of designing a variety of hybrid mesoporous composites. On adding by solution, the reaction species travel to pores and rearrange in a specific structure in a dynamic reaction. The primary driving forces are as follows:

(a) their concentration in the solution,
(b) the rate of their incorporation into pores,
(c) the rate of their nucleation and growth in a stable crystallite within pores,
(d) the morphology and size in the pores, and
(e) $P_r$ over the particle growing into a stable structure.

This results in a specimen structure that is modified with respect to that obtained by bulk reaction under equilibrium conditions. The method is extended

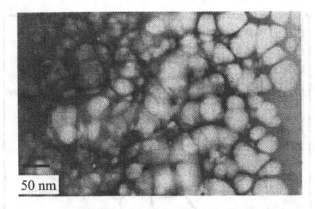

**FIGURE 11.** TEM images in a porous $CrO_2$ of empty-shell structure.

successfully in deriving kinds of mesoporous composites such as doped $Al_2O_3$, $ZrO_2$, or $CrO_2$ using metals (copper, silver, or gold) as well as molecules, e.g., PVA, polyvinyl pyrolidone, benzonitrile, and benzaldehyde, and rare earth doped $Al_2O_3$. The doping causes surface enhanced plasmon bands in cermets while surface enhanced $\pi \rightarrow n\pi^*$ electronic transitions in the organics, with optical density improved by as much as 1-2 orders of magnitude.[151,154]

   A typical TEM micrograph in Fig. 11 presents a porous $CrO_2$ of empty shells. The sample was derived by heating a polymer precursor at 250°C in air for 1 h. The major axis in vowel shape of shells is 30-50 nm. The shell is as thin as a few nm. Otherwise, it wouldn't be reflected in empty shells of whitish contrast. It sounds that the shells are created and filled-up in situ with air while heating the precursor in air. One can argue that the air, which persists in shells over these temperatures, supports the $CrO_2$ phase stability, i.e., a kind of effectively $O_2$-partial pressure assisted $CrO_2$ synthesis. Otherwise, $CrO_2$ hardly keeps the oxygen stoichiometry over early growth stages in ambient air pressure. This provides a new route for pressureless synthesis of phase-stabilized $CrO_2$ ceramics by doping of oxygen rich and high-temperature oxides such as $ZrO_2$, $TiO_2$, or $HfO_2$ in ambient air.

## 3. STABILITY AND CONTROLLED TRANSFORMATION IN PHASE-STABILIZED PARTICLES

Virgin $CrO_2$ is a metastable compound. At atmospheric pressure in air, it reacts with moisture, forming a topotactic $CrO(OH)$ surface layer. TEM images clearly reveal core-layer in such $CrO_2$ particles of characteristically different structure.[102,126] If not controlled, the layer grows in $\beta$-$CrO(OH)$ polymorph in a distorted orthorhombic or hexagonal crystal structure.[126,138] Essig *et al.*[126] simulated $\beta$-$CrO(OH)$ growth over $CrO_2$ along (110) and (010) planes. As a result, $CrO_2$ surface degrades into $Cr_2O_3$ (antiferromagnetic) at an early temperature as 200°C.[94,97] A complete $CrO_2 \rightarrow Cr_2O_3$ conversion lies at temperatures above 400°C according to

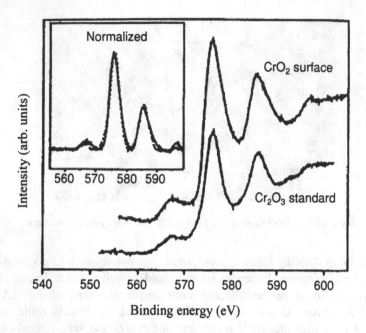

**FIGURE 12.** XPS Cr 2p bands of $CrO_2$ and $Cr_2O_3$ standard (dashed line in the inset) (Reprinted with permission from J. Dai, J. Tang, H. Xu, L. Spinu, W. Wang, K. Wang, A. Kumbhar, M. Li, and U. Diebold, Characterization of the natural barriers of intergranular tunnel junctions: $Cr_2O_3$ surface layers on $CrO_2$ nanoparticles, Appl. Phys. Lett. 77(18), 2840-2842. Copyright [2000], American Institute of Physics).

microstructure. In CVD grown $CrO_2$ films, Liang et al.[111] observed such conversion reaction near the film surface at room temperature. X-ray diffraction $Cr_2O_3$ peaks occur in superposition of $CrO_2$ peaks after 100 days of annealing at room temperature.

For the above reasons, the structural defects, which lead to deteriorate $M_s$ and other useful properties very sensitively, are not easy to heal in a permanent basis by annealing in $O_2$ gas.[94,108] Occurrence of $Cr_2O_3$ in a surface barrier layer in $CrO_2$ particles is confirmed more unambiguously with XPS (photoelectron spectroscopy) studies. In Fig. 12, a close similarity in the $Cr 2p$ band shapes between $CrO_2$ and $Cr_2O_3$ standard confirms $Cr^{3+}$ to be dominating the near-surface region of the $CrO_2$ particles. Notice XPS is sensitive within 1-2 nm into the particle surface. Invariably, the 1-3 nm thick top layer in $CrO_2$ particles is not $CrO_2$ but a $Cr^{3+}$ compound. The $Cr_2O_3$, if not stabilized and controlled, grows on at $CrO_2$ expense. A dense and uniform $Cr_2O_3$ layer (stable) serves a tunnel barrier in the spin-dependent intergranular tunnel junctions in so-called phase-stabilized $CrO_2$ particles.

Now, we are convinced ourselves with the realization of a surface coating of a phase-stabilized $CrO_2$ of particles. A stable surface interface, which occurs in the case of dispersing and embedding $CrO_2$ particles in a matrix in a hybrid composite, serves also a similar purpose of phase-stabilizing $CrO_2$ particles. Such samples

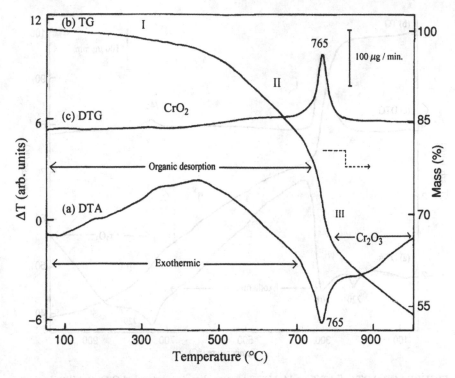

**FIGURE 13.** (a) DTA, (b) TG, and (c) DTG curves showing extended thermal stability in $CrO_2$ particles against transformation to $Cr_2O_3$ in a polymer precursor. The data were measured at 10°C/min heating rate in air.

have better stability in ambient atmosphere. Figure 13 shows (a) DTA (differential thermal analysis), (b) TG (thermogravimetric), and (c) DTG (derivative of TG) curves measured during heating a $Cr^{4+}$-PVA/sucrose composite powder derived by the solution chemistry in Sec. 2.2.9. The TG curve yields a total of $\sim 38\%$ mass loss in three successive steps as marked over the curve by symbols I, II, and III. The I and II regions (impart an $\sim 22\%$ partial mass loss) describe disintegration and combustion in the organics, with an exothermic DTA signal over 100-700°C. Part of carbon (after the organic combustion), which still adheres to the $CrO_2$ surface, burns out mainly in region II as oxides. The region III infers the $CrO_2 \rightarrow Cr_2O_3$ conversion (adding $\sim 10\%$ mass loss with respect to the III-set-point), with an endothermic peak in DTA or DTG of extended temperature 765°C. A reduced chemical process,

$$CrO_2 \rightarrow \frac{1}{2}Cr_2O_3 + \frac{1}{4}O_2 \uparrow, \tag{5}$$

has as much mass loss as 9.5%. The releasing $O_2$ gas results inevitably in the decrease of mass of the specimen.

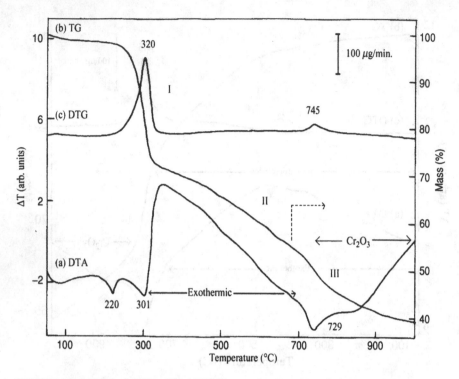

**FIGURE 14.** (a) DTA, (b) TG, and (c) DTG curves showing extended $CrO_2$ stability against transformation to $Cr_2O_3$ in 2.0 wt% $CrO_2$-PVA composite films. The data were measured at 10°C/min heating rate in air.

Figure 14 is an example of a reinforced polymer composite film with separately prepared $CrO_2$ particles. The mass loss process, starting at ~200°C, terminates in the three parts at ~329, 729 and 851°C, respectively. The step I imparts ~28% mass loss due to the matrix decomposition in two endothermic DTA peaks of 220 and 301°C. Combustion of the surface carbon over $CrO_2$ particles starts at 329°C in step II, adding ~19% mass loss before the step III sets in. That is manifested as an exothermic signal in DTA over 301-688°C. The endothermic $CrO_2 \rightarrow Cr_2O_3$ transformation occurs in a peak at 729°C in DTA while at 745°C in DTG, with 9.0% mass loss after the step set-in. In bulk $CrO_2$, the mass loss attributes to $CrO_2$ heat decomposition over 425-580°C, i.e., a topotactic transformation from Rutile phase $CrO_2$ to corundum phase $Cr_2O_3$. The crystallographic relations between two phases are $\{001\}_{Cr2O3} \| \{100\}_{CrO2}$ and $\{100\}_{Cr2O3} \| \{001\}_{CrO2}$.[6,118] Evidently, the carbon supports the extended temperature (relative to 450°C in bulk $CrO_2$[118]) stability in such $CrO_2$ particles.

## 4. ELECTRONIC BAND STRUCTURE

A half-metal such as $CrO_2$ is a solid with an unusual electronic structure. Self-consistent spin-polarized band structure calculations describe that for electrons

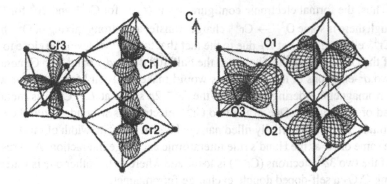

**FIGURE 15.** Angular distribution of electron-spin-density for the nearest $Cr^{4+}$ neighbours (xy (left) and $yz + zx$ (right)). Cr1 and Cr2 are $Cr^{4+}$ at the corners of the unit cell and Cr3 is in the body centre position. Solid circles denote $O^{2-}$. The arrow shows the direction of crystallographic c-axis (Reprinted figure with permission from M. A. Korotin, V. I. Anisimov, D. I. Khomskii, and G. A. Sawatzky, Phys. Rev. Lett, 80(19), 1998. Copyright (1998) by the American Physical Society).

of one spin it is a metal with a Fermi surface while for the opposite spin there is a gap in spin-polarized density of states, like a semiconductor or insulator.[9] In $CrO_2$, the prevalent metallic character arises for the majority spin ↑ electrons while the $E_F$ level falls in a semiconductor gap for the spin ↓ electrons.[9,97] Presupposing a magnetically ordered state defines the spin quantization axis. The responses of such a solid to an electric and magnetic field at zero temperature are quite different. There is electric conductivity, but no high field magnetic susceptibility. Nevertheless, photoemission studies showed the system to be an insulator for both the kinds of spins.[94,155] $CrO_2$ is one of the few oxides that has metallic conductivity and ferromagnetic ordering, $T_C = 391$ K and spin quantized magnetic moment $2\mu_B$.

The crystal field ($\sim$2.5 eV) leads to split the $Cr^{4+}$ $d^2$ orbitals into a $t_{2g}$ triplet and an $e_g$ doublet.[48,156,157] In the Rutile $CrO_2$ structure, the Cr atoms form a body-centered tetragonal lattice with distorted oxygen octahedron. The octahedra surrounding $Cr^{4+}$ at the body's center and corner positions differ by a 90° rotation about the c-axis. The simplest $CrO_6$ cluster calculations show that this kind of distortion of the octahedra (elongation along c-axis) leads to the new natural basis for the $t_{2g}$ orbitals: $xy$, $yz + zx$ and $yz - zx$ in a local coordinate system (LCS) for every octahedron (Fig. 15). Korotin et al.[96] interpreted ferromagnetic ordering in $CrO_2$ in terms of a double exchange mechanism[158] by performing LSDA + U (local Coulomb and exchange interactions) calculations in the linearized muffin-tin orbitals (LMTO) approach and argued that this distorted structure of the octahedra is responsible for the peculiar properties of $d^2$ bands in $CrO_2$. As a consequence, the $t_{2g}$ orbitals are split further into a nonbonding $d_{xy}$ orbit (which lies in the equatorial plane of the oxygen octahedron) and a $d_{yz}$, $d_{zx}$ doublet, which form an antibonding $d_{yz} \pm d_{zx}$ ($\pi^*$) combinations with respect to the $O^{2-}$ p-orbital perpendicular to the $Cr_3O$ triangles.[48,156,157]

Thus, the formal electronic configuration is $(t_{2g}^2)^{\uparrow}$ for $Cr^{4+}$ and $2p^6$ for $O^{2-}$ although there is some $O^{2-} \rightarrow Cr^{4+}$ charge transfer and strong mixing of $O^{2-}$ hole and $Cr^{4+}$ electron states at $E_F$ due to the fact that the $Cr^{4+}$ $d^2$ levels lie close to the top of the $O^{2-}$ $2p$ band and $E_F$ lies in the half-full $d_{yz} \pm d_{zx}$ band.[48,96] Generally, the two $d^2$-electrons in the $t_{2g}$ orbitals would make $CrO_2$ an Mott insulator with antiferromagnetic ordering but due to the $O^{2-}$ $2p$ states at $E_F$, $CrO_2$ is metallic instead of an insulator. One of the two $Cr^{4+}$ electrons is $\pi$ bonded with the $O^{2-}$ $2p$ orbitals, forming a partially filled narrow band at $E_F$. The width of this band is of the same order as the Hund's rule interatomic exchange interaction. As a result, one of the two $3d^2$ electrons ($Cr^{4+}$) is localized where as the other one is itinerant making $CrO_2$ a self-doped double exchange ferromagnet.

In the double exchange mechanism, the itinerant electrons transfer not only charge but also magnetic order. The spin of these electrons interacts with the localized $Cr^{4+}$ spins and puts them in a ferromagnetic order. As a result, conduction electrons are more mobile when the localized spins are aligned parallel than without magnetic order- favors a metal-insulator transition.[97] This mechanism clearly explains the metallicity and ferromagnetism in $CrO_2$ in spite of large U values and relatively narrow $d$-bands, and shows that $CrO_2$ belongs to a class of solids of small or even negative charge-transfer gap leading to a self-doping.[96,97]

Starting from the band structure calculations by augmented spherical wave (asw) method by Williams,[159] self-consistent spin-polarized calculations with the *asw* by Schwarz,[9] LSDA, LSDA + U, and GCA calculations have a refined $CrO_2$ energy band diagram. Almost every calculation confirms the spin split band structure of $CrO_2$ like a $I_A$-type half-metal as given in Fig. 16a, with a spin gap

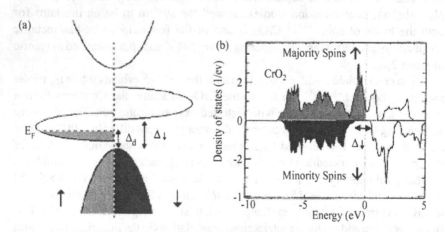

**FIGURE 16.** (a) Schematic density of states for a IA type half-metal with only spin-up electrons at the EF level and (b) spin polarization of the density of states in $CrO_2$ (a) (Reprinted with permission from J. M. D. Coey and M. Venkatesan, Half-metallic ferromagnetism: example of $CrO_2$, J. Appl. Phys. 91(10), 8345-8350. Copyright 2002, American Institute of Physics). b) (Reprinted figure with permission from S. P. Lewis, P. B. Allen, and T. Sasaki, Band structure and transport properties of $CrO_2$, Phys. Rev. B 55(16), 10253-10260, 1997. Copyright (1997) by the American Physical Society).

$\Delta_\downarrow > 1$ eV and a spin-flip gap $\Delta_{sf}$ of a few tenths of an eV.[48] Figure 16b shows the spin polarization of the density of states of $CrO_2$ with a $t_{2g}$ bandwidth of 2.5 eV with a trident structure including a narrow peak in the density of states due to the $d_{xy}$ electrons. The $E_F$ level lies in a pseudogap between the $d_{yz} \pm d_{zx}$ bands, but its position relative to the minimum is rather sensitive to details of the calculation.[48] The half-metallic character is maintained up to the surface of the $CrO_2$ particles.[48,155]

# 5. ELECTRONIC PROPERTIES

## 5.1. Dielectric Properties

The values for dielectric constant ($\varepsilon$) and dielectric power loss (tan $\delta$) for $CrO_2$ in forms of bulk crystals, thin films, bulk powder compacts, nanopowder compacts, or composites at room temperature are given in Table 3. Analogous to ferrites, PZTs [i.e. $Pb(Zr_{1-x}Ti_x)O_3$], and similar ceramics, the $CrO_2$ ceramics in general are supposed to assume improved $\varepsilon$-values, with usefully decreased tan $\delta$ values, in ideal single-domain particles in granular films, nanopowders, or hybrid composites. Furthermore, the value for $\varepsilon$ as well as tan $\delta$ highly depends on frequency ($f$) applied to measuring them with one of the conventional methods. For instance, Fig. 17 presents the data for a cold pressed $CrO_2$ nanopowder compact (as a thin pellet or billet) as a function of $f$, in the 100 Hz-1 MHz range, at a logarithmic scale. As usual in such ceramics,[165-168] both the kinds of the characteristic values are found to be decreasing monotonically with increasing $f$-value over the range. As reflected by the large values, effectively, all types of possible polarizations (i.e., interfacial, atomic, dipolar, or ionic and electronic ones[166,167]) are active at early $f$-values.

**TABLE 3.** Dielectric constant $\varepsilon$, dielectric power loss tan $\delta$, and electrical resistivity $\sigma$ in $CrO_2$ available in selected forms

| Samples | $\varepsilon$ | tan $\delta$ | dc $\sigma(\Omega$-cm) (Nyquist diagram) | References |
|---|---|---|---|---|
| $CrO_2$ nanopowder | 300 (100 Hz) 56 (1 MHz) | 5.57(100 Hz) 0.01(1 MHz) | 11.4 | Unpublished work |
| Bulk powder | — | — | — | — |
| Thin films | — | — | — | — |
| 50 wt% $CrO_2$ - $TiO_2$ | — | — | $1.02 \times 10^2$ | 160 |
| 20 wt% $CrO_2$ - $TiO_2$ | — | — | $3.90 \times 10^6$ | 160 |
| PZT films | 400 (10 kHz) | 4 (10 kHz) | — | 161 |
| Gd doped PLZT | 3750 (10 kHz) | 0.004 (10 kHz) | — | 162 |
| Fe doped PLZT | 1008 (10 kHz) | 0.05 (10 kHz) | $0.78 \times 10^6$ | 163 |
| $BaTiO_3$ films | 192 (1 kHz) 173 (1 MHz) | 0.03 (1 kHz) 0.20 (1 MHz) | — | 164 |

The values are reported at room temperature. The powders were studied as pellets. PLZT: lead lanthanum zirconium titanate, Fe doped PLZT: $Pb_{0.9}$ $(La_{0.5} Fe_{0.5})_{0.1}(Zr_{0.65}Ti_{0.35})_{0.975}O_3$.

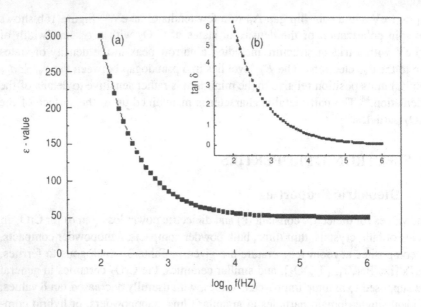

**FIGURE 17.** Frequency dependence of (a) $\varepsilon$ and (b) tan $\delta$ in CrO$_2$ nanopowder compacts at room temperature in the 100 Hz to 1 MHz frequency range.

Usually, the electronic polarization is observed at very high $f$-values ($>10^{13}$ Hz). As a function of increasing $f$-value, a decrease in the tan $\delta$ value together with the $\varepsilon$-value, as observed in this example, implies that the active component (Ohmic) of the current is practically independent of $f$-value and the reactive component (capacitative) increases proportional to the $f$-value. CrO$_2$ thin films or nanocomposites have more or less similar nonlinear $f$-dependence of $\varepsilon$ and tan $\delta$ values. A similar trend in $\varepsilon$ or tan $\delta$ variation with a function $f$ is observed in PZTs, modified PZTs, or ferrites.[169-172] Unfortunately, not many reports are available on the kinds of HMF compounds in order to present a better comparison of the values among other kinds of half-metallic ferromagnets.

Figure 18 portrays the impedance $\{Z(f) = R(f) + iX(f) = Z \cos \theta + iZ \sin \theta\}$ spectrum (Nyquist diagram) for the CrO$_2$ nanopowder compacts, where the imaginary part of impedance $\{X(f)\}$ is plotted against the real part of impedance $\{R(f)\}$ in the same $f$-range as in Fig. 17. The plot consists of a single semicircle. Figure 19 displays the real and imaginary parts of complex impedance against the $f$-values. The specific $f$-value, at which the maximum $X(f)$ occurs, is found to be 2.014 kHz. The observed features in $R(f)$ and $X(f)$ are associated with the strength of anisotropy, which acts as a result of ordering in the CrO$_2$ particles. There are only few reports so far on the high-frequency transport properties of CrO$_2$.

Fu et al.[173] studied the high-frequency magneto-impedance for polycrystalline CrO$_2$ thin films and demonstrated specific feature of impedance spectra,

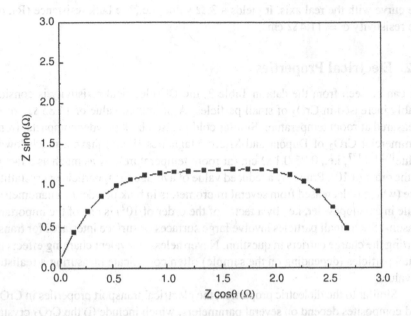

**FIGURE 18.** The Nyquist diagram of complex impedance in $CrO_2$ nanopowder compacts at room temperature in the 100 Hz to 1 MHz frequency range.

which are distinct from those of a usual metallic ferromagnet and are attributed to the nature of the half-metallic ferromagnets.[174,175] Further analysis has shown the impedance to be related to the interplay of magnetization response and dielectric relaxation under alternative electromagnetic fields. As described elsewhere,[119,160,175] in Fig.18, the resistance of the sample can be determined from the interception of

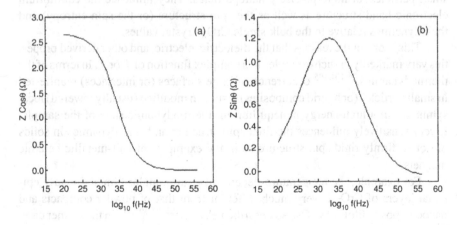

**FIGURE 19.** (a) The real and (b) imaginary parts of the complex impedance in $CrO_2$ nanopowder compacts at room temperature in the 100 Hz to 1 MHz frequency range.

the curve with the real axis. It yields a 3 $\Omega$ value, i.e., the bulk resistance (R), or the resistivity $\sigma = 11.4\,\Omega$-cm.

## 5.2. Electrical Properties

As can be seen from the data in Table 3, the DC electrical resistivity is considerably increased in $CrO_2$ of small particles. A promoted value of 14.83 $\Omega$-cm is measured at room temperature. Similar cold pressed bulk powder compacts, from commercial $CrO_2$ of Dupont and Micro Magnetics (USA), have a much lower value[102,176,177], i.e., 0.05-0.1 $\Omega$-cm (at room temperature), by as much as a factor of the order of $10^2$. Primarily a reduced value of average $CrO_2$ particle or crystallite size (which is decreased from several micrometers in bulk powder to a nanometric scale in a nanopowder, i.e., by a factor of the order of $10^2$) is one of the important reasons. Such small particles involve large surfaces or surface interfaces for transporting the charge carriers in question. Nevertheless, prevalent charging effects in small particles (depending on the sample) often complicate measuring a realistic $\sigma$-value.

Similar to the dielectric properties, the electrical transport properties in $CrO_2$ and composites depend on several parameters, which include (i) the $CrO_2$ crystal structure and local symmetry, (ii) the artificial interstitial impurities and defects, (iii) the basic $CrO_2$ size and morphology or topology, (iv) the surface interface or surface layer, if any, (v) the distribution of $CrO_2$ of ordered lattice in a specific microstructure, i.e., homogeneous or heterogeneous [in combination to the factors (iii) and (iv)], (vi) the exchange coupling between $CrO_2$ of particles, (vii) the intergranular tunneling of charge carriers, or in particular in composite, (viii) the $\alpha$ - $\beta$ macroscopic interactions in the basic $CrO_2$ particles or crystallites ($\alpha$) and the matrix ($\beta$). The last four or five factors become extremely important in the case of thin films, powder compacts, and hybrid composites, involving especially small particles, or more precisely nanoparticles. They influence the equilibrium electronic band structure as well as the spin structure (or the spin entropy) and their dynamics relative to the bulk single $CrO_2$ crystal values.

This is one of the reasons that the dielectric, electric, and other derived properties vary markedly in such examples in a complex function of $T$ or $f$. In terms of the thermodynamics,[76,104,105] occurrence of large surfaces (or interfaces) manifested in small particles (or hybrid composites) results in modified (usually lowered) local symmetry in a high-energy nonequilibrium thermodynamic state of the sample. It very sensitively influences the final spin structure and spin-dynamics in solids of not so firmly rigid spin structure as in the examples of half-metallic ceramic magnets.

For example, the temperature-dependent $\sigma$ in granular film and nearly epitaxial layers of $CrO_2$ is very much different from that of powder compacts and nanocomposite films of $CrO_2$. As described elsewhere,[97,107,111] in the former case, the $\sigma$-value increases monotonically with increasing $T$, a typical metallic innateness, while decreases in the others.[100,102,109] The processing techniques (affect the

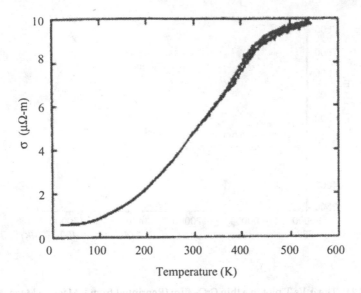

**FIGURE 20.** The σ Vs T plot in single crystal CrO$_2$ film on an Al$_2$O$_3$ substrate (Reprinted with permission from L. Ranno, A. Barry, and J. M. D. Coey, Production and magnetotransport properties of CrO$_2$ films, J. Appl. Phys. 81(8), 5774-5776. Copyright [1997], American Institute of Physics).

microstructure) used to prepare the specimen affect the electro-transport properties, e.g., a typical cold pressed CrO$_2$ powder compact has a negative slope of $R$ vs. $T$ plot in the range $150 < T < 300$ K, on the other hand, a hot pressed pellet of the same CrO$_2$ sample exhibits a positive slope in this range. The $R$ increases sharply in both samples with decreasing $T$ below 150 K.[177] Moreover, the σ-value in single-crystal CrO$_2$ films shows residual resistivity $\sigma_o$ of a few $10^{-8}$ Ω m[48,98,121,178] and a resistivity ratio $\sigma_T$ (300 K)/$\sigma_o$ that is found to be as high as 140.[48,179] For example, Fig. 20 portrays a typical σ vs. $T$ plot for such CrO$_2$ films ($t = 0.6$ μm) grown on Al$_2$O$_3$ or TiO$_2$ single-crystal substrates.

As evident from Fig. 20, the σ-value in CrO$_2$ is metallic over the measured temperature range. There is a decrease in slope of σ (T) at the $T_C$-value but the slope remains positive, indicating no sign of electron localization. At room temperature, σ($T$) is ∼4.8 × 10$^{-6}$ Ωm and the residual value can be as low as 2 × 10$^{-7}$ Ωm. The spread in the absolute values of resistivity may originate from the film roughness or defects in stoichiometry. After subtracting the residual value, the data follow a power law σ α $T^n$, where the exponent ($2.4 < n < 3.4$) is sample dependent. The σ-value is interpreted as being the sum of a scattering due to electronic correlation and a magnetic scattering. Figure 21 describes a similar $T$-dependence of σ in a CVD grown epitaxial CrO$_2$ thin film on Al$_2$O$_3$ (0001) substrate.[97] The σ-value retains a metallic $T^2$ behavior in the 150-330 K range.

The σ-$T^2$ dependence is an indication for a dominating $e$-$e$ scattering mechanism in this temperature range. At higher temperatures, a deviation of the $T^2$

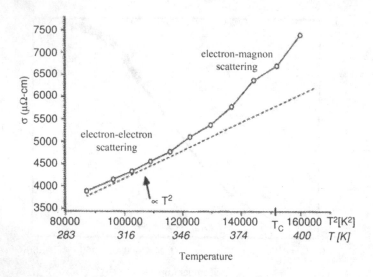

**FIGURE 21.** The $\sigma$ Vs T plot in a thin $CrO_2$ film (Reprinted from J. Magn. & Magn. Mater., 211, M. Rabe, J. Dreßen, D. Dahmen, J. Pommer, H. Stahl, U. Rüdiger, G. Güntherodt, S. Senz, and D. Hesse, Preparation and characterization of thin ferromagnetic $CrO_2$ films for applications in magnetoelectronics, 314-319, 2000, with permission from Elsevier).

dependence of $\sigma$-value indicates that an additional scattering mechanism contributes to the total $\sigma$-value. The e-m scattering (m:magnon) with a $T^{4.5}$ dependence possibly contributes to the value in the vicinity of $T_C$. At lower temperatures, e-m scattering is suppressed due to the semiconducting behavior of the minority spin channel. In another example of a CVD grown granular $CrO_2$ film (Fig. 22), $\sigma$-value decreases with increasing temperature up to about $T^{ADS} = 125$ K ($T^{PAS} = 165$ K), then turns on to increase with an increase in temperature above $T^{ADS}$ or $T^{PAS}$ (ADS: as-deposited sample, PAS: post-annealing sample).[111] The metallic nature of the film at the higher temperature range could, nevertheless, not be described consistently by a normal expression in powers of $T$, but rather by $\sigma \sim T^2 \exp(-\Delta_1/T)$, in which the metallic majority and semiconducting minority electrons are assumed to be acting in parallel.[111,180]

The dash lines in Fig. 22 show the best fit to the data with $\Delta_1^{ADS} = 407$ K and $\Delta_1^{PAS} = 413$ K. At low temperatures ($T < 6$ K), the conductance decreases according to $\exp(-\Delta_2/T)^{1/2}$ with decreasing temperature, where $\Delta_2$ is proportional to the Coulomb charging energy and barrier thickness. The observation infers that the intergranular tunneling plays a dominating role in this specific range of temperature.[100,102,109]

Coey et al.[176] studied the conduction mechanism of $CrO_2$ powder compacts of acicular shaped particles of an average 300 nm length and an aspect ratio ~8:1. The mechanism is associated with the spin-dependent intergranular tunneling across the grain boundaries. Figure 23 shows the R-value as a function of $T$ for a typical sample. The resistance $R$ has been normalized to the value at $T = 300$ K and in

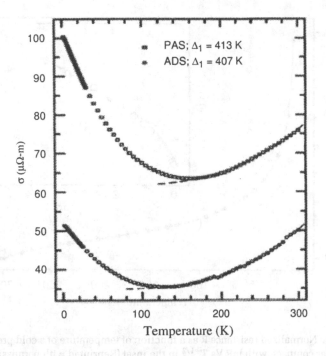

**FIGURE 22.** The $\sigma$ Vs T plot at low temperatures for a CVD grown granular $CrO_2$ film (Reprinted from J. Magn. & Magn. Mater., 239, J. J. Liang, S. F. Lee, Y. D. Yao, C. C. Wu, S. G. Shyu, and C. Yu, Magnetotransport study of granular chromium dioxide thin films prepared by the chemical vapor deposition technique, 213-216, 2002, with permission from Elsevier).

the inset ln $R$ is plotted against $T^{-1/2}$. Clearly, the ln $R$ is linear to $T^{-1/2}$ at $T$ below 35 K, typical of the intergranular tunneling.

In a specimen of thin film or powder compact of $CrO_2$, the $s$-dependent tunneling of charge carriers from one particle to other in nearest neighbors is described in terms of "intergranular tunneling." The process operates through the surface interface in $CrO_2$ particles of a Coulomb gap $\Delta$.[100,102] In an empirical relation,[181,182] the tunneling resistance from particle to particle is expressed in a simple function of $T$ as follows

$$R_{tun} \propto (1 + P^2 m^2)^{-1} \exp(\Delta/T)^{1/2} \qquad (6)$$

with $P$ the spin polarization and $m = M/M_s$ the relative magnetization. The $\Delta$-value is considered proportional to the Coulomb charging energy and the barrier thickness.

A linear relationship between ln R and $1/T^{1/2}$ in a $CrO_2$ powder compact has been reported only at low temperatures in prevalent intergranular tunneling effects.[109,183] Phenomenologically, the observation also suggests that at high temperatures, $s$-dependent tunneling conductance ($G_{SDT}$) is no longer the major

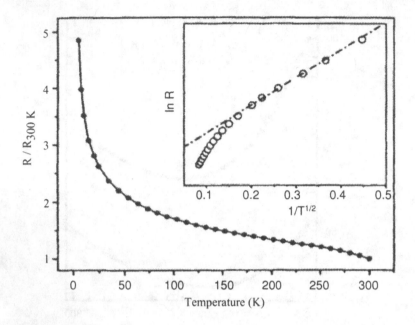

**FIGURE 23.** Normalized resistance R as a function of temperature of a cold pressed bulk CrO$_2$ powder compact, with lnR Vs $T^{-1/2}$ in the inset (Reprinted with permission from J. Dai, J. Tang, H. Xu, L. Spinu, W. Wang, K. Wang, A. Kumbhar, M. Li, and U. Diebold, Characterization of the natural barriers of intergranular tunnel junctions: Cr$_2$O$_3$ surface layers on CrO$_2$ nanoparticles, Appl. Phys. Lett. 77(18), 2840-2842. Copyright [2000], American Institute of Physics).

contribution. Other mechanisms become predominant. The suppression of $s$-dependent contribution results in a decrease for the MR-value at high temperatures when a $s$-independent channel (inelastic hopping conductance) becomes operative with increasing temperatures.[109] As a mater of fact, a power law temperature dependence of conductance ($G$), which is characteristic of the higher-order inelastic hopping, is being observed. Thus, the conduction mechanism in CrO$_2$ powder contains two channels, namely, the $s$-dependent tunneling (SDT) channel with an exponential $T^{-1/2}$ dependence and the $s$-independent (SI) hopping channel that follows a power-law.[109,183] As a result, phenomenologically, the total $G$-value is expressed in a simple form as[184]

$$G = G_{\text{SDT}} + G_{\text{SI}} = C_1 \exp[-(\Delta/T)^{1/2}] + C_2 T^{\gamma} \qquad (7)$$

where $C_1$ and $C_2$ are free parameters. The second term $C_2 T^{\gamma}$ denotes the inelastic hopping conductance and $\gamma = N - [2/(N + 1)]$. The mechanism of $G_{\text{SI}}$ results from the hopping along quasi one-dimensional chains of $N$-localized states ($N$-Ls). Glazman and Matveev[185] proposed a hopping conductance model with $e$-$p$ interaction (EPI), indicating that, in the inelastic tunneling process, strong EPI with the impurity leads to the appearance of inelastic scattering channels.

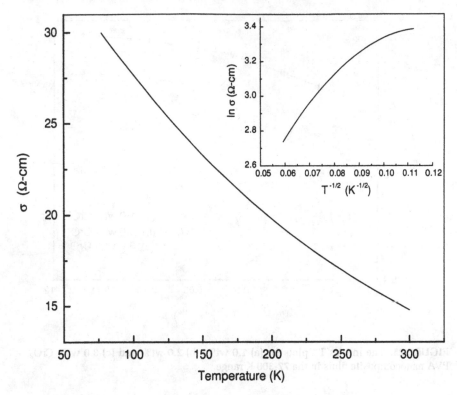

**FIGURE 24.** $T$-dependence of $\sigma$-value in $CrO_2$ nanoparticles (a cold pressed powder) prepared with polymer precursor. The inset plots $\ln\sigma$ as a function of $T^{-1/2}$.

The $T$-dependence of $\sigma$-value in $CrO_2$ nanoparticles (cold pressed as a pellet), prepared with the polymer precursor method, is shown in Fig. 24. As usual, the $\sigma$-value decreases with increasing $T$-value in a typical semiconductor behavior. A linear relationship of $\ln\sigma$ with $T^{-1/2}$ below 110 K is implying the $s$-dependent tunneling to be effective at such low values of temperature. Similar $\ln R$ vs. $1/T^{1/2}$ plots (Fig. 25) are obtained in the $CrO_2$-PVA nanocomposite films. Below 150 K, the plots are linear in the prevalent effects of the $s$-dependent tunneling associated with a Coulomb gap.

Following Eq. 6, the $\Delta$-values determined from the slopes of the linear parts of the plots in Fig. 25 are found to be 0.0062 K, 0.0093 K and 0.0130 K for the three samples of $CrO_2$-PVA nanocomposite films; (a) 1.0, (b) 2.0 and (c) 3.0 wt% $CrO_2$, respectively. Small differences in the $\Delta$-values infer small variations in the Coulomb charging energy and the barrier thickness in these samples. Developing a model empirical relation could be helpful in order to analyze the partial contributions $\Delta_C$ and $\Delta_b$ from the respective control parameters.

Figure 26 presents the current—voltage (I-V) characteristics and $G$-value as a function of $V$ at room temperature for a cold pressed powder compact from $CrO_2$ nanoparticles. In the low-bias voltage region (up to 5 V used here), the $G$-values

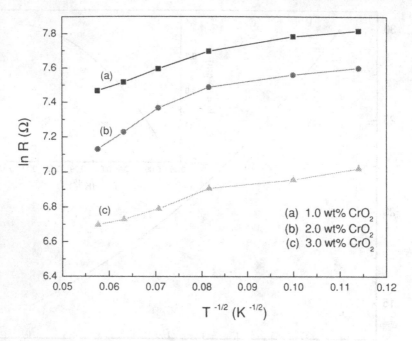

**FIGURE 25.** The lnR Vs T⁻ plots for (a) 1.0 wt%, (b) 2.0 wt% and (c) 3.0 wt% $CrO_2$ – PVA nanocomposite films in the 77-300 K range.

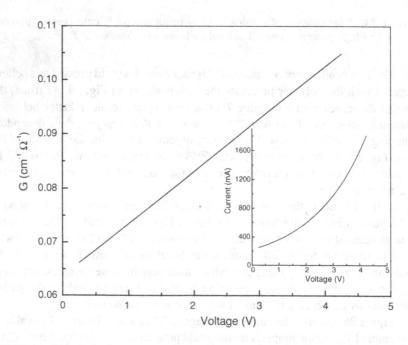

**FIGURE 26.** The $G$-$V$ plot at room temperature for a cold pressed compact of $CrO_2$ nanopowders. The inset presents the $I$-$V$ characteristics.

remain to be linear with a function of $V$. At higher $V$-values (not included in this figure), the plot deviates from the ideal linearity primarily due to the rise of the higher order inelastic hopping conductance. This has been studied and analyzed in bulk compact $CrO_2$ powder.[109,176] In an empirical relation, $G$-value is expressed as[109]

$$G = \sigma_0 + \sigma_1 V^{1.33} + \sigma_2 V^{2.5} \qquad (8)$$

where $\sigma_0$, $\sigma_1$ and $\sigma_2$ are the free parameters. Although the power law dependence of the inelastic hopping conductance on the bias $V^\gamma$ is predicted under condition $eV \gg k_B T$, the present model theory explicitly states that an increasing value of either the temperature or the bias voltage favors the hopping along the chains with more localized states ($N > 2$) resulting in promoted nonlinear dependence of the conductance on temperature and voltage.

# 6. MAGNETIC PROPERTIES

As mentioned above in section 1.2, as a ferromagnet or optoelectronic material, the $CrO_2$ ceramics and composites offer a variety of functional properties of direct applications in several kinds of devices. A low value of $T_C$, i.e. $\sim$391 K in pure $CrO_2$ crystals, is one of the basic limitations in miniaturizing the strategies. Depending on crystallite size (ranging from a few tens of nanometers to a submicrometer scale in the range of the single magnetic domains), $CrO_2$ as such is a soft magnetic compound, with significant values for $H_c$, remanence $M_r$, and derived parameters. As summarized in Table 4, it has a huge value of the magnetic

TABLE 4. Magnetic properties in $CrO_2$ ceramics and composites

| Sample | $M_s$ (emu/g) | $T_C$ (K) | $H_c$ (Oe) | $M_r$ (emu/g) | References |
|---|---|---|---|---|---|
| Bulk powder | 115 (5K) | — | 600 | | DuPont (102) |
| | | | 1000 (5K) | 55 (5K) | 102 |
| | 110 | — | 10 | — | 117 |
| | 135 (5K) | — | 50 (5K) | — | 117 |
| Aligned powder* | 110 (5K) | — | 1000 (5K) | 75 (5K) | 100 |
| Granular film | — | — | — | — | — |
| Epitaxial thin film* | 95 | 390 | 130 | 80 | 113 |
| | 132 (6K) | — | 185 (6K) | 115 (6K) | 113 |
| Polycrystalline thin | 96 | 391 | — | — | 41 |
| film | 132 (0K) | — | — | — | 41 |
| Nanopowder | 95.3 | — | 1050 | 74.2 | Unpublished work |
| Polystyrene – $CrO_2$ composite | — | — | — | — | — |
| $CrO_2$-PVA films | — | — | — | — | Unpublished work |

The values other than at room temperature are given with temperature in parentheses.
*Applied magnetic field is parallel to the easy axis of magnetization.

moment, i.e., $2\mu_B$ per formula unit (which is comparable to the $2.21\mu_B$ value per Fe atom in Fe metal) or $4\mu_B$ per crystal unit cell according to the $z = 2$ lattice number. As a result, the saturation magnetic polarization $P_s$ is found to be as high as $0.65\ T$, as can be derived by the relation $P_s = \mu_o M_s = 4\mu_B\mu_o/a^2c$, with $V = a^2c \equiv 0.0570$ nm$^3$ as the lattice volume. The intrinsic magnetocrystalline anisotropy $K_1 = 27$ kJ/m$^3$ (reported in epitaxial CrO$_2$ films[48,186]) at room temperature, with the easy axis of the magnetization lying along the crystallographic $c$-axis.

Burdett et al.[187] reported spin canting in CrO$_2$ powder. According to it, the easy axis of the magnetization in this case is tilted by an angle $\theta \sim 20°$ from the $c$-axis. This results in a reduced value of $P_s$ by a factor of cos $\theta$. For acicular CrO$_2$ particles, being used in the magnetic recording, the required shape anisotropy is as high as $\mu_o M_s^2/4$, implying a typical value of surface anisotropy $K_s = 50$ kJ/m$^3$ at room temperature.[48] Also, there is report of magnetostriction coefficient $\lambda_s$, i.e., $5 \times 10^{-6}$, in CrO$_2$ particles.[188] In fact, as in other identified classes of magnetic materials (Table 1), i.e., selected metal oxides, mixed oxides, metals, alloys, or intermetallics, all of the intrinsic magnetic properties strongly depend on size, morphology, topology, and preferred orientation of growth in a specific specimen of CrO$_2$ or hybrid composite of thin films or powder compacts.

Assuming a maximum $\mu = 4\mu_B$ value per crystal unit cell, a pure CrO$_2$ in the ferromagnetic structure would have a maximum $M_s$ value of 132.57 emu/g, i.e., equal to the spontaneous magnetization $M_0$. This is possible to be observed in whole at low temperature, precisely near absolute zero value. Against this theoretical value, Bajpai and Nigam[117] reported an $M_s = 135$ emu/g value at 5 K in a highly pure CrO$_2$ powder. All other reports on granular CrO$_2$ envisage occurrence of a much lower $M_s$ value in the range of 100-115 emu/g only.[102,176] This is quite feasible in $s$-canting structure, especially at the surface, and then it becomes increasingly important in small particles, which involve relatively large surfaces or surface interfaces in the surface stabilized CrO$_2$ particles or the CrO$_2$ reinforced polymer or ceramic composites.

For instance, Fig. 27 portrays the magnetization curves at 5 K of magnetic field aligned (alignment is parallel to the easy axis) acicular shaped CrO$_2$ particles.[100] Notice, when the applied field is parallel to the easy axis (alignment direction), the hysteresis loop is close to a near square with a high remanence and shows bit large value of $H_c \sim 1000$ Oe, with $M_s \sim 110$ emu/g. When the magnetic field is applied perpendicular to the easy axis, the hysteresis loop becomes more gradual and rounded with a much-reduced value of remanence as shown in the figure. There is no much change in $H_c$ value, possible in imperfect alignment of single magnetic $H_c$ domains. A perfect alignment is feasible only in a specimen of monodispersed single crystallites.

A CrO$_2$ nanopowder (of microstructure as given in Fig. 7), obtained from a polymer precursor, presents a rectangular hysteresis loop, which yields $M_s = 95.3$ emu/g; remanence $M_r = 74.2$ emu/g and $H_c = 1050$ Oe as measured at room temperature. The values were measured of sample annealed at 350°C for 2 h

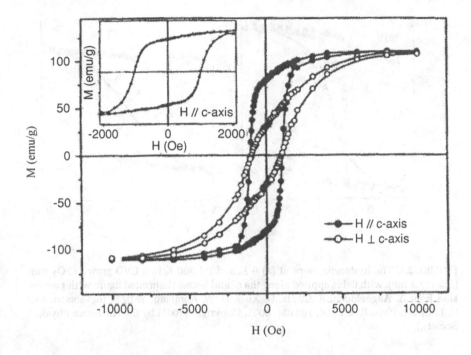

**FIGURE 27.** The hysteresis loops in magnetic field aligned $CrO_2$ powders at 5 K; field parallel to the easy axis (•) and perpendicular to the easy axis (o). The inset is a close-up of the loop (•) (Reprinted figure with permission from J. Dai and J. Tang, Phys. Rev. B, 63, 054434, 2001. Copyright (2001) by the American Physical Society.)

and then washed in benzene under heating conditions (with ultrasonic stirring) at 50-70°C for 1-2 h in order to recover from due carbon by-product impurities. Bulk $CrO_2$ powder of DuPont has bit larger $M_s \sim 115$ emu/g with a smaller $H_c = 600$ Oe value (1000 Oe at 5 K) at room temperature.[102] Increased surface or surface anisotropy seems to favor spin canting in small $CrO_2$ particles according to the reduced $M_s$ value. Assuming an arbitrary $\theta = 20°$ value, as used earlier by Burdett et al.[187], successfully reproduces a $M_s$ value of 89.6 emu/g. Possible that significant part of $Cr^{4+}$ at the surface is converted as $Cr^{3+}$ and that supports the canting. This is responsible for improving the $H_c$ value (with superior MR, mechanical, and other properties) of $Cr^{3+}$ surface stabilized $CrO_2$ of small particles.

Figure 28 presents an example of a thin $CrO_2$ film ($t = 137.5$ nm). It had been developed by a CVD process with a $CrO_2Cl_2$ liquid precursor on $TiO_2$ substrate.[113] The two hysteresis loops are measured at (a) 6 K and (b) 300 K in order to evaluate the thermal effects on the s-dynamics. It can be seen that the easy-axis (in-plane c-axis) hysteresis loops are rectangular like irrespective of the temperature. At the low temperature, an improved $M_s = 650$ emu/cc (or 132 emu/g)) value lies along with an improved $H_c = 185$ Oe value relative to the room temperature value $M_s = 465$ emu/cc ($H_c = 130$ Oe). No significant $H_c$ value appears in measuring

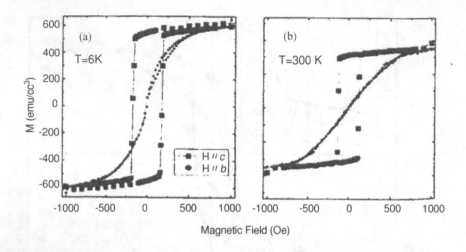

**FIGURE 28.** The hysteresis loops at (a) 6 K and (b) 300 K in a CVD grown $CrO_2$ film ($t = 137.5$ nm), with fields applied along the c- and b-axes (Reprinted figure with permission from A. Anguelouch, A. Gupta, G. Xiao, D. W. Abraham, Y. Ji, S. Ingvarsson, and C. L. Chien, Phys. Rev. B, 64, 180408, 2001. Copyright (2001) by the American Physical Society.)

with fields applied along the $b$-axis (the hard axis of the magnetization). The results infer that preferred orientations of magnetic domains with respect to the applied filed plays a determining role in engineering rectangular hysteresis loops, with useful $H_c$-values, of such $CrO_2$ films.

One of the standard methods for determining the value of $T_C$ in such examples of ferro or ferrimagnetic materials involves the measurement of thermomagnetogram by heating the specimen as a function of $T$ under a specific value of applied magnetic field. As we described elsewhere,[189–191] a low value of field, i.e., a few hundreds of Oe (insufficient to reach the $M_s$ value), is often selected for allocating precisely the $T_C$ point at which the induced magnetic moment drops to almost a zero value as $T \rightarrow T_C$. Otherwise, as the $T_C$ approaches, the curve no longer undergoes so sharp transition. This is demonstrated with a typical $M_s$ vs $T$ plot in Fig. 29 for a sample of thin $CrO_2$ films ($t = 1.3$ μm). The value of $M_s$, which is 650 emu/cm$^3$ at $T \sim 0$ K, has decreased stiffly to 471 emu/cm$^3$ at room temperature and then dropped rather rapidly to almost a zero value at $T_C = 391$ K as marked over the curve. The thermomagnetogram well below $T_C$ follows a Bloch $T^{3/2}$ law with the spin-wave stiffness constant $D \sim 1.8 \times 10^{-40}$ J m$^2$, indicating that normal ferromagnetic spin waves are excited.[48,180] They are studied and confirmed with ferromagnetic resonance, where the Gilbert damping parameter is very small.[192]

Figure 30 shows typical transverse Kerr loops for $CrO_2$ thin films ($t = 0.1$ to1 μm) grown from $CrO_3$ on two different substrates (a) $Al_2O_3$ (110) and (b) $TiO_2$ (110). As described by Ranno et al.,[107] both the loops were recorded at

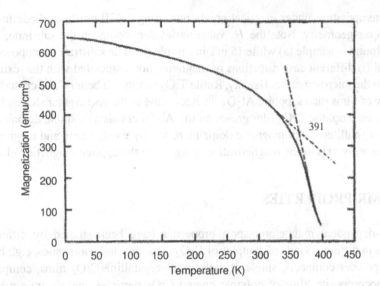

**FIGURE 29.** The variation of MS in a CrO$_2$ film measured during heating with a SQUID magnetometer (Reprinted figure with permission from H. Brändle, D. Weller, S. S. P. Parkin, J. C. Scott, P. Fumagalli, W. Reim, R. J. Gambino, R. Ruf, and G. Güntherodt, Phys. Rev. B, 46, 13889-13895, 1992. Copyright (1992) by the American Physical Society.)

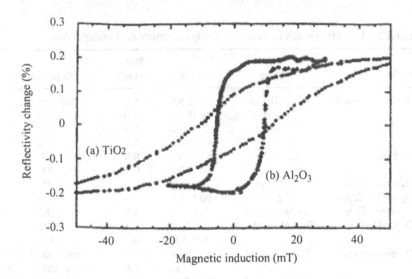

**FIGURE 30.** Transverse Kerr loops for CrO$_2$ films grown on Al$_2$O$_3$ (110) and TiO$_2$ (110) substrates (Reprinted with permission from L. Ranno, A. Barry, and J. M. D. Coey, Production and magnetotransport properties of CrO$_2$ films, J. Appl. Phys. 81(8), 5774-5776 1997. Copyright [1997], American Institute of Physics.)

room temperature under identical conditions using a 670 nm laser diode in the transverse geometry. Note the $H_c$ value varies depending on the substrate, i.e., 7 mT found in sample (a) while 15 mT in sample (b). The substrate is supposed to control (i) different easy directions of magnetization associated with the textures and (ii) the microstructure. Usually, Rutile $TiO_2$ supports a better $CrO_2$ crystalline quality of films than sapphire $Al_2O_3$.[107] According to the rectangular shape of the hysteresis loop, the $CrO_2$ film grown on the $Al_2O_3$ consists of single domains of $CrO_2$ crystallites. Such magnetic domains relatively easily rotate and magnetize with the easy axis of the magnetization lying along the applied magnetic field.

## 7. GMR PROPERTIES

The s-dependent magnetotransport properties have been studied by different groups in different kinds of samples of $CrO_2$ ceramics and composites, e.g., bulk $CrO_2$ powder compacts, single crystal or polycrystalline $CrO_2$ films, compacts or nanocomposite films of polymer coated $CrO_2$ particles, and ceramic matrix composites (CMC), mainly with $TiO_2$ and $Cr_2O_3$ as the matrix.[97,111,118,176,177,183] Selected materials of representative MR values are given in Table 5. A common strategy had been developing (i) the basic understanding of the tunneling effects in such materials, (ii) the MR-value for potential use in novel spintronic devices, (iii) expertise and knowledge of devising GMR devices and sensors,

**TABLE 5.** The MR values in selected $CrO_2$ ceramics and composites

| Sample | Nature | MR (%) | Field (kOe)* | References |
|---|---|---|---|---|
| Bulk $CrO_2$ powder | Pellet | 28 (5 K) | 1.0 | DuPont (102) |
| | Pellet | 29 (5 K) | 1.0 | 176 |
| | Pellet | 42 (4.2 K) | 1.5 | 177 |
| | Aligned | 41 (5 K) | 1.0 | 100 |
| $CrO_2$ film | Granular | 5 (70 K) | 0.8 | 97 |
| | | 26 (7 K) | 0.8 | 97 |
| $CrO_2$-$Cr_2O_3$ film | Granular | 27 (5K) | 1.0 | 95 |
| $CrO_2$ – 75% $Cr_2O_3$ | Cold pressed | 50 (5 K) | — | 176 |
| $CrO_2$ –78% $TiO_2$ | Cold pressed | 7 (77 K) | 1.2 | 183 |
| $CrO_2$-$Cr_2O_3$/Co | Tunnel junction | 1.5 (150 K) | 0.1 | 193 |
| | | 8 (5 K) | 0.1 | 193 |
| $CrO_2$ nanopowder | Cold pressed | 0.2 (300 K) | 1.0 | Unpublished work |
| | | 3.0 (77 K) | 1.0 | Unpublished work |
| $CrO_2$ – polystyrene | Cold pressed | 14 (77 K) | 1.5 | MicroMagnetics(118) |
| | | 8.2 (300 K) | 1.1 | 118 |
| $CrO_2$-PVA films | Granular | 11 (77 K) | 1.5 | Unpublished work |
| | | 6.8 (300 K) | 1.1 | Unpublished work |

The temperature at which the MR value is measured is given in parentheses.
* The magnetic field at which the included MR value is reached.

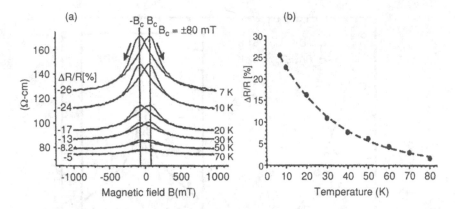

**FIGURE 31.** Intergranular tunneling MR in a polycrystalline $CrO_2$ film on $Al_2O_3$ (0001) substrate at selected temperatures as indicated over the curves and (b) exponential decrease of the MR value with temperature (Reprinted from J. Magn. & Magn. Mater., 211, M. Rabe, J. Dreßen, D. Dahmen, J. Pommer, H. Stahl, U. Rüdiger, G. Güntherodt, S. Senz, and D. Hesse, Preparation and characterization of thin ferromagnetic $CrO_2$ films for applications in magnetoelectronics, 314-319, (2000), with permission from Elsevier.)

(iv) newer GMR materials of superior properties and applications and (v) miniaturizing the applications of the kinds of GMR or magnetic sensors. As already mentioned above, a natural surface barrier oxide layer of $Cr_2O_3$ ($t$ = 2-3 nm) on $CrO_2$ serves extremely well as the tunnel barrier.[95,102,176] The $CrO_2$ grain boundaries or surface layers often impart an amorphous $CrO_2$ or CrOOH structure.[102,126] The surface interface in a hybrid $CrO_2$ composite interfaces hybridized properties of interest in a composite electronic structure by modulating the $s$-dynamics.

For example, a $CrO_2$/native oxide/Co junction hardly supports a MR value, i.e., 1.5% at 150 K and 8% at 5 K.[193] Pronounced MR effects occur in granular $CrO_2$ of films or powder compacts. Several research groups studied magnetotransport properties in thin $CrO_2$ films grown on selected ceramic substrates with different methods. In Fig. 31a, a typical CVD grown polycrystalline $CrO_2$ film on $Al_2O_3$ (0001) substrate, after Rabe et al.,[97] displays an example of MR primarily arising by the mechanism of the intergranular tunneling. At 7 K, the MR value reaches to be as large as −26%. As given in Fig. 31b, an exponential decrease of the MR value is found with raising the temperature in this example. A maximum value of resistivity (also MR) is found at an applied field $B_c = \pm$ 80 mT, i.e., equal to the $H_c$ value.

As another example, Fig. 32 shows the temperature dependence of MR value in a CVD grown granular $CrO_2$ film on a Si (100) substrate. As much MR value as $\sim$ −20% occurs at 2 K. As plotted in Fig. 32, the MR value decreases rapidly as $\exp\left(-T/T_{mr}\right)$ to roughly 1.2% at 240 K with $T_{mr}^{ADS} = 61$ K and $T_{mr}^{PAS} = 62$ K.[111] In this regard, it is argued that in granular $CrO_2$ films, the MR value is independent of the relative orientation of the current and magnetization. The variation of the

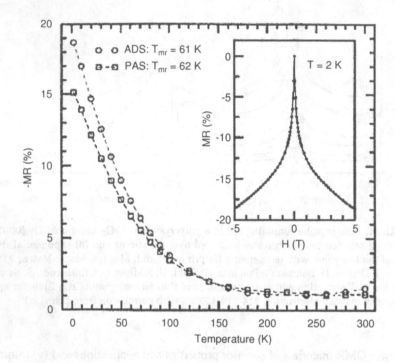

**FIGURE 32.** The temperature dependent variation of MR value in a granular $CrO_2$ film on a Si (100) substrate. The inset shows the field dependence behaviour of the MR value at 2 K (Reprinted from J. Magn. & Magn. Mater., 239, J. J. Liang, S. F. Lee, Y. D. Yao, C. C. Wu, S. G. Shyu, and C. Yu, Magnetotransport study of granular chromium dioxide thin films prepared by the chemical vapor deposition technique, 213-216, (2002), with permission from Elsevier.)

low-temperature MR value as function of the magnetic field is shown in the inset. The suppression of MR over high temperatures is believed to occur primarily due to the increase of the $s$-independent hopping channel at the expense of the $s$-dependent contribution to the conductance. Notice, in this specific example, a thermal annealing did not affect the MR value much, except leading to improve the room temperature $\sigma$-value from roughly 50 to above 75 $\mu\Omega$m.[111] In usual ferromagnetic metal conductors, a thermal annealing often adversely leads to reduce the final $\sigma$-value owing to effective grain-growth and expense of defects and imperfections if any present.

In Fig. 33, a cold-pressed powder compact from the Dupont $CrO_2$ powder has a similar variation of the low-temperature MR value as a function of the magnetic field as shown above in granular $CrO_2$ films. The powder had single crystal needle-shaped particles of 400 nm length and 9:1 aspect ratio. As large MR value as $\sim$ $-28\%$ has been observed at an applied field of $\sim$ 1.0 kOe. In the inset in Fig. 33 are shown the low-field MR and magnetic hysteresis loop which present a very good correlation between the two kinds of the interdependent properties.

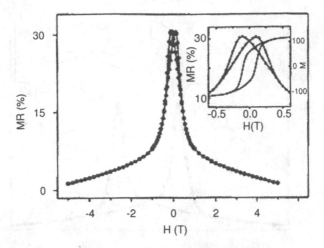

**FIGURE 33.** The field dependence of MR in a cold-pressed bulk powder compact at 5 K. Inset shows the low field MR and hysteresis loop (Reprinted with permission from J. Dai, J. Tang, H. Xu, L. Spinu, W. Wang, K. Wang, A. Kumbhar, M. Li, and U. Diebold, Characterization of the natural barriers of intergranular tunnel junctions: $Cr_2O_3$ surface layers on $CrO_2$ nanoparticles, Appl. Phys. Lett. 77(18), 2840-2842 2000. Copyright [2000], American Institute of Physics.)

Furthermore, Dai *et al.*[100] studied the MR in field-aligned needle-shaped $CrO_2$ particles of Dupont, with the magnetic field applied along the easy axis (lying along the needle) of magnetization [001]. Figure 34 shows the optimal value of MR to be $\sim -41\%$ at 5 K and is characterized by two well-separated peaks near the $H_c = \pm 1000$ Oe value. Unlike nonaligned powder, whose MR curves are typically "butterfly" shaped, the aligned powder has sharp switching of the resistance between the high and low states in a small field region in the vicinity of the $H_c$ field. Notice there is no change in the MR when the direction of the current is changed from the [001] direction to perpendicular one and then back to the [001] position. As suggested by Dai *et al.*,[100] this kind of magnetic tunneling junction switching behavior can be attributed to the fact that there is a narrow distribution of the switching field of the magnetic moments of the aligned particles.

As can be seen from the hysteresis loop of the aligned $CrO_2$ powders in Fig. 27, when decreasing the field from the saturation to a zero value, nearly all moments remain aligned in the original direction and result in the low-resistance state near the zero field. The situation is different in the case of nonaligned $CrO_2$ particles in which a significant part of the moments remains to be misaligned as soon as the field is reduced to a near zero value, leading to a rather large increase in the resistance under identical conditions of other parameters. A comparison of the results with those in the inset in a merely cold pressed sample clearly implies that the LFMR value and the sharpness of the switching effect in the nonaligned sample are much lower than those in the aligned sample.

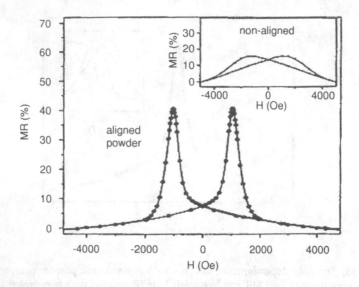

**FIGURE 34.** The MR in magnetic field aligned $CrO_2$ particles at 5 K, with that of cold pressed and nonaligned sample in the inset (Reprinted figure with permission from J. Dai and J. Tang, Phys. Rev. B , 63, 054434, 2001. Copyright (2001) by the American Physical Society.)

Figure 35 shows the temperature dependence of MR in a $CrO_2$ nanopowder (processed at 350°C from a polymer precursor) compact under an applied pressure of 500 MPa at room temperature followed by annealing the pellet at 375°C in air. At 77 K, the MR value is a maximum of −3.0%, which is decreasing exponentially with the increase of temperature and attains a reduced value of −0.2% at room temperature. A derived $CrO_2$-PVA nanocomposite of films ($t = 125\,\mu m$) from such nanopowder involve a dramatically improved value of room temperature MR as depicted in Fig. 36. A maximum value of MR of 6.8% is found in such films of 3.0 wt% $CrO_2$ at an applied magnetic field of 1.1 kOe. A relatively small value of MR of −6.3% is found in the sample of 2.0 wt% $CrO_2$ while a still lower value of −5.8% in that of 1.0 wt% $CrO_2$. This specific kind of hysteretic MR as observed in this example is associated with the alignment of the magnetization of single domain $CrO_2$ particles since the maximum resistance practically coincides with the $H_c$ of the sample.

In general, there are reports of MR value as high as ∼ −50% (Table 5) but at very low temperatures (∼5 K). Only very few reports are available for MR at room temperature. Manoharan et al.[177] observed a value of ∼ −42% at 4.2 K whereas Coey et al.[176] observed ∼ −29% value at 5 K in a cold pressed bulk $CrO_2$ powder. An improved value ∼ −50% appeared in a dilute magnetic composite of $CrO_2$ −75% $Cr_2O_3$ but still at low temperatures only. Recently, Chen et al.[118] reported a room temperature MR value 8.2% in a cold pressed sample of polystyrene coated $CrO_2$ particles (of needle shape of ∼ 40 nm diameter from the Micro Magnetics

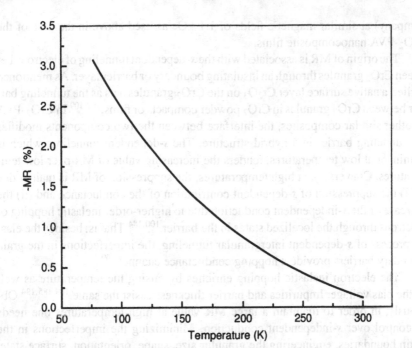

**FIGURE 35.** Temperature-dependent variation of MR in a cold pressed CrO₂ nanopowder.

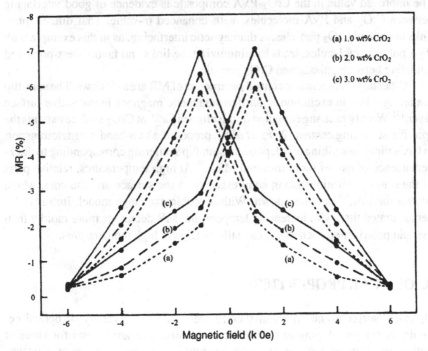

**FIGURE 36.** The field dependence of MR in CrO₂-PVA hybrid nanocomposite films at room temperature; (a) 1.0, (b) 2.0, and (c) 3.0 wt% CrO₂ contents.

company) at similar magnetic fields of 1.1 kOe as used above in the case of the $CrO_2$-PVA nanocomposite films.

The origin of MR is associated with the $s$-dependent tunneling of electrons between $CrO_2$ granules through an insulating boundary or barrier layer. As mentioned earlier, a native surface layer $Cr_2O_3$ on the $CrO_2$ granules acts as the tunneling barrier between $CrO_2$ granules in $CrO_2$ powder compacts or films.[102,109] In $CrO_2$-PVA or other similar composites, the interface between the two components modifies the tunneling barrier in a hybrid structure. The $s$-dependent tunneling, which is dominant at low temperatures, renders the increasing value of MR over low temperatures. Conversely, at high temperatures, the suppression of MR is mainly due to (i) the suppression of $s$-dependent contribution of the conductance and (ii) the increase of the $s$-independent conduction due to higher-order inelastic hopping of electrons through the localized states in the barrier.[100,184] That is, besides the elastic process of $s$-dependent intergranular tunneling, the imperfections in the grain boundary barriers provide a hopping conductance channel.[109]

The electron inelastic hopping enriches by raising the temperature as well as the bias voltage. Impurities and barrier thickness assist the same.[109,185,194] Obviously, in order to maintain a large MR value at high temperatures one needs to control over $s$-independent conduction. Minimizing the imperfections in the grain boundaries, engineering the granular size, shape, orientation, surface state, and adjusting the $\Delta$-value can be used to engineer a room temperature MR.[109,176] The improved value in the $CrO_2$-PVA composite is evidence of good interfacing between $CrO_2$ and PVA molecules, with enhanced $\sigma$-value. That dilutes interlinks in divided $CrO_2$ particles. A diamagnetic interfacing, as in this example with PVA polymer molecules, leads to minimizing the links and favors the $s$-polarized tunneling between misaligned $CrO_2$ granules.

Other high-temperature control parameters of MR are as follows. The spin-flip scattering arises in excitation of antiferromagnetic magnons in the native surface layer.[176] Weakly exchange coupled or misaligned $Cr^{3+}$ at $CrO_2$ surface extends the spin-flip scattering centres. Watts et al.[178] proposed a two-band magnetotransport in $CrO_2$ films, describing a $T$-dependent spin flip scattering corresponding to the $T$-dependence of carrier mobilities and MR.[109] At high temperatures, relative parts of the spin polarizations being suppressed from the surface and the core, which control the MR,[100] are not known. With a double-exchange model, Itoh et al.[195] demonstrated that, with increasing temperature, MR decreases more rapidly than the spin polarization such that it can still be large at high temperatures.

## 8. OPTICAL PROPERTIES

Optical absorption and emission spectra are studied in a variety of optical ceramics containing doping of $Cr^{4+}$ cations. Identified ceramic hosts for kinds of optical applications include sapphire, yttrium-iron garnets, forsterite, silicates, and silicate glasses.[196-200] There are not many reports on the optical properties

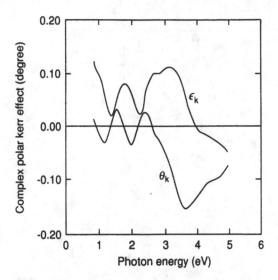

**FIGURE 37.** Magneto-optical, polar Kerr rotation $\theta_k$, and the Kerr ellipticity $\epsilon k$ of $CrO_2$ film, in the 0.8-5 eV range, measured in $\pm20$ kOe magnetic-field at room temperature (Reprinted figure with permission from H. Brändle, D. Weller, S. S. P. Parkin, J. C. Scott, P. Fumagalli, W. Reim, R. J. Gambino, R. Ruf, and G. Güntherodt, Phys. Rev. B, 46, 13889-13895, 1992. Copyright (1992) by the American Physical Society.)

of $CrO_2$ ceramics or composites. Brändle *et al.*[41] reported the magneto-optical properties of polycrystalline $CrO_2$ films ($t = 1.3$ μm) grown on a sapphire (0001) substrate. The films were obtained with CVD method by decomposing $CrO_3$ at 200°C in a closed reactor. Figure 37 shows the magneto-optical polar Kerr rotation $\theta_k(\hbar\omega)$ and the Kerr ellipticity $\varepsilon_k(\hbar\omega)$ spectra measured at room temperature in the $0.8 \leq \hbar\omega \leq 5$ eV range. The Kerr rotation shows a maximum value $|\theta_k| = 0.154°$ at 3.7 eV. It is significantly below 0.1° in the energy range between 1.5 and 3 eV (826-413 nm), which is the interesting region for optical data storage and applications of $CrO_2$ as an optical material. The energy levels in 3d→3d electronic transitions in $Cr^{4+}$ ($3d^2$) system are well described with the Tanabe-Sugano diagram and that is available in the textbooks.[201-203] According to the diagram, the $Cr^{4+}$ cations in octahedral symmetry, as in the $CrO_2$ crystals, have three spin-allowed transitions. In the case of the absorption spectrum, they occur in terms of the transition energy in the sequence as follows:

$$^3T_1(F) \longrightarrow {}^3T_2(F),$$
$$\longrightarrow {}^3T_1(P), \text{ and}$$
$$\longrightarrow {}^3A_2(F),$$

where $^3T_1(F)$ is the ground electronic state. Unfortunately, no such absorption spectra are studied in a systematic manner in $CrO_2$ or similar $Cr^{4+}$-compounds. Nevertheless, $Cr^{3+}$-compounds are studied extensively with absorption as well

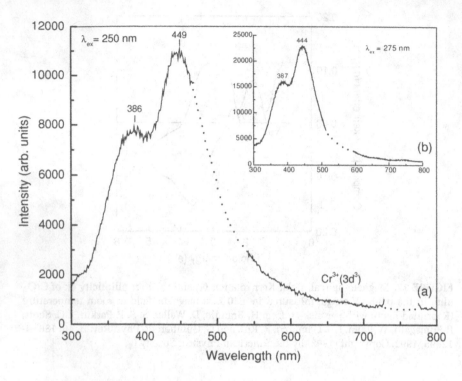

**FIGURE 38.** Photoluminescence spectra in $CrO_2$ nanoparticles measured by irradiating the specimen at (a) 250 nm and (b) 275 nm wavelengths from a xenon lamp.

as emission excitations.[202-206] This is because of the fact that $Cr^{3+}$ is the most common oxidation state of chromium cations.

The photoluminescence observed in 3d→3d electronic transitions in $Cr^{4+}$ ($3d^2$) in $CrO_2$ nanoparticles before and after a thin surface coating of Ag-metal are given in Fig. 38 and 39. The spectra were measured by irradiating the specimens at two selected wavelengths $\lambda_{exc}$ = 250 and 275 nm from a standard xenon lamp in order to confirm their origin of the electronic transitions. There is no significant change in average position of the spectrum on measuring at varied $\lambda_{exc}$ values in the uv-visible range. One of the obvious differences noticed is that the integrated intensity assumes selective values with the changing of $\lambda_{exc}$ value. This occurs in minor variation in the conditions of the excitation process followed by the radiative emission from the radiative electronic states, i.e., identified to be the $^3T_2(F)$ and $^3T_1(P)$ states. As a result, the photoluminescence spectrum consists of two distinct bands in the $^3T_1(F) \leftarrow^3 T_2(F)$ and $^3T_1(F) \leftarrow^3 T_1(P)$ electronic transitions, as given in Table 6, irrespective to the $\lambda_{exc}$ values. There is a signature of a very weak characteristic band due to the trace of $Cr^{3+}$ ($3d^3$) impurities (i.e., less than 1%) at ~680 nm.

**TABLE 6.** Average peak positions and relative peak intensities I in photoluminescence bands in virgin and Ag-coated $CrO_2$ nanoparticles

| Band | $CrO_2$ Position (nm) | $CrO_2$ I | Coated $CrO_2$ Position (nm) | Coated $CrO_2$ I | Transitions |
|---|---|---|---|---|---|
| Band I | 444 | 100 | 488 | 22 | $^3T_1(F) \leftarrow {}^3T_2(F)$ |
| Band II | 387 | 69 | 356 | 35 | $^3T_1(F) \leftarrow {}^3T_1(P)$ |

The I-values are given assuming a normalized value 100 in the most intense band.

As can be seen from the spectra in Fig. 39, one of the obvious effects of the surface Ag-coating of $CrO_2$ nanoparticles is that it diminishes the integrated intensity of the photoluminescence spectrum, i.e., roughly by an order of magnitude. Presumably, it cures the surface defects by interbridging a kind of ideal surface interference in associating with the native $CrO_2$ surfaces. In other words, the Ag-metal, being a diamagnetic of nature, extends a ferromagnetic—diamagnetic interfacing of a specific robust shell structure. This influences the $Cr^{4+}$ ($3d^2$) electronic excitation process and the intermediate cross relaxations by capping the ferromagnetic $CrO_2$ domains as observed. Several experiments are demanded in order to identify the specific processes and to resolve the underlying mechanisms.

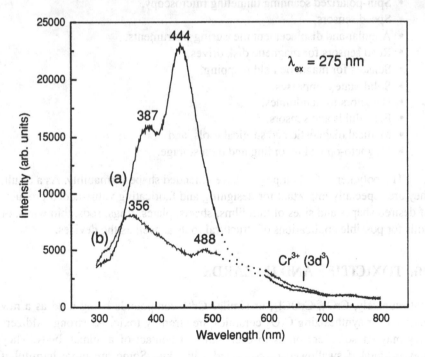

**FIGURE 39.** Modification in photoluminescence in (a) $CrO_2$ nanoparticles by (b) surface coating of Ag metal. The two spectra were measured by irradiating the specimens under identical conditions using a xenon lamp at 275nm wavelength.

## 9. APPLICATIONS

The efforts of developing $CrO_2$ ceramics and hybrid nanocomposites of controlled microstructure stem from unusually half-metallic ferromagnetic behavior of them as a new generation of spintronic materials. Identified applications include magnetic, electrical, GMR, or optical devices and sensors. The availability of nearly 100% spin polarization make such specific materials especially suitable for spin-polarized electron injectors, spin valves, and other magneto-electronic and magneto-optical applications.[41,42,48,79,94-100] $CrO_2$ nanoceramics and composites are emerging as the basic materials of spintronics. It is suggested that in tunneling junctions made of half-metals, like $CrO_2$ ceramics on both sides of an insulating barrier and with ideal interfaces, the GMR effect can be as high as 50%, and current across the structure would be either on or off.[102] Even with a resistance drop much lower than that, $CrO_2$ ceramics would still provide a much greater signal-to-noise ratio than conventional metals, making it possible to build devices to run at low voltage and high speed.

To be more specific, the GMR properties in the kinds of $CrO_2$ ceramics and hybrid composites offer widespread applications in numerous disciplines such as

- Magnetic random access memory (MRAM) in computers,
- Spin-polarized scanning tunneling microscopy,
- Speed sensors,
- Angular and displacement measuring instruments,
- Read sensors for magnetic disk drives,
- Sensors for magnetic field mapping,
- Solid-state compasses,
- Detectors for landmines,
- Ferrofluids and sensors,
- Medical diagnostic and surgical tools, and
- Magneto-optical recording and data storage.

The polymer $CrO_2$ composites have extended shape formability. As a result, they are especially important for designing and fabricating value-added products of desired shapes and sizes of thin films, sheets, plates, rings, rods, thin wires, or coils for possible applications of structural tools as well as the devices.

## 10. TOXICITIES AND HAZARDS

Unfortunately, $CrO_3$, $CrOCl_2$, or similar $Cr^{6+}$-compounds being used as a raw material for synthesizing $CrO_2$ ceramics are strongly toxic. As strong oxidizers they may cause severe burns to every area of contact of a human body. They may be fatal if swallowed or contacted with skin. Some are more harmful if inhaled, inviting severe damage of lungs and other sensitive parts of the human body or other creatures. Synthesizing $CrO_2$ in form of powder often involves an aqueous precursor solution of $CrO_3$ or in general a $Cr^{6+}$-compound. In the proposed polymer precursor method, the $Cr^{6+}$ cations ultimately are reduced to $Cr^{4+}$ cations as soon as admixing to a polymer, which reacts with $Cr^{6+}$ to form a metal ion-polymer complex capping in part of the polymer molecules. The resulting polymer complex as such is not as much toxic as the free $CrO_3$ (or other $Cr^{6+}$-compounds), which as such is highly hygroscopic and may spread in moist air (as a carrier gas) in the open atmosphere if not controlled. The involved experiments are not hazardous if carried out in a controlled manner in a specific chamber.

A chemical $Cr^{6+}$-vapor is used in CVD and other similar methods in the $CrO_2$ synthesis of films. It may be more hazardous if by chance it leaks out from the apparatus and spreads in the open air. Necessary care and precautions are advised while dealing with such experiments and the raw material, especially in industries.

# ACKNOWLEDGEMENTS

This work has been financially supported by the Ministry of Human Resource and Development (MHRD) and the Defense Research and Development Organization (DRDO), Government of India.

# REFERENCES

1. J. Jung, J. P. Franck, W. A. Miner, and M. A.-K. Mohamed, Effect of substitution of Bi, Ga, and Fe on the structure and superconducting transition of $Y_1Ba_2Cu_3O_{6.5+\delta}$, *Phys. Rev. B* **37**(13), 7510-7515 (1988).
2. Z. Iqbal, H. Eckhardt, F. Reidinger, A. Bose, J. C. Barry, and B. L. Ramakrishna, Microstructure and properties of the ~90-K superconductor $Bi_2Sr_{3-x}Ca_xCu_2O_{8+\delta}$, *Phys. Rev. B* **38**(1), 859-862 (1988).
3. S. Ram and K. A. Narayan, Effects of Bi(x) additives on microstructure and superconductivity in $YBa_{2-x}Bi_xCu_3O_{7-\delta}$, *Phys. Rev. B* **42**(13), 8627-8629 (1990).
4. V. P. S. Awana, A. Gupta, H. Kishan, M. Karppinen, H. Yamauchi, A. V. Narlikar, E. Galstyan, and I. Felner, Micro-structure and magnetization of the 80-K superconductor, $TbSr_2Cu_{2.7}Mo_{0.3}O_{7+\delta}$, *Physica C* **415**, 69-73 (2004).
5. A. Knizhnik, G. M. Reisner, and Y. Eckstein, Increase of $T_c$ in the 1:2:3 superconductors YBCO and $(Ca_xLa_{1-x})(La_uBa_{1-u})_2Cu_3O_y$ after slow cooling or low temperature annealing, *J. Phys. Chem. Solids* **66**, 1137-1144 (2005).
6. E. P. Wohlfarth, *Ferromagnetic Materials* (North Holland, New York, 1982).
7. R. S. Tebble and D. J. Craik, *Magnetic Materials* (Wiley-Interscience, London, 1969).
8. W. D. Kingery, H. K. Bowen, and D. R. Uhlmann, *Introduction to Ceramics* (John Wiley & Sons, New York, 1975).
9. K. Schwarz, $CrO_2$ predicted as a half-metallic ferromagnet, *J. Phys. F: Met. Phys.* **16**, L211-L215 (1986).
10. F. Heusler, *Verh. Dtsch. Phys. Ges.* **5**, 219 (1903).
11. R. A. de Groot, F. M. Mueller, P. G. van Engen, and K. H. J. Buschow, New class of materials: half-metallic ferromagnets, *Phys. Rev. Lett.* **50**(25), 2024-2027 (1983).
12. K. H. J. Buschow, *Handbook of Magnetic Materials* (North-Holland, Amsterdam, 1991).

13. H. Munekata, H. Ohno, S. von Molnar, A. Segmüller, L. L. Chang, and L. Esaki, Diluted magnetic III-V semiconductors, *Phys. Rev. Lett.* **63(17)**, 1849-1852 (1989).

14. H. Munekata, T. Penny, and L. L. Chang, Diluted magnetic III-V semiconductor structures, *Surf. Sci.* **267(1-3)**, 342-348 (1992).

15. H. Ohno, A. Shen, F. Matsukura, A. Oiwa, A. Endo, S. Katsumoto, and Y. Iye, (Ga,Mn)As: A new diluted magnetic semiconductor based on GaAs, *Appl. Phys. Lett.* **69(3)**, 363-365 (1996).

16. A. Oiwa, S. Katsumoto, A. Endo, M. Hirasawa, Y. Iye, H. Ohno, F. Matsukura, A. Shen, and Y. Sugawara, Nonmetal-metal-nonmetal transition and large negative magnetoresistance in (Ga, Mn)As/GaAs, *Solid State Commun.* **103(4)**, 209-213 (1997).

17. S. Koshihara, A. Oiwa, M. Hirasawa, S. Katsumoto, Y. Iye, C. Urano, H. Takagi, and H. Munekata, Ferromagnetic order induced by photogenerated carriers in magnetic III-V semiconductor heterostructures of (In,Mn)As/GaSb, *Phys. Rev. Lett.* **78(24)**, 4617-4620 (1997).

18. J. P. Bucher, D. C. Douglass, and L. A. Bloomfield, Magnetic properties of free cobalt clusters, *Phys. Rev. Lett.* **66(23)**, 3052-3055 (1991).

19. A. J. Cox, J. G. Louderback, and L. A. Bloomfield, Experimental observation of magnetism in rhodium clusters, *Phys. Rev. Lett.* **71(6)**, 923-926 (1993).

20. S. E. Ampsel, J. W. Emmert, J. Deng, and L. A. Bloomfield, Surface-enhanced magnetism in nickel clusters, *Phys. Rev. Lett.* **76(9)**, 1441-1444 (1996).

21. S. Ram and H. J-Fecht, Millimeter sized ferromagnetic Fe-clusters: formation by mechanical attrition, microstructure and magnetic properties, *Mater. Trans., JIM*, **41(7)**, 754-760 (2000).

22. S. Ram, Surface structure and surface-spin induced magnetic properties and spin-glass transition in nanometer Co-granules of fcc crystal structure, *J. Mater. Sci.* **35**, 3561-3571 (2000).

23. S. Ram, Allotropic phase transformations in hcp, fcc and bcc metastable structures in co-nanoparticles, *Mater. Sci. & Eng. A* **304-06**, 923-927 (2001).

24. S. Ram and G. P. Johari, Glass-liquid transition in hyperquenched metal alloys, *Phil. Mag. B* **61**, 299-310 (1990).

25. S. Ram, Calorimetric investigation of structural relaxation in supercooled $Ni_{75}Al_{22}Zr_2B$ amorphous alloy, *Phys. Rev. B* **42(15)**, 9582-9586 (1990).

26. S. Ram and J. C. Joubert, Crystallization of small and separated magnetic particles of $Nd_2Fe_{14}B$ alloy, *J. Appl. Phys.* **72**, 1164-1171 (1992).

27. S. Ram and J. C. Joubert, Production of substantially stable Nd-Fe-B hydride (magnetic) powders using chemical dissociation of water, *Appl. Phys. Lett.* **61(5)**, 613-615 (1992).

28. P. S. Frankwicz, S. Ram, and H. J. Fecht, Enhanced microhardness in $Zr_{65.0}Al_{7.5}Ni_{10.0}Cu_{17.5}$ amorphous rods on coprecipitation of nanocrystallites through supersaturated intermediate solid phase particles, *Appl. Phys. Lett.* **68(20)**, 2825-2827 (1996).

29. S. Ram and P. S. Frankwicz, Granular GMR sensors of Co-Cu and Co-Ag nanoparticles synthesized through a chemical route using $NaBH_4$, *Phys. Stat. Sol. (a)* **188(3)**, 1129-1140 (2001).

30. G. Kumar, J. Eckert, S. Roth, W. Loser, S. Ram, and L. Schultz, Magnetic properties of Nd-Fe-Co(Cu)-Al-B amorphous alloys prepared by nonequilibrium techniques, *J. Appl. Phys.* **91**, 3764-3768 (2002).

31. M. Sagawa, S. Hirosawa, H. Yamamoto, S. Fujimura, and Y. Matsuura, Nd-Fe-B permanent magnet materials, *Jpn. J. Appl. Phys.* **26(6)**, 785-800 (1987).

32. E. Claude, S. Ram, I. Gimenez, P. Chaudouet, D. Boursier, and J. C. Joubert, Evidence of a quantitative relationship between the degree of hydrogen intercalation and the coercivity of the two permanent magnet alloys $Nd_2Fe_{14}B$ and $Nd_2Fe_{11}Co_3B$, *IEEE. Trans. Magn.* **29**, 2767-2769 (1993).

33. S. Ram, Kinetics of the desorption of interstitial hydrogen in stable $Nd_2Fe_{14}BH_x$, $x \leq 5$, *Phys. Rev. B* **49(14)**, 9632-9638 (1994).

34. S. Ram, E. Claude, and J. C. Joubert, Synthesis, stability against air and moisture corrosion, and magnetic properties of finely divided loose $Nd_2Fe_{14}BH_x$, $x \leq 5$, hydride powders, *IEEE. Trans. Magn.* **31**, 2200-2208 (1995).

35. S. Ram, M. Febri, H. J. Fecht, and J. C. Joubert, Synthesis of $Nd_2Fe_{14}B$ nanocrystals using interstitial hydrides, *Nanostruct. Mater.* **6**, 473-476 (1995).

36. 36. S. Haldar, S. Ram, P. Ramachandrarao, and H.D. Banerjee, Synthesis of high-energy-density $Pr_2Fe_{14-x}Co_xB$, $x \leq 3$, magnets for practical applications, *Bull. Mater. Sci.* **18**, 963-974 (1995).

37. S. Ram, Synthesis, magnetic properties, and formalism of magnetic properties of high-quality refined $Nd_2Fe_{14}B$ powders for permanent magnet devices, *J. Mater. Sci.* **32**, 4133-4148 (1997).

38. S. Ram, H. J-Fecht, S. Haldar, P. Ramachandrarao, and H. D. Banerjee, Calorimetric study of the desorption of the interstitial hydrogen atoms in ferromagnetic $Nd_2Fe_{14}BH_x$ ($x < \sim 5$) microcrystals, *Phys. Rev. B* **56(2)**, 726-737 (1997).

39. S. Ram, H. D. Banerjee, S. Haldar, and P. Ramachandrarao, Formation of $Nd_2Fe_{14}BH_x$, $x \leq 5$, hydride by milling of anhydride particles in toluene in a closed reactor, *Bull. Mater. Sci.* **20**, 1049-1058 (1997).

40. J. L. Tsai, T. S. Chin, and S. K. Chen, Coercivity mechanism and microstructure study of sputtered Nd-Fe-B/X/Si(111) (X = W, Pt) films, *Jpn. J. Appl. Phys.* **38(10)**, 5879-5884 (1999).

41. H. Brändle, D. Weller, S. S. P. Parkin, J. C. Scott, P. Fumagalli, W. Reim, R. J. Gambino, R. Ruf, and G. Güntherodt, Magneto-optical properties of $CrO_2$, *Phys. Rev. B* **46(21)**, 13889-13895 (1992).

42. S. P. Lewis, P. B. Allen, and T. Sasaki, Band structure and transport properties of $CrO_2$, *Phys. Rev. B* **55(16)**, 10253-10260 (1997).

43. I. I. Mazin, Robust half metallicity in $Fe_xCo_{1-x}S_2$, *Appl. Phys. Lett.* **77(19)**, 3000-3002 (2000).

44. R. Yamamoto, A. Machida, Y. Moritomo, and A. Nakamura, Electronic structure of half-metallic materials: comparison of pyrite with doped manganite, *Physica B* **281-282**, 705-706 (2000).

45. I. S. Elfimov, S. Yunoki, and G. A. Sawatzky, Possible path to a new class of ferromagnetic and half-metallic ferromagnetic materials, *Phys. Rev. Lett.* **89(21)**, 216403 (2002).

46. I. Galanakis, P. H. Dederichs, and N. Papanikolaou, Slater-Pauling behavior and origin of the half-metallicity of the full-Heusler alloys, *Phys. Rev. B* **66**, 174429 (2002).

47. C. M. Fang, G. A. de Wijs, and R. A. de Groot, Spin-polarization in half-metals, *J. Appl. Phys.* **91(10)**, 8340-8344 (2002).

48. J. M. D. Coey and M. Venkatesan, Half-metallic ferromagnetism: example of $CrO_2$, *J. Appl. Phys.* **91(10)**, 8345-8350 (2002).

49. Z. Szotek, W. M. Temmerman, A. Svane, L. Petit, and H. Winter, Electronic structure of half-metallic double perovskites, *Phys. Rev. B* **68**, 104411 (2003).

50. M. Zhang, X. Dai, H. Hu, G. Liu, Y. Cui, Z. Liu, J. Chen, J. Wang, and G. Wu, Search for new half-metallic ferromagnets in semi-Heusler alloys NiCrM (M = P, As, Sb, S, Se and Te), *J. Phys.: Condens. Matter* **15**, 7891-7899 (2003).

51. B. Nadgorny, I. I. Mazin, M. Osofsky, R. J. Soulen, P. Broussard, R. M. Stroud, D. J. Singh, V. G. Harris, A. Arsenov, and Y. Mukovskii, Origin of high transport spin polarization in $La_{0.7}Sr_{0.3}MnO_3$: direct evidence for minority spin states, *Phys. Rev. B* **63**, 184433 (2001).

52. 52. S. Picozzi, A. Continenza, and A. J. Freeman, $Co_2MnX$ (X = Si, Ge, Sn) Heusler compounds: an *ab initio* study of their structural, electronic, and magnetic properties at zero and elevated pressure, *Phys. Rev. B* **66**, 094421 (2002).

53. R. Vidya, P. Ravindran, A. Kjekshus, and H. Fjellvåg, Huge magneto-optical effects in half-metallic double perovskites, *Phys. Rev. B* **70**, 184414 (2004).

54. Y. Sakuraba, J. Nakata, M. Oogane, Y. Ando, H. Kato, A. Sakuma, T. Miyazaki, and H. Kubota, Magnetic tunnel junctions using B2-ordered $Co_2MnAl$ Heusler alloy epitaxial electrode, *Appl. Phys. Lett.* **88**, 022503 (2006).

55. H. Y. Hwang, S.-W. Cheong, N. P. Ong, and B. Batlogg, Spin-polarized intergrain tunneling in $La_{2/3}Sr_{1/3}MnO_3$, *Phys. Rev. Lett.* **77(10)**, 2041-2044 (1996).

56. P. Bach, A. S. Bader, C. Rüster, C. Gould, C. R. Becker, G. Schmidt, L. W. Molenkamp, W. Weigand, C. Kumpf, E. Umbach, R. Urban, G. Woltersdorf, and B. Heinrich, Molecular-beam epitaxy of the half-Heusler alloy NiMnSb on (In,Ga)As/InP (001), *Appl. Phys. Lett.* **83(3)**, 521-523 (2003).

57. Y. S. Dedkov, U. Rudiger, and G. Guntherodt, Evidence for the half-metallic ferromagnetic state of $Fe_3O_4$ by spin-resolved photoelectron spectroscopy, *Phys. Rev. B* **65(6)**, 064417 (2002).

58. Z. Szotek, W. M. Temmerman, A. Svane, L. Petit, G. M. Stocks, and H. Winter, *Ab initio* study of charge order in $Fe_3O_4$, *Phys. Rev. B* **68**, 054415 (2003).

59. J. S. Parker, P. G. Ivanov, D. M. Lind, P. Xiong, and Y. Xin, Large inverse magnetoresistance of $CrO_2$/Co junctions with an artificial barrier, *Phys. Rev. B* **69**, 220413 (2004).

60. L. Wang, K. Umemoto, R. M. Wentzcovitch, T. Y. Chen, C. L. Chien, J. G. Checkelsky, J. C. Eckert, E. D. Dahlberg, and C. Leighton, $Co_{1-x}Fe_xS_2$: a tunable source of highly spin-polarized electrons, *Phys. Rev. Lett.* **94**, 056602 (2005).

61. G. Banach, R. Tyer, and W. M. Temmerman, Study of half-metallicity in LSMO, *J. Magn. & Magn. Mater.* **272-276**, 1963-1964 (2004).

62. K. -I. Kobayashi, T. Kimura, H. Sawada, K. Terakura, and Y. Tokura, Room-temperature magnetoresistance in an oxide material with an ordered double-perovskite structure, *Nature* **395**, 677-680 (1998).

63. R. Weht, and W. E. Pickett, Half-metallic ferrimagnetism in $Mn_2VAl$, *Phys. Rev. B* **60(18)**, 13006-13010 (1999).

64. D. J. Singh, Magnetoelectronic effects in pyrochlore $Tl_2Mn_2O_7$: role of Tl-O covalency, *Phys. Rev. B* **55(1)**, 313-316 (1997).

65. H. Y. Hwang, and S.-W. Cheong, Low-field magnetoresistance in the pyrochlore $Tl_2Mn_2O_7$, *Nature* **389**, 942-944 (1997).

66. P. Mavropoulos, I. Galanakis, and P. H. Dederichs, Multilayers of zinc-blende half-metals with semiconductors, *J. Phys.: Condens. Matter* **16**, 4261-4272 (2004).

67. M. R. Castell, P. L. Wincott, N. G. Condon, C. Muggelberg, G. Thornton, S. L. Dudarev, A. P. Sutton, and G. A. D. Briggs, Atomic-resolution STM of a system with strongly correlated electrons: NiO (001) surface structure and defect sites, *Phys. Rev. B* **55(12)**, 7859-7863 (1997).

68. H. Akai, Ferromagnetism and its stability in the diluted magnetic semiconductor (In, Mn)As, *Phys. Rev. Lett.* **81(14)**, 3002-3005 (1998).

69. N. A. Hill, Why are there so few magnetic ferroelectrics?, *J. Phys. Chem. B* **104(29)**, 6694-6709 (2000).

70. M. M. Kumar, A. Srinivas, and S. V. Suryanarayana, Structure property relations in $BiFeO_3$/$BaTiO_3$ solid solutions, *J. Appl. Phys.* **87(2)**, 855-862 (2000).

71. J. Y. Son, B. G. Kim, C. H. Kim, and J. H. Cho, Writing polarization bits on the multiferroic $BiMnO_3$ thin film using Kelvin probe force microscope, *Appl. Phys. Lett.* **84(24)**, 4971-4973 (2004).

72. J. Zhai, N. Cai, Z. Shi, Y. Lin, and C. W. Nan, Coupled magnetodielectric properties of laminated $PbZr_{0.53}Ti_{0.47}O_3$/$NiFe_2O_4$ ceramics, *J. Appl. Phys.* **95(10)**, 5685-5690 (2004).

73. S. B. Majumder, S. Bhattacharyya, R. S. Katiyar, A. Manivannan, P. Dutta, and M. S. Seehra, Dielectric and magnetic properties of sol-gel-derived lead iron niobate ceramics, *J. Appl. Phys.* **99**, 024108 (2006).

74. C. Tien, E. V. Charnaya, V. M. Gropyanov, I. S. Mikhailova, C. S. Wur, and A. A. Abramovich, Magnetic properties of a cermet on the base of $Al_2O_3$, *J. Magn. & Magn. Mater.* **220**, 147-151 (2000).

75. S. Ram, D. Ghosh, and S. K Roy, Microstructure and topological analysis of $Co:Al_2O_3$ nanocermets in new FCC and BCC metastable Co-structures, *J. Mater. Sci.* **36**, 3745-3753 (2001).

76. S. Ram, Self-confined dimension of thermodynamic stability in Co-nanoparticles in fcc and bcc allotrope structures with a thin amorphous $Al_2O_3$ surface layer, *Acta Mater.* **49**, 2297-2307 (2001).

77. S. Rana, S. Ram, S. Seal, and S. K. Roy, Surface structure and topology in surface stabilized Co-nanoparticles with a thin $Al_2O_3$ amorphous layer, *Appl. Surface Sci.* **236**, 141-154 (2004).

78. S. Rana, and S. Ram, X-ray diffraction and x-ray photoelectron spectroscopy studies of stabilized cobalt nanoparticles with a thin $Al_2O_3$ surface layer, *Mater. Sci. & Tech.* **21(2)**, 243-249 (2005).

79. S. Biswas, and S. Ram, (a) Morphology and stability in a half-metallic ferromagnetic $CrO_2$ compound of nanoparticles synthesized via a polymer precursor, *Chem. Phys.* **306**, 163-169

(2004) and (b) Synthesis of shape-controlled ferromagnetic $CrO_2$ nanoparticles by reaction in micelles of $Cr^{6+}$-PVA polymer chelates, *Mater. Chem. Phys.* (in press).

80. J. H. Kim, H. Kim, D. Kim, Y. E. Ihm, and W. K. Choo, Magnetoresistance in laser-deposited $Zn_{1-x}Co_xO$ thin films, *Physica B* **327**, 304-306 (2003).

81. S.-J. Han, B. Y. Lee, J.-S. Ku, Y. B. Kim, and Y. H. Jeong, Magnetic properties of $Zn_{1-x}Co_xO$, *J. Magn. & Magn. Mater.* **272-276**, 2008-2009 (2004).

82. A. Dinia, G. Schmerber, V. Pierron-Bohnes, C. Meny, P. Panissod, and E. Beaurepaire, Magnetic perpendicular anisotropy in sputtered $(Zn_{0.75}Co_{0.25})O$ dilute magnetic semiconductor, *J. Magn. & Magn. Mater.* **286**, 37-40 (2005).

83. H.-J. Lee, S. H. Choi, C. R. Cho, H. K. Kim, and S.-Y. Jeong, The formation of precipitates in the ZnCoO system, *Europhys. Lett.* **72(1)**, 76-82 (2005).

84. H. Ndilimabaka, S. Colis, G. Schmerber, D. Muller, J. J. Grob, L. Gravier, C. Jan, E. Beaurepaire, and A. Dinia, As-doping effect on magnetic, optical and transport properties of $Zn_{0.9}Co_{0.1}O$ diluted magnetic semiconductor, *Chem. Phys. Lett.* **421**, 184-188 (2006).

85. D. Rodic, V. Spasojevic, A. Bajorek, and P. Onnerud, Similarity of structure properties of $Hg_{1-x}Mn_x S$ and $Cd_{1-x}Mn_x S$ (structure properties of HgMnS and CdMnS), *J. Magn. & Magn. Mater.* **152**, 159-164 (1996).

86. D.-S. Chuu, Y. -C. Chang, and C.-Y. Hsieh, Growth of CdMnS films by pulsed laser evaporation, *Thin Solid Films* **304**, 28-35 (1997).

87. L. Levy, D. Ingert, N. Feltin, and M. P. Pileni, $Cd_{1-y}Mn_y S$ nanoparticles: absorption and photoluminescence properties, *J. Crystal Growth* **184/185**, 377-382 (1998).

88. Q. Pang, B. C. Guo, C. L. Yang, S. H. Yang, M. L. Gong, W. K. Ge, and J. N. Wang, $Cd_{1-x}Mn_x S$ quantum dots: new synthesis and characterization, *J. Crystal Growth* **269**, 213-217 (2004).

89. J. P. Woods, B. M. Patterson, A. S. Fernando, S. S. Jaswal, D. Welipitiya, and D. J. Sellmyer, Electronic structures and Curie temperatures of iron-based rare-earth permanent-magnet compounds, *Phys. Rev. B* **51(2)**, 1064-1072 (1995).

90. J. M. D. Coey, Interstitial intermetallics, *J. Magn. & Magn. Mater.* **159**, 80-89 (1996).

91. J. M. D. Coey and P. A. I. Smith, Magnetic nitrides, *J. Magn. & Magn. Mater.* **200**, 405-424 (1999).

92. R. Q. Wu, G. W. Peng, L. Liu, and Y. P. Feng, Possible *graphitic*-boron-nitride-based metal-free molecular magnets from first principles study, *J. Phys.: Condens. Matter* **18**, 569-575 (2006).

93. H. Eschrig and W. E. Pickett, Density functional theory of magnetic systems revisited, *Solid State Commun.* **118**, 123-127 (2001).

94. K. P. Kämper, W. Schmitt, G. Güntherodt, R. J. Gambino, and R. Ruf, $CrO_2$—a new half-metallic ferromagnet?, *Phys. Rev. Lett.* **59(24)**, 2788-2791 (1987).

95. H. Y. Hwang and S.-W. Cheong, Enhanced intergrain tunneling magnetoresistance in half-metallic $CrO_2$ films, *Science* **278(5343)**, 1607-1609 (1997).

96. M. A. Korotin, V. I. Anisimov, D. I. Khomskii, and G. A. Sawatzky, $CrO_2$ : a self-doped double exchange ferromagnet, *Phys. Rev. Lett.* **80(19)**, 4305-4308 (1998).

97. M. Rabe, J. Dreßen, D. Dahmen, J. Pommer, H. Stahl, U. Rüdiger, G. Güntherodt, S. Senz, and D. Hesse, Preparation and characterization of thin ferromagnetic $CrO_2$ films for applications in magnetoelectronics, *J. Magn. & Magn. Mater.* **211**, 314-319 (2000).

98. A. Gupta, X. W. Li, and G. Xiao, Magnetic and transport properties of epitaxial and polycrystalline chromium dioxide thin films, *J. Appl. Phys.* **87(9)**, 6073-6078 (2000).

99. A. Barry, J. M. D. Coey, and M. Viret, A $CrO_2$-based magnetic tunnel junction, *J. Phys.: Condens. Matter.* **12**, L173-L175 (2000).

100. J. Dai and J. Tang, Junction-like magnetoresistance of intergranular tunneling in field-aligned chromium dioxide powders, *Phys. Rev. B* **63**, 054434 (2001).

101. W. F. McClume, JCPDS X-ray powder diffraction file **9-332,** Joint Committee on powder diffraction standards (International Center for Diffraction Data, Swarthmore, PA USA, 1979).

102. J. Dai, J. Tang, H. Xu, L. Spinu, W. Wang, K. Wang, A. Kumbhar, M. Li, and U. Diebold, Characterization of the natural barriers of intergranular tunnel junctions: $Cr_2O_3$ surface layers on $CrO_2$ nanoparticles, *Appl. Phys. Lett.* **77(18)**, 2840-2842 (2000).

S. Ram, S. Biswas, and H. J-Fecht

58

103. R. Cheng, B. Xu, C. N. Borca, A. Sokolov, C.-S. Yang, L. Yuan, S.-H. Liou, B. Doudin, and P. A. Dowben, Characterization of the native $Cr_2O_3$ oxide surface of $CrO_2$, *Appl. Phys. Lett.* **79(19)**, 3122-3124 (2001).

104. H. Gleiter, Nanocrystalline materials, *Prog. Mater. Sci.* **33(4)**, 223-315 (1989).

105. R. W. Siegel, Cluster-assembled nanophase materials, *Annu. Rev. Mater. Sci.* **21**, 559-578 (1991).

106. B. Kubota, Decomposition of higher oxides of chromium under various pressures of oxygen, *J. Am. Ceram. Soc.* **44(5)**, 239-248 (1961).

107. L. Ranno, A. Barry, and J. M. D. Coey, Production and magnetotransport properties of $CrO_2$ films, *J. Appl. Phys.* **81(8)**, 5774-5776 (1997).

108. M. A. K. L. Dissanayake and L. L. Chase, Optical properties of $CrO_2$, $MoO_2$, and $WO_2$ in the range 0.2-6 eV, *Phys. Rev. B* **18(12)**, 6872-6879 (1978).

109. J. Dai and J. Tang, Temperature dependence of the conductance and magnetoresistance of $CrO_2$ powder compacts, *Phys. Rev. B* **63**, 064410 (2001).

110. S. Ishibashi, T. Namikawa, and M. Satou, (a) Epitaxial growth of $CrO_2$ on sapphire in air, *Jpn. J. Appl. Phys.* **17(1)**, 249-250 (1978) and (b) Epitaxial growth of ferromagnetic $CrO_2$ films in air, *Mater. Res. Bull.* **14(1)**, 51-57 (1979).

111. J. J. Liang, S. F. Lee, Y. D. Yao, C. C. Wu, S. G. Shyu, and C. Yu, Magnetotransport study of granular chromium dioxide thin films prepared by the chemical vapor deposition technique, *J. Magn. & Magn. Mater.* **239**, 213-216 (2002).

112. W. J. DeSisto, P. R. Broussard, T. F. Ambrose, B. E. Nadgorny, and M. S. Osofsky, Highly spin-polarized chromium dioxide thin films prepared by chemical vapor deposition from chromyl chloride, *Appl. Phys. Lett.* **76(25)**, 3789-3791 (2000).

113. A. Anguelouch, A. Gupta, G. Xiao, D. W. Abraham, Y. Ji, S. Ingvarsson, and C. L. Chien, Near-complete spin polarization in atomically smooth chromium-dioxide epitaxial films prepared using a CVD liquid precursor, *Phys. Rev. B* **64**, 180408 (2001).

114. P. G. Ivanov, S. M. Watts, and D. M. Lind, Epitaxial growth of $CrO_2$ thin films by chemical-vapor deposition from a $Cr_8O_{21}$ precursor, *J. Appl. Phys.* **89(2)**, 1035-1040 (2001).

115. P. A. Dowben, Y. G. Kim, S. Baral-Tosh, G. O. Ramseyer, C. Hwang, and M. Onellion, Fabrication of ferromagnetic and antiferromagnetic chromium oxides by organometallic chemical vapor deposition, *J. Appl. Phys.* **67(9)**, 5658-5660 (1990).

116. R. C. DeVries, Epitaxial growth of $CrO_2$, *Mater. Res. Bull.* **1(2)**, 83-93 (1966).

117. A. Bajpai and A. K. Nigam, Synthesis of high-purity samples of $CrO_2$ by a simple route, *Appl. Phys. Lett.* **87**, 222502 (2005).

118. Y. J. Chen, X. Y. Zhang, and Z. Y. Li, Enhanced room-temperature magnetoresistance in half-metallic $CrO_2$/polymer composites, *Chem. Phys. Lett.* **375**, 213-218 (2003).

119. Y. J. Chen, X. Y. Zhang, T. Y. Cai, and Z. Y. Li, Temperature dependence of ac response in diluted half-metallic $CrO_2$ powder compact, *J. Alloy. Compd.* **379**, 240-246 (2004).

120. T. J. Swoboda, P. Arthur, N. L. Cox, J. N. Ingraham, A. L. Oppegard, and M. S. Sadler, Synthesis and properties of ferromagnetic chromium oxide, *J. Appl. Phys.* **32(3)**, S374-S375 (1961).

121. B. L. Chamberland, (a) Crystal growth of $CrO_2$, *Mater. Res. Bull.* **2(9)**, 827-835 (1967) and (b) The chemical and physical properties of chromium (IV) oxide and tetravalent chromium oxide derivatives, *Crit. Rev. Solid State Mater. Sci.* **7(1)**, 1-31 (1977).

122. T. S. Kannan and A. Jaleel, *Indian Patent No.* 627/Del/87 (1987).

123. R. A. McCurrie, *Ferromagnetic Materials: Structure & Properties* (Academic Press, New York, 1994).

124. K. Ramesha and J. Gopalakrishnan, A new method for the synthesis of chromium (IV) oxide at ambient pressure, *Chem. Commun.* 1173-1174 (1999).

125. N. L. Cox, *US Patent* 3 278 263 (1966).

126. M. Essig, M. W. Müller, and E. Schwab, Structural analysis of the stabilization layer of chromium dioxide particles, *IEEE Trans. Magn.* **26(1)**, 69-71 (1990).

127. T. Yu, Z. X. Shen, J. He, W. X. Sun, S. H. Tang, and J. Y. Lin, Phase control of chromium oxide in selective microregions by laser annealing, *J. Appl. Phys.* **93(7)**, 3951-3953 (2003).

128. E. Gaffet, M. Abdellaoui, and N. Malhouroux-Gaffet, Formation of nanostructural materials induced by mechanical processings (Overview), *Mater. Trans., JIM*, **36(2)**, 198-209 (1995).

129. J. Y. Huang, Y. K. Yu, and H. Q. Ye, Allotropic transformation of cobalt induced by ball milling, *Acta mater.* **44(3)**, 1201-1209 (1996).

130. V. R. Palkar, P. Ayyub, S. Chattopadhyay, and M. Multani, Size-induced structural transitions in the Cu-O and Ce-O systems, *Phys. Rev. B* **53(5)**, 2167-2170 (1996).

131. B. S. Murty and S. Ranganathan, Novel materials synthesis by mechanical alloying/milling, *Inter. Mater. Rev.* **43(3)**, 101-141 (1998).

132. M. Qi and H. J-Fecht, Structural transition of zirconia during mechanical attrition, *Mater. Sci. Forum* **269-272**, 187-192 (1998).

133. T. Tsuzuki and P. G. McCormick, Synthesis of $Cr_2O_3$ nanoparticles by mechanochemical processing, *Acta mater.* **48(11)**, 2795-2801 (2000).

134. O. K. Tan, W. Cao, Y. Hu, and W. Zhu, Nanostructured oxides by high-energy ball milling technique: application as gas sensing materials, *Solid State Ion.* **172**, 309-316 (2004).

135. S. Ram and S. Biswas, *Indian Patent* 198130 (2005).

136. M. P. Klug and L. E. Alexander, *X-ray Diffraction Procedure for Polycrystalline and Amorphous Materials* (Wiley, New York, 1974).

137. Y. Shibasaki, F. Kanamaru, and M. Koizumi, The conversion from $CrO_2$ into orthorhombic CrOOH, *Mater. Res. Bull.* **8(5)**, 559-564 (1973).

138. W. Krakow, H. Colign, and O. Müller, *Phil. Magn. A* **41(3)**, 369-384 (1980).

139. Z. C. Wang, T. J. Davies, N. Ridley, and A. A. Ogwu, Superplasticity of ceramic materials— II. Effect of initial porosity and doping on the superplastic behaviour of alumina, *Acta. Mater.* **44(11)**, 4301-4309 (1996).

140. H. Yang, A. Kuperman, N. Coombs, S. M. Afara, and G. A. Ozin, Synthesis of oriented films of mesoporous silica on mica, *Nature* **379**, 703-705 (1996).

141. S. H. Joo, S. J. Choi, I. Oh, J. Kwak, Z. Liu, O. Terasaki, and R. Ryoo, Ordered nanoporous arrays of carbon supporting high dispersions of platinum nanoparticles, *Nature* **412**, 169-172 (2001).

142. S. Inagaki, S. Guan, T. Ohsuna, and O. Terasaki, An ordered mesoporous organosilica hybrid material with a crystal-like wall structure, *Nature* **416**, 304-307 (2002).

143. M. E. Davis, Ordered porous materials for emerging applications, *Nature* **417**, 813-821 (2002).

144. T. Sun, and J.Y. Ying, Synthesis of microporous transition-metal-oxide molecular sieves by a supramolecular templating mechanism, *Nature* **389**, 704-706 (1997).

145. W. Cai, Y. Zhang, J. Jia, and L. Zhang, Semiconducting optical properties of silver/silica mesoporous composite, *Appl. Phys. Lett.* **73(19)**, 2709-2711 (1998).

146. O. Jessensky, F. Muller, and U. Gosele, Self-organized formation of hexagonal pore arrays in anodic alumina, *Appl. Phys. Lett.* **72(10)**, 1173-1175 (1998).

147. S. Kondoh, Y. Iwamoto, K. Kikuta, and S. Hirano, Novel processing for mesoporous silica films with one-dimensional through channels normal to the substrate surface, *J. Am. Ceram. Soc.* **82(1)**, 209-212 (1999).

148. H. Gleiter, Nanostructured materials: basic concepts and microstructure, *Acta Mater.* **48(1)**, 1-29 (2000).

149. Y. Sakamoto, M. Kaneda, O. Terasaki, D.Y. Zhao, J. M. Kim, G. Stucky, H. J. Shin, and R. Ryoo, Direct imaging of the pores and cages of three-dimensional mesoporous materials, *Nature* **408**, 449-453 (2000).

150. S. Ram and S. Rana, Synthesis of porous $Al_2O_3$ ceramic clusters by surface hydrolysis of a thin Al-metal plate in water, *Mater. Sci. & Eng. A* **304-306**, 790-795 (2001).

151. P. Mohanty and S. Ram, (a) Confined growth in $Eu_2O_3$ nanocrystals in a new polymorph in an amorphous mesoporous $Al_2O_3$, *Phil. Magn. B* **82**, 1129-1144 (2002), and (b) Enhanced photoemission in dispersed $Eu_2O_3$ nanoparticles in amorphous $Al_2O_3$, *J. Mater. Chem.* **13**, 3021-3028 (2003).

152. S. Ram, Dynamics of formation of self-organized mesoporous $AlO(OH).\alpha H_2O$ structure in Al-metal surface hydrolysis in humid air, *J. Am. Ceram. Soc.* **86(12)**, 2037-2043 (2003).

153. A. Mondal and S. Ram, Synthesis of nanoparticles of monolithic zirconia in a new polymorphs in orthorhombic crystal structure, *Chem. Phys. Lett.* **382**, 297-306 (2003).

154. P. Mohanty and S. Ram, Multiple light emission associated with the $^5D_0 \rightarrow ^7F_3$ forbidden transition in $Eu^{3+}$ cations dispersed in a $Eu^{3+}:Al_2O_3$ mesoporous structure, *Phil. Magn. Lett.* (in press).

155. H. van Leuken and R. A. de Groot, Electronic structure of the chromium dioxide (001) surface, *Phys. Rev. B* **51(11)**, 7176-7178 (1995).

156. J. B. Goodenough, Metallic oxides, *Prog. Solid State Chem.* **5**, 145-399 (1971).

157. P. I. Sorantin and K. Schwarz, Chemical bonding in rutile-type compounds, *Inorg. Chem.* **31(4)**, 567-576 (1992).

158. C. Zener, Interaction between the d-shells in the transition metals. II. Ferromagnetic compounds of manganese with perovskite structure, *Phys. Rev.* **82(3)**, 403-405 (1951).

159. A. R. Williams, J. Kübler, and C. D. Gelatt Jr., Cohesive properties of metallic compounds: augmented-spherical-wave calculations, *Phys. Rev. B* **19(12)**, 6094- 6118 (1979).

160. Y. J. Chen, X. Y. Zhang, and Z. Y. Li, Role of grain boundaries on magnetoresistance of $CrO_2$-$TiO_2$ composites: impedance spectroscopy study, *J. Magn. & Magn. Mater.* **267**, 152-160 (2003).

161. Q. Zou, H. E. Ruda, B. G. Yacobi, K. Saegusa, and M. Farrell, Dielectric properties of lead zirconate titanate thin films deposited on metal foils, *Appl. Phys. Lett.* **77(7)**, 1038-1040 (2000).

162. H. B. Park, C. Y. Park, Y. S. Hong, K. Kim, and S. J. Kim, Structural and dielectric properties of PLZT ceramics modified with lanthanide ions, *J. Am. Ceram. Soc.* **82(1)**, 94-102 (1999).

163. R. Rai, S. Sharma, and R. N. P. Choudhary, Dielectric and piezoelectric studies of Fe doped PLZT ceramics, *Mater. Lett.* **59**, 3921-3925 (2005).

164. S. Canulescu, G. Dinescu, G. Epurescu, D. G. Matei, C. Grigoriu, F. Craciun, P. Verardi, and M. Dinescu, Properties of $BaTiO_3$ thin films deposited by radiofrequency beam discharge assisted pulsed laser deposition, *Mater. Sci. & Eng. B* **109**, 160-166 (2004).

165. I. S. Zheludev, *Physics of crystalline dielectrics* (Plenum Press, New York, 1971).

166. R. N. P. Choudhary, R. Palai, and S. Sharma, Structural, dielectric and electrical properties of lead cadmium tungstate ceramics, *Mater. Sc. & Engg. B* **77**, 235-240 (2000).

167. R. Palai, R. N. P. Choudhary, and H. S. Tewari, Structural and dielectric properties of $Ba_4R_2Ti_4Nb_6O_{30}$ (R = Y, Sm and Dy) ferroelectric ceramics, *J. Phys. Chem. Solids* **62**, 695-700 (2001).

168. M. Thirumal and A. K. Ganguli, Studies on dielectric oxide materials containing niobium and tantalum, *Prog. Crystal Growth and Charact.* **44**, 147-154 (2002).

169. A. Wu, P. M. Vilarinho, I. M. M. Salvado, and J. L. Baptista, Sol-gel preparation of lead zirconate titanate powders and ceramics : Effect of alkoxide stabilizers and lead precursors, *J. Am. Ceram. Soc.* **83(6)**, 1379-1385 (2000).

170. A. M. Abo El Ata and M. A. Ahmed, Dielectric and AC conductivity for $BaCo_{2-x}Cu_xFe_{16}O_{27}$ ferrites, *J. Magn. & Magn. Mater.* **208**, 27-36 (2000).

171. K. Bouayad, S. Sayouri, T. Lamcharfil, M. Ezzejari, D. Mezzane, L. Hajji, A. E. Ghazouali, M. Filalil, P. Dieudonne, and M. Rhouta, Sol–gel processing and dielectric properties of $(Pb_{1-y}La_y)(Zr_{0.52}Ti_{0.48})O_3$ ceramics, *Physica A* **358**, 175-183 (2005).

172. P. H. Xiang, X. L. Dong, C. D. Feng, R. H. Liang, and Y. L. Wang, Dielectric behavior of lead zirconate titanate/silver composites, *Mater. Chem. Phys.* **97**, 410-414 (2006).

173. C. M. Fu, C. J. Lai, J. S. Wu, J. C. A. Huang, C.-C. Wu, and S.-G. Shyu, High frequency impedance spectra on the chromium dioxide thin film, *J. Appl. Phys.* **89(11)**, 7702-7704 (2001).

174. C. M. Fu, C. J. Lai, H. S. Hsu, Y. C. Chao, J. C. A. Huang, C. -C. Wu, and S.-G. Shyu, Characterization of magnetoimpedance on polycrystalline and amorphous chromium oxides bilayered thin films, *J. Appl. Phys.* **91(10)**, 7143-7145 (2002).

175. C. M. Fu, Y. C. Chao, S. H. Hung, C. P. Lin, and J. Tang, Impedance spectra of field-aligned $CrO_2$ needle-shape powders, *J. Magn. & Magn. Mater.* **282**, 283-286 (2004).

176. J. M. D. Coey, A. E. Berkowitz, L. Balcells, F. F. Putris, and A. Barry, Magnetoresistance of chromium dioxide powder compacts, *Phys. Rev. Lett.* **80(17)**, 3815-3818 (1998).

177. S. S. Manoharan, D. Elefant, G. Reiss, and J. B. Goodenough, Extrinsic giant magnetoresistance in chromium (IV) oxide, $CrO_2$, *Appl. Phys. Lett.* **72(8)**, 984-986 (1998).

178. S. M. Watts, S. Wirth, S. von Molnar, A. Barry, and J. M. D. Coey, Evidence for two-band magnetotransport in half-metallic chromium dioxide, *Phys. Rev. B* **61(14)**, 9621-9628 (2000).

179. P. A. Stampe, R. J. Kennedy, S. M. Watts, and S. von Molnar, Strain effects in thin films of $CrO_2$ on rutile and sapphire substrates, *J. Appl. Phys.* **89(11)**, 7696-7698 (2001).

180. A. Barry, J. M. D. Coey, L. Ranno, and K. Ounadjela, Evidence for a gap in the excitation spectrum of $CrO_2$, *J. Appl. Phys.* **83(11)**, 7166-7168 (1998).

181. J. Inoue and S. Maekawa, Theory of tunneling magnetoresistance in granular magnetic films, *Phys. Rev. B* **53(18)**, 11927-11929(1996).

182. T. Zhu and Y. J. Wang, Enhanced tunneling magnetoresistance of $Fe-Al_2O_3$ granular films in the coulomb blockade regime, *Phys. Rev. B* **60(17)**, 11918-11921 (1999).

183. Y. J. Chen, X. Y. Zhang, T. Y. Cai, and Z. Y. Li, Study of the conductance and magnetotransport of $CrO_2-TiO_2$ composites, *Mater. Lett.* **58**, 262 (2004).

184. C. H. Shang, J. Nowak, R. Jansen, and J. S. Moodera, Temperature dependence of magnetoresistance and surface magnetization in ferromagnetic tunnel junctions, *Phys. Rev. B* **58(6)**, 2917-2920 (1998).

185. L. I. Glazman and K. A. Matveev, Inelastic tunneling across thin amorphous films, *Sov. Phys. JETP* **67 (6)**, 1276-1282 (1988).

186. F. Y. Yang, C. L. Chien, E. F. Ferrari, X. W. Li, G. Xiao, and A. Gupta, Uniaxial anisotropy and switching behavior in epitaxial $CrO_2$ films, *Appl. Phys. Lett.* **77(2)**, 286-288 (2000).

187. J. K. Burdett, G. J. Miller, J. W. Richardson, Jr., and J. V. Smith, Low-temperature neutron powder diffraction study of $CrO_2$ and the validity of the Jahn-Teller viewpoint, *J. Am. Chem. Soc.* **110**, 8064-8071 (1988).

188. X. W. Li, A. Gupta, and G. Xiao, Influence of strain on the magnetic properties of epitaxial (100) chromium dioxide ($CrO_2$) films, *Appl. Phys. Lett.* **75(5)**, 713-715 (1999).

189. S. Ram, Development of planar hexagonal $Fe_2-Y$ ferrite particles for millimeter wave devices, *J. Magn. & Magn. Mater.* **72**, 315-318 (1988).

190. S. Ram and J. C. Joubert, Variation in particle morphology and Curie temperature of $SrZn_2Fe_{16}O_{27}$ ceramic powders, *Phys. Rev. B* **44**, 6825-6831 (1991).

191. S. Ram and J. C. Joubert, Development of high quality ceramic powders of $Sr_{0.9}Ca_{0.1}Zn_2$-W type hexagonal ferrite for permanent magnet devices, *IEEE. Trans. Magn.* **28**, 15-20 (1992).

192. P. Lubitz, M. Rubinstein, M. S. Osofsky, B. E. Nadgorny, R. J. Soulen, K. M. Bussmann, and A. Gupta, Ferromagnetic resonance observation of exchange and relaxation effects in $CrO_2$, *J. Appl. Phys.* **89(11)**, 6695-6697 (2001).

193. A. Gupta, X. W. Li, and G. Xiao, Inverse magnetoresistance in chromium-dioxide-based magnetic tunnel junctions, *Appl. Phys. Lett.* **78(13)**, 1894-1896 (2001).

194. Y. Xu, D. Ephron, and M. R. Beasley, Directed inelastic hopping of electrons through metal-insulator-metal tunnel junctions, *Phys. Rev. B* **52(4)**, 2843-2859 (1995).

195. H. Itoh, T. Ohsawa, and J. Inoue, Magnetoresistance of ferromagnetic tunnel junctions in the double-exchange model, *Phys. Rev. Lett.* **84(11)**, 2501-2504 (2000).

196. V. Petricevic, S. K. Gayen, and R. R. Alfano, Laser action in chromium-activated forsterite for near-infrared excitation: Is $Cr^{4+}$ the lasing ion?, *Appl. Phys. Lett.* **53(26)**, 2590-2592 (1988).

197. N. V. Kuleshov, V. P. Mikhailov, V. G. Scherbitsky, B. I. Minkov, T. J. Glynn, and R. Sherlock, Luminescence study of $Cr^{4+}$- doped silicates, *Opt. Mater.* **4**, 507-513 (1995).

198. X. Wu, H. Yuan, W. M. Yen, and B. G. Aitken, Compositional dependence of the luminescence from $Cr^{4+}$-doped calcium aluminate glass, *J. Lumin.* **66 & 67**, 285-289 (1996).

199. B. Lipavsky, Y. Kalisky, Z. Burshtein, Y. Shimony, and S. Rotman, Some optical properties of $Cr^{4+}$-doped crystals, *Opt. Mater.* **13**, 117-127 (1999).

200. X. Feng and S. Tanabe, Spectroscopy and crystal-field analysis for Cr(IV) in alumino-silicate glasses, *Opt. Mater.* **20**, 63-72 (2002).

201. Y. Tanabe and S. Sugano, The absorption spectra of complex ions, *J. Phys. Soc. Jpn.* **9**, 753-766 (1954).

202. D. F. Shrirer, P. W. Atkins, and C. H. Langford, *Inorganic Chemistry* (Wiley, New York, 1990).

203. A. Paul, *Chemistry of Glasses* (Chapman and Hall, London, 1990).

204. S. Ram, K. Ram, and B. S. Shukla, Optical absorption and EPR studies of borate glasses with $PbCrO_4$ and $Pb_2CrO_5$ microcrystals, *J. Mater. Sci.* **27**, 511-519 (1992).

205. R. Bhatt, S. Kar, K. S. Bartwal, and V. K. Wadhawan, The effect of Cr doping on optical and photoluminescence properties of $LiNbO_3$ crystals, *Solid State Commun.* **127**, 457-462 (2003).

206. S. Ram, Synthesis and structural and optical properties of metastable $ZrO_2$ nanoparticles with intergranular $Cr^{3+}/Cr^{4+}$ doping and grain surface modification, *J. Mater. Sci.* **38**, 643-655 (2003).

# QUESTIONS

1. What are typical ferromagnetic ceramics? Enumerate their important applications in electronic industries.
2. What is a hybrid composite? How does it present modified electronic structure and properties of the basic components? Support your arguments with representative examples.
3. What are half-metallic ferromagnetic compounds of ceramics? How do they assume a metal-like structure of magnetic spins in a semiconductor in terms of the electronic band structure? Enumerate special advantages of such specific features in spintronic devices.
4. Explain the basic features of spintronics and spintronic materials? Use selective examples of different kinds of such materials in order to support your comments. Enumerate important applications in electronic devices and components.
5. What is spin polarization? List practically useful ceramics of a reasonably large value of spin polarization. How it is useful in developing novel spin tunneling or optoelectronic devices?
6. What is magnetoresistance MR? Classify typical series of different kinds of ceramics or granular hybrid ceramic-composites of such solids. How does microstructure control such specific transport properties in a heterogeneous solid?
7. Name typical solids that assume a significant MR value at room temperature. Enumerate specific applications of such materials in electronic industries.
8. What is a low-field MR (often called LFMR) or a high-field MR (often called HFMR)? Describe physical characteristics to distinguish them unambiguously. Enumerate their individual applications.
9. How does spin tunneling result in a tunneling magnetoresistance (TMR)? Give your comments with specific examples. How does this specific class of MR materials differ from the GMR or CMR ones? Give technically useful ceramics of a reasonably large value of TMR at room temperature.
10. Name selected half-metallic ferromagnets that involve good optical as well as other useful semiconductor properties for practical applications in optical and magnetic data storage systems or optoelectronic devices.
11. Give classifications of two series of Heusler and semi-Heusler alloys or compounds with selected examples. Describe important electrical transport and optoelectronic properties in the two classes of the solids or the hybrid composites.
12. Argue the statement that the Heusler alloys represent a specific series of dilute magnetic semiconductors or semimagnetic semiconductors. Enu-

merate applications of such a specific class of solids in magnetic sensors and biosensors.

13. Enumerate applications of half-metallic ceramics as novel structural materials.
14. How do porous ceramics of such a specific class of compounds extend the applications as gas sensor, gas storage, hot filters and selective gas separators?
15. List biological and tribological activities and applications of such a specific class of magnetic compounds.

# 2

# Functional Nanostructured Thin Films

## Hare Krishna and Ramki Kalyanaraman

Department of Physics, Washington University in St. Louis, MO, 63130
Center for Materials Innovation, Washington University in St. Louis, MO, 63130.
Email: ramkik@wuphys.wustl.edu.

Functional nanostructured surfaces comprised of spatially ordered features, like clusters, wires or thin films, are important in various applications, including nonlinear photonics, magnetic data storage and gas sensing. The primary challenge to realizing the potential of this area is to develop cost-effective fabrication techniques and reliable characterization of such nanostructures. In this chapter we discuss the potential towards controlling the spatial arrangement of surface structures by modification to thin film nucleation and growth processes during physical vapor deposition. We also review structures resulting from surface instabilities generated by energetic ion beam irradiation—another active research area that promises simple and reliable self-assembly of nanostructures. Careful characterization of spatial order and thin film morphological properties like size, shape and connectedness are very important towards understanding the role of processing parameters, as well as in structure-property correlations. In this regard, image analysis can be a simple but vital step. We review the techniques of Fourier analysis and Minkowski functional analysis to understand the length scales and morphologies of surface nanostructures. Finally, examples of applications from the areas of nonlinear optics, magnetism and gas sensing are provided which highlight the need for control of size and spatial order in nanostructures.

## 1. INTRODUCTION

Three broad categories of functional nanostructures, i.e., nanostructures leading to devices, can be considered: (1) those that work at the molecular level and are built by tailoring at the molecular level[1,2]; (2) those that involve large-scale structures ($\sim\mu$m) comprised of nanometer-scale components arranged in three dimensions[3]; and (3) those based on nanostructured surfaces involving nanometer scale (1-1000 nm) lateral features and/or ultrathin films of nanometer scale thickness.[4-6]

In this chapter, we will concentrate on this third category of functional nanostructured thin films wherein a surface is comprised of nanoscale features, for example, nanodots or nanowires, whose spatial arrangement and other attributes, like size and shape, can lead to various applications. There are numerous promising applications in this area of nanostructured surfaces, including: nanostructured oxide semiconductors for high-sensitivity gas sensors[7]; regular arrays of discrete nm scale magnets for potential memory storage devices[8,9]; surface plasmon waveguides made from linear chains of metal nanoclusters[10,11]; and spatially ordered metal nanocatalyst seeds to make high-efficiency flat-panel displays from carbon-nanotube arrays.[12] In these applications, the control of ordering, including spatial arrangement of the nanofeatures, control of its size, shape, composition, and crystallographic orientation is important. In most instances, the ability to control ordering over large length scales, i.e., to have long-range ordering is also necessary.[13] *These requirements summarize one of the outstanding challenges in the area of nanoscience and nanotechnology, which is the need for cheap and cost-effective nanomanufacturing processes that can make ordered surfaces with the desired attributes.*

Thin films of virtually any solid material can be deposited on a variety of surfaces. The atomic-level precision of film thickness control, wide range of microstructure and electronic properties, various characterization techniques to correlate structure and property, and the wide variety of deposition techniques available, make the thin film area a powerful materials physics arena. In fact, thin film science is at the heart of major industries, including the semiconductor, magnetic and optical areas. Therefore, it is reasonable to expect that our fundamental understanding of thin film processes and the available infrastructure can be used to innovate nanoscale processing for functional nanostructured surfaces. In the past, the emphasis of fundamental thin film studies has been on lateral homogeneity, i.e., to prepare films with uniform properties such as thickness, crystal structure, chemical composition, etc., over short (nanoscale) as well as large (device scale) length scales within the plane of the wafer.[14] Consequently, experimental and theoretical studies focused on film nucleation and growth in macroscopically homogeneous systems. For example, where the incident atom deposition rate (or the atom flux) and surface temperatures were considered (and desired) to be uniform over the area of interest. For these conditions a sound basic understanding was achieved (Sec. 1.2).

Now, the need for nanoscale order has shifted the emphasis towards understanding how processing parameters need to be changed in order to make surfaces that have ordered arrangements of nanostructures over large areas. *The fundamental difference here is that now the properties of the film or the surface must vary in the nanometer scale, but must have long-range order, i.e., where different regions of the surface have identical nanoscale properties.* Some recent examples emphasize this change in approach. One is the use of lasers to create spatially periodic patterns of nanofeatures by manipulation of the depositing film on the substrate[15,16] or by manipulation of the atom flux prior to deposition on the substrate.[17,18] Another example is self-assembly by creating special conditions for film growth on the substrate.[19]

Our goal in this chapter is introduce readers to topics that are currently active research areas. We attempt this by providing some basic information along with examples from exciting research areas with the hope that they will pursue further reading using the numerous references provided here. We first review techniques for nanostructure processing (Sec. 1.1) with an emphasis on self-assembly approaches. In this broad area we restrict the discussion to self-assembly by manipulation of nucleation and growth conditions for thin film deposition from the vapor phase (e.g., by evaporation, sputtering, and MBE) in Sec. 1.2 and by the processing of the thin film or a surface by energetic ion beams in Sec. 1.3. We follow this in Sec. 1.4 by a discussion of two important computer image analysis techniques that are important for characterization of nanoscale features like spatial order (Fourier analysis) and morphology (Minkowski analysis). The chapter ends with examples of some important applications requiring ordered surface nanostructures (Sec. 1.5).

## 1.1. Fabricating Nanostructured Surfaces

Two broad approaches, lithography and self-assembly, are currently utilized to make nanostructured surfaces. While both are extremely powerful, some problematic issues persist making research to discover new manufacturing techniques an ongoing activity. A brief review of lithography and self-assembly is presented below.

### 1.1.1. Lithography

Lithography is the most successful approach to make surface structures with long-range periodicity. Here, a surface is subdivided into smaller regions down to 10's of nanometers using lithographically determined patterns.[20,21] Patterns can be defined by masks, interference patterns and beam scanning (e.g., electron beams). The final patterned surface is obtained by etching and removal of the exposed (or unexposed) areas. In some instances a deposition step is performed after the first etching step. A typical lithography set-up will consist of the exposure optics, the etching station and the deposition system. In optical lithography, the achievable feature sizes are a combination of the wavelength of light and the sensitivity of the etching process. Typical feature sizes with optical lithography are on the order of 100 nm.

The key detrimental issues with using lithography to make ordered surface nanostructures lie in: (a) the high cost and complexity to achieve sizes in the 1-100 nm range; (b) the need for etchants with higher sensitivity and material specificity; and (c) the relative complexity of the overall process. Presently, complex and expensive technologies like e-beam and X-ray lithography are capable of manufacturing sub-100 nm scale features. However, the future of these approaches is unclear because of the number of processing steps involved and the expensive infrastructure required.

### 1.1.2. Self-assembly

Self-assembled structures result from the evolution of a system based on kinetic and thermodynamic considerations. The primary example of self-assembly is crystal growth. In crystal growth, the periodic arrangement of atoms results from energy and entropy considerations. Proper control of experimental variables like temperature and concentration lead to high-quality crystals of large size. While there is as yet no analogous example of similar high-quality crystals containing periodically arranged nanoscale features like nanoclusters or nanowires, the essential concept is that either internal or external forces will control the energetics and kinetics to form nanostructure arrays. One example of self-assembly that can potentially lead to nanometer-sized clusters is thin film growth. In film growth from the vapor phase, atom adsorption on the surface and subsequent adatom diffusion and binding lead to nucleation and growth of clusters. The rate of formation of the cluster and its stable size is determined by energy and kinetic considerations. By a proper choice of experimental parameters, the smallest stable cluster ranges from a few atoms in size (even as small as a single adatom at low temperatures) and upwards. However, one important aspect of film nucleation and growth is the inherent spatial randomness in the various processes involved, including location of atom adsorption and the subsequent randomness in diffusion and binding. *Eventually this leads to features with a random spatial distribution as well as a wide distribution in feature size, crystallographic orientation, and so on.*

However, self-assembly in the presence of internal or external driving forces that inherently bias the system to follow certain thermodynamic or kinetic pathways is a potential means to make ordered nanostructures.[22,23] One example for thin film growth is the case of Ge quantum dots grown on Si substrates, where the cluster size and distribution are a result of epitaxial strain.[24] The achievable cluster sizes are in the nanometer scale. Another example is the modification of an atomically smooth surface into one containing nanostructures following ion irradiation.[25] Self-assembly based approaches are exciting because no true limitation to the smallest size of clusters is seen. More importantly, there are a variety of natural processes that tend to form ordered arrays of nanostructures which suggests the potential to harvest natural processes in nanomanufacturing.[26] Another advantage is that self-assembly approaches explored thus far have generally been simple to implement and are cost-effective. However, one present drawback is the spatial extent of ordering. Most of these processes are capable of generating order over small spatial length scales, but are presently unable to generate long-range order, like that provided by lithography. *However, the benefits provided by the low-cost and ease of such techniques makes it one of the most important and exciting areas of current research.*

In the following sections we discuss some important concepts in film nucleation and growth that can be then applied to create nanostructures with order, as well as another growing area, that of energetic beam irradiation of surfaces, that can also result in ordered surface nanostructures.

## 1.2. Self-assembly of Nanostructures by Film Nucleation and Growth

In order to manipulate experimental parameters to influence film nucleation and growth thereby leading to ordered nanostructures, it is first necessary to understand the basic concepts of nucleation and growth phenomenon. A number of excellent reviews and books have been written on the subject of nucleation and growth in thin films.[14,27,28] Here we present the capillary model of film growth from the vapor phase to highlight the inherent disorder present in film growth. This understanding will allow us to propose changes that lead to more order in nucleation and growth.

### 1.2.1. Disorder During Vapor Phase Film Nucleation and Growth

The capillary model for vapor-based film growth[27] can be used to identify the origin of disorder in the spatial arrangement and size of nanoclusters during film growth. The discussion below emphasizes the role of the two critical experimental parameters—the atom deposition rate $R(x, y)$ (usually expressed in units of *# of atoms/cm$^2$-s*), and the substrate temperature $T_S(x, y)$—in determining film nucleation and growth characteristics. Here $x$ and $y$ designate coordinates in the plane of the substrate. We restrict the discussion to physical adsorption on a defect-free and isotropic surface where the adsorbed atom (adatom) diffuses by a random-walk.[28] The important physical processes taking place following impingement of atoms from a vapor phase are shown in Fig. 1 and can be described as follows:

- An incident atom adsorbs at *random positions* and at *random times* on the surface. This surface adatom has a thermally activated lifetime for re-evaporation (or desorption) $\tau_a$ which is related to its surface binding energy $E_b$ and the substrate temperature $T_S$ as:

$$\tau_a = \tau_a^o \exp(E_b/kT_s) \tag{1}$$

where, $\tau_a^o$ is a pre-exponential factor dependent on the material system and $k$ is the Boltzmann constant. Consequently, the larger is the binding

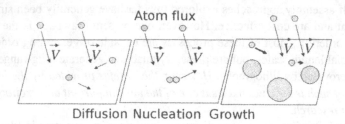

**FIGURE 1.** Schematic illustration of the important physical processes during vapor phase deposition of a thin film.

energy and/or smaller the substrate temperature, the longer is the desorption time.

- These adatoms diffuse *randomly* over a time scale $\tau_d$ before attachment to other adatoms to form clusters. $\tau_d$ is thermally activated and related to the adatom migration or diffusion energy $E_d$ as well as the areal concentration of adatoms (or clusters) which is determined by $R$. One immediate constraint on achieving film deposition is that the adatom must have a finite lifetime on the surface, i.e., the binding energy must be larger than the diffusion energy or $E_b \geq E_d$.

These physical processes result in two important regimes of film formation:

1. *Nucleation.* Nucleation is the formation of a cluster that is more likely to grow than shrink. In the capillary model, this occurs for critical cluster sizes determined by deposition rate, $R(x, y)$, and the substrate surface temperature, $T_S(x, y)$. Under conditions of low adatom concentration (or supersaturation), the critical size is achieved primarily by adatom surface diffusion and binding, rather than by direct impingement from the vapor. The critical cluster size can range from a single stable atom (e.g., at extremely low temperatures) to those containing hundreds of atoms. Therefore, nanometer-sized stable clusters are readily accessible in film deposition. Continued deposition results in a saturated density of stable clusters, $N^{\max}(R, T_s)$, which is a strong function of $R$ and $T_S$. Now the adatom deposition occurs at random locations on the surface $(x, y)$ and at random times. The adatoms then undergo diffusion via random-walk and meet other atoms (or clusters) and bind to form clusters at random positions and times on the surface.[29,30] *Consequently, the nucleation under these conditions is spatially homogeneous and occurs over a finite time scale. Therefore, the nuclei are randomly distributed on the surface and at any instant have a broad distribution in size.*

2. *Growth.* Once a stable cluster is nucleated, continued deposition leads to the growth of the nuclei. However, since the stable nuclei form at different times, the growth of each cluster occurs over different time scales, resulting in a broad distribution of size. In general the nuclei formed early on during film deposition grow to larger average sizes than the nuclei which formed

later in time. The growth eventually stops when each cluster impinges with other clusters. The growth stage is important as it can be used to control the average size of clusters. *However, in the absence of coalescence (or other driving forces, for example, strain relaxation[24]), the grains will be disordered in spatial location as well as having a large size spread.*[31,33]

A more quantitative description of the role of $R$ and $T_S$ on film deposition can be made by further evaluating the consequence of the two characteristic time scales. It can be shown that these time scales actually result in a characteristic length scale or critical capture radius ($r_c$) surrounding each adatom. To understand the origin of this $r_c$, visualize any adatom or cluster on the surface. With continuing deposition, new adatoms attach to the surface some distance away from this cluster (or adatom). If the distance of this adatom is smaller than $r_c$ then the adatom is more likely to diffuse and attach to the cluster (or adatom). On the other hand, if the adatom is more than $r_c$ away from any other cluster or other adatom, it is more likely to re-evaporate. This critical region is known as the Burton-Cabrera-Frank (BCF) zone[30,34,35] and is expressed by:

$$r \propto e^{\frac{E_b - E_d}{2kT_s}} \tag{2}$$

where k is the Boltzmann constant. This critical radius can be experimentally determined by evaluating the maximum number of nuclei $N^{\max}(R, T_s)$ that form on the surface as:

$$r_c = \frac{1}{\sqrt{(N^{\max})}} \tag{3}$$

On the basis of this understanding, we can distinguish between two important conditions for film nucleation:

1. *Nucleation under uniform substrate temperature $T_S(x, y)$.*

   For a fixed temperature, the most important experimental variable is the deposition rate $R$ as it determines the average spacing between adatoms. For instance, if $R$ is small such that $\tau_d > \tau_a$, then the adatom is more likely to re-evaporate than attach. However, if $R$ is large such that the average separation between adatoms (and/or clusters) is decreased, hence making $\tau_d < \tau_a$, then attachment leading to cluster nucleation and/or growth becomes more likely.

2. *Nucleation under constant R.*

   The $T_s$ strongly influences the critical radius $r_c$. When the $T_s$ is decreased for a given $R$, the capture zone increases in size and hence adatom attachment leading to increased cluster nucleation (or growth) becomes more probable. However, if the $T_s$ is increased $r_c$ decreases in size and so for the same $R$, re-evaporation becomes important and nucleation is decreased.

The above conditions form the basis for thin film science. However, the exact quantitative relation between $R$ and $T_s$ to nucleation characteristics like cluster density, size and size distribution is of considerable interest and debate. Typically, the relationship of nucleation to $R$ and $T_s$ is best described in terms of scaling laws.[36,37] Nevertheless, qualitatively at least, the above conditions remain true.

## 1.2.2. Introducing Order During Film Nucleation and Growth

From the discussion above, in order to achieve spatial order and control over size spread, one can manipulate the important experimental parameters controlling nucleation and growth, i.e., the deposition rate $R(x, y)$ and substrate temperature $T_s(x, y)$. We discuss below various approaches that can result in the formation of ordered surface nanostructures.

### 1.2.2.1. Spatially Varying Deposition Rate $R(x, y)$.
In the earlier discussion, the deposition rate $R$ was assumed to be spatially uniform over the substrate surface. This resulted in a homogeneous but spatially random nucleation and growth. However, if it were possible to use atom fluxes that were periodic, then it is conceivable that the resulting nucleation and growth would reflect this periodicity. Such experiments have indeed been done by using optical forces to steer atoms and create a spatially periodic atom flux. It was first shown by Timp and co-workers[17] that when a well-collimated beam of Na atoms was passed through a optical standing wave created by interference of two laser beams, the resulting Na atom flux had the spatial periodicity of the standing wave. Subsequently, McClelland and co-workers showed that chromium lines of 85 nm width and approximately 200 nm separation can be fabricated using the optical forces in the standing wave.[18] Figure 2(A) gives a schematic representation of the experiment while Fig. 2(B) is an atomic force microscope (AFM) representation of the resulting Cr film surface morphology.

### 1.2.2.2. Spatially Varying Substrate Temperature $T_s(x, y)$.
Analogous to the above situation of spatially varying the atomic flux, another theoretically possible way to obtain periodic nanostructures is by spatial modulation of $T_s(x, y)$. Consider depositing a film onto a surface with a periodically varying $T_s$ distribution in one-dimension, i.e., $T_s(x, y) = T_s(x + L, y)$, where L is the periodicity of the temperature in the $x$-direction. The average diffusion distance $l$ in some time scale $\tau$ is usually given by:

$$l = \sqrt{D_o \exp(-E_d/kT_s).\tau} \tag{4}$$

where $D_o$ is the pre-exponential factor of the diffusivity. Therefore, adatoms in the hot regions, i.e., larger $T_s$, will have larger average diffusion lengths then those in the cold, i.e., smaller $T_s$. Consequently, these adatoms are more likely to move out of the hot and into the cold zones. In addition, the thermal gradient from hot to cold will introduce an anisotropic diffusion flux to move atoms from hot to cold. The result should be preferential material accumulation dictated by the temperature distribution.

In order to verify this possibility, we have performed computational studies[38] as well as experiments[15,16,39] of film deposition under periodic but time-varying

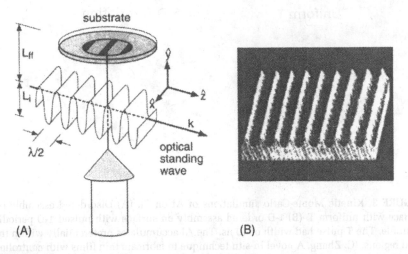

**FIGURE 2.** (A) Schematic representation of the steering experiment where the laser interference standing wave deflects the atomic beam. (Reprinted figure with permission from G. Timp, R. Behringer, D. Tennant, J. Cunningham, M. Prentiss, and K. Berggren, Physical Review Letters, 69, 1636-39, 1992. Copyright (1992) by the American Physical Society.) (B) Experimental observation of ordered lines of Cr atoms achieved by atom steering. (J. McClelland, R. Scholten, E. Palm, and R. Celotta, Laser focused atomic deposition, Science, 262, 877-880 (1993).

surface temperature gradients. The periodic temperature gradients were created by using pulsed laser interference heating of the substrate. The laser pulse width was 9 ns in the computations and experiments performed. Below we summarize some key results of simulation and experiments.

1. *Kinetic Monte-Carlo (KMC) simulations of Al film patterning under periodic surface temperature variations.* A KMC code called ADEPT, developed at Bell laboratories,[40,41] was used to simulate sputter deposition of Al metal onto single crystal Si surfaces. Its output provides information on the film morphology formed under various conditions, such as substrate temperature and deposition rate. At low to moderate temperatures, the model treats dimers (a pair of adatoms) as a stable cluster. All quantities for nucleation and diffusion were derived from molecular dynamics calculations of atomic level energetics and kinetics. User-controlled parameters in this code include: number of atoms deposited, substrate temperature, deposition rate, etc. To observe the morphology changes under spatially varying surface temperatures, the code was suitably modified so as to allow a sinusoidal temperature variation on the surface along with the deposition process. The local temperature changes were incorporated into adatom behavior via changes in the hopping rates based on an Arrhenius relation. The time scale of the temperature pulse could be arbitrarily varied to represent a real time scale. The short length scale (20 nm) and

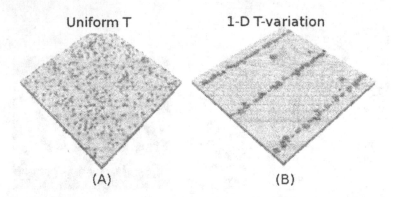

**FIGURE 3.** Kinetic Monte-Carlo simulations of Al on Si. (A) Disordered assembly on surface with uniform T. (B) 1-D ordered assembly on surface with pulsed 1-D periodic T profile. The T-pulse had width of 10 ns. The Al accumulates preferentially within the cold regions. (C. Zhang, A novel in-situ technique to fabricate thin films with controlled lateral-thickness modulations. PhD thesis, Department of Physics, Washington University, St. Louis, MO 63130, June 2004. Advisor: Ramki Kalyanaraman.)

large deposition rate (1.5 nm/s) are extreme values, chosen partly due to substrate size limitations in the code and to allow practical computation times on a PC. Al was sputter deposited onto a Si(100) surface at: (i) a uniform temperature of 150 K; or (ii) a 1-D periodic sinusoidal surface temperature (T) with peak T at 800 K and valley at 150 K. 500 Al atoms were deposited at a rate of 1.5 nm/s. The sinusoidal temperature was maintained in 10 ns long pulses, while a uniform T of 150 K was maintained between the pulses. The pulses were repeated at 50 Hz to mimic the laser used in the experiments.[15] Figure 3(A) shows the random arrangement of Al clusters on the uniform-T surface, while Fig. 3(B) shows semi-continuous lines periodically arranged along the cold temperature zones.[38] This result, i.e., that the Al metal preferentially accumulates in the regions of lower temperature, is consistent with the simple argument presented earlier on the role of surface temperature on nucleation.

2. *Experimental results on film growth under pulsed-laser heating.* We have performed experiments to study the change in film morphology under application of pulsed laser-interference heating of the surface simultaneous with film growth. The experiment typically consisted of evaporation of a metal like Co at deposition rates between 0.1 to 1 nm/min onto Si substrates.

During the deposition, the substrate was exposed to two-beam interference heating using a 266-nm UV laser with a pulse width of 9 ns and a repetition frequency of 50 Hz.[15,16] The interference fringe spacing, which is determined by the angle between the interfering beams, was 400 nm. Figure 4 compares the morphology of the film deposited with (A) and without (B) the interference pattern. Clearly, the film grown on the substrate

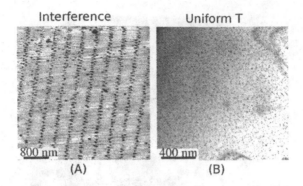

**FIGURE 4.** Morphology of a thin Co film deposited with and without two-beam laser interference irradiation. (A) AFM topographic image showing the one-dimensional ordering of 400 nm spaced arrays of Co nanodots deposited under two-beam interference; (B) Transmission electron micrograph of the Co film deposited on a substrate without laser irradiation. The disorder in the nanoparticle arrangement is apparent. (W. Zhang, C. Zhang, and R. Kalyanaraman, Dynamically ordered thin film nanoclusters, J. Vac. Sci. Tech. B, 23, L5-L9 (2005).

with the interference irradiation is regularly patterned. Careful analysis of the spacing between the rows of particles showed it to be consistent with the interference fringe spacing of 400 nm. While these preliminary results suggest that pattern formation is possible, further work is required to understand the actual changes to film nucleation and growth under such experimental conditions.

### 1.2.3. Selective Nucleation

In the discussion of Sec. 1.2.1, we had assumed that nucleation and growth occurred on a defect-free and isotropic surface. The importance of this assumption is that it leads to homogeneous nucleation on the entire surface with each surface site being equally likely to act as a nucleation site. Under most conditions of film growth, especially on single crystal surfaces, this assumption is rarely valid. Various defects, such as, dislocations, kinks, impurities and vacancies act as a favorable nucleation site that enhances nucleation.[28] Often, other driving forces, such as strain energy minimization during epitaxial growth, can also modify the nucleation and growth process.[24] From the perspective of fabrication of spatially ordered nanostructures, film deposition on a surface comprising spatially ordered defects that serve as nucleation sites can lead to ordering. Below we give one example of such an approach.

*1.2.3.1. Preferential Nucleation at Ordered Defects.* When a crystal is cleaved along a certain crystallographic plane under ultrahigh vacuum conditions, the exposed surface undergoes a rearrangement of atomic positions to achieve the most stable structure. This phenomenon is known as reconstruction and

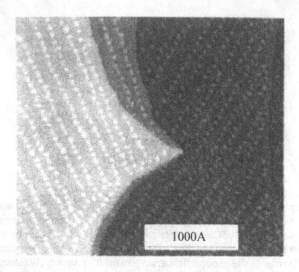

**FIGURE 5.** The herringbone arrangement of Ni nanoclusters on Au due to preferential nucleation and growth on dislocations on the Au(111) surface. (Reprinted figure with permission from D. Chambliss, R. Wilson, and S. Chiang, Phys. Rev. Lett., 66, 1721-24, 1991. Copyright (1991) by the American Physical Society.)

is the relaxation of the top layers of the crystal.[42] Until the invention of the scanning probe microscope, evidence for surface reconstructions were few and controversial. Now there are numerous examples, including the well-studied reconstruction of the $Si(1 \times 1)$ surface.[43,44] Another such reconstruction is that of the Au(111) face which results in periodic zig-zag arrangement of surface-lattice dislocations.[22] These dislocations can act as preferential nucleation sites, as observed by various groups.[45,46]

Figure 5 shows the self-assembly of ordered Ni nanoclusters on this periodic dislocation array on Au. The rows of Ni nanoclusters are spaced by 14 nm with a 7 nm spacing within the rows. Similar results of preferential nucleation have been reported for Fe[47] and Co[48] deposition on Au(111). This preferential nucleation was explained in terms of an increased binding energy of adatoms at these preferred sites. It was explained that the individual atoms diffusing on the surface can bind at these dislocations sites with low sticking probability and act as nuclei for further aggregation. However, for some other metal systems such as Al[49], Cu[50], Ag[51] and Au[52], preferential nucleation was not observed, indicating that the model of simple trapping of diffusing adatoms may not be adequate. In this respect, Meyer et al.[53] tried to explain the mechanism behind these type of film growth processes using Ni/Au(111) system as two-step process. In the first step, Ni exchanges the place with Au atoms at the elbow of the herringbone reconstruction, while the second step involves the nucleation of Ni islands on top of these substitutional Ni atoms.

In conclusion, controlling nucleation and growth by actively modifying growth parameters such as deposition rate and substrate temperature, or via modification of the surface, is an important area of research towards fabrication of nanostructured surfaces.

## 1.3. Self-assembly by Ion Irradiation

Another approach that appears to be promising towards creation of ordered nanostructures is irradiation of surfaces and thin films with energetic ion or photon beams. The irradiation results in a variety of nanoscale patterns. Since most of the experimental and theoretical work in this area has been to understand nanostructure formation by ion irradiation, we discuss this area in some detail.

Ion irradiation is the bombardment of solid surfaces by means of energetic (1-100 keV) ionized particles, which enhance surface diffusion and affect the surface morphology. It can be used to modify surfaces as well as to analyze surfaces and thin films. Some of the common applications are sputter depth profiling,[54] ion beam etching,[55] ion beam assisted deposition (IBAD),[56] surface texturing[57] and ion beam machining.[58] In addition, ion irradiation followed by annealing is frequently used to prepare chemically cleaned[59] and structurally ordered crystal surfaces for the use in surface sciences. Apart from these applications, it has attracted much attention recently due to its ability to pattern surfaces on the nanometer scale. Nanometer scale uniform periodic structures can be formed without application of an external mask or template.

A large variety of self-assembled nanostructures under ion irradiation have been reported for metals, semiconductors and amorphous materials. The synthesis of two-dimensional nanostructured arrays of metallic and semiconductor islands with at least short-range order using ion irradiation techniques is one of the exciting features with respect to nanostructured surfaces.[60,65] The surface morphology of these ordered patterns have been investigated in various material systems, including metals like Co,[66] Cu,[67,68] Ag,[69] Au,[61] and Pt,[63] and in semiconductors such as Si,[62,70] Ge,[71] and InP.[64] However, despite the extensive research, a comprehensive theory that predicts the final morphology under given irradiation conditions is lacking. This is primarily because of the strong influence of various experimental parameters, such as ion beam incidence angle, substrate temperature, ion beam energy, and surface defects. In addition, it also requires an understanding of surface processes like diffusion and kinetics of the ion/surface interaction, which essentially defines the morphology in ion-irradiation based nanostructures. Therefore, the mechanism involved in self-assembly of ordered nanostructures via ion-irradiation is an area of ongoing intense research today. Despite the complexities involved in developing a comprehensive theory, there are important theoretical understandings that form the basis of ion beam irradiation induced pattern formation and are discussed below.

### 1.3.1. Mechanism of Ion Beam Patterning

Bombarding a solid surface with an ion beam leads to the spontaneous formation of periodic structures (ripple or wave-like). A typical ion beam irradiation geometry

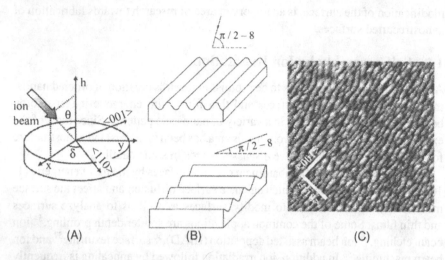

**FIGURE 6.** (A) Schematic of the ion beam patterning geometry showing the incident ion direction and important angles and surface directions. (G. Costantini, S. Rusponi, F. B. d. Mongeot, C. Boragno, and U. Valbusa, Periodic structures induced by normal-incidence sputtering on Ag(110) and Ag(001): flux and temperature dependence, J. Phys: Condens. Matter, 13, 5875-5891 (2001). (B) the relation between ripple orientation and ion beam direction (R. M. Bradley and J. M. E. Harper, Theory of ripple topography induced by ion bombardment, J. Vac. Sci. Technol. A, 6(4), 2390-2395 (1988). (C) Rippling patterns observed on a Ag(110) surface following irradiation by ions. (G. Costantini, S. Rusponi, F. B. d. Mongeot, C. Boragno, and U. Valbusa, Periodic structures induced by normal-incidence sputtering on Ag(110) and Ag(001): flux and temperature dependence, J. Phys: Condens. Matter, 13, 5875-5891 (2001).

is shown in Fig. 6(A) with the typical orientation of ripples in relation to the ion beam direction shown in Fig. 6(B). The self-assembled ripple patterns arise as a result of balance between multiple surface processes, such as roughening and smoothening. A typical example of a pattern formed by ion beams and exhibiting short range order (SRO) is shown in Fig. 6(C). This shows the ripple structure for Ag(110)[69] by irradiation of ions using the geometry shown in Fig. 6. The primary model to explain this pattern formation is the Bradley-Harper model.[72]

### 1.3.1.1. Bradley-Harper Model for Rippling.
Surface ripple under ion irradiation is linked to a surface instability that is caused by the competition between surface roughening and surface smoothening processes.[73] This instability model was first proposed by Bradley and Harper (BH)[72] and is the most successful model for ion beam rippling. This model was based on the theory proposed by Sigmund,[74] in which the removal of atoms on a surface by sputtering due to ion irradiation was extremely sensitive to the local surface curvature. Importantly, the BH model explained why ion-bombardment produces periodic height modulations (ripple pattern) on solid surfaces. The primary consequence of ion irradiation of surfaces at appropriate energies is removal of atoms from the surface by sputtering and the creation of near-surface defects like vacancies which leads to surface diffusion and

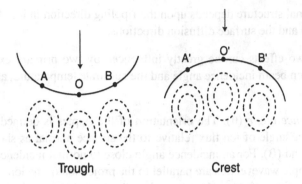

Trough                                        Crest

**FIGURE 7.** Origin of the difference in sputter rates between troughs and crests during ion beam irradiation leading to rippling in the Bradley-Harper model. The contours show regions of constant implanted ion energy. (R. M. Bradley and J. M. E. Harper, Theory of ripple topography induced by ion bombardment, J. Vac. Sci. Technol. A, 6(4), 2390-2395 (1988).)

interlayer diffusion. These consequences can be described broadly as erosion and surface diffusion:

1. *Erosion*: A well-collimated beam of heavy ions with sufficient kinetic energy bombarding a flat substrate with precise angles with respect to the normal of the uneroded surface leads to sputtering and thereby to surface erosion within some finite distance from the surface. The ion penetrates inside the solid and induces a cascade of collisions among the atoms of the substrate. As a result of this, atoms located at the surface may be affected by this collision and acquire enough energy to leave the surface.[74] However Sigmund showed[74] that the tendency for atoms to be removed depended upon the local curvature of the surface, as shown in Fig. 7. It was shown that atoms that are located at the valley (or troughs) can gain more energy and be preferentially removed as compared to those near the peaks (or crests).[72] This was because the energy deposited by the ions at the troughs was larger as compared to the crests, as can be understood from Fig. 7. As Bradley and Harper subsequently pointed out, this preferential sputtering leads to a surface *instability*, because once this local roughness was initiated, the peak-to-valley amplitude would continue to grow. This in essence was the BH model for ion-induced surface rippling. An alternate way of visualizing this effect was by assigning the ion irradiated surface a negative surface tension which tended to maximize the surface area through roughness.

2. *Surface diffusion*: The *instability* or the negative surface tension caused by ion-irradiation can be stabilized by surface diffusion, which is always present at nonzero temperatures. At sufficient temperatures, atoms and defects on or near the surface tend to diffuse and look for the most stable surface positions.[73] This effect is especially important in crystalline surfaces where there is large anisotropy in the direction of diffusion. Hence

the final structure depends upon the rippling direction induced by the ion beam and the surface diffusion directions.

These two effects can be greatly influenced by two primary experimental parameters, ion beam incidence angle and the substrate temperature, as discussed below.

- *Incidence angle* ($\theta$): The orientation of the ripples is defined by the incidence angle of ion flux relative to the surface normal as shown in Fig. 6(A) and (B). For an incidence angle close to normal incidence ($\theta \sim 0°$), the ripple wavevectors are parallel to the projection of the ion beam direction on the surface.[72] However, when the angle between the local normal to the surface and the incidence beam approaches 90°, i.e., close to grazing incidence, the ripple wavevector is perpendicular to the projection of the ion beam direction on the surface. At normal incidence several different orientations are possible. Finally at grazing angles, when there is an increase in the reflection of the ions by the surface, the rate of erosion diminishes. In addition, rotating the ion beam around the surface normal (i.e., the azimuthal angle, $\delta$) also rotates the pattern direction.
- *Substrate temperature* ($T_s$): Since diffusion of atoms and defects is a thermally activated process, the surface temperature plays a significant role on the patterning of the surface.[60,75] In general. at low temperatures (typically $\leq 200$ K), the interlayer mass transport, i.e., between surface layers and the bulk, is inhibited or less effective and the surface morphology does not show significant ordering. In the temperature range of $200 K < T_s < 300 K$, the interlayer diffusion becomes important and lead to interaction of point defects (vacancies and interstitials) with the surface atoms. This results in appearance of hill and valley structures on the surface. In this temperature range, surface mass transport is usually along a surface direction determined by the crystallographic properties of the surface. This is because specific crystallographic directions have low activation barrier energies for diffusion and so atom motion along these directions predominate. Finally, at temperature $T_s > 320$ K the interlayer mass transport along the perpendicular direction is also activated and the surface leads to ripple morphology. However, very high temperatures ($T_s \gg 320$ K) can degrade the surfaces and therefore understanding of the temperature window for ordered nanopatterning is extremely important.

**1.3.1.2. Present Status of Rippling Models.** The BH model has been very successful in predicting the ripple orientation and its dependence on the angle of incidence of the ion beam. Ripples produced by off-normal ion sputtering have been observed on $SiO_2$, diamond, Si, GaAs and Ge. However, this model did not agree well with the experimental observations on metallic surfaces[60,61,63,69] and some compound semiconductor surfaces.[62,64,76] As mentioned, this was primarily because of the important role of surface diffusion on the final ripple morphology, which was not accounted for in the BH model. Recently, Cuerno and Barabasi modified the BH model and developed a continuum equation which describes the time evolution of the surface height during sputter erosion. Referring to Fig. 6, the continuum equation can be written as

$$\frac{\partial h}{\partial t} = -v_o + \gamma \frac{\partial h}{\partial x} + v_x \frac{\partial^2 h}{\partial x^2} + v_y \frac{\partial^2 h}{\partial y^2} + \frac{\lambda_x}{2} \left( \frac{\partial h}{\partial x} \right)^2$$

$$+ \frac{\lambda_y}{2} \left( \frac{\partial h}{\partial y} \right)^2 - k \nabla^2 \left( \nabla^2 h \right) + \eta \tag{5}$$

In Eq. 5, $\frac{\partial h}{\partial t}$ represent the rate of evolution of the surface height $h(x, y, t)$ with time, $v_o$ is the erosion rate of the surface and the term containing $\gamma$ accounts for the uniform motion of the surface features along the x-direction. The terms containing $v_x$ and $v_y$ represent ion-induced surface tension terms that determine the rate of erosion of the valley and crests in terms of their curvature, i.e., the second derivative term of the local height. The terms containing $\lambda_x$ and $\lambda_y$ characterize the slope dependence of the erosion rate, while $k \nabla^2 (\nabla^2 h)$ is the surface self-diffusion term, where $k$ is a temperature dependent positive constant. Finally, $\eta$ is a white noise term, which accounts for the stochastic arrival of the ions.

In conclusion, self-assembled patterns can be formed using ion-irradiation without applying any mask or template. But it requires understanding the fundamental processes of ripple formation based on the interaction of surface processes like diffusion with ion/surface interaction.

## 1.4. Characterization

While fabrication of functional nanostructured surfaces is an outstanding challenge, the ability to characterize nanostructure morphology, spatial order, feature size, shape, etc., and correlate these characteristics to physical properties is an equally important activity. The primary goal of research into characterization techniques is to achieve a nanometer scale understanding of the chemical and physical attributes of the nanostructures so that size-dependent effects on electronic properties may be understood. In order to do this, various experimental techniques that offer high-spatial and elemental resolution are often used, including microscopy (electron microscopy, optical microscopy), spectrometry (X-ray spectrometry, mass spectrometry), scattering (X-ray diffraction, Rutherford backscattering, etc.) and scanning probe analysis (scanning tunneling microscopy, atomic force microscopy, etc). The fundamental science and technology behind these and numerous other techniques have been detailed in various books and review articles. In this section, rather than emphasizing these techniques, we will review characterization from the perspective of the physical attributes of nanostructures that can be obtained from a quantitative analysis of information obtained from techniques like atomic force microscopy and scanning electron microscopy. These techniques readily provide information on morphology, characteristic length scales, nature of spatial order and the average size and shape of surface and thin film nanostructures and often serve as the most basic characterization tools in the area of nanostructured surfaces.

The analysis of the images taken from different microscopy technique is a big challenge in the field of nanomaterials. There are number of techniques that researchers use to extract the maximum information out of the images. Here we will emphasize two techniques: Fourier analysis to obtain information on characteristic length scales and the nature of ordering; and Minkowski analysis to obtain information on morphology.

### 1.4.1. Spatial Order in Nanostructures

How can one obtain characteristic length scales and quantify order in nanostructures? As emphasized in earlier parts of this chapter, the spatial ordering in nanostructures is important in various applications. Therefore, we must have reliable techniques to quantify the nature and degree of order in nanostructures. The spatial nature of arrangement of nanostructures can be classified into three broad categories:

1. *Long-range order (LRO).* In this form, the nanostructures are spatially periodic over the entire length scale of the active area of the nanostructure in one or two dimensions. In other words, the physical properties in this structure, designated say by $T_k(x, y)$, can be expressed mathematically in 2-D by the relation:

$$T_j(\vec{x}, \vec{y}) = T_i(\vec{x} + n_1\lambda_x, \vec{y} + n_2\lambda_y) \text{ for all } x \leq L_x \quad \text{and} \quad y \leq L_y \quad (6)$$

Here, $\lambda_x$ and $\lambda_y$ are the periodic or characteristic length scales in the $\vec{x}$ and $\vec{y}$ directions, respectively; $L_x$ and $L_y$ define the maximum dimensions of the area of interest, for example, the areal size of the device; and $n_1$ and $n_2$ are integers characterizing the magnitude of the separation between the two points. Equation 1.6 describes a 2-D lattice with positions that are identical to each other. In other words, a nanostructure whose spatial arrangement follows Equation 6 is said to have 2-D translation symmetry and the properties of the nanostructure are identical at any of these related locations. Similarly, a nanostructure can have translation symmetry in 1-D only, and then any two identical positions in this nanostructure are related to each other as

$$T_1(\vec{x}, \vec{y}) = T_1(\vec{x} + n_1\lambda_x, \vec{y}) \qquad (7)$$

2. *Short-range order (SRO).* Unlike LRO where periodicity exists over the entire active or useful area of the nanostructure, in short-range order, there is no spatial periodicity. Instead, there is a characteristic length scales that appears because of the high probability of a specific spatial arrangement of features on the surface. For instance, a collection of nanoparticles on a surface can be arranged in a manner analogous to atoms in an amorphous

crystal.[77] Here, the nearest-neighbor (NN) spacing between the features is well defined, but there is no long-range regularity in the spatial position. In such a situation, the mathematical description of the structure is not straightforward, as for the LRO given by Eq. 1.6. However, a graphical representation of the spatial variation of the properties readily shows that certain length scales predominate in the system. Figure 10 is an example of average spacing, often referred to as $G(k)$, the radial distribution function (*RDF*), between nanoparticles in a laser treated film. The well-defined peak in $G(k)$ is evidence that certain spacings are more predominant than others. These spacings are the characteristic length scales in such SRO systems.

3. *Disorder*. The third category is a highly disordered system where no peaks or periodicity are visible in the $G(k)$, i.e., there is no *SRO* or *LRO*. In this situation, the nanostructured system does not have a characteristic length scale. An example is a completely random arrangement of nanoparticles on the surface, as shown in Fig. 11. Here, the $G(k)$ does not show any peak. An example of such a situation is a system whose processes are dictated by random stochastic processes.

With this introduction to characteristic length scales and the nature of order, we look at Fourier analysis and the autocorrelation function, which are the primary mathematical techniques to quantitatively evaluate the nature of ordering and the characteristic length scales in a system.

**1.4.1.1. Fourier Transforms.** Fourier analysis is a mathematical technique that allows us to decompose a signal in terms of sinusoidal signals with various frequency and amplitude.[78] The original signal can be of any type, including temporal (i.e., time varying) and spatial (i.e., varying in space). The Fourier transform (FT) of a temporal signal gives the characteristic frequencies present in the signal as well as their amplitudes. Likewise, the FT of a spatial signal, such as an image of surface nanostructures, can give information on the characteristic length scales in the system and the effective amplitudes of the features giving these length scales. The basic mathematical steps involved in Fourier analysis of a 1-D signal $T(x)$ are as follows:

- The signal is assumed to comprise a number of sine and cosine components with spatial frequencies of the form:

$$T(x) = A + \sum_{m=1}^{\infty} B_m \cos(mkx) + \sum_{m=1}^{\infty} C_m \sin(mkx) \tag{8}$$

where, $mk$ are the characteristic spatial frequency or wavenumber in the signal and $A$, $B_m$ and $C_m$ are the amplitudes of the various wavenumbers. The wavenumbers are related to real-space wavelengths by:

$$k = \frac{2\pi}{\lambda} \tag{9}$$

The analogous equation in integral form is given as:

$$T(x) = \frac{1}{2\pi} \int\limits_{-\infty}^{+\infty} T(k) e^{-ikx} dk \tag{10}$$

- The amplitudes and wavenumbers in the signal are evaluated by the Fourier integral given by:

$$T(k) = \frac{1}{2\pi} \int\limits_{-\infty}^{+\infty} T(x) e^{ikx} dx \tag{11}$$

where, $T(k)$ is known as the 1-D Fourier transform of the signal $T(x)$, or:

$$T(k_x, k_y) = \int\limits_{-\infty}^{+\infty} \int\limits_{-\infty}^{+\infty} T(x, y) e^{i(k_x x + k_y y)} dx dy \tag{12}$$

where, $T(k_x, k_y)$ is the 2-D FT.

### 1.4.1.2. Power Spectrum and the Autocorrelation Function.
The result of performing the FT, as in Eq. 12, is a complex-valued function $T(k_x, k_y)$ that provides wavenumber and phase information of the harmonic components of the signal. However, in most situations of digital image processing, one is interested in the characteristic wavenumbers $k$ and their amplitudes, rather than phase information. The amplitude information can be obtained from analyzing the power spectrum of the signal, which is defined as[79]:

$$P(k_x, k_y) = |T(k_x, k_y)|^2 \tag{13}$$

The power spectrum is also related to an important quantity known as the autocorrelation function, which quantitatively relates the similarity between any two positions $T(x, y)$ and $T(x - x', y - y')$ in the image by evaluating the function:

$$A(x', y') = \int\limits_{-\infty}^{+\infty} \int T(x, y) T^*(x - x', y - y') dx dy \tag{14}$$

Because of the convolution property of FTs, the FT of the autocorrelation function is identical to the power spectrum. Thus, it is common in image analysis to evaluate the power spectrum of the image and use it to calculate the characteristic length scales as well as the amplitudes of the various components. In the next section, we provide examples of Fourier analysis of nanostructure morphologies comprising varying degrees of order.

### 1.4.1.3. Fourier Analysis of Ordering in Nanostructures.
Figures 1.8, 1.9 and 1.10 show various images of nanoscale patterns in ultrathin cobalt metal films following laser melting while Fig. 11 shows a pattern following deposition by e-beam evaporation. Besides each figure is the corresponding power spectrum

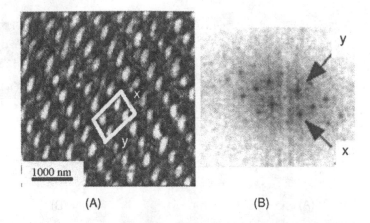

(A)                                    (B)

**FIGURE 8.** (A) 2-D arrangement of nanoscale surface features following laser irradiation of Co films on Si. (B) the power spectrum showing 2-D symmetry from which the characteristic lengths scales can be evaluated. (C. Zhang, A novel in-situ technique to fabricate thin films with controlled lateral-thickness modulations. PhD thesis, Department of Physics, Washington University, St. Louis, MO 63130, June 2004. Advisor: Ramki Kalyanaraman.)

and (in some cases) the radial distribution function $G(k)$. Note that the image is in real space, but the power spectrum and $G(k)$ are expressed in reciprocal space in terms of the wavenumber $k$.

1. *LRO*. Figure 8(A) shows a atomic force microscope image of Co nanoparticles on Si surface. The nanoparticles are arranged in a fairly periodic 2-D

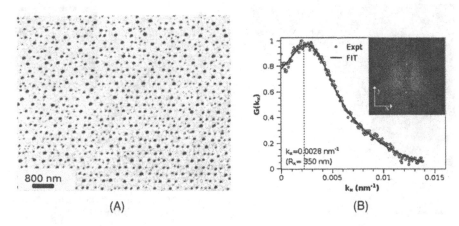

(A)                                    (B)

**FIGURE 9.** (A) Co nanoparticles on Si formed by two-beam laser interference irradiation showing 1-D arrangement of row containing the nanoparticles. (B) the radial distribution function in the x-direction showing a peak in the G(k) indicating that short-range order exists between particles within the row. The inset shows the power spectrum. (C. Favazza, J. Trice, H. Krishna, R. Sureshkumar, and R. Kalyanaraman, Laser induced short and long-range ordering of Co nanoparticles on $SiO_2$, App. Phys. Lett., 88, 153118, (2006).

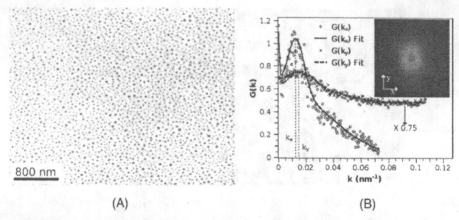

(A)                                              (B)

**FIGURE 10.** (A) Co nanoparticles on Si formed by laser irradiation with a single beam showing an apparently random distribution. (B) The G(k) in the x- and y-directions indicate peaks implying short-range order in both directions. The power spectrum in the inset shows this short range order as a annular diffraction ring. (C. Favazza, J. Trice, H. Krishna, R. Sureshkumar, and R. Kalyanaraman, Laser induced short and long-range ordering of Co nanoparticles on SiO2, App. Phys. Lett., 88, 153118, (2006).

array. The power spectrum (Fig. 8(B)) was obtained using the fast-Fourier transform technique[80] available in the open-source program ImageJ.[81] The power spectrum shows spots in the $k_x$ and $k_y$ directions that correspond to the long-range periodic spacings of nanoparticles in the $x$ and $y$ directions of the real image. This periodic spacing is analogous to information

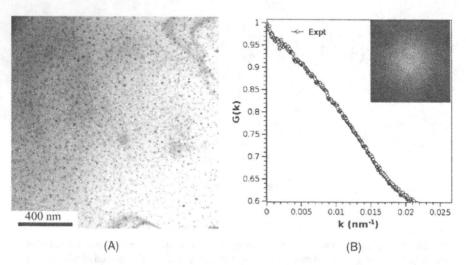

(A)                                              (B)

**FIGURE 11.** (A) Co nanoparticles formed by film deposition onto Si at room temperature. (B) The G(k) does not show a peak and the power spectrum (inset) is also featureless indicating a completely random arrangement of particles with no characteristic length scale. (W. Zhang, C. Zhang, and R. Kalyanaraman, Dynamically ordered thin film nanoclusters, J. Vac. Sci. Tech. B, 23, L5-L9 (2005).

contained in the diffraction pattern obtained from crystalline material and in fact the power spectrum is the computer generated diffraction pattern of the nanoparticle array.

2. *LRO and SRO*. Figure 9(A) shows a scanning electron microscope image of periodically spaced rows of Co nanoparticles along the y-direction of the surface. The pattern was formed by two-beam laser-interference heating of an ultrathin Co film on $SiO_2$[82]. The power spectrum (inset of Fig. 9(B)) shows spots along $k_y$ corresponding to the 1-D periodicity, while it shows arcs in the $k_x$ direction perpendicular to the spots. To quantitatively understand these arcs, the RDF, $G(k_x)$, is plotted in Fig. 9(B). This shows a maximum in the $G(k_x)$ at $k = 0.0028$ nm$^{-1}$ with no other peaks. This is an example of *SRO* in the positioning of particles. The physical interpretation of the *SRO* is that the nearest-neighbour spacing between particles in each row is not completely random but has a well-defined length scale given by, $\lambda_x = \frac{1}{k_x} \sim 357$ nm (here the scaling relation between $\lambda$ and $k$ in Fig. 9(B) is $\lambda_x = \frac{1}{k_x}$). This SRO cannot readily be discerned from the real space image of Fig. 9(A) and emphasizes the importance of performing power spectrum analysis on such images.

3. *SRO*. Figure 10(A) shows Co nanoparticles formed by single beam laser heating on $SiO_2$ surface.[82] No immediate evidence for ordering is visible from the image. However, a better understanding of the spatial arrangement of the particles can be seen in the power spectrum (inset of Fig. 10(B)), which shows the presence of a ring. The $G(k_x)$ in two perpendicular directions, $k_x$ and $k_y$ are plotted in Fig. 10(C). Peaks are visible in the $G(k_x)$ in each direction indicating *SRO* of the nanoparticles due to a characteristic nearest-neighbor spacing.

4. *Disorder*. Figure 11(A) shows Co nanoparticles formed by electron-beam deposition of a 0.5 nm thick Co film onto Si substrate. The power spectrum in Fig. 11(B) does not show any distinctive feature other than a monotonic decrease in intensity away from the origin. This effect is captured more clearly in the plot of the $G(k_x)$. Clearly no distinctive peak is seen, but rather there is a gradual decay in intensity. This type of diffraction signal is characteristics of a disordered or random arrangement of particles on the surface. In this example, there is no characteristic length scale in the spatial ordering of nanoparticles.

From these examples, the utility of the Fourier or power spectrum analysis is apparent, especially in situations where SRO is present but difficult to evaluate from the real-space images.

## 1.4.2. Minkowski Analysis

Patterns occur in many systems and images of these patterns consist of regions of varying morphology, including changes in contrast, size, shape, connectivity and spatial order over various length scales. Understanding the morphology quantitatively is often important in understanding of physical processes. For instance, the classic patterns exhibited by diffusion-limited aggregation, spinodal dewetting, or other hydrodynamic instabilities give important clues to the underlying physical processes involved in structure formation. While Fourier analysis and autocorrelation functions provide vital information about the nature of order and characteristic length scales, they are incapable of quantitatively describing morphology of irregular structures, including the vertical shape, curvature and connectedness. Therefore, it is useful to apply techniques that quantitatively characterize the morphology of spatially varying patterns. In this respect, Minkowski functionals analysis could be considered as a potential morphological measure for patterns. This analysis is well known in image analysis of astronomical data, and in mathematical morphology and integral geometry. However, it has only recently attracted attention as a tool for analysis in materials science.[83]

### 1.4.2.1. Minkowski Functionals.

The Minkowski measurement allows for a novel morphological characterization, which is somewhat independent of the image size. All the available conventional image analysis are done either in direct real space or in Fourier space. In the direct $(x-y)$ space, the brightness of the image or the height of the sample surface, for example in scanning probe microscopy (SPM) images, usually undergo some local averaging or ranking operation taking place in a limited neighborhood, called a kernel around the considered pixel. On the other hand, in the Fourier analysis of image spatial frequencies the treatment is not local but global and the periodic features are identified by averaging its local morphological properties. The Minkowski functionals described below, describe global geometric characteristics of structures in *real* space.

The functionals are estimated by thresholding a given image into two different regions, $H$ and $L$, of high and low intensity pixel levels and subsequently estimating different pixel characteristics as a function of the threshold level $\rho$. In this case the threshold level $\rho$ can be thought to behave like a third co-ordinate, i.e., the $z$-coordinate, representative of the height of features on the surface. These functionals give information on the vertical shape, boundary length and connectivity of characteristic features. Below we describe these functionals based on the work of Mecke and co-workers[84,85] and the article by Salerno and co-workers.[83] To perform the calculations the following quantities (with reference to Fig. 12) must be evaluated as a function of the threshold $\rho$:

- the total number of pixels in the image $N$. Obviously this quantity is independent of $\rho$;
- the number of white pixels $N_{white}(\rho)$, i.e., pixels above threshold $\rho$;
- the number of black pixels $N_{black}(\rho)$, i.e., pixels below threshold $\rho$;
- the number of boundary pixels, $N_{bound}(\rho)$, which are defined as the number of adjacent black-white pixels.
- the number of discrete white ($C_{white}(\rho)$) and black ($C_{black}(\rho)$) domains, which denote the number of regions enclosed by the opposite pixel type.

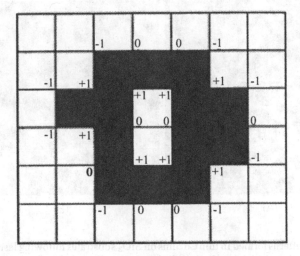

**FIGURE 12.** A black and white image constructed from square pixels that is used to explain the different definitions used in the Minkowski functionals analysis. (Reprinted figure with permission from K. R. Mecke, Phys. Rev. E, 53(5), 4794-4800, 1996. Copyright (1996) by the American Physical Society.)

From these quantities the following Minkowski functionals can be estimated:

a. Area fraction: $V(\rho) = \dfrac{N_{\text{white}}(\rho)}{N}$

b. Boundary length: $S(\rho) = \dfrac{N_{\text{bound}}(\rho)}{N}$

c. Connectivity: $\chi(\rho) = \dfrac{C_{\text{white}}(\rho) - C_{\text{black}}(\rho)}{N}$

### 1.4.2.2. Application of Minkowski Analysis to Thin Film Patterns.

Here we provide an example of the application of the above functionals to characterize the morphology. In order to highlight the information provided by each functional we have chosen two images of distinctly different morphologies. Figure 13 shows typical microstructures observed at different stages in the dewetting process of Co ultrathin films from an $SiO_2$ surface. Figure 13(A) shows a cellular pattern made by rings of Co while Fig. 13(B) is a nanoparticle morphology. The total Co concentration in both images is comparable. Below we explain the information contained in each functional for the two images.

1. *Area fraction.* Shown in Fig. 14(A) is the rate of change $V(\rho)$ for the two samples. Some important differences can be noted from a careful evaluation of the variation of $V(\rho)$ for the two figures. For values of $\rho$ near 0, $V(\rho)$ decreases slowly for the particles as compared to the rings, while for larger values of $\rho$, i.e., near 50, the particle drops rapidly as compared to the rings. As mentioned earlier, $\rho$ can be visualized as an

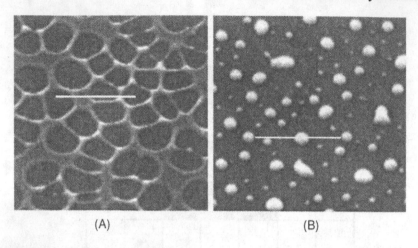

(A)                                                      (B)

**FIGURE 13.** Patterns formed in thin Co films on SiO$_2$ substrates following laser irradiation for (A) a short time which results in cellular structures and (B) a long time leading to nanoparticles. Both pattern have identical length scales. (C. Favazza, R. Kalyanaraman, R. Sureshkumar, Nanotechnology, 17, 4229–4234, (2006).)

effective height scale in the z-direction. Hence, the rate of change of $V(\rho)$ with $\rho$ gives information on how the feature shape changes with height. To explain this functional in simpler terms, Fig. 15(A) shows two hypothetical 1-D feature shapes, one a rectangle and another some fraction of a circle (dashed line). In this example, $V(\rho)$ at any value of $\rho$ is simply the length of the feature in the $x$-coordinate. Therefore, the plot of $V(\rho)$ vs. $\rho$, shown in Fig. 15(B) will be a rectangle in one case while a parabolic curve for the circle, representing the shape of the sidewalls of the features. Therefore, the $V(\rho)$ for the rings and particles, which is a cumulative result for all the features on the image suggests that the particles have a larger rate of height variation compared to the ring patterns. This is consistent with the fact that since the mass of Co is conserved, the smaller particles must have a larger height as compared to the ring-like patterns.

2. *Boundary length $S(\rho)$.* In comparison to the area fraction, the $S(\rho)$ (Fig. 14(B)) is a measure of the total curvature of features in the image. This can be explained from the line scans across typical features in the ring and particle images. Figure 13 shows the location of the line scans, while Fig. 16 shows the gray value plotted vs. length for the scans in the ring and particle images. Since $S(\rho)$ estimates the total length of the boundary between black and white pixels, one can note the following: for the particle, at $\rho = 0$, $S(\rho) = 0$ and then there is a rapid increase with a boundary length equivalent to the projected perimeter of the particle on the surface. As $\rho$ continues to increase, the projected perimeter decreases because of the particles curvature, and finally $S(\rho)$ drops of to 0 as $\rho$ becomes $>\rho_c$

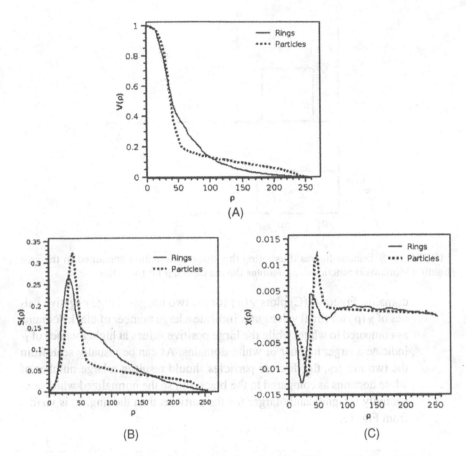

**FIGURE 14.** Minkowski functionals for the two patterns of Fig. 13: (A) the area fraction $V(\rho)$; (B) the boundary length $S(\rho)$; (C) the connectivity $\chi(\rho)$.

(Fig. 15(A)) for the particle. A similar behavior is seen for the rings, but with one difference. Since the ring patterns comprise two distinct feature, i.e., the narrow ring wall and the broad film region between two merging rings, therefore the line scan of Fig. 16(A) shows two characteristic features: a single peak at the ring wall; and a double peak in the region between merging rings. Consequently, $S(\rho)$ also results in two peaks as visible in Fig. 14(C) which can also be easily visualized by varying $\rho$ in the line scan figure of Fig. 16. Therefore the boundary length provides important information on the type of features present in the image.

3. *Connectivity* $\chi(\rho)$. In contrast to $V(\rho)$ and $S(\rho)$ which provide area and length information, the connectivity functional provides a quantitative estimate of the number of domains of black or white regions. Since a domain is an isolated region of black or white, this parameter quantifies the discreteness of the features or in other words the connectedness of the various

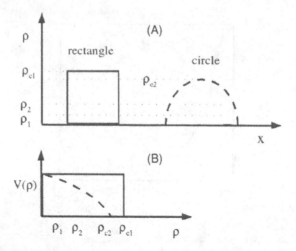

**FIGURE 15.** Schematic figures illustrating the shape information contained in the area fraction Minkowski functional. $\rho_c$ denotes the maximum $\rho$ for the feature.

domains. Figure 14(C) plots $\chi(\rho)$ for the two images. Large negative values of $\chi(\rho)$ for small values of $\rho$ indicate a large number of black domains as compared to white, while the large positive values at higher values of $\rho$ indicate a larger number of white domains. As can be visually seen from the two images, the discrete particles should result in a large number of white domains as compared to the black and so the normalized white connectivity is significantly larger for the particles than the ring, as is visible from Fig. 13.

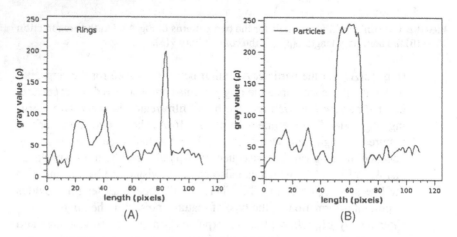

**FIGURE 16.** Linescans taken from the patterns of Fig. 13. (A) Line scan for the cellular pattern showing the strong peak from the ring wall and a smaller double peak from the region between merging rings. (B) Line scan from the particle morphology showing the rapid change in height.

In this section we have reviewed two important image processing techniques: the Fourier transform, which provides information on spatial ordering, and the Minkowski functionals, which provides information on morphology. Below, in the final section of this chapter we provide some typical application areas that are strongly connected to control of nanoparticle size and their ordering.

## 1.5. Applications

The final section of this chapter provides examples from cutting-edge applications of nanostructured surfaces and thin films. These examples highlight the importance of particle size in improving the sensitivity of gas sensors, the potential for coherent optical response from sub-wavelength sized structures based on plasmon waveguiding effects in metallic nanoparticle chains, and the need for order in size and size distribution for magnetic data storage applications.

### 1.5.1. Gas Sensors

Semiconductor resistive metal oxide (MOx) gas sensing is an essential part of the various sensing approaches for monitoring and detection of various biological and environmental toxins. Recent trends[87] indicate that nanotechnology is impacting this field primarily via the enhanced properties observed for nanostructured morphologies.[88] Since the first reports in 1962 by Seiyama and co-workers[89] that a semiconductor can be used to detect gases, this field has progressed significantly. The general requirement for gas sensors is that they be sensitive to gas absorption while being chemically and electrically stable at the relatively high temperatures of operation (200°C to 500°C). This is usually satisfied by materials with large band-gaps, like the MOx semiconductors $SnO_2$, $ZnO$ and $TiO_2$, while ruling out the more conventional semiconductors like Si and GaAs. Table 1 lists the common materials used as semiconductor gas sensors and the detected gases.

There is a continuous need for more energy-efficient, sensitive, selective, stable, fast and reliable sensors which are cheap and easy to manufacture. This

**TABLE 1.** The commonly used metal oxide sensing material and the gases they are responsive to. (From Ref. [97].)

| Oxide type | Detectable gas |
|---|---|
| $SnO_2$ | $H_2$, CO, $NO_2$, $H_2S$, $CH_4$ |
| $TiO_2$ | $H_2$, $C_2H_5OH$, $O_2$ |
| $Fe_2O_3$ | CO |
| $Cr_{1.8}Ti_{0.2}O_3$ | $NH_3$ |
| $WO_3$ | NO, $NH_3$ |
| $In_2O_3$ | $O_3$, $NO_2$ |
| $LaFeO_3$ | $NO_2$, $NO_x$ |

has been typically approached by combining miniaturization[90] using Si processing techniques and modifications to oxide film properties, including increased porosity and metal doping for chemical and electronic effects. A very popular and highly evolved gas sensing concept has become the Electronic Nose,[91] where an array of sensors[92] with varying electrical characteristics[93] is simultaneously monitored and selective gas sensing achieved through complex electronics and algorithms.[94] Presently, the focus is clearly shifting to the implementation of nanotechnology, either via morphological changes[95] or a combination of miniaturization and nanostructured morphology.[96]

### 1.5.1.1. Role of Nanostructure in Thin Film MOx Gas Sensors.

MOx gas sensors detect surface-absorbed gases via a change in resistance.[98,99] The widely accepted model for the resistance change is due to variation in concentration of absorbed oxygen in the near surface regions.[100] Absorbed oxygen species ($O^-$, $O^{2-}$, etc.) capture conduction electrons (in the case of $n$-type semiconductors) in the near surface region leading to a change in the concentration of charge carriers and hence the resistance. When a reducing gas like $H_2$ is introduced into the system, it interacts with the adsorbed O species, which in turn results in a change in the carrier concentration and therefore a change in resistance. The volume over which charge carrier is depleted is known as the space charge layer and is quantitatively approximated by the Debye length,[97,101] which can be written as:

$$L_D = \left(\frac{\varepsilon_o kT}{n_o e^2}\right)^{\frac{1}{2}}$$

(15)

Where $\varepsilon_o$ is the static dielectric constant, $n_o$ is the total carrier concentration, $e$ is the carrier charge, $k$ is the Boltzmann constant, and $T$ is the absolute temperature.

When the particle size is decreased, at least two effects play a role:

1. The number of surface sites available for gas adsorption increases with decreasing grain size because of the increase in the surface-to-volume ($s/v$) ratio as the grain size decreases. This ratio for a spherical particle of radius $r$ can be expressed as:

$$\frac{s}{v} = \frac{3}{r}$$

(16)

2. As the size of the grain becomes comparable to the depletion layer or the Debye length, the sensitivity of gas detection is found to increase significantly. The sensitivity ($S$) can be expressed as the change in conductivity ($\Delta G$) due to change in the carrier concentration ($\Delta n$) and is given by:

$$S = \Delta G/G_o = (\Delta n/n_o)$$

(17)

where $G_o$ and $n_o$ are the conductivity and concentration, respectively, following adsorption of the oxygen species but prior to adsorption to reducing gases. Experimental and theoretical work shows that the sensitivity is maximum when the particle size is about half the Debye length. The size of this space-charge layer, which depends on the intrinsic carrier concentration, is typically of the order

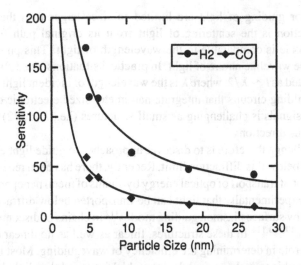

**FIGURE 17.** The dramatic increase in sensitivity in measurement of $H_2$ and CO gas concentrations as a function of decreasing metal oxide semiconductor nanoparticle size. (F. Cosandey, G. Skandan, and A. Singhal, Materials and processing issues in nanostructured semiconductor gas sensors, JOM-e, 52, 1-6 (2000). http://www.tms.org/pubs/journals/JOM/0010/Cosandey/Cosandey-0010.html.)

of 1 to 100 nm for the important gas-sensing oxides.[102] A typical value for this length in the case of $SnO_2$ has been measured to be ~3 nm.[103] Because of the above effects, scaling of grain and film dimensions into the nm scale enhances the sensitivity to gas detection and also speeds the response time of these sensors. Both experiments,[96,104,106] and theory,[107–110] support this enhanced performance with reduced grain size or film thickness. Figure 17 shows that the sensitivity increases significantly with decreasing grain size for detection of $H_2$ and CO by $SnO_2$ sensors.

### 1.5.2. Optics Below the Diffraction Limit

The miniaturization of optical devices and its integration to sub-micron electronic components is currently a major research effort in optoelectronics and photonics.[111] In this area, the fundamental challenge is the need to confine optical modes to small dimensions and to exchange and route optical energy into different directions in the circuits. For example, it will be necessary to transport optical energy through corners, elbows and tees, something that is presently not available in length scales ≤ μm.

Optical and photonic structures, such as high density photonic circuits, must overcome optical cross-talk and high losses that are common in present photonic devices. To eliminate cross-talk it is necessary to develop structures that either work in the strong confinement regime (high index differential between the waveguide and the surrounding media) or work below the diffraction limit. Currently the key technologies are planar waveguides and photonic crystals. However the device size scales for guiding of light are limited to sizes larger than the diffraction limit. Diffraction is the scattering of light from its original path from objects whose size scale is comparable to the wavelength of light. This phenomenon is inherent to the wavelike nature of light. In practical situations, the diffraction limit is approximated as $l \sim \lambda/2$, where $\lambda$ is the wavelength of incident light. Because of this limit, building circuits that integrate nanometer sized electronic components with optical signals is challenging as small structures (i.e., $l < \lambda/2$) will scatter light in various directions.

The challenge therefore is to discover approaches to guide light efficiently in length scales below this diffraction limit. Recently, there has been renewed interest in the behavior of transport of optical energy by chains of metal nanoparticles. It has been shown experimentally that light can be transported below diffraction limited length scales by surface plasmon guiding modes[112] in chains of disconnected metal nanoparticles.[10,113,114] In these structures, linear as well as nonlinear effects play an important role in determining the efficiency of waveguiding. Most importantly, it appears possible to make optical devices which can modulate light based on the efficient non-linear response of waveguides made from such metal nanoparticle chains.[11,115] However, some of the key challenges in realizing optical devices based on such metal-nanoparticle array architectures is to reduce transmission losses and find innovative fabrication processes that allow superior control over physical attributes that influence linear and non-linear effects.

The basic physical phenomenon that causes the waveguiding is the production of surface waves or collective charge oscillations known as plasmons in the individual nanoparticles. Simply put, the oscillation in an individual nanoparticle induces dipole moments in neighboring particles via near-field electromagnetic interactions and this process leads to transfer of optical energy. Below, we describe plasmon oscillations, which are important towards achieving sub-micron waveguides.

### 1.5.2.1. Plasmons.
A collection of free moving charges in an otherwise neutral system can be treated as a plasma. In condensed matter, one example of such a plasma is the free electron gas in a metal. Let us consider a metal slab containing $n$ free electrons per unit volume, as shown in Fig. 18(A). Imagine that the electron gas is displaced by a distance $x$ with respect to the fixed positive ion positions. This will lead to a net surface charge of $-nex$, where $n$ is the electron charge volume density and $-e$ is the charge of the electron. Consequently, the long-range Coulomb force tries to restore the electron gas to equilibrium. This force can be expressed through an internal electric field expressed using Gauss's law as $E(x) = nex/\varepsilon_o$. *The result is that a collective longitudinal oscillation of the electron gas is excited, with the electric field in the direction of the oscillation.* These longitudinal oscillations are known as plasma oscillations. The 1-D equation of motion of this electron gas under this restoring force can be written as:

$$nm\frac{d^2x}{dt^2} = -neE(x) = -ne\left(\frac{nex}{\varepsilon_o}\right) = \frac{-n^2e^2x}{\varepsilon_o} \tag{18}$$

Neutral

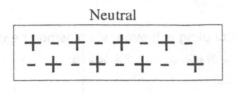

Displaced Charge

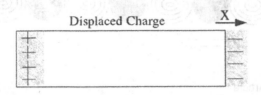

FIGURE 18. Schematic illustrating the collective displacement of the electron gas in a metal that gives rise to the plasmon oscillation.

where $m$ is the mass of the electron, $\varepsilon_o$ is the permittivity of free space, and $x$ is its displacement due to the electric field. This equation describes a harmonic oscillator with a natural plasma frequency given by:

$$\omega_p = \sqrt{\frac{ne^2}{\varepsilon_o m}} \qquad (19)$$

The plasma frequency is a very useful parameter as it determines a characteristic length scale, i.e., the wavelength $\lambda_p = 2\pi c/\omega_p$ of light, that will propagate through the metal; In general, when the wavelength of light is less than $\lambda_p$ it will propagate through the metal, otherwise it will be reflected. This frequency is also used to define a quantum of bulk plasma oscillation known as the *plasmon*. A plasmon has energy given by $E_p = \hbar\omega_p$. Plasmons can be excited by interaction of the metal with charged particles like electrons or electromagnetic light waves.

- *Surface plasmons.* Another form of longitudinal oscillation is observed when the electron gas is confined to small length scales, for instance, in ultrathin films or nanoparticles. Analogous to the bulk, a small perturbation in the local electron density produces a restoring force, resulting in oscillations that are localized to the near surface regions. Solutions of the electromagnetic Maxwell's equation at the surface shows that these longitudinal oscillations have a characteristic frequency that is related to the bulk plasma frequency but also dependent on the shape and geometry of the structure.[116] For example, the frequency of the surface plasmon at a planar metal/vacuum interface is $\omega_s = \omega_p/\sqrt{2}$, while it is $\omega_s = \omega_p/\sqrt{3}$ for a spherical metal nanoparticle.

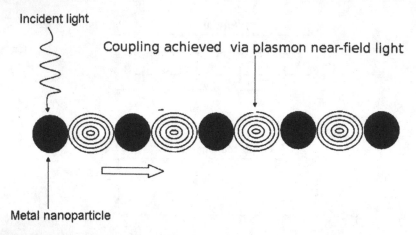

**FIGURE 19.** Illustration of the concept of waveguiding with structures having size below the diffraction limit. Incident light is coupled into one end of an appropriate nanoparticle chain. Surface plasmon oscillations in the metal nanoparticle generates a electric field whose near-field component couples to adjoining nanoparticles thus resulting in electromagnetic energy transfer. (Reprinted by permission from Macmillan Publishers Ltd.:, Nature Materials, J. Krenn, Nanoparticle waveguides: Watching energy transfer, 2, 210-211, (2003).)

### 1.5.2.2. Transport of Energy by Metal Nanoparticle Chains.

When light of frequency comparable to the surface plasmon frequency is incident on a metal nanoparticle, there is a strong coupling leading to enhanced scattering of this light. This strong interaction of the individual nanoparticles with light can be used to fabricate waveguides. The dipole field resulting from the light-induced plasmon oscillations in a nanoparticle induces an oscillation in a closely spaced neighbouring particle due to near field electrodynamic interactions.[10] When metal nanoparticles are closely spaced the distance dependent near field term in the expansion of the electric dipole interaction dominates. The phase and the interaction strength of the electric field in neighbouring particles are both polarization and frequency dependent. Under certain conditions, this type of interaction leads to coherent oscillating modes and the wavevector $k$ of the oscillation is along the nanoparticle arrays. As shown by Maier and co-workers,[11] such effects can lead to propagation of light along the metal nanoparticle chain with particle sizes and spacings much smaller than the wavelength of light. Figure 19 depicts how a nanoparticle waveguide is locally excited by light, typically by NSOM (near-field scanning optical microscope). Then the electromagnetic energy is transferred to the neighboring nanoparticles by near-field coupling of electromagnetic energy due to polarization by the surface plasmon oscillations. This type of interaction between metallic nanoparticles with light promises to be an exciting area of research for achieving subwavelength sized optical devices.

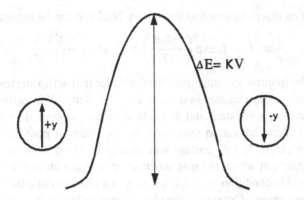

**FIGURE 20.** The activation energy barrier determined by the nanoparticle volume V and magnetic anisotropy constant K that determines the lifetime of magnetization direction and therefore information storage in magnetic materials.

### 1.5.3. Nanoparticles for Magnetic Data Storage

One of the current outstanding materials challenges is the goal of achieving high-density magnetic storage of 1 Tbit/in.$^2$ or greater.[8,9] This areal density corresponds roughly to regularly spaced nanomagnets every 25 nm in a square array. Clearly, for magnetic data to be stored-at and read-from such size scales, it is critical that the magnetic bits be arranged in some regular manner such that the read-write process can be programmed. However, there are some fundamental materials challenges with regards to the behavior of such small-sized magnetic materials that also need to be addressed.

In present high-density magnetic recording media with densities of 10-100 Gbit/in.$^2$, a bit is stored in a single magnetic domain that is comprised of numerous grains. Typically the grain size is a small fraction of the domain. However, as recording densities increase, the grain size becomes an increasing fraction of the domain and eventually, a single domain or bit is a single grain with size in the range 1-100 nm. In this regime a new problem is encountered, that of superparamagnetism.

#### 1.5.3.1. Superparamagnetism and Ordering of Nanoparticles. In general, ferromagnetic crystals below the Curie temperature have an easy axis of magnetization that is related to the crystal structure. Consider such a single grain crystal, shown in Fig. 20, whose easy axis lies along the y-direction. Then, the direction of magnetization is equally likely to point along the +y or −y directions. Once the magnetization direction is set by an external magnetic field, then to reverse direction an energy barrier must be overcome. This energy barrier $\Delta E$, as depicted in Fig. 20, is related to the volume of the grain V and a material anisotropy constant K. The lifetime or decay time t of magnetization, i.e., the time to reversal

of magnetization direction was first derived by Neel and can be expressed as:

$$\frac{1}{t} = f = f_o \exp\left(\frac{-\Delta E}{kT}\right) = f_o \exp\left(\frac{-KV}{kt}\right) \tag{20}$$

where, $f_o$ is the frequency of attempts ($\sim 10^9$ s$^{-1}$) for magnetization reversal, $K$ is the material anisotropy constant in units of energy/volume, $V$ is particle volume, $k$ is the Boltzmann s constant and $T$ is absolute temperature. The exponential nature Eq. 20 of the relaxation time on $V$ and $T$ makes it possible to define a blocking temperature $T_b$ (at constant volume), or blocking volume $V_b$ (at constant temperature), at which the magnetization goes from an unstable condition ($t \ll t_c$) to a stable condition ($t \gg t_c$), where $t_c$ is a characteristic time scale relevant to the application of interest. Consider an application such as magnetic data storage where the stability of magnetization must be significantly larger than the expected lifetime of the storage media. The hard-disk industry has decided that for a magnetic material to be a candidate as a storage medium, the minimum value of the exponential term, $KV/kT$ must be about 50 at room temperature.[8] As shown by White,[8] this sets a characteristic lifetime for the magnetization reversal to be:

$$t_c = \frac{1}{f_o} e^{50} = 5.18 \times 10^{12} s \sim 17k \text{ years}$$

However, the exponential nature of the behavior results in dramatic changes in the lifetime as a function of the volume of the particle. Consider a collection of magnetic nanoparticles with a large size distribution such that particle volumes range from $V_{min} \leq V_{av} \leq V_{max}$, where $V_{min}$ is of the order of $V_{av}/2$. For this minimum volume in the collection, the characteristic time for reversal can be expressed as:

$$\frac{t}{t_c} = \frac{e^{50/2}}{e^{50}} = e^{-25} \quad \text{or} \quad t = t_c e^{-25} = 5.07 \times 10^{11} \times e^{-25}$$

or

$$t = t_c e^{-25} = 5.07 \times 10^{11} \times e^{-25} \sim 70s$$

This dramatic drop in the lifetime of the magnet is known as superparamagnetism, i.e., when the magnetic material cannot retain magnetism over time scales of interest.

From the above discussion it is clear that besides the materials challenge of finding magnets with appropriate values of the material anisotropy constant, *ordering* in nanoparticle size by having a very narrow size distribution is critical to realizing high-density storage at densities beyond 1 Tbit / in.$^2$

## 1.6. Conclusion

In this chapter we introduced functional surface nanostructures that have potential for use in various areas of nanotechnology, including gas sensing, photonics and magnetism. We focused on self-assembly techniques to making such surface nanostructures. One promising area is the manipulation of thin film nucleation and growth that can lead to spatially ordered nanostructures. Another was the use of ion irradiation of surfaces that can lead to periodic features in the nm length scale. Since the characterization of order and morphology is a critical part of understanding the properties of these structures, we discussed with examples the information that Fourier analysis and Minkowski functionals analysis can give on order and morphology. Finally, applications of ordered nanostructures were presented, with examples from areas of gas sensing, surface plasmon waveguiding and magnetic data storage.

## ACKNOWLEDGMENTS

RK acknowledges support by the National Science Foundation through a CAREER grant #DMI- 0449258. The authors would also like to acknowledge useful discussion with Christopher Favazza on Minkowski functionals analysis.

## REFERENCES

1. K. Drexler, Molecular Engineering: An approach of the development of general capabilities for molecular manipulation, in *Proceedings of the National Academy of Sciences*, **78**, 5275-78 (1981).
2. K. Drexler and Foster. J.S., Synthetic tips, *Nature*, **343**, 600-604 (1990).
3. H. Gleiter, Nanostructured materials: basic concepts and microstructure, *Acta. Mater.*, **48**, 1-29 (2000).
4. K. Inomata and Y. Saito, Spin-dependent tunneling through layered ferromagnetic nanoparticles, *Appl. Phys. Lett.*, **73**, 1143-45 (1998).
5. J. Stahl, M. Debe, and P. Coleman, Enhanced bioadsorption characteristics of a uniquely nanostructured thin film, *J. Vac. Sci. Tech. A*, **14**, 1761-64 (1996).
6. D. Shtanski, S. Kulinich, E. Levashov, and J. Moore, Structure and physical-mechanical properties of nanostructured thin films, *Phys. Sol. St.*, **45**, 1177-84 (2003).
7. S. Shukla and S. Seal, Room temperature hydrogen gas sensitivity of nanocrystalline pure tin oxide, *J. Nanosci. Nanotech.*, **4**, 141-149 (2004).
8. R. White, The physical boundaries to high-density magnetic recording, *J. Mag. Magn. Mat.*, **209**, 1-5 (2000).
9. C. Black, C. Murray, R. Sandstrom, and S. Sun, Spin-dependent tunneling in self-assembled conanocrystal superlattices, *Science*, **290**, 1131-34, (2000).
10. M. Quinten, A. Leitner, J. Krenn, and F. Aussenegg, Electromagnetic energy transport via linear chains of silver nanoparticles, *Optics Lett.*, **23**(17), 1331-33 (1998).
11. S. A. Maier, P. G. Kik, H. A. Atwater, S. Meltzer, E. Harel, B. E. Koel, and A. G. Requicha, Local detection of electromagnetic energy transport below the diffraction limit in the metal nanoparticle plasmon waveguides, *Nature Materials*, **2**, 229 (2003).
12. S. Fan, M. Chapline, N. Franklin, T. Tombler, A. Cassell, and H. Dai, Self-oriented regular arrays of carbon nanotubes and their field emission properties, *Science*, **283**, 512-514 (1999).
13. J. Lodder, M. Haast, and L. Abelman, Patterned magnetic thin films for ultra high density recording, in *Proceedings of NATO Advanced Study Institute on Magnetic Systems Beyond 2000* (G.C. Hadjipannayis, ed.), pp. 117-145 (Dordrecht, Netherlands: Kluwer Academic Publishers, 2002).

14. J. Matthews, *Epitaxial Growth: Parts A and B*. (Academic Press, New York, 1975).
15. W. Zhang, C. Zhang, and R. Kalyanaraman, Dynamically ordered thin film nanoclusters, *J. Vac. Sci. Tech. B*, **23**, L5-L9 (2005).
16. C. Zhang and R. Kalyanaraman, In situ nanostructured film formation during physical vapor deposition, *App. Phys. Lett.*, **8323**, 4827-29 (2003).
17. G. Timp, R. Behringer, D. Tennant, J. Cunningham, M. Prentiss, and K. Berggren, Using light as a lens for submicron, neutral-atom lithography, *Phys. Rev. Lett.*, **69**, 1636-39 (1992).
18. J. McClelland, R. Scholten, E. Palm, and R. Celotta, Laser-focused atomic deposition, *Science*, **262**, 877-880 (1993).
19. L. Huang, S. Chey, and J. Weaver, Buffer-layer-assisted growth of nanocrystals: Ag-Xe-Si(111), *Phys. Rev. Lett.*, **80**, 4095-98 (1998).
20. S. Wolf and R. Tauber, *Silicon Processing for the VLSI Era-1 Process Technology*. (Lattice Press, Sunset Beach, California, 1986).
21. J. Sheats and B. Smith, *Microlithography Science and Technology*. (Marcel Dekker, New York, 1998).
22. D. Chambliss, R. Wilson, and S.Chiang, Nucleation of ordered Ni island arrays on Au(111) by surface-lattice dislocations, *Phys. Rev. Lett.*, **66**, 1721-24 (1991).
23. S. Sun, C. Murray, D. Weller, L. Folks, and A. Moser, Monodisperse Fe-Pt nanoparticles and ferromagnetic Fe-Pt nanocrystal superlattices, *Science*, **287**, 1989- 1992 (2000).
24. F. Ross, J. Tersoff, and R. Tromp, Coarsening of self-assembled Ge quantum dots on Si(001), *Phys. Rev. Lett.*, **80**, 984-87 (1998).
25. S. Facsko and H. Kurz, Energy dependence of quantum dot formation by ion sputtering, *Phys. Rev. B*, **63**, 165329-1-165329-5 (2001).
26. M. Sarikaya and I. Aksay, *Design and Processing of Materials by Biomimicking*. (AIP, New York, 1994).
27. L. Maissel and R. Glang, *Handbook of Thin Film Technology*, Ch. 8. (McGraw Hill, New York, 1970).
28. J. Venables, *Introduction to Surface and Thin Film Processes*, Ch. 5. (Cambridge University Press, Cambridge, 2000).
29. D. Walton, Nucleation of vapor deposits, *J. Chem. Phys.*, **37**, 2182-88 (1962).
30. B. Lewis and D. Campbell, Nucleation and initial-growth behavior of thin-film deposits, *J. Vac. Sci. Tech.*, **4**, 209-218 (1967).
31. Y. Mo, J. Kleiner, M. Webb, and M. Lagally, Activation energy for surface diffusion of Si on Si(001): A scanning-tunneling-microscopy study, *Phys. Rev. Lett.*, **66**, 1998-2001 (1991).
32. J. Venables, G. Spiller, and M. Hanbucken, Nucleation and growth of thin films, *Rep. Prog. Phys.*, **47**, 339-459 (1984).
33. P. Mulheran and J. Blackman, Capture zones and scaling in homogeneous thin-film growth, *Phys. Rev. B*, **53**, 10261-68 (1996).
34. W. Burton, N. Cabrera, and F. Frank, The growth of crystals and the equilibrium structure of their surfaces, *Phil. Trans. R. Soc. Lond. A*, **243**, 299-346 (1951).
35. R. Sigsbee and G. Pound, Heterogeneous nucleation from the vapor, *Advan. Coll. Interf. Sci.*, **1**, 335-390 (1967).
36. S. Stoyanov and D. Kaschiev, *Current Topics in Materials Science*, **7**. (North- Holland, Amsterdam, 1981).
37. M. Bartelt and J. Evans, Scaling analysis of diffusion-mediated island growth in surface adsorption processes, *Phys. Rev. B*, **46**, 12675-88 (1992).
38. C. Zhang, A novel in situ technique to fabricate thin films with controlled lateral-thickness modulations. PhD thesis, Department of Physics, Washington University, St. Louis, MO 63130, June 2004. Advisor: Ramki Kalyanaraman.
39. C. Zhang and R. Kalyanaraman, In situ lateral patterning of thin films of various materials deposited by physical vapor deposition, *J. Mat. Res.*, **19**(2), 595-599 (2004).
40. G. Gilmer, H. Huang, and C. Roland, Thin film deposition: fundamentals and modeling, *Comp. Mat. Sci.*, **12**, 354-380 (998).

41. H. Huang, G. Gilmer, and T. Diaz de la Rubia, An atomistic simulator for thin film deposition in three dimensions, *J. Appl. Phys.*, **84**, 3636-49 (1998).

42. Hudson. J.B., *Surface Science: An Introduction*. (Wiley Interscience, New York, 1998).

43. R. M. Tromp, R. J. Hamers, and J. E. Demuth, Atomic and electronic contributions to Si(111)— (7 × 7) scanning-tunneling-microscopy images, *Phys. Rev. B (Condensed Matter)*, **34**(2), 1388-1391 (1986).

44. R. J. Hamers, R. M. Tromp, and J. E. Demuth, Surface electronic structure of Si(111)—(7 × 7) resolved in real space, *Physi. Rev. Lett.*, **56**(18), 1972-1975 (1986).

45. D. D. Chambliss, R. J. Wilson, and S. Chiang, Ordered nucleation of Ni and Au islands on Au(111) studied by scanning tunneling microscopy, *J. Vac. Sci. Technol.*, **9**, 933 (1991).

46. F. A. Moller, O. M. Magnussen, and R. J. Behm, Overpotential-controlled nucleation of Ni island arrays on reconstructed Au(111) electrode surfaces, *Phys. Rev. Lett.*, **77**(26) 5249-5252 (1996).

47. J. A. Stroscio, D. T. Pierce, R. A. Dragoset, and P. N. First, Microscopy aspects of the initial growth of metastable fcc iron on Au(111), *J. Vac. Sci. Technol. A*, **10**, 1981 (1992).

48. B. Voigtlander, G. Meyer, and N. M. Amer, Epitaxial growth of thin magnetic cobalt films on Au(111) studied by scanning tunneling microscopy, *Phys. Rev. B*, **44**(18), 10354-10357 (1991).

49. B. Fischer, H. Brune, J. V. Barth, A. Fricke, and K. Kern, Nucleation kinetics on inhomogeneous substrate: Al/Au(111), *Phys. Rev. Lett.*, **82**(8), 1732-1735 (1999).

50. M. B. Hugenschmidt, M. Ruff, A. Hitzke, and R. J. Behm, Rotational epitaxial vs. missing row reconstruction: Au/Cu/Au(110), **388**, L1100-L1106 (1997).

51. M. M. Dovek, C. A. Lang, J. Nogami, and C. F. Quate, Epitaxial growth of Ag on Au(111) studied by scanning tunneling microscopy, *Phys. Rev. B*, **40**(17), 11973-11975 (1989).

52. C. A. Lang, M. M. Dovek, J. Nogami, and C. F. Quate, Au(111) epitaxial study of scanning tunneling microscopy, *Surface Science*, **224**, L947 (1989).

53. J. A. Meyer, I. D. Baikie, E. Kopatzki, and R. J. Behm, Preferential island nucleation at the elbows of the Au(111) herringbone reconstruction through place exchange, *Surface Science*, **365**, L647-L651 (1996).

54. S. Hofmann, Sputter depth profile analysis of interfaces, *Rep. Prog. Phys.*, **61**, 827-888 (1998).

55. R. E. Lee, Microfabrication by ion-beam etching, *J. Vac. Sci. Technol.*, **18**(2), 164-170 (1979).

56. J. M. Choi, H. E. Kim, and I. S. Lee, Ion-beam-assisted deposition (IBAD) of hydroxyapatite coating layer on Ti-based metal substrate, *Biomaterials*, **21**, 469-473 (2000).

57. L. Zhou, K. Kato, N. Umehara, and Y. Miyake, Nanometer scale island-type texture with controllable height and area ratio formed by ion-beam etching on hard-disk head sliders, *Nanotechnology*, **10**, 363-372 (1999).

58. J. G. Pellerin, D. P. Griffis, and P. E. Russell, Focussed ion beam machining of Si, GaAs and InP, *J. Vac. Sci. Technol. B*, **8**(6), 1945-1950 (1990).

59. J. C. Kim, J. Y. Ji, J. S. Kline, J. R. Tucker, and T. C. Shen, Preparation of atomically clean and flat Si(100) surfaces by low-energy ion sputtering and low-temperature annealing, *Applied Surface Science*, **220**, 293-297 (2003).

60. S. Rusponi, G. Costantini, and Boragno, Ripple wave vector rotation in anistropic crystal sputtering, *Phys. Rev. Lett.*, **81**(13), 2735-2738 (1998).

61. M. V. Ramamurty, T. Curcic, and Judy, X-ray scattering study of the surface morphology of Au(111) during $Ar^+$ ion irradiation, *Phys. Rev. Lett.*, **80**(21), 4713-4716 (1998).

62. J. Erlebacher, M. Aziz, and Chason, Spontaneous pattern formation on ion bombardment Si(001), *Phys. Rev. Lett.*, **82**(11), 2330-2333 (1999).

63. T. Michely, M. Kalff, and Cosma, Step edge diffusion and step atom detachment in surface evaluation: Ion erosion of Pt(111), *Phys. Rev. Lett.*, **86**(12), 2589-2592 (2001).

64. F. Frost, A. Schlinder, and F. Bigl, Roughness evolution of ion sputtered rotating InP surfaces: pattern formation and scaling laws, *Phys. Rev. Lett.*, **85**, 19, 4116-4119 (2000).

65. G. Costantini, F. B. Mongeot, and Boragno, Is ion sputtering always a "negative homoepitaxial deposition"? *Phys. Rev. Lett.*, **86**(5), 838-841 (2001).

66. O. Malis, J. D. Brock, and Headrick, Ion-induced pattern formation on Co surfaces: an x-ray scattering and kinetic Monte Carlo study, *Phys. Rev. B*, **66**, 035408 (2002).

67. H. J. Ernst, The pattern formation during ion bombardment of Cu(001) investigated with helium atom beam scattering, *Surface Science*, **383**, L755-L759 (1997).

68. M. Ritter, M. Stindtmann, M. Farle, and K. Baberschke, Nanostructuring of the Cu(001) surface by ion bombardment: a STM study, *Surface Science*, **348**, 243-252 (1996).

69. G. Costantini, S. Rusponi, F. B. d. Mongeot, C. Boragno, and U. Valbusa, Periodic structures induced by normal-incidence sputtering on Ag(110) and Ag(001): flux and temperature dependence, *J. Phys: Condens. Matter*, **13**, 5875-5891 (2001).

70. G. Carter and V. Vishnyakov, Roughening and ripple instability on ion-bombarded Si, *Phys. Rev. B*, **54**(24), 17647-17653 (1996).

71. S. J. Chey, J. E. V. Nostrand, and D. G. Cahill, Surface morphology of Ge(001) during etching by low-energy ions, *Phys. Rev. B*, **52**(23), 16696-16701 (1995).

72. R. M. Bradley and J. M. E. Harper, Theory of ripple topography induced by ion bombardment, *J. Vac. Sci. Technol. A*, **6**(4), 2390-2395 (1988).

73. E. Chason, T. M. Mayer, B. K. Kellerman, D. T. McIlroy, and A. J. Howard, Roughening instability and evolution of the Ge(001) surface during ion sputtering, *Phys. Rev. Lett.*, **72**(19), 3040-3043 (1994).

74. P. Sigmund, A mechanism of surface micro-roughening by ion-bombardment *J. Mat. Sci.*, **8**, 1545-1553 (1973).

75. S. Rusponi, C. Boragno, and U. Valbusa, Ripple structure on Ag(110) surface induced by ion sputtering, *Phys. Rev. Lett.*, **78**(14), 2795-2798 (1997).

76. S. Facsko, T. Dekorsy, C. Koerdt, C. Trappe, H. Kurz, A. Vogt, and H. L. Hartnagel, Formation of ordered nanoscale semiconductor dots by ion sputtering, *Science*, **285**, 1551-1553 (1999).

77. B. Warren, *X-ray Diffraction*. (Dover, New York, 1990).

78. E. Hecht, *Optics*, Ch. 5. Fourth ed. (Addison-Wesley, Reading, 2002).

79. A. Rosenfeld and A. Kak, *Digital Picture Processing. Computer Science and Applied* Mathematics (Academic, New York, 1976).

80. A. Jain, *Fundamentals of Digital Image Processing*. (Prentice-Hall, Englewood Cliffs, NJ, 1989).

81. W. Rasband, ImageJ, tech. rep., U. S. National Institutes of Health, Bethesda, MD, USA (1997), http://rsb.info.nih.gov/ij/.

82. C. Favazza, J. Trice, H. Krishna, R. Sureshkumar, and R. Kalyanaraman, Laser-induced short and long-range ordering of Co nanoparticles on $SiO_2$, *App. Phys. Lett.*, **88**, 153118 (2006).

83. M. Salerno and M. Banzato, Minkowski measures for image analysis in scanning probe microscopy, *Microscopy and Analysis*, **19**(4), 13-15 (2005).

84. K. Mecke, T. Buchert, and H. Wagner, Robust morphological measures for large-scale structure in the universe, *Astron. Astrophys.*, **288**, 697-704 (1994).

85. K. R. Mecke, Morphological characterization of patterns in reaction-diffusion systems, *Phys. Rev. E*, **53**(5), 4794-4800 (1996).

86. C. Favazza, R. Kalyanaraman, R. Sureshkumar, *Nanotechnology*, **17**, 4229-34 (2006).

87. J. Watson and K. Ihokura, Gas-sensing materials, in MRS Bull. (J. Watson and K. Ihokura, eds.), *Mat. Res. Soc.*, 24 (1999).

88. W. Gopel, Nanostructured sensors for molecular recognition, *Phil Trans.*, **353**, 333-354 (1995).

89. T. Seiyama, A. Kato, K. Fujishi, and M. Nagatani, A new detector for gaseous components using semiconductive thin film, *Anal. Chem.*, **34**, 1052-1953 (1962).

90. E. van Setten, T. M. Gür, D. H. A. Blank, J. C. Bravman, and M. R. Beasley, Miniature Nernstian oxygen sensor for deposition and growth environments, *Rev. Sci. Instr.*, **73**, 156-161 (2002).

91. H. Nagle, R. Gutierrez-Osuna, and S. Schiffman, The how and why of electronic noses, *IEEE Spect.*, **35**(9), 15 (1998).

92. P. Althainz, J. Goschnick, S. Ehrmann, and Ache H.J., Multisensor microsystem for contaminants in air, *Sens. Actuat. B*, **33**, 72-76 (1996).

93. B. Yang, M. Carotta, G. Faglia, M. Ferroni, V. Guidi, G. Martinelli, and G. Sberveglieri, Quantification of H2S and NO2 using gas sensor arrays and an artificial neural network, *Sens. Actuat. B*, **43**, 235-238 (1997).

94. J. Anglesea, P. Corcoran, and W. Elshaw, The application of genetic algorithms to multisensor array optimization, *Proc. 12th Europ. Conference on Sold-State Tranducers*, **2**, 1103-07 (1998).

95. G. Martinelli, M. Carotta, E. Traversa, and G. Ghiotti, Thick-film gas sensors based on nano-sized semiconducting oxide powders, *MRS Bull.*, **24**, 30-35 (1999).

96. M. Kennedy, F. Kruis, H. Fissan, B. R. Mehta, S. Stappert, and G. Dumpich, Tailored nanoparticle films from monosized tin oxide nanocrystals: particle synthesis, film formation, and size-dependent gas-sensing properties, *J. App. Phys.*, **93**, 551-560 (2003).

97. F. Cosandey, G. Skandan, and A. Singhal, Materials and processing issues in nanostructured semiconductor gas sensors, JOM-e, **52**, 1-6 (2000). http://www.tms.org/pubs/journals/JOM/0010/Cosandey/Cosandey-0010.html.

98. G. Heiland, Homogeneous semiconductor gas sensors, *Sens. Actuat.*, **2**, 343-361 (1982).

99. K. Ihokura and J. Watson, The Stannic oxide gas sensors principles and applications. (CRC Press, Boca Raton, FL, 1994).

100. N. Yamazoe and T. Seiyama, Sensing mechanism of oxide semiconductor gas sensors, in Solid-state sensors and actuators, 3rd Int. Conf., Philadelphia, PA, pp. 376-379 (1985).

101. D. Williams, *Solid-State Gas Sensors* (Adam Hilger, Philadelphia, 1987).

102. H. Ogawa, M. Nishikawa, and A. Abe, Hall measurement studies and electrical conductivity model of tin oxide ultrafine particle films, *J. Appl. Phys.*, **53**, 4448-4454 (1982).

103. C. Xu, J. Tamaki, N. Miura, and N. Yamazoe, Grain size effects on gas sensitivity of porous $SnO_2$ based element, *Sens. Actaut. B*, **3**, 147-155 (1991).

104. G. Williams and G. Coles, The gas-sensing potential of nanocrystalline tin dioxide produced by laser ablation techniques, *MRS Bull.*, **24**, 25-27 (1999).

105. G. Korotchenkov, V. Brynzari, and S. Dmitriev, $SnO_2$ films for thin film gas sensor design, *Mat. Sci. and Eng. B*, **63**, 195-204 (1999).

106. T. Yang, H. Lin, B. Wei, C. Wu, and C. Lin, UV enhancement of the gas sensing properties of nano-TiO2, *Rev. Adv. Mater. Sci.*, **4**, 48-54 (2003).

107. H. Windischmann and P. Mark, A model for the operation of a thin-film SnOx conductance modulation carbon monoxide sensor, *J. Elec. Soc.*, **126**, 627-633 (1979).

108. A. Brailsford, M. Yussouff, and E. M. Logothetis, Theory of gas sensors, *Sens. Actuat. B*, **13**, 135-138 (1993).

109. S. Gulati, N. Mehan, D. P. Goyal, and A. Mansingh, Electrical equivalent model for $SnO_2$ bulk sensors, *Sens. Actuat. B*, **87**, 309-320 (2002).

110. F. Hossein-Babaei and M. Orvatinia, Analysis of thickness dependence of the sensitivity in thin film resistive gas sensors, *Sens. Actuat. B*, **89**, 256-261 (2003).

111. S. Koehl and M. Paniccia, The quest to siliconize photonics, *Photonics Spectra*, pp. 53-60 (Nov. 2005).

112. U. Fischer, A. Dereux, and J.-C. Weeber, Near-field optics and surface plasmon polaritons. In: *Topics in Applied Physics, Controlling Light Confinement by Excitation of Localized Surface Plasmons*, (Springer, New York, 2001), pp. 49-69.

113. J. Krenn, J. C. Weeber, A. Dereux, B. Schider, A. Leitner, F. R. Aussenegg, and C. Girard, Direct observation of localized surface plasmon coupling, *Phys. Rev. B*, **60**, 5029-5033 (1999).

114. M. L. Brongerman, J. W. Hartman, and H. A. Atwater, Electromagnetic energy transfer and switching in nanoparticle chain arrays below the diffraction limit, *Phys. Rev. B*, **62**, R16356 (2000).

115. S. A. Maier, M. L. Brongersma, P. G. Kik, and H. A. Atwater, Observation of near-field coupling in metal nanoparticle chains using far-field polarization spectroscopy, *Phys. Rev. B.*, **65**, 193408-1 (2002).

116. U. Kreibig and M. Vollmer, *Optical Properties of Metal Clusters.* (Springer, Berlin, 1994).

117. J. Krenn, Nanoparticle waveguides: Watching energy transfer, *Nature Materials*, **2**, 210-211, (2003).

## QUESTIONS

1. What is the necessary constraint on the time scales of re-evaporation ($\tau_a$ and $\tau_d$) for film deposition to occur?
2. What are the differences between self-assembly and lithography?
3. What competing processes result in the surface instability during ion-irradiation?
4. Assume a characteristic volume and depth for energy deposition by each ion in the ion-irradiation process. Show that the effective energy deposited is influenced by local curvature and is higher for the troughs as compare to the crests.
5. What is the working principle for gas sensors?
6. Calculate the Debye length ($L_D$) at room temperature when the total carrier concentration ($n_o$) is $1 \times 10^{17}$ cm$^{-3}$.
7. Compare and contrast the behavior of the radial distribution function $G(k)$ for disordered and ordered (short-range and long-range) spatial arrangements.

# 3

# MEMS for Nanotechnology: Top-down Perspective

## Ghanashyam Londe[1], Arum Han[2], Hyoung J. Cho[3]

[1]Electrical and Computer Engineering, University of Central Florida, Orlando, FL 32816-2450

[2]Electrical and Computer Engineering, Texas A&M University, College Station, TX 77843-3128

[3]Mechanical, Materials and Aerospace Engineering, University of Central Florida, Orlando, FL 32816-2450

[x]To whom correspondence should be addressed. Hyoung J. (Joe) Cho, Mechanical, Materials and Aerospace Engineering, University of Central Florida, Orlando, FL 32816-2450. E-mail: joecho@mail.ucf.edu

## 1. INTRODUCTION

Since the invention of the transistor at Bell Labs in 1947, the semiconductor industry has achieved tremendous success in mass production of integrated circuits via planar batch processing (Brinkman et al. 1997). As predicted by Dr. R. Feynman in his classic talk in 1959 (Feynman), we have already gained the capability of "writing by putting atoms down in a certain arrangement" by exploring "new kind of forces and new kinds of possibilities at the atomic scale." For nanoscale devices, he envisioned photolithography and biologically inspired chemical assembly, which are now considered top-down and bottom-up approaches in nanotechnology, respectively. Massive parallel replication of electronic components using photolithography has realized the era of digital electronics in the past century. Based upon this success, revolutionary adaptation of microfabrication technology in various applications has resulted in an unprecedented amount of scientific and engineering feats in many arenas.

In addition to the commercially driven technical progress in solid-state electronics, with the U.S. government's support since the 1990s, Micro-Electro-Mechanical-Systems (MEMS) research has emerged as a novel scientific and engineering methodology for the integration of miniaturized sensors, actuators and subsystems. Meanwhile, in chemistry and biology, the idea of self-assembly, photolithography and precision engineering techniques have been adapted for making massive arrays of microscale conduits and vessels. The freedom in defining and controlling volumes at the nano- and picoliter level as well as the favorable use of diffusion-dominant microflows in analysis, has made Lab-on-a-Chip (LOC) concepts into real devices.

Nanotechnology, from its own definition, utilizes any functional component at the scale of 1-100 nm and includes tools and instruments with the same scale of precision. This chapter will focus on the "top-down approach," and more specifically, on efforts by the MEMS community, and while covering the subject of nano/micro interfacing and integration for practical applications. Various micromachining techniques and consequential nanofabrication methods will be reviewed. Integration and interfacing nano into microsystems will be discussed. Most of the successful implementations of nanotechnology in functional systems comes from the creative adaptation of various interdisciplinary research approaches. We will review the exemplary devices and systems for practical applications in this chapter.

## 2. MICROMACHINING TECHNIQUES

Micromachining techniques, by which MEMS devices are constructed, have been derived from silicon microfabrication. However, the adaptation of these methods is not restricted to semiconductor materials only. Micromachining combines traditional precision techniques for 3-D structure construction along with planar fabrication to realize massive replication of small parts using various materials.

### 2.1. Photolithography

Although the definition of micromachining techniques is more inclusive nowadays, most of such processes start with the top-down approach of photolithography. Photolithography involves a writing process by the use of light on a substrate. The typical process is illustrated in Fig. 1. (Schellenberg 2003). Silicon wafers are commonly used due to their commercial availability and well-characterized electrical and mechanical properties. Typically, layers of oxide, nitride, metal or any additive film are grown or deposited and/or photoresists are coated on top. By exposure to a light source, the patterns predefined on the mask are transferred to the photoresist. The exposed portion is either removed (positive photoresist) or left behind (negative photoresist) after a developing process, dissolving away the unwanted area(s). Either an etching or another deposition step follows. The etching process removes films on the open areas selectively while the photoresist-covered areas remain protected. Alternatively, the additional layers can be deposited through the open areas of the photoresist film. This entire process is repeated, layer by layer,

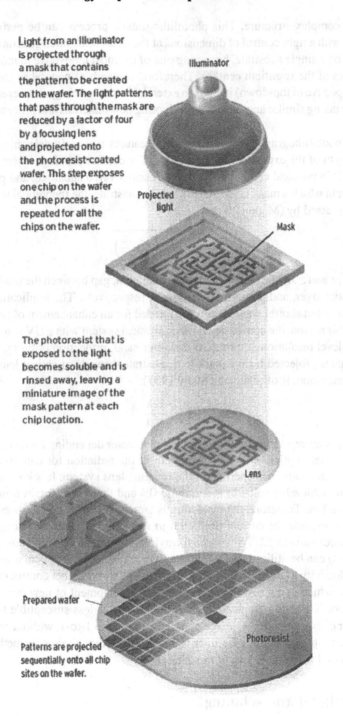

Light from an illuminator is projected through a mask that contains the pattern to be created on the wafer. The light patterns that pass through the mask are reduced by a factor of four by a focusing lens and projected onto the photoresist-coated wafer. This step exposes one chip on the wafer and the process is repeated for all the chips on the wafer.

Illuminator

Projected light

Mask

The photoresist that is exposed to the light becomes soluble and is rinsed away, leaving a miniature image of the mask pattern at each chip location.

Lens

Prepared wafer

Photoresist

Patterns are projected sequentially onto all chip sites on the wafer.

**FIGURE 1.** Schematic illustration of photolithography. Schellenberg, A little light magic, IEEE Spectrum 34-39. (© [2003] IEEE)

for any complex structure. This photolithographic process can be performed in parallel with a tight control of dimensions at the small scale, thus generating many devices on a single substrate, becoming one of the most innovative manufacturing processes of the twentieth century. Therefore, research of nanofabrication (from the perspective of top-down) intends to extend this methodology to an even smaller scale by taking similar approaches and adopting new materials and processes along the way.

In photolithography, the resolution of features is mainly determined by the wavelength of the exposure radiation beam. For contact printing, where the photomask is in physical contact with a photoresist-covered substrate, and proximity printing, in which a mask is located off from a substrate, the theoretical resolution, $R$, is expressed by (Madou 2002):

$$R = 1.5\sqrt{\lambda\left(s + \frac{z}{2}\right)}$$

where $\lambda$, $s$ and $z$ are the wavelength of the radiation, gap between the mask and the photoresist layer, and photoresist thickness, respectively. The implication of the equation is that shorter wavelengths are needed for an enhancement of resolution. Due to this reason, the conventional lithographical system with a UV source yields micron-level resolution at most. On the other hand, projection printing, in which the image is projected from a mask to a substrate through an optical lens system, yields resolution, R of (Chiu and Shaw 1997):

$$R = \frac{k_1\lambda}{NA}$$

where $k_1$ is an experimentally determined parameter depending on resist, process and instrument optics, $\lambda$ is the wavelength of the radiation for pattern exposure and NA is the numerical aperture of the imaging lens system. It is known that the minimum achievable value of $k_1$ is around 0.4 and there is a certain limitation in increasing NA. Therefore, the resolution is largely dependent on the wavelength.

To overcome the current limitation in resolution, shorter wavelength exposure sources such as EUV (extreme ultraviolet lithography, $\lambda \sim 10$ nm) or X-ray ($\lambda \sim 1$ nm) can be utilized. EUV lithography requires special optical materials for high reflectivity at the specified wavelength and careful defect control on optical coatings, which is not trivial and requires the development of new materials and optics. For X-ray lithography, although the process is less susceptible to defects and their diffraction effects, image transfer is limited to 1-to-1, without any reduction, due to its lack of optical reduction capability. Other alternative methods will be discussed in the latter part of this chapter.

## 2.2. Bulk Micromachining

After the discovery of piezoresistivity in Si and Ge in the 1950s (Smith 1954), the commercial silicon pressure sensors using electromechanical properties were first developed by Honeywell in 1963 (Sanchez 1963). For the sensor fabrication, a movable structure such as a diaphragm must be produced with controlled geometry. In order to make mechanical components that can move out of plane, different types of fabrication techniques have been developed. These can be categorized into bulk micromachining and surface micromachining.

*Bulk micromachining* is a subtractive process by which a structure is made out of a substrate. From the structure defined by the photolithographic step, the etching process removes unwanted parts out of the substrate, leaving behind a thickness-controlled region. For example, by anisotropic etching along crystalline orientation, an etched pit and a remaining diaphragm with a unique profile can be fabricated. A wide variety of structures are possible using this technique as shown in Fig. 2 (Kovacs, et al. 1998). The bulk micromachining intends to obtain thin membranes, trenches, holes or undercuts depending on the characteristics of etching chemistry and crystallographic orientation of the monocrystalline silicon substrate. Rounded isotropic pits (Fig. 2(a)), pyramidal pits (Fig. 2(b)) and an undercut beam with a pyramidal pit can be made using this method (Fig. 2(c)).

Due to its well-defined crystal structure, a silicon substrate shows different etch rates along its individual plane directions in commonly used etchant solutions such as KOH (potassium hydroxide) and TMAH (tetramethylammonium hydroxide). For example, the ratio of etch rates of (100) plane over (111) is typically a few hundreds in KOH solutions due to the variation in atomic densities of the planes and strengths of the bonds. As a result, the slowest etching planes are exposed at the end while the fastest, e.g., high index planes, disappear during the etching process. Usually, the (100) and (111) planes contain the etched final structure and the angle between these planes is $54.74°$.

For an isotropic etching profile, HNA mixture solutions are commonly used. The solution is made of hydrofluoric acid (HF), nitric acid ($HNO_3$), and acetic acid ($CH_3 COOH$). More detailed information can be found from references (Kovacs 1998; Zurbel 1998; Zurbel and Barycka 1998). The etching process can be done either wet or dry using liquid or gaseous phase etchants. Plasma etching or RIE (reactive ion etching) is used for dry etching. RF power energizes and activates ion species, which react with the silicon substrate and form into a volatile complex. These reaction by-products are transported out from the chamber by a carrier gas. In common practice, sulfur hexafluoride ($SF_6$), hydrochlorocarbon ($CHCl_3$), chlorocarbon ($CCl_4$) and chlorine gases are used. For example, when $SF_6$ is used, it reacts with silicon and forms a silicon fluoride compound vapor, and the chemistry is thought to be isotropic. However, a mixture of $SF_6$ and $O_2$ with a magnetically confined plasma configuration achieves a highly anisotropic deep trench (Christopher et al. 1992).

During the process, oxygen forms silicon oxide, passivates side walls, prevents further etching of silicon while the $SF_6$ bombards the trench bottom and takes away silicon atoms. This shows the careful selection of process parameter matrix that can yield a controlled etch profile. A more widely commercialized etch process

ISOTROPIC WET ETCHING: AGITATION

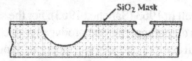

ISOTROPIC WET ETCHING: NO AGITATION

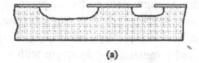

(a)

ANISOTROPIC WET ETCHING: (100) SURFACE

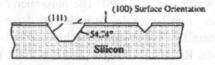

ANISOTROPIC WET ETCHING: (110) SURFACE

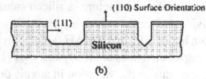

(b)

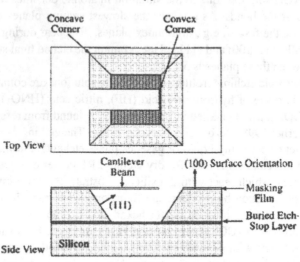

(c)

**FIGURE 2.** Illustration of possible bulk micromachined structures: (a) rounded, isotropically etched pits in a silicon substrate (b) pyramidal pits etched into (100) and (110) silicon using anisotropic wet etchants, bounded by (111) crystal planes (c) a pyramidal pit etched down to a buried etch-stop layer in (100) silicon, with an undercut cantilever beam. Kovacs, Maluf et al., Bulk Micromachining of Silicon, Proc.IEEE 86(8): 1536-1551. (© [1998] IEEE)

is known as the Bosch process (Learmer et al. 1999). The Bosch process was initially developed to overcome the limitations of the chlorine-based plasma etching process such as low etch rates and poor mask selectivity. In contrast to the process using the $SF_6 + O_2$ mixture, the Bosch process utilizes Teflon-like sidewall films for passivation to achieve an anisotropic sidewall. The Bosch approach is based on a variation of the Teflon-film sidewall passivation technique which avoids recombination of active species in the gas phase: deposition and etch steps are performed subsequently, as is illustrated in Fig. 3 (Learmer et al. 1999). As the etching proceeds, polymer film on sidewalls for passivation is removed and deposited continually, which enables localized anisotropic etching to progress at the bottom of the trench. This can be performed at room temperature so that higher selectivity over standard photoresist masks can be achieved as high as 200:1. For better selectivity, a variety of hard masking materials can be used including $SiO_2$ and $Si_3N_4$. Etch rates of several microns per minute can be readily achieved with a good control of the side profile with this process.

## 2.3. Surface Micromachining

*Surface micromachining* is an additive process that utilizes deposited thin films for structural construction. The device is made with overlaying films on top of the substrate. Often a sacrificial layer is used to produce a beam or membrane for moving parts after freeing films attached on top. A resonant gate sensor was demonstrated using this concept first in the mid-1960s (Nathanson et al. 1967). For surface micromachining, as a structural material, polysilicon (polycrystalline silicon) or metal is commonly used. Due to fatigue issues with metals, polysilicon is preferred for mechanical applications. However, some problem areas for polysilicon as a MEMS material include residual stress, stress gradients through the film thickness and statistical variations of the effective Young's modulus in the polycrystalline material (Bustillo et al. 1998).

Many of these issues have been solved under precise control of process parameters. Polysilicon is deposited from silane ($SiH_4$) by LPCVD (low-pressure vapor deposition) at around 600°C. Silane decomposes into silicon and hydrogen in the reaction chamber. It is known that the crystalline phase becomes dominant at higher temperatures (T > 600°C). In situ doping (during deposition) can be performed using arsine ($AsH_3$), phosphine ($PH_3$) or diborane ($B_2H_3$) gases if the control of resistivity is needed. In order to release the residual stress of the film, an annealing step often follows at around 900°C. During the etch release process, due to the high surface-to-volume ratio and surface tension of liquid, liquid-phase chemicals are trapped between layers and cause "stiction." This could easily lead to structural damage. To avoid structural damage, the release methods including the addition of a physical structure for preventing stiction (i.e., stand-off bumps on the underside of a poly-Si plate), supercritical drying, and plasma etching, have been suggested (Bustillo et al. 1998).

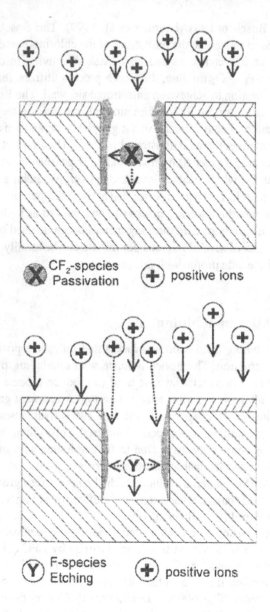

**FIGURE 3.** Sidewall passivation mechanism of the Bosch deep silicon etching process. The protecting film deposited in the passivation step is extending deeper into the trenches during the etch step to yield a smooth sidewall. Learmer, Schilp et al., Bosch deep silicon etching: improving uniformity and etch rate for advanced MEMS applications. Proc. MEMS, Orlando, FL. (© [1999] IEEE)

For a metal structure, a typical process starts with a seed layer deposition on a thermally grown silicon oxide layer. Oxide serves as an electrical insulator over semiconductor silicon. Cr or Ti is usually used for the underlayer for conductive metallic films such as Au, Cu or Ni. Photoresist is spun and patterned by photolithography, and the open area, where photoresist is removed, is exposed to an electroplating solution. Then an electric current is supplied from outside for forming metal films out of source ions in an electroplating bath. (There are a variety of commercially available plating and etching baths, e.g., http://www.technics.com) For a mechanical structure, Ni and Ni alloys are used in general. Stand-alone structures need control of residual stress, which is dependent on the process conditions of electroplating, such as the chemical composition of bath, temperature, pH, current density and duration. More references can be found in traditional electroplating books (Lowenheim 1974).

## 2.4. Combined Method

Bulk and surface micromachining techniques have their own advantages and disadvantages (French and Sarro 1998). Since bulk micromachining uses single crystal silicon, its electrical and mechanical characteristics are superior to and more well-defined than those of surface micromachined components in most cases. However, crystallographic orientation restricts the etching profile and final structure achievable. In comparison, although surface micromachining is free from this, slight variations in deposition process conditions can affect the properties of polysilicon seriously. The process often needs a post process to release the residual stress. Stiction is also problematic and control of mechanical properties such as the level of residual stress and Young's modulus, are known to be very difficult. The combination of these two techniques has been attempted to overcome technical constraints imposed by only one side.

The surface bulk micromachining (SBM) process was demonstrated with (111) silicon wafers for making free-standing structures by applying both methods synergistically (Lee et al. 1999). First, a (111) silicon wafer is reactive ion etched, leaving the structural patterns to be released later, and the bulk silicon under the patterns is etched in an aqueous alkaline etchant to release the patterns. In the developed technology, the thickness of the structural layer as well as the thickness of the sacrificial gap is defined by deep silicon RIE. The release of structure is accomplished by aqueous alkaline etching, utilizing the slow etching planes as the etch stop. This results in very smooth and flat bottom surfaces. The required release etch time in a KOH solution, a commonly used alkaline etchant, is comparable to that of etching the sacrificial oxide in HF. Figure 4 illustrates the process flow and Fig. 5 shows a resultant structure made by the SBM process. The SBM process demonstrated its capability of duplicating almost all surface micromachined microsensors and microactuators. Material properties of bulk silicon crystal could be utilized while a thin layer of a free-standing structure could be made with

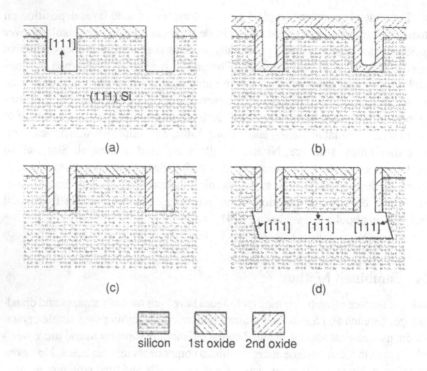

**FIGURE 4.** Detailed process flow: (a) silicon etch using RIE (b) passivation nitride and oxide deposition (c) oxide and nitride etch (d) silicon wet etch in aqueous alkaline etchant. Lee, Park et al., The surface/bulk micromachining (SBM) process: a new method for fabricating released MEMS in single crystal silicon, J. Microelectromech. Syst. 8(4): 409-416. (© [1999] IEEE)

a large sacrificial gap. This method effectively removes the problem associated with stiction.

Another example of a combined fabrication method was demonstrated in the micro accelerometer (Yazdi and Najafi 2000). Figure 6 shows the process flow. The process starts with a thin, heavily boron-doped area on double-polished p-type (100) Si wafer using thermal oxide as a mask. The boron doping is used for an etch stop, where the bonding strength between boron and silicon is much stronger than silicon-silicon (Peterson 1982). Both sides of the wafer are patterned, and the patterns are aligned to each other. A shallow boron diffusion is performed to define the beams, the proof mass, and the supporting rim. Then, LPCVD nitride is deposited and patterned to form the polysilicon electrode anchors and isolation dielectric under the polysilicon electrode dimples. The next masking step is etching trenches to define the vertical electrode stiffeners. The trenches are then refilled completely using sacrificial LPCVD oxide and then LPCVD polysilicon.

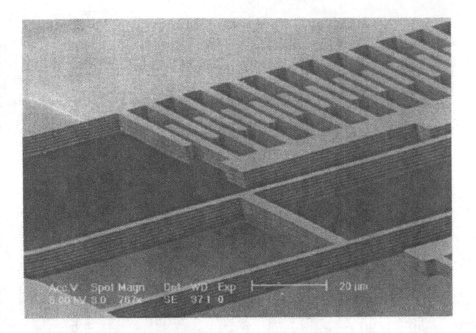

**FIGURE 5.** Surface bulk micromachined comb-drive resonator for electrostatic actuation. Lee, Park et al., The surface/bulk micromachining (SBM) process: a new method for fabricating released MEMS in single crystal silicon, J. Microelectromech. Syst. 8(4): 409-416. (© [1999] IEEE)

The polysilicon is etched back using a blanket RIE etch, exposing the sacrificial oxide and leaving polysilicon plugs in the trenches. In this manner, the step height due to the trench etch is reduced, and sacrificial oxide at the bottom of the trenches is protected. After a series of deposition and etching steps to define electrodes, the final release structure is obtained. For prevention of stiction, before the wet etching process, self-assembled-monolayer is used for surface coating on the release structure.

## 3. NANOFABRICATION

Various efforts are being made toward the development of nanofabrication techniques. Some of them can be considered as an extension of the micromachining techniques. Others are being tried based on novel ideas.

According to International Technology Roadmap for Semiconductors updated in 2005 (Takekawa et al. 2005) (ITRS 2005), EUV is viewed as the most likely to achieve 32 nm and 22 nm half-pitch patterning. However, ITRS 2005 predicts

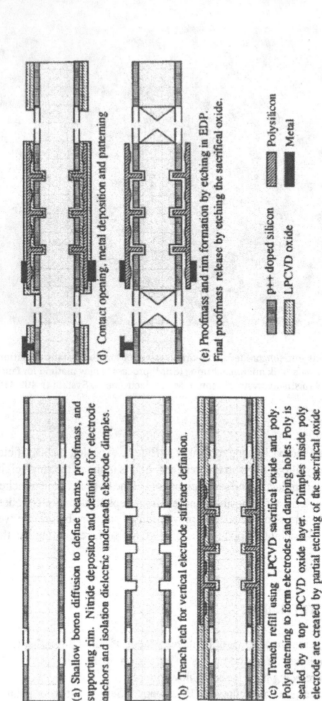

(a) Shallow boron diffusion to define beams, proofmass, and supporting rim. Nitride deposition and definition for electrode anchors and isolation dielectric underneath electrode dimples.

(b) Trench etch for vertical electrode stiffener definition.

(c) Trench refill using LPCVD sacrifical oxide and poly. Poly patterning to form electrodes and damping holes. Poly is sealed by a top LPCVD oxide layer. Dimples inside poly electrode are created by partial etching of the sacrifical oxide before poly deposition.

(d) Contact opening, metal deposition and patterning

(e) Proofmass and rim formation by etching in EDP. Final proofmass release by etching the sacrifical oxide.

Polysilicon

Metal

p++ doped silicon

LPCVD oxide

FIGURE 6. Combined surface and bulk micromachining process for micro accelerometer. Yazdi and Najafi, An all-silicon sigle-wafer micro-g accelerometer with a combined surface and bulk micromachining process, J. Microelectromech. Syst. 9(4): 544-550. (© [2000] IEEE)

that due to inherent problems of optical lithography methods such as diffraction and aberration, the post-optical alternatives are in demand and need to be considered:

> Maskless lithography has been applied to niche applications in development for prototyping and transistor engineering and to low volume application specific integrated circuit (ASIC) production, but its role could be expanded. Breakthroughs in direct-write technologies that achieve high throughput could be a significant paradigm shift, eliminating the need for masks and resulting in cost and cycle-time reduction. Maskless lithography for application beyond prototyping is currently in the research phase, and many significant technological hurdles will need to be overcome for maskless lithography to be viable for cost-effective semiconductor manufacturing. Imprint lithography has the potential to be a cost-effective solution, but there are a number of problems that need to be solved for this to happen, including the difficulties associated with 1× templates, defects, template lifetime, and overlay. The introduction of nonoptical lithography will be a major paradigm shift that will be necessary to meet the technical requirements and complexities that are necessary for continued adherence to Moore's Law at DRAM 32 nm half-pitch and beyond.

For production scale top-down nanofabrication methods, concurrent efforts are needed in developing exposure tools, masks, and materials.

## 3.1. Electron Beam Lithography (EBL)

Electron beam lithography is one example of a maskless lithography method. As discussed earlier, since the resolution of photolithography is limited by the wavelength of the radiation source, the logical solution is to use a short-wavelength beam for exposure. For this purpose, radiation sources such as EUV and X-ray can be used. However, due to the inherent diffraction and lens aberration in optical system (EUV) and limitation in image reduction ratio and difficulty in generating mask patterns (X-ray), a direct beam-writing method such as electron-beam (e-beam) or ion beam lithography has been also considered as an alternative (Chiu and Shaw 1997). For nanoscale feature production, electron beam lithography (EBL) is preferred to ion beam. It has been developed based on the same principle as scanning electron microscopy. This method utilizes high-energy charged particles as a writing tool on the substrate with a controlled dose. The advantages and disadvantages of this method are listed in Table 1 (Madou 2002).

Due to technical restrictions, mainly associated with low throughput, the use of EBL has been limited to photomask generation initially. Recent research progress discusses an improvement of throughput using projection approaches (Liddle et al.; Chang et al. 2001; Dhaliwal et al. 2001; Sohda et al. 2003). Figure 7 illustrates the evolution of electron-beam lithography from the early scanning-electron-microscopy-type systems, which expose integrated-circuit (IC) patterns one pixel at a time, to mass parallel projection of pixels in electron projection lithography (EPL) systems targeted at exposing ten million pixels per shot (Dhaliwal et al. 2001). Scattering with Angular Limitation Projection Electron-beam

**TABLE 1.** Advantages and disadvantages of EBL [Tabulated from (Madou 2002)]

| Advantages | Disadvantages |
|---|---|
| Precise control of the energy and dose | Scattering of electrons in solid, which limits practical resolution |
| Modulation of e-beams with speed and precision by electromagnetic fields | Requirement of high vacuum chamber for fabrication which makes production system complex |
| Imaging of electrons to form a small point of <100A (as opposed to 5000A for light beam) | Slow exposure speed |
| No need for a physical mask | High cost |
| Ability to register accurately over small areas of wafer | |
| Lower defect densities and large depth of focus | |

Lithography (SCALPEL) research started at Bell Labs in 1989 with the initial invention, and continued at the level of an initial feasibility study for several years (Liddle et al.). In 1994, a proof-of-concept program was begun, with the successful results from the first prototype exposure tool, making the transition to a proof-of-lithography phase in 1997.

In each previous case, the design utilizes a step-and-repeat writing strategy. That is, the electron-optical system is capable of illuminating a mask area containing the full pattern of at least one circuit level. An image of the circuit is projected onto a resist-coated wafer and brought into registration with previously defined levels. After exposure of the image, the wafer is moved, or stepped, to the next

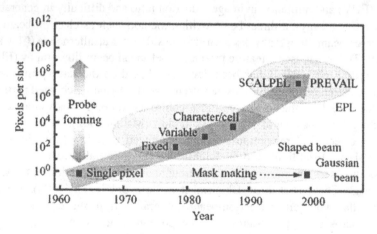

**FIGURE 7.** Evolution of electron beam lithography. Dhaliwal, R. S., W. A. Enichen, et al. (2001). "PREVAIL—Electron projection technology approach for next-generation lithography." IBM J. Res. Develop. 45(5): 615-638.

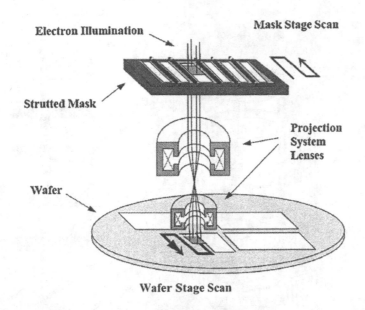

**FIGURE 8.** Step and scan operation of SCALPEL. Waskiewicz, Berger et al., Electron-Optical Design for the SCALPEL Proof-of Concept Tool." Proc. SPIE: pp. 13-22, 1995.

chip site and the full pattern exposure is repeated (Waskiewicz et al. 1995). This approach employs a scattering (transmission) mask and a step-and-scan writing strategy. The use of a scattering mask overcomes the problems of stencil masks. Here, the illuminated mask area is substantially smaller than the full circuit pattern to be printed on the wafer. Due to this diminished optical field size, the mask and wafer are both placed on stages and mechanically scanned during the exposure, thereby reaching all patterned portions of the mask. The requirements of the electron-optics are therefore independent of the size of the chip being fabricated.

Another advantage from the greatly reduced field size requirement for the step-and-scan technique is that short focal-length lenses can be employed, thus achieving a compact overall optical column length. The step and scan SCALPEL is schematically illustrated in Fig. 8. The mask consists of a membrane and a cover layer. The mask membrane is made of low atomic number materials while the pattern is prepared on a cover layer of high atomic number materials. Due to the low scattering, an incident beam (with a typical energy of 100 keV) passes through the membrane at a smaller angle compared to one through the patterned layer. Contrast is obtained by the difference in electron scattering characteristics between the membrane and the patterned materials. An aperture in the back-focal (pupil) plane of the projection optics blocks the strongly scattered electrons, forming a high-contrast aerial image at the wafer plane (Waskiewicz et al. 1995). The functions of contrast generation and energy absorption are thus separated between the mask and the aperture, which contributes to minimal absorption of incident energy by the mask, preventing thermal instabilities.

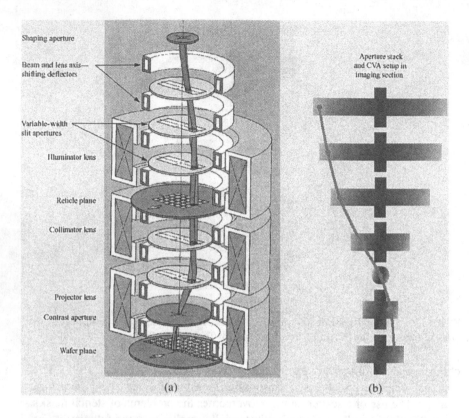

**FIGURE 9.** Schematic of PREVAIL. Reprinted with permission from Dhaliwal, Enichen et al., PREVAIL—Electron projection technology approach for next-generation lithography, IBM J. Res. Develop., 45(5): 615-638, 2001. Copyright [2001], American Institute of Physics.

Based on the research progress of shaped beam lithography at IBM, Projection Exposure with Variable Axis Immersion Lenses (PREVAIL) was proposed in 1995 as a way of improving throughput in EBL (Pfeiffer and Stickel 1995). In PREVAIL, a different electro-optical system is used as depicted in Fig. 9. The PREVAIL concept enables stitching through a combination of high-speed electronic beam scanning and moderate-speed mechanical scanning of both reticle and wafer (Pfeiffer et al. 1999).

The electron beam column consists of the following building blocks: an electron source or gun; a first lens system to generate a square-shaped beam; a second lens system to provide illumination of individual square sections—subfields—of the reticle with the square beam of essentially the same size; deflectors and correctors to scan the beam over multiple subfields with minimal loss of image quality and to control exposure timing; and a third lens system to project the reticle subfields reduced in size onto the wafer, including the means to maintain image

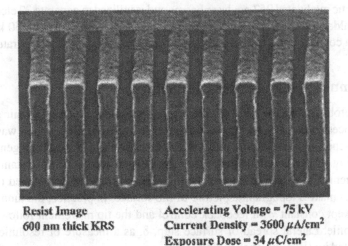

Resist Image
600 nm thick KRS

Accelerating Voltage = 75 kV
Current Density = 3600 μA/cm²
Exposure Dose = 34 μC/cm²

FIGURE 10. Resist cross section micrograph of 100 nm lines and spaces. Reprinted with permission from Pfeiffer, Dhaliwal et al., Projection reduction exposure with variable axis immersion lenses: Next generation lithography, J. Vac. Sci. Technol. B, 17(6): 2840-2846, 1999. Copyright [1999], American Institute of Physics.

quality and accurate stitching. The basic optics is depicted in Fig. 9, together with a schematic cross-section of the column. The lens systems most favorably employed are telecentric antisymmetric doublets (TADs) known to inherently have a minimum of optical aberrations. Lenses, deflectors and correctors all use magnetic fields to shape and position the beam, whereas the exposure is controlled with high-speed electric deflectors moving the beam on and off a beam stop with a pass-through aperture. Located between the illumination and imaging section is the reticle mounted on a precision movable stage in a chamber; the wafer is mounted on a similar stage below the imaging section. Column and stage chambers are under vacuum, requiring appropriate reticle and wafer load/unload systems to connect to the outside. Except for the vacuum requirements, the column and mechanical system closely resembles optical scanners in widespread use today. The electronic control system, however, is much more sophisticated (Dhaliwal et al. 2001). Figure 10 shows the pattern of 100 nm lines and spaces exposed in 600 nm thick resist at a 6:1 aspect ratio in a single layer of KRS photoresist (Pfeiffer et al. 1999).

Hitachi developed a cell projection EBL system (Nakayama et al. 1990; Sakitani et al. 1992; Matsuzaka and Soda 1999) in which an array of cells were exposed simultaneously. Increasing the number of cells increase the throughput (Sohda et al. 2003). The cell projection combined with continuously moving wafer stage was demonstrated for sub-180-nm lithography with alignment accuracy of 50 nm. Recent research efforts include multiple e-beam generation within a lithography system using a photocathode, in which a photocathode is controlled by acousto-optic modulated 257 nm laser beams and is utilized to generate 32 electron beams (Maldonado et al. 2003). The multiple beams are accelerated at 50 kV in an electron column, demagnified and focused on the mask or wafer substrate.

## 3.2. Scanning Probe Lithography (SPL)

Scanning probe lithography (SPL) is the term used for describing an atomic scale writing process based on scanning tunneling microscopy (STM), which was initially developed at the IBM Zurich Lab in the early 1980s. The STM can generate images using a sharpened conducting tip over the surface with a constant tunneling current. The displacements of the tip given by the voltages applied to the piezo drives make a topographic picture of the surface. In practice, the tunneling current is kept constant with feedback control and the tip movement follows the surface profile. Lateral spread of surface step, $\delta$, as a measure of resolution, is approximated by (Binning et al. 1982):

$$\delta = 3 \left( \frac{2R}{A\varphi^{1/2}} \right)^{1/2}$$

where R is the tip radius, A is a constant whose value is around $1.0 \, \text{Å}^{-1} \, \text{eV}^{-1/2}$ and $\varphi$ is a planar tunnel barrier. As shown in the equation, a reduction of tip radius can improve the lateral resolution. A typical STM has sub-angstrom vertical resolution and sub-nano lateral resolution.

An extensive review of SPL can be found in (Tseng et al. 2005). AFM was developed later based on STM principles and by the same research group (Binnig et al. 1986). The atomic force microscope (AFM) is made of a cantilever beam with a sharp tip at its free end which scans over the surface. The deflection of the beam varies, depending on the attractive and repulsive interaction between the tip and the surface. By measuring this deflection, topographical images can be obtained. For nanofabrication, the same tools are used to deposit, remove and modify the materials on the surface.

One advantage of STM over EBL is its use of low electron energy (<100 eV) compared to EBL (10-100 keV). Since lower-energy electrons have more confined lateral scattering, STM, in general, has a better resolution than EBL. A comparative study (Wilder et al. 1998) shows that it is possible to print sub-50-nm features using both systems, but SPL has a wider exposure latitude (the percent change in line width for a percent change in exposure dose) at these small feature sizes (Fig. 11). SPL requires a significantly higher incident electron dose for exposure than does EBL. In EBL, lithography control is most limited by proximity effects which arise from backscattered electrons whose range is considerably larger than the forward scattering range in the resist film. As a result, the exposed feature dimension depends strongly on the local feature density and size, leading to unacceptable line width variations across a wafer. These limitations are alleviated in the case of SPL exposures.

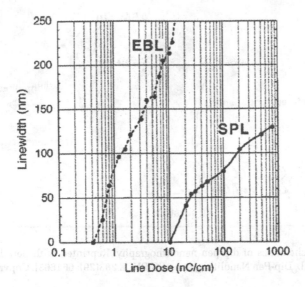

**FIGURE 11.** Patterned line width dependence on exposure dose for EBL and SPL single-pass exposures of SAL601 resist. Reprinted with permission from Wilder, Quate et al., Electron beam and scanning probe lithography: A comparison, J. Vac. Sci. Technol. B, 16(6): 3864-3873, 1998. Copyright [1998], American Institute of Physics.

Considering the advantages of ultrahigh-resolution capability and other unique features, including inexpensive equipment and relatively easy operation, SPM-based nanofabrication technologies have proliferated with diverse approaches and with varying maturity, from concept to commercialization (Tseng et al. 2005). One thing to be noted is that the development of SPL has come about from the progress in MEMS fabrication techniques. By anisotropic chemical etching, a pyramidal tip could be fabricated at the end of a silicon cantilever beam.

However, SPL has the significant drawback of a low throughput. Scanning probe is an inherently serial writing tool and its application is limited by its slow speed and low productivity. Feasible solutions have been suggested such as multiple tip operation and scan speed enhancement by integrated control (Minne et al. 1998; Lutwyche et al. 2000; Bullen et al. 2004).

Dip pen nanolithography (DPN) is a branch technique of SPL. DPN uses an atomic force microscope (AFM) tip as a "nib," a solid-state substrate (in this case, Au) as "paper," and molecules with a chemical affinity for the solid-state substrate as "ink." (Piner et al. 1999). Molecules are transported via capillary action from the AFM tip to the solid substrate as shown in Fig. 12. The migrated molecules are then immobilized by self-assembly. The first DPN experiment demonstrated the writing capability of 1-octadecanethiol (ODT) to Au surface utilizing thiol bonding (Piner et al. 1999). ODT formed a patterned self-assembled monolayer (SAM) following the writing track. Affinity of the ink molecules to the substrate is an important parameter in DPN. With the optimized substrate condition and

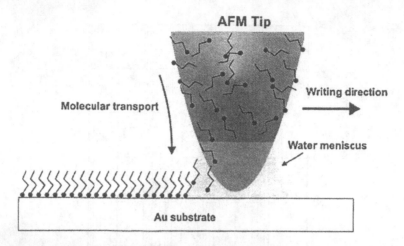

**FIGURE 12.** Schematics of dip-pen nanolithography. Reprinted with permission from [Piner, Zhu et al., Dip-Pen Nanolithography SCIENCE 283(29): 661663]. Copyright [1999] AAAS.

experimental variables such as humidity, temperature, contact force and scan speed, DPN can achieve a lateral resolution in the order of 10 nm.

DPN is a good example that shows how top-down (manipulation by a scanning probe) and bottom-up (self organization of molecules) approaches are interfaced to realize toolkits and methodology for nanoscale manufacturing.

## 3.3. Soft Lithography

Soft lithography is a printing method that uses physical contact of a master stamp pattern to transfer the image on the substrate using a special ink. Soft lithography represents a nonphotolithographic strategy based on self-assembly and replica molding for carrying out micro- and nanofabrication. It can generate patterns and structures with feature sizes ranging from 30 nm to 100 μm. For the preparation of the master stamp itself, other lithography methods are used. Once the pattern is generated, soft lithography is used to replicate features faithfully as a complementary technique.

Although the history of contact printing is quite long in human history, it has been rediscovered by Whitesides group in 1993 for micro- and nanofabrication (Kumar and Whitesides 1993). They initially demonstrated that an elastomer stamp inked with an alkanethiol could transfer a contact image on a gold surface by self-assembly. A good review in this area can be found from (Xia and Whitesides 1998; Gates, Xu et al. 2004) and (Michel et al. 2001). Five techniques have been demonstrated: microcontact printing (CP), replica molding (REM), microtransfer molding (TM), micromolding in capillaries (MIMIC), and solvent-assisted micromolding (SAMIM) (Xia and Whitesides 1998). Figure 13 illustrates these techniques.

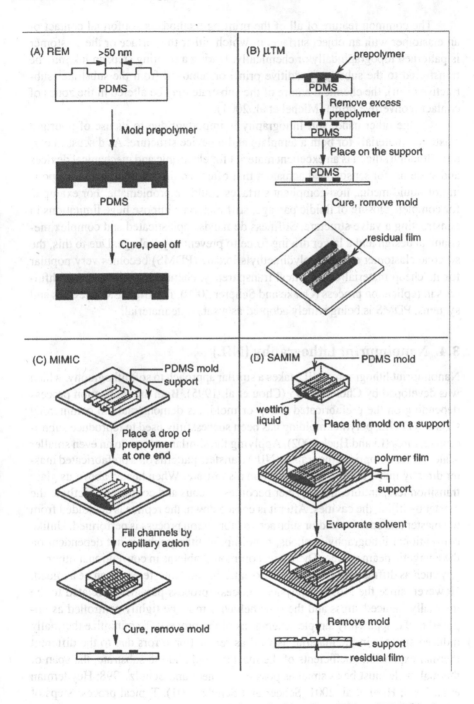

**FIGURE 13.** Schematic illustration of procedures for (a) replica molding (REM) (b) micro-transfer molding (TM) (c) micromolding in capillaries (MIMIC), and (d) solvent-assisted micromolding (SAMIM). Xia and Whitesides, Soft Lithography, Annu. Rev. Mater. Sci., 28: 153-184, 1998.

The common feature of all of the printing methods is conformal contact of an elastomer with an object surface, in which either the surface or the elastomer is patterned topographically or chemically. During a printing step, an ink may be transferred to the substrate (additive print) or removed from the substrate (subtractive print), the chemical nature of the substrate may be altered in the zones of contact (convertive print) (Michel et al. 2001).

On the other hand, soft lithography is important due to its use of nonrigid elastomer materials for both a template and a device structure. As discussed earlier, although silicon is an excellent material for electronic and mechanical devices and systems, for sensors and actuators that often require sampling and transporting of liquid media, non-compliant surfaces could be problematic. For example, for complete closure of fluidic passage, stiff materials impose huge limitations in constructing a valve structure. Stiffness demands sophisticated and complex mechanical design and a larger driving force to prevent any leakage. Due to this, the silicone elastomer such as polydimethylsiloxane (PDMS) becomes very popular for its cheap material cost, optical transparency, chemical stability and faithfulness in replication process (Quake and Scherer 2000). For BioMEMS devices and systems, PDMS is being widely adopted as a substrate material.

## 3.4. Nanoimprint Lithography (NIL)

Nanoimprint lithography (NIL) takes a similar approach to soft lithography, which was developed by Chou in 1995 (Chou et al. 1995). Both are replication process depending on the prefabricated stamp or mold. As demonstrated in commercial CDs and DVDs, polymer molding has been successfully used to reproduce submicron features (Li and Huck 2002). Applying the similar concept to an even smaller scale, the nanoimprint lithography (NIL) transfers patterns on a prefabricated master directly into a thermoplastic resist on a substrate. When heated above its glass transition temperature, the polymer becomes viscous and conforms exactly to the master by filling the cavities. After it is cooled down, the replica is demolded from the master and additive and/or subtractive fabrication process is performed. Unlike conventional lithography methods, nanoimprint lithography is not dependent on the energetic beam source. Therefore, common problems in conventional lithography such as diffraction, scattering/backscattering and interference can be avoided. However, since the NIL is a physical process, process parameters related to the thermally induced stress and the flow behavior must be tightly controlled as opposed to the apparently simple schematics of the process. To minimize thermally induced stresses in the material as well as replication errors due to the different thermal expansion coefficients of the master mold and the substrate, the span of thermal cycle must be as small as possible (Scheer and Schulz 1998; Heyderman et al. 2000; Hirai et al. 2001; Scheer and Schulz 2001). Typical process steps of NIL are illustrated in Fig. 14 (Torres et al. 2003).

NIL has been demonstrated with 10-100 nm resolution (Chou et al. 1995; Chou 1996; Chou and Krauss 1997; Xia and Whitesides 1998; Xia and Whitesides

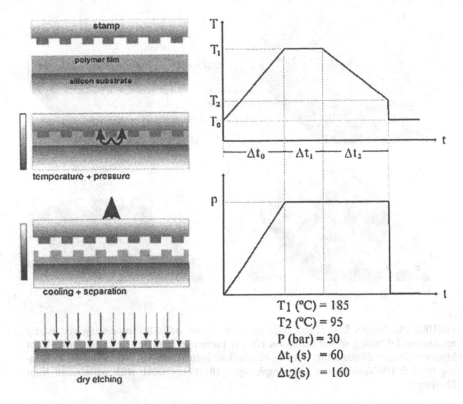

**FIGURE 14.** Schematics of the NIL process (left) and of the temperature–pressure temporal sequence (right). The inset shows typical process parameters. Reprinted from Materials Science and Engieering, C 23, Torres, Zankovych et. al., Nanoimprint lithography: an alternative nanofabrication approach, pp. 23-31, (2003), with permission from Elsevier.

1998; Michel et al. 2001; Zankovych et al. 2001; Chou et al. 2002). To address the issues described above, research efforts in this field involve the use of a short-pulse laser or the combination of NIL and photolithography to reduce thermal cycle time and the residual stress. In the technique of laser-assisted nanoimprint (Xia et al. 2003; Grigaliunas et al. 2004), a thin surface layer of polymer is molten by a single excimer laser pulse, and a master mold is embossed into the low viscous polymer in a very short time of less than 500 ns. In the nanoimprint system with UV-photolithography capabilities (Bender et al. 2000; Bender et al. 2002; Bender et al. 2004; Chen and Guo 2004; Plachetka, Bender et al. 2004), a master mold is made from a UV transparent material and low-viscosity UV-curable resists are used for faithful pattern definition. Relatively high aspect ratio structures can be generated by this technique.

Combinations of two techniques—UV-NIL and soft lithography—were also tried for micro- and nanofabrication (Bender et al. 2000; Bender et al. 2004;

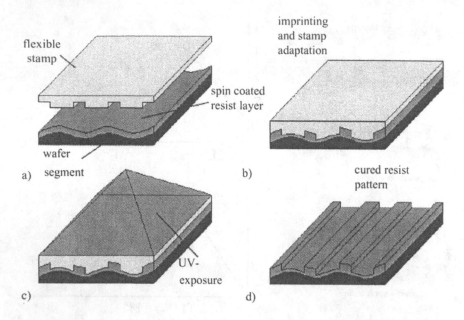

**FIGURE 15.** Process flow of soft NIL: (a) spin coated wafer (b) imprinting and stamp adaptation (c) curing of the resist via UV (d) patterned high contrast. Reprinted from Microelectronic Engineering, 73-74, Plachetka, Bender et. al., Wafer scale Patterning by Soft UV-Nanoimprint Lithography, pp. 167-171, (2004), with permission from Elsevier.

Plachetka Bender 2004). For example, starting with a hard surface such as Si or GaAs, patterns are generated, then transferred to a soft and transparent elastomer stamp by casting. The stamp is then used to regenerate the original structure on the desired surface by UV-NIL. The schematic of this soft NIL is shown in Fig. 15 (Plachetka et al. 2004).

It should be noted that NIL is a complementary technique that can be combined with other production method for batch fabrication. There have been continuous efforts toward low-cost mass production in the area of nanofabrication. One of the most frequently used fabrication methods in the lab scale production is EBL or SPL. Although progress is being made, slow serial processing limits the use of this technology in mass production. While EUV or X-ray nanolithography offers parallel batch fabrication, it requires huge initial investment on equipment and toolkits. The development of practical methods for generating sub-100 nm resolution on a variety of substrates has been a great technical challenge. As an alternative solution, prefabrication of patterns on a stamp or mold with high-precision techniques such as EUV, EBL and SPL and replication of those with NIL is suggested. Although there are other issues related to viscoelastic behavior of resist materials, multiple layer alignment as well as print and uniformity across a large area wafer, technical progress is being made.

## 4. INTEGRATION AND INTERFACE

In order to generate signals and motions that can communicate with the macro world, novel materials assembled from a bottom-up approach need to be integrated and interfaced with larger-scale devices and systems. Top-down nanofabrication can offer a testing platform for functional materials developed from bottom-up methodology. In such sense, two approaches are complementary. As discussed by Fujita (Fujita 2003), micromachined devices provide not only infrastructures for nanotechnology but also tools for examining nanomaterials. Thus, for engineering applications, nano/micro integration and interface needs to be conceived from the design stage. In this, the effort made toward integration and interface concerns more about methodology than about scale itself. A few examples will be discussed regarding integration and interface.

### 4.1. Carbon Nanotube (CNT) Manipulation with Microelectrode

Carbon nanotubes are considered to be a very promising candidate material for nanodevices due to their unique chemical, mechanical and electrical characteristics. Synthesis of CNT has been widely studied while the integration of CNT into devices is still elusive (McEuen et al. 2002). For CNT integrated devices, chemical vapor deposition with an activated catalytic site can be used. However, high-temperature processing is not very desirable due to its incompatibility with semiconductor batch fabrication. Alternatively, CNTs can be fabricated separately and suspended in liquid media, then assembled into a device. For manipulation of CNTs in this method, microfabricated tools such as a scanning probe and microelectrodes can be used.

In the first approach, the scanning probe is used to simply control the shape and position of individual CNT (Hertel et al. 1998) and to further fabricate a single-electron transistor (Roschier et al. 1999). One major drawback, however, is its lack of capability to make multiple devices concurrently with reproducibility. To overcome the yield problem for manufacturing, selective alignment of CNTs from liquid media with micro- and nanoelectrode using dielectrophoresis has been drawing attention (Yamamoto et al. 1998; Nagahara et al. 2002; Chung and Lee 2003; Chung et al. 2003; Suehiro et al. 2003; Chan et al. 2004). Yamamoto et. al. (1996) initially discovered CNTs in isopropyl alcohol (IPA) could be separated from other particles and aligned towards an electrode due to mobility difference under dielectrophoresis. Dielectrophoresis is the translational motion of neutral matter caused by polarization effects in a nonuniform electric field. This force results from differences in polarizability between the particle and the medium. Polarizability is the ease with which an external field can distort the charge distribution in a molecule. The electrical force on any object undergoing dielectrophoresis is directly related to its volume and polarizability, which is a function of its complex permittivity. Integration of CNTs into the electrode was reported (Nagahara et al. 2002; Chung and Lee 2003) and was utilized for a thermal sensor (Chan et al. 2004) and an ammonia gas sensor (Suehiro et al. 2003). A more sophisticated alignment method was developed with a nanogap electrode using MEMS fabrication techniques (Chung and Lee 2003; Chung et al. 2003).

Passive integration without resorting to any active power source was also suggested as an integration method. Rao used organic molecular marks on a substrate to guide the self-assembly of individual single-walled CNTs. Using microcontact printing, two distinct surface areas were created with polar chemical (amino- or carboxyl- ) groups and nonpolar (methyl-) group, then CNTs were aligned along polar region due to electrostatic interaction (Rao et al. 2003). A large area patterning has been demonstrated. This is also an example of combined methods of chemical interaction and the top-down approach of soft lithography.

## 4.2. Nanoparticle Interface with Microelectrode

Out of recent progress in chemistry and materials science, a wide variety of sub-10 nm scale engineering materials in wire, tube or particle form have been produced. Meanwhile, lithographic techniques generating reliable micrometer scale features have been utilized in electronics. The interfacing those two principles needs to be explored for successful implementation of nanomaterials into working devices and systems. Among others, nanoparticles are prevalent in materials development: the organized patterning of nanoparticles and connecting those with microelectrodes is a crucial task in achieving any tangible function in small devices.

The strategy for the selective creation of covalent or electrostatic binding sites is employed for collective pattern generation of nanoparticles (Shipway et al. 2000). A standard photolithography process can be used to make a defined structure that protects the surface from chemical treatment. The exposed area holds (or loses) specific chemical functional groups which become preferential assembling (or distracting) sites of nanoparticles. UV exposure during the photolithography itself can also be used for patterning purpose. For example, irradiation on thiol groups leads to the destruction of active surface groups, making them unable to bind gold nanoparticles (Liu et al. 1998).

Soft lithography is also an obvious choice for integration. Using a stamp, microcontact printing of an affinity layer for nanoparticles or direct application of nanoparticles in a solution with guided microchannels could generate patterns.

Interfaced with microelectrodes, collective characteristics of nanoparticles in the gas sensor structure have been studied by one of the authors (Rajnikant et al. 2004; Shukla et al. 2005; Shukla et al. 2005). Compared to the conventional gas sensors using the equivalent material system, very high sensitivity can be obtained. Figure 16 shows the device and its cross-sectional view. In this sensor, the nanoparticle layer consists of trivalent $In_2O_3$-doped $SnO_2$ semiconductor nanocrystalline particles with Pt-clusters as the catalyst. Hydrogen molecules are first adsorbed on the sensor surface, and then are dissociated into protons and electrons on the Pt-catalyst surface. While the protons are involved in forming water molecules

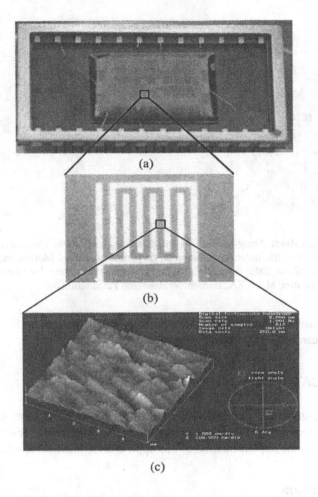

(a)

(b)

(c)

**FIGURE 16.** Fabricated nanoparticle-MEMS sensor: (a) packaged sensor (b) interdigitated microelectrode (c) AFM image of nanoparticles. Rajnikant, Shukla et al., A nanoparticle-based microsensor for room temperature hydrogen detection, Proc. IEEE Sensors 24-27 Oct. 2004, pp 395-398, vol-1. (© [2004] IEEE)

via reaction with the surface-adsorbed oxygen ions, the electrons contribute to an enhanced conduction between the interdigitated electrodes. The sensor interface was made by micromachining electrodes and a sol-gel dip coating of nanoparticles. Scaling-down from a bulk sensor to a microsensor leads to increased mass transport rate of the analytes, small volume requirements, precise control over the geometry and dimensions, while the nanoparticles provide increased area of space charge region and a total number of reactive sites for a given volume. Overall sensor performance can be dramatically improved from this nano-micro interface. The high sensitivity of $10^5$ was observed even at room temperature, which had not been reported before (Shukla et al. 2005; Shukla et al. 2005).

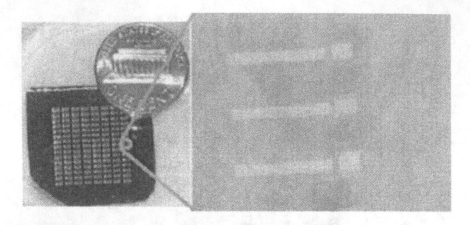

**FIGURE 17.** Cantilever Array Sensor. Biswal, Raorane et al. 2005, Thermally Induced Phase Transitions of Biomolecules Observed Via Nanomechanical Motion from Micro-cantilevers, MicroTAS 2005: 9th Intl. Conf. on Miniaturized Systems for Chemistry and Life Sciences, Boston, MA, USA, Transducer Research Foundation.

An extensive review on the subject of nanoparticle patterning for practical applications can be found in (Shipway et al. 2000).

## 5. APPLICATIONS

MEMS structures and fabrication methodology provides infrastructure for nanoscale materials to facilitate interactions with the outer world. The following sections discuss the implementation of this approach for practical applications.

### 5.1. Nanobeam

In the cantilever-based sensor, the adsorption/desorption of nanoscale molecules and surface reorganization induce a stress that causes the cantilever to bend. The binding event can be observed with an amplified signal output, i.e., beam deflection. This phenomenon has been utilized to study phase transitions in macromolecules (Biswal et al. 2005). A bimorph cantilever array shown in Fig. 17 can be fabricated with a 200 $\mu$m silicon nitride beam and a 20 nm gold layer. Optical detection is used to monitor the bending of the bimorph cantilever due to a change in temperature. The relation between the surface stress in the thin gold layer to the cantilever tip rotation and deflection is given by the Stoney's formula:

$$\Delta h = \frac{3\Delta\gamma(1 - v_1)L^2}{E_1 t_1^2}$$

where $\Delta\gamma$ is the change in surface stress, $v_1$ is the Poisson's ratio of the thick layer, $E_1$ is the Young's modulus of the thick layer, $t_1$ is the thickness of the thick layer and $L$ is the length of the cantilever. DNA strands were immobilized and hybridized on the cantilever and then the temperature was ramped up to induce DNA melting. The result is in good agreement with the temperature at which the double helix separation is expected.

By scaling down further, aluminum nanocantilever-based dynamic sensors for highly sensitive mass sensors have been reported (Davis and Boisen 2005). When mass sensitivity ($\delta m / \delta f$) is calculated for cantilevers of different materials, it is observed that Al has the best mass sensitivity followed by poly-Si and then single crystal Si, due to its low density and low Young's modulus compared to the other cantilever materials. The fabrication of the Al nanocantilevers is based on a lift-off technique. The devices are realized on 4-in. Si wafers. First, a resist mold is formed on the Si substrate using UV lithography. Then, Al is deposited using electron beam evaporation. Next, the Al deposited on the resist is removed by lift-off. Finally, the metal structure is released by $SF_6$ dry etching of the underlying Si layer. An important aspect of this process is that the release step is performed by dry etching, which alleviates stiction problems that are often observed in wet release techniques. The fabricated nano cantilever is shown in Fig. 18. On the fabricated sensor, AC and DC voltages are applied between the driver electrode and the Al nanocantilever, which drive vertical motion. The frequency responses of the nano cantilever reveal frequency shifts (6.5 kHz) in response to mass of the carbon deposit (14 fg). The mass sensitivity of the device is found to be around 2 ag/Hz.

Another example of microcantilever sensor with nano interdigitated electrodes (IDEs) has been reported for DNA binding protein analysis (Wu et al. 2001; Lee et al.). In this sensor, a 0.5 μm thick low-stress $SiN_x$ thin film is grown by LPCVD on the front and back side of a silicon wafer. Back side $SiN_x$ is etched by reactive ion etching (RIE) and then Si is anisotropically etched from the backside by TMAH. Au/Cr layer is deposited on the front side of the wafer and the nano IDEs are patterned by e-beam lithography. Finally, the $SiN_x$ layer is released by RIE to form the microcantilever. The 100 nm line and spaced nano IDEs are located at the clamped end of the 15 μm wide microcantilever. As shown in Fig. 19, the nano-IDEs maximize the microcantilever deflection induced by conformal change of DNA bridges due to interaction of biomolecules from DNA hybridization and DNA protein binding.

A nano cantilever with spherical tips for dynamic (oscillating cantilever) noncontact AFM has been reported (Hoummady et al. 1997). The fabrication process consists of a thin platinum mesh containing an electrolyte liquid (KOH) that is maintained by surface tension. A wire made of tungsten (W), can be etched by inserting it into the mesh by means of an $X$-$Y$-$Z$ nano-translator stage. A microscope and a charge-coupled device (CCD) camera are used to control the position of the wire in the electrolyte as well as to visually control the size of the wire during the etching procedure. First, the diameter of the W wire is reduced using rapid AC etching followed by a DC electro-polishing to attain the shape desired. Fig. 20(a) and (b) show an SEM picture of a perfect ball tip fabricated by this technique.

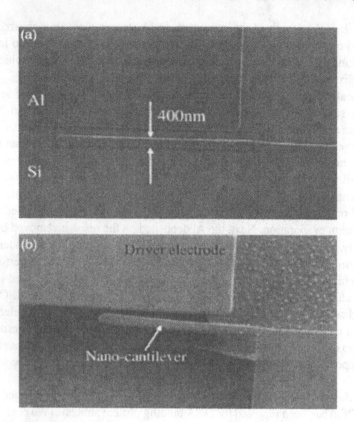

**FIGURE 18.** SEM images of Al nanocantilever before (a) and after (b) dry underetching. Reprinted with permission from Davis and Boisen, Aluminum nanocantilevers for high sensitivity mass sensors. Applied Physics Letters 87(1): 013102-3, 2005. Copyright [2005], American Institute of Physics.

Fig. 20(c) shows a simple procedure by which to make a ball tip. After reducing the size of the wire, the ball tip is obtained by a combination etching of the selected region and scanning the wire through the electrolyte thin film in order to obtain a smooth surface. By controlling the etching time and the region to be etched, it is possible to make any structure with cylindrical symmetry. Due to the small size of the nano cantilever (neck) and the small mass of the oscillating ball, the natural frequency typically lies in the range of 100 MHz–1 GHz. With the fabricated device at the length of 200 nm, mass of $10^{-14}$ g can be measured at 300 MHz.

A tweezers-type AFM probe device using micromachining technology in order to combine the function of nano objects manipulation with AFM observation has been developed (Takekawa et al.). Out of a SOI wafer, after front and back side etching, the free standing area of the structure is formed. Final $SiO_2$ etching by an HF solution results in the AFM tweezers as shown in Fig. 21. The main shape of the AFM tweezers is determined by deep RIE, while the triangular cross-section probes are formed before the deep RIE etching. Two thin probes with a triangular

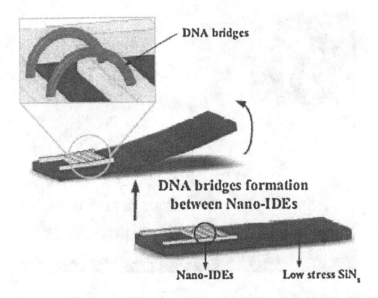

**FIGURE 19.** DNA bridges between the nano-IDEs maximizing the microcantilever deflection. Lee, Yun et al., 2005, Fabrication of Microcantilever with Nano-interdigitated Elecetrodes (IDEs) for DNA Binding Protein Detection, MicroTAS 2005: 9th Intl. Conf. on Miniaturized System for Chemistry and Life Sciences, Boston, MA, USA, Transducer Research Foundation.

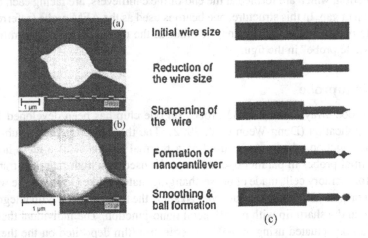

**FIGURE 20.** SEM image of the ball tip and procedure to fabricate the ball tip of the nano cantilever. (a) Low magnification inspection. The diameter of the ball is 1.5 mm. (b) High magnification of the nano cantilever. The diameter is around 50 nm. (c) Procedure by which to make a ball tip supported by a nano-cantilever. Reprinted with permission from Hoummady, Farnault et al., New technique for nanocantilever fabrication based on local electrochemical etching: Applications to scanning force microscopy, The fourth international conference on nanometer scale science and technology, Beijing (China), AVS, 1997. Copyright [1997], American Institute of Physics.

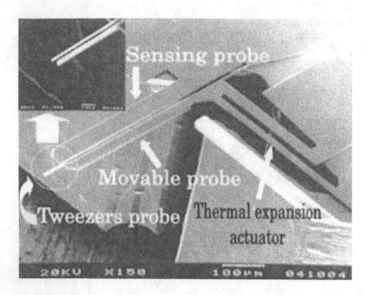

**FIGURE 21.** A SEM photograph of the fabricated AFM tweezers. The insert figure is a magnified view of the triangular probes. Takekawa, Nakagawa et al., The AFM Tweezers: Integration of a Tweezers Function With an AFM Probe. Transducers'05, Seoul, Korea, IEEE Electron Devices Society. (© [2005] IEEE)

cross-section, which are formed at the end of the cantilevers, are facing each other with a 2 μm gap. In this structure, one beam is used as the AFM probe (referred as "sensing probe" in the figure), and the other as the manipulation probe (referred as "movable probe" in the figure).

## 5.2. Nanoprobe

A nanoprobe array with 32 × 32 probes on one chip has been developed for a sensor application (Dong-Weon et al. 2002). The thermal probe has a sub-100-nm nano junction which can be used as a thermal sensor or as a nano heater. The thermal probes in parallel operation can be used as a high-rate data transfer device for memory cells made of phase changing material like GeSbTe. The whole system consists of Pyrex glass plate with Ni stick, the thermal probe with integrated actuator, and a sharp tip with metal–metal nano-junction. The individual thermal probe can be actuated using an AlN piezoelectric film deposited on the thermal probe. The designed thermal probe has a fundamental resonance frequency of around 212 kHz, and the spring constant is designed under 2 N/m for contact operation between the tip and a medium. The thermal probes in the array are arranged in 140 μm pitch horizontally and in 100 μm pitch vertically. The nano thermal probes are fabricated by a combination of conventional micromachining techniques, photolithography, directional fast atom beam (FAB) dry etching to pattern metals and inductively coupled plasma reactive ion etching (ICP-RIE) to

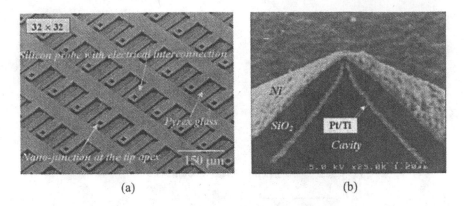

**FIGURE 22.** SEM views of (a) The thermal probe array (32 × 32) with nano-junction. (b) A cross sectional SEM view of the SiO₂ tip made by FIB cutting. Dong-Weon, Ono et al., Microprobe array with electrical interconnection for thermal imaging and data storage, Microelectromechanical Systems, Journal of 11(3): 215. (© [2002] IEEE)

etch silicon. An SEM view of the fabricated probe array is shown in Fig. 22(a). A cross section of the thermal tip with metal–metal junction is shown in Fig. 22(b), which is obtained by FIB (focused ion beam) cutting.

In another example, nano-probes were batch fabricated for near-field photo-detection optical microscopy (NPOM) (Davis et al. 1995). The nano-probes are fabricated on a silicon wafer. A sandwich of $SiO_2$ and $Si_3N_4$ is grown on the Si wafer. A lithography step is performed and the exposed $Si_3N_4$ and $SiO_2$ layers are etched through leaving an array of $10 \times 10$ μm pads of the sandwich layers on the substrate. The wafer is then placed in a quasi-isotropic silicon etch consisting of nitric acid, $H_2O$, ammonium fluoride and hydrofluoric acid. The silicon is etched until a broad tip is formed as shown in Fig. 23(a). The silicon is thermally oxidized (100 nm thickness) and the oxide is then removed with buffered HF (BHF). The resulting structure is again oxidized as shown in Fig. 23(b). This re-oxidation process further sharpens the tip. The photoresist is spun for only 0.2 s, to get a very thick coat. The xylene in the photoresist is allowed to evaporate and the wafer is then spun for 30 s at 3000 rpm. By this modified photoresist process, only the end of the tip protrudes through the photoresist as shown in Fig. 23(c). Then the oxide is etched from the exposed tip region with BHF and the resist is removed as shown in Fig. 23(d). Al is then deposited over the entire structure. Photoresist is again applied in the manner described above and the Al in the tip region is removed with a standard Al etchant solution. The photoresist is removed and the Al–Si contacts are sintered at 420°C as shown in Fig. 23(e). An SEM image shows the resultant nano-probe in Fig. 24. Hence, this simple fabrication method meets all the NPOM probe requirements i.e., an ultrasharp tip (nanometer radius), a small Schottky contact area, and a very small (sub-micrometer) optically sensitive region. The diode has a maximum optical sensitivity at a reverse bias of $-1.0$V where it can

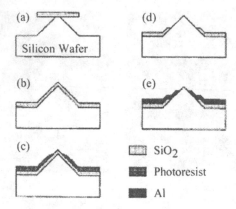

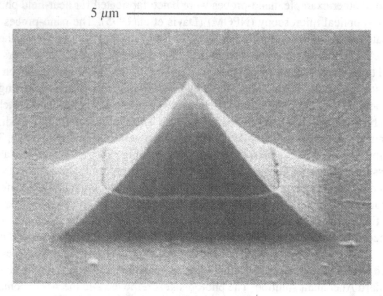

**FIGURE 23.** Probe fabrication. (a) A pyramid is Micromachined on a silicon wafer. (b) The tip is sharpened through oxidation. (c) A special application of photoresist exposes the oxide near the end of the tip. (d) The oxide is etched off the exposed region. (e) Al is deposited over the entire structure and then etched off the end of the tip. Reprinted with permission from Davis, Williams et al., Micromachined submicrometer photodiode for scanning probe microscopy, Applied Physics Letters 66(18): 2309-2311, 1995. Copyright [1995], American Institute of Physics.

**FIGURE 24.** A SEM image of the nano-probe detector. The opening in the aluminum is 0.7 × 0.7 μm and the tip radius is less than 50 nm. Reprinted with permission from Davis, Williams et al., Davis, Williams et al., Micromachined submicrometer photodiode for scanning probe microscopy, Applied Physics Letters 66(18): 2309-2311, 1995. Copyright [1995], American Institute of Physics.

detect optical variations as small as 150 fW. It has a spectral response similar to bulk silicon detectors.

In the area of nanosensors for biological analysis, the development of dynamic intracellular sensing probes is crucial to understand interactions between cell signaling mechanisms and between networks of cells. Nanotips and nanoprobes have been developed to be capable of electrochemically analyzing the cell membrane surface and the cell interior (Fasching et al. 2005). A cantilever transducer system with platinum tip electrodes in submicro range was developed using etching, shadow masking, and focused ion beam technology. A sharpened tip with a tip radius smaller than 50 nm was fabricated with an AFM cantilever so that the tip stands out vertically on top of the cantilever. These probes are capable of piercing the cell with minimum damage and also conducting electrochemical detection.

## 5.3. Nanopore and Nanogap

In biology and chemistry, many efforts have been focused on molecular-level detections. Nanopores are one of the most successful structures with the capability of single molecular level detection. Electrical detection of translocation of single-stranded DNA and RNA molecules through a voltage-biased $\alpha$-haemolysin protein pore (14 Å at the narrowest constriction) within a lipid bilayer was first reported in 1996 (Kasianowicz et al. 1996). Each translocation molecule either partially or fully blocks the normal ionic current through the pore in a lipid bilayer that can be detected by an electrical signal, where the signal depends on the characteristics of the target molecule. Using this technology, detection of single nucleotide differences, distinction between polycytosine and polyadenine molecules, and characterization of the hybridization of individual DNA strands has been demonstrated (Akeson et al. 1999; Meller et al. 2000; Howorka et al. 2001; Meller et al. 2001; Vercoutere et al. 2001). Although these pores work great, the sizes are fixed and their stability and noise characteristics are restricted by chemical, mechanical, electrical, and thermal constraints. Also, the use of a fragile lipid bilayer membrane makes this approach limited in practice.

The use of solid-state nanopores overcomes these limitations and provides a wide variety of advantages. The size of the nanopore can be easily controlled and the pores are chemically and mechanically robust under a wide variety of pH, temperature, and electrical conditions. A solid-state nanopore can be integrated with various types of other sensing mechanisms, can have the inner surface functionalized for different applications, and can be integrated into a larger and more complex system. The advances in nano and microfluidic technologies make possible a fully integrated molecular level sensing platform for various applications. Solid-state pores have been used to detect DNA contour lengths, DNA hybridization, DNA folding, drag on individual DNA molecules, single porphyrin molecules, and to filter single stranded DNA (ssDNA) from a solution containing ssDNA and double stranded DNA (dsDNA) (Li et al. 2001; Li et al. 2003; Storm et al. 2003; Chen et al. 2004; Fologea et al. 2005; Heins et al. 2005; Storm et al.

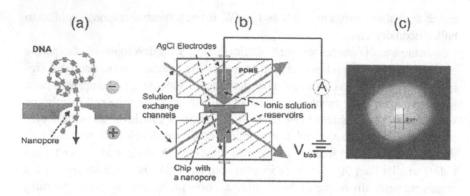

**FIGURE 25.** (a) Schematic illustration of solid state nanopores for DNA molecule translo-cating (b) experimental setup for single molecule measurements with a nanopore detector (c) TEM of a silicon nanopore with a 4 nm diameter. Reprinted with permission from Fologea, Gershow et al.Detecting Single Stranded DNA with a Solid State Nanopore." Nano Letters 5(10): 1905-1909. Copyright (2005) American Chemical Society.

2005). Techniques such as e-beam lithography, ion milling, etching and deposition have been used to fabricate nanopores in silicon nitride and silicon dioxide. The sizes of the nanopores are typically in the tens of nanometer range. To detect ds-DNA for example, a pore diameter smaller than the molecule persistence length of 50 nm and larger than the 2 nm cross-sectional size of the molecule is required (Li et al. 2003). Typically, the molecular detection is performed by placing a nanopore-containing chip between two separated fluidic chambers. These two chambers are electrically connected only by the ionic solution through the nanopore. By apply-ing a voltage, charged molecules can pass through the nanopore, causing a change in the ionic current being measured. Figure 25 shows solid-state nanopores used to detect DNA (Li et al. 2003).

    It has been reported recently that the nanopores not only operate as a coulter-counter where the ionic current measured across the pore decreases when a molecule passes through the pore, but that the ionic current can actually increase due to the interaction between the surface and the molecule passing through the pore (Chang et al. 2004). During the translocation of dsDNA molecules through a nanopore "channel," the ionic current increased due to electrical gating of surface current in the channel resulting from the charge on the DNA itself. The salt depen-dence of ion transport and DNA translocation through solid-state nanopores has been also studied over a wide range of concentrations. The result shows that the conductance increases as the salt concentration is lowered due to the salt-dependent surface charge of the pore. It is expected that the nanopore single molecule de-tection technology can greatly improve the high-speed DNA sequencing (Fologea et al. 2005).

    The nano-porous electrode systems enhance sensitivity of electrochemical biosensors by employing redox cycling between submicron spaced electrodes (Muller et al.). First basic electrodes (counter and reference) and an insulating layer

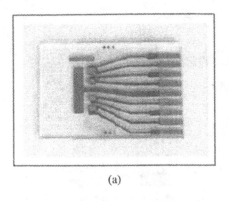
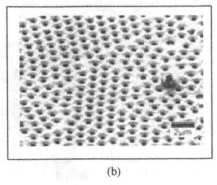

(a)                                              (b)

**FIGURE 26.** (a) Sensor chip with four sensor positions, a large counter electrode & connecting pads in the periphery. (b) SEM image of the nano-porous top electrode. One non removed particle is still visible, demonstrating the down-scaling effect when using particles as shadow masks during metallization. Muller, Kentsch et al., 2005, Sub-micrometer Spaced Nano-Porous Electrode Systems: Fabrication, Properties and Application to Sensitive Electrochamical Detection, MicroTAS 2005: 9th Intl. Conf. on Miniaturized Systems for Chemistry and Life Sciences, Boston, MA, USA, Transducer Research Foundation.

are deposited on glass. An integrated gold electrode and a formatted Ag/AgCl-wire serve as counter and reference electrodes respectively. Figure 26(a) shows four sensor positions and a centrally positioned counter electrode. A monolayer of BSA adsorbed as an adhesive agent on top of the insulator helps in the self-assembly of micro- or nanospheres in a closely packed layer. Etching in oxygen plasma results in a regular pattern of isolated particles which act as shadow masks during deposition of the top electrode layer. Gold is sputter deposited on to the self-assembled surface and then the nano particles are removed. Pores are etched in $CF_4$ plasma employing the porous top electrode as the etching mask. Figure 26(b) shows the porous top electrode.

A simple device based on electrostatic actuation with a controllable nanogap for electromechanical characterization of nano-scale objects is fabricated (Gel et al., 2005; Park et al. 1999). Figure 27(a) shows the schematic view of the in-plane bending of the device. When voltage is applied to the left actuation electrodes, bending towards the actuation electrode results in approach of the tips. Retraction of the tips can be realized when voltage is applied to the right actuation electrode. Figure 27(b) shows the fabricated device after oxide release etching. At the tip of each cantilever a metal electrode is separated by an initial gap of approximately 30 nm from the other tip.

## 5.4. Channel and Needle

With the emergence of nano-biotechnology, there is a need to further downsize microfluidic channels that are widely used in liquid handling in biology and chemistry for sample manipulation. Fluid transport and molecular behavior at

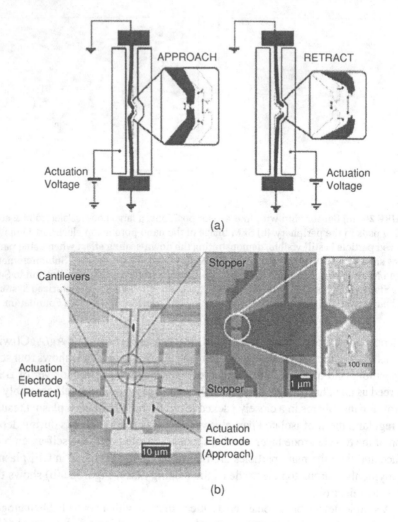

**FIGURE 27.** (a) Working principle for device based on in-plane bending. (b) Fabricated tests sample after oxide release etching. Gel, Edura et al., 2005, Controllable Nano-Gap Mechanism For Characterization Of Nanoscale Objects, MicroTAS 2005: 9th International Conference on Miniaturized Systems for Chemistry and Life Sciences, Boston, Massachusetts, USA, Transducer Research Foundation.

extremely small dimensions can be used in applications such as single molecule manipulation and detection for biosensing, chemical analysis, and medical diagnosis (Turner et al. 2002). Advances in nanoimprint lithography (NIL) greatly resolved the issue of low-cost fabrication while maintaining high accuracy (Guo 2004). Nanofluidic channels with cross-sections of 700 nm and 300 nm were used to stretch DNA, allowing the study of the statics and dynamics of DNA molecules in a confined geometry (Guo et al. 2004). If a DNA molecule is forced through a nanofluidic channel with a cross-section comparable to the persistence length of the molecule, it is favorable for the DNA molecule to be in the stretched state. This phenomenon can be applied to detect, analyze, and separate the DNA molecules. Austin's group reported the use of such nanofluidic channel to directly measure the contour length of single DNA molecules confined in the channels and the statistical analysis of the dynamics of the polymer in the nanochannel (Tegenfeldt et al. 2004). This technique has advantages over conventional flow stretching or stretching using a tethered molecule since the presence of a known external force is not required while allowing for continuous measurement of length.

Nanochannels have been also used for drug delivery applications (Sinha et al. 2004). Nanofluidic channels with 60-nm channel height were fabricated in silicon by selectively growing oxide and then etching the sacrificial oxide. Glucose flow through a 60-nm channel showed a zero-order release rate.

DNA stretching experiments have been successfully demonstrated by using nanofluidic channels fabricated by nanoimprinting of hydrophilic hydrogen silsesquioxane (HSQ) (Guo et al. 2004; Cheng et al. 2005). HSQ is an inorganic hydrophilic polymer with a repeating unit of $HSiO_{3/2}$ and has the ability to retain its solvent in a spin-coated film. First, a 390 nm thick layer of HSQ is spin coated on a silicon substrate and then imprinted by using a surfactant coated grating mold under a pressure of 1000 psi at room temperature. A surfactant coated grating mold has a period of 700 nm, 50% duty cycle and a depth of 500 nm prepared by interference lithography and dry etching (Cao et al. 2002). For sealing the channels, HSQ is spin coated on an $O_2$ treated PDMS stamp and put in physical contact with HSQ channels for 5 minutes. The same sealing process can be used to seal nanochannels etched in glass as well. Figure 28 shows the

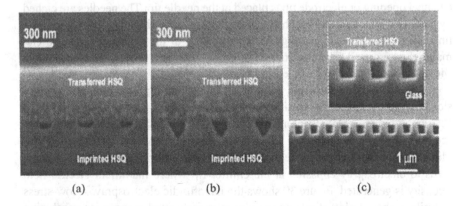

(a)    (b)    (c)

**FIGURE 28.** (a) & (b) SEM images of cross section of HSQ nanofluidic channels with the HSQ sealing. (c) SEM image of the HSQ sealing on glass nanochannels. Cheng, Chang et al., 2005, Nanoimprint of Nanofluidic Channels by Using Hydrophilic Hydrogen Silsesquioxane (HSQ), MicroTAS 2005: 9th Intl. Conf. on Miniaturized Systems for Chemistry and Life Sciences, Boston, MA, USA, Transducer Research Foundation.

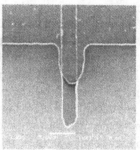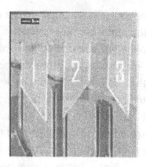

**FIGURE 29.** SEM images of planar nanoneedles. (a) Hollow nanoneedle. The channel is clearly visible in the middle of the needle. Inner channel dimensions: 150 nm high, 2 μm wide. Needle dimensions: 0.7 μm high, 3 μm wide & 5 μm long. (b) Electrode nanoneedle. The electrode tip is 2 μm long, 2 μm wide & 100 nm high.. (c) Three plane nanoneedles marked 1, 2 and 3. Needle dimensions: 260 nm high, 3 μm wide and 10 μm long. Emmelkamp, Gardeniers et al., 2005, Planar Nanoneedles On-Chip for Intracelluler Measurements, MicroTAS 2005: 9th Intl. Conf. on Miniaturized Systems for Chemistry and Life Sciences, Boston, MA, USA, Transducer Research Foundation.

cross section of the SEM images of HSQ channels and HSQ sealed glass channels, respectively.

Needles with nanosized inner channels or electrodes have been fabricated (Emmelkamp et al.). Excellent penetration with diminished leakage from HL60 cells after insertion and release of the nanoneedle has been reported. Figure 29(a) and Fig. 29(b) shows the hollow needle and the needle with electrode, respectively. In these types of needles the channels or electrodes are encapsulated between a 300 nm LPCVD-nitride bottom layer and a 300 nm PECVD-nitride top layer. Channel opening or electrode tip is placed at the needle tip. The needles are etched by reactive ion etching and the channels are made by a sacrificial layer of chromium line etched by ion beam etching. After deposition of the top layer the sacrificial material is removed. The electrodes are formed by wet etching. The plane needle shown in Fig. 29(c) only has the LPCVD layer.

A nanofluidic electrospray emitter tip incorporating a nanofluidic capillary slot has been realized for use in the electrospray-mass spectrometry (ESI-MS) (Descatoire et al. 2005) (Brinkmann et al. 2004). A liquid placed into the macroscopic reservoir spontaneously fills the microfluidic slot and then the nanofluidic slot and is presented at the extremity of the emitter tip. An applied voltage leads to electrospray emission at the emitter tip where maximum electric field intensity is generated. Figure 30 shows the nanofluidic electrospray. A low-stress polysilicon-based slotted silicon cantilever is fabricated using micromachining techniques involving low-pressure chemical vapor deposition of polysilicon, optical photolithography and pattern transfer using a He-Cl based reactive ion etch. Then nanofabrication of the nanofluidic slot is carried out by using a Ga-ion-based Focused Ion Beam (FIB) (Le Gac et al. 2004). The emitter tip has width of

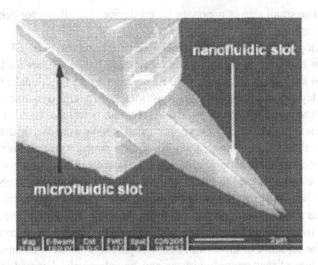

**FIGURE 30.** SEM image of the nanofluidic electrospray. C. Descatoire, D.Troadec et al., 2005, A Nanofluidic Electrospray Source Fabricated Using Focussed Ion Beam Etching, MicroTAS 2005: 9th Intl. Conf. on Miniaturized Systems for Chemistry and Life Sciences, Boston, MA, USA, Transducer Research Foundation.

200 nm and height of 100 nm implying the liquid volume in the nanofluidic part is less than 1pl. The effective electrospray is around 0.02 $\mu m^2$.

## 5.5. Nanowire and Nanotube

The sensing properties of nanowires and nanotubes have been widely studied recently due to their nanoscale dimensions and huge surface-to-volume ratio. Metallic nanowires have been used to measure thiol and amine molecules (gold nanowire) and hydrogen (palladium nanowire) (Li et al. 2000; Favier et al. 2001). Nanowires can be configured as field-effect transistors, which change conductance upon binding of molecules to receptors linked to the nanowire surfaces. It has been shown that the conductance of metallic nanowires is sensitive to molecular adsorption onto the nanowires. Field-effect transistors (FETs) based on individual $In_2O_3$ nanowires were used to investigate the chemical gating effect of organic molecules and biomolecules with amino or nitro groups (Li et al. 2003). The electron-donating capability of amine groups and electron-withdrawing capability of nitro groups showed significant shifts in the gate threshold voltages. Single-crystalline $In_2O_3$ nanowire with a diameter around 10 nm were used to bridge the source/drain electrodes with a channel length of 3 $\mu m$. This device was also used to sense low-density lipoprotein (LDL) cholesterol. The advantage of such chemically sensitive FETs is that molecular binding can be monitored by a direct change in conductance or related electrical properties. Using this device, sensitive, label-free, real-time detection of a wide range of chemical and biological species is possible. The small size of nanowire-based device also allows for dense arrays of such sensors.

Boron-doped silicon nanowires (SiNWs) were also used in a FET configuration to create highly sensitive, real-time electrically based sensors for biological and chemical samples (Cui et al. 2001). Amine and oxide-functionalized SiNWs showed pH-dependent conductance that was linear over a large dynamic range. The dopant type and concentration of semiconductor nanowires can be controlled, enabling the sensitivity to be tuned. Biotin-modified SiNWs were used to detect streptavidin down to at least a picomolar concentration range. Antigen-functionalized SiNWs showed reversible antibody binding and concentration-dependent detection and the device was also used to detect the reversible binding of the metabolic indicator $Ca^{2+}$. The same group recently showed direct, real-time electrical detection of single virus particles using a similar device (Patolsky et al. 2004). The nanowire arrays were modified with antibodies for influenza A and showed conductance changes in the presence of influenza A but not paramyxovirus or adenovirus. The group also showed that multiple viruses can be selectively detected in parallel. Figure 31 shows the concept of the device. This device has been further used to detect cancer markers at femtomolar concentration with high selectivity and simultaneous incorporation of control nanowire to discriminate against false positives (Zheng et al. 2005).

Carbon nanotubes have also been used for biosensors in a FET configuration as the conducting channel (Star et al. 2003). Single-walled carbon nanotubes (SWNTs) are molecular scale wires where the electrical properties are sensitive to surface charge transfer and changes in the surrounding environment (Shim et al. 2002). Simple adsorption of certain molecules can change the electrical properties of SWNTs dramatically. Biotin-streptavidin binding has been detected by changes in the device characteristics. Other groups have reported the use of single-stranded DNA (ss-DNA) as the chemical recognition site and single-walled carbon nanotube field effect transistors as the electronic readout component (Staii and Johnson 2005). Different gases showed differences in signs and magnitudes that can be tuned by choosing the base sequence of the ss-DNA. Arrays of carbon nanotubes placed on a silicon chip have been also used as a DNA chip (Li et al. 2003). DNA molecules attached to each carbon nanotube binds to the target DNA, resulting in conductance increase.

Most of the nanowires and nanotubes aforementioned measure the electrical signal of the nanostructure to detect molecules. Since many sensing materials are not conductive, mechanical properties can be also used to detect molecules. Microfabricated cantilevers have been recently used to detect bending or changes in the oscillation of the cantilever due to the binding of analyte molecules onto the cantilever surface (Wu et al. 2001). Polymer nanowires have been used to bridge two prongs of a microfabricated tuning fork to detect the change in mechanical properties of the nanowires (Boussaad and Tao 2003). A combination of mechanical stretching and focused ion beam (FIB) has been used to fabricate a 100 nm thick nanowire. When the nanowire is exposed to organic vapors, the stress in the

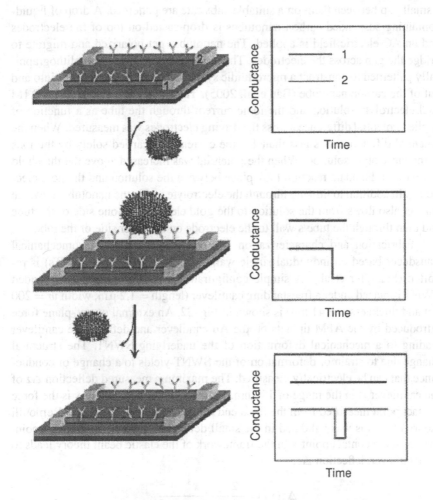

**FIGURE 31.** Nanowire based detection of single viruses. (Left) schematic shows two nanowire devise, 1 and 2, where the nanowires are modified with different antibody receptors. Specific binding of a single virus to the receptors on nanowire 2 produces a conductance change (Right) characteristic of the surface charge of the virus only in nanowire 2. When the virus unbinds from the surface the conductance returns to the baseline value. Patolsky, Zheng et al., Electrical Detection of Single Viruses, Proceedings of the National Academy of Sciences of the United States of America 101(39): 14017-14022. Copyright (2004) National Academy of Sciences, U.S.A.

polymer nanowire reduces, which can be detected from the changes in the resonance of the tuning fork. Other simple polymer nanowire fabrication techniques such as electrospinning can also be used (Kameoka and Craighead 2003).

A method of fabricating a novel device consisting of a carbon nanotube connecting two wells and filling up the nanotube with various liquids or suspensions, has been developed (Kim and Bau 2005; Kim et al. 2004). First, electrodes with a small gap between them on a suitable substrate are patterned. A drop of liquid-containing suspended carbon nanotubes is drop-casted on top of the electrodes and an AC electric field is applied. The nanotubes get polarized and migrate to bridge the gap across the electrodes. Then, SU-8 photoresist is photolithographically patterned to construct a microfluidic system to facilitate liquid flow into and out of the carbon nanotube (Bau et al. 2005). The nanotube was filled with 0.1M KCl electrolyte solution and the ionic current through the tube as a function of applied potential difference across the driving electrodes was measured. When the potential difference was less than 1V, the current was carried solely by the ions in the electrolyte solution. When the potential was increased above the threshold of ($\sim$1V), a Faradaic reaction took place between the solution and the gold electrodes. In addition to flowing through the electrolyte inside the nanotube, now, the current also flows from the solution to the gold electrode on one side of the tube and then through the tube's wall to the electrode on the other side of the tube.

Fabrication and characterization of a nanometer-scale electromechanical transducer based on individual single-walled carbon nanotubes (SWNTs) is reported (Stampfer et al.). A simple configuration where a contacted suspended SWNT is placed under a freestanding cantilever (length = 1.2 $\mu$m, width $w = 200$ nm and thickness $t = 41$ nm) is shown in Fig. 32. An external out-of-plane force introduced by the AFM tip acts on the Au cantilever and deflects the cantilever leading to a mechanical deformation of the underlying SWNT. The structural change due to strain or deformation of the SWNT yields to a change of conductance that can be electrically measured. The maximum measured deflection $\Delta z$ of the cantilever is in the range of 150 nm. Below $x = 800$ nm, where $x$ is the force contact point measured from the fixed end of the cantilever, The Euler- Bernoulli theory of beams was validated in the small deflection regime. Assuming a point load $F$ at the contact point $x$ in the framework of the elastic beam theory leads to the cantilever deflection $\Delta z$,

$$\Delta z(x) = \frac{12}{3} \frac{x^3}{F \cdot E \cdot w \cdot t^3}$$

where $E$ is the Young's modulus, w the width, $t$ the thickness of the cantilever and $x$ is the contact point.

Figure 33 shows the electrical response of a SWNT as a function of a mechanical deformation due to deflecting the Au cantilever (by AFM tip) at a constant contact point $x$ (see inset in Fig. 33). The resistance increases from $R(0) \approx$ 5 $M\Omega$, by a factor of approximately 3 up to 15 $M\Omega$ when deflecting the cantilever leading to a deflection of the SWNT by 40 nm. It is important to note that the resistance change is reversible, making the proposed system suitable for sensing purposes.

Passive CNT based gas sensors have been reported (Keat Ghee et al. 2002). Remote query detection of carbon dioxide, oxygen and ammonia is possible using multiwalled carbon nanotubes (MWNT) based upon the change in MNWT permittivity and conductivity with gas exposure. The transduction platform is a planar inductor- capacitor (LC) resonant circuit. The general sensor structure is

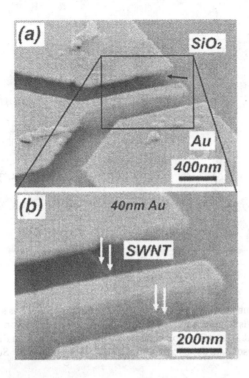

**FIGURE 32.** SEM image of the released SWNT based structure. (a) image under an angle where the under etching of the freestanding structures is highlighted (balck arrow). (b) close up of the same device: the suspended SWNT is indicated by white arrows. Stampfer, Jungen et al., Nano Electromechanical Transducer Based on Single Walled Carbon Nanotubes, Transducers'05, Seoul, Korea, IEEE Electron Devices Society, (© [2005] IEEE).

shown in Fig. 34(a). A planar inductor-interdigital capacitor pair is photolithographically defined upon a copper-clad PCB. The printed LC resonant circuit is then coated with an electrically insulating layer of $SiO_2$, followed by a second layer of gas responsive MWNT- $SiO_2$ mixture with the $SiO_2$ matrix acting to physically bind the MWNTs to the sensor. The cross-sectional view of the sensor is shown in Fig. 34(b). As the sensor is exposed to various gases, the relative permittivity of the MWNT varies, changing the effective complex permittivity of the coating and hence the resonant frequency of the sensor. The frequency spectrum of the sensor is obtained by wirelessly measuring the impedance spectrum of a sensor monitoring loop antenna. By modeling the sensor with an RLC circuit, the complex permittivity $\varepsilon_r' - j\varepsilon_r''$ of the MNWT-$SiO_2$ and $SiO_2$ layers together can be calculated. The permittivity and conductivity changes are reversible, without hysteresis between increasing and decreasing $CO_2$ concentrations. The resistance of the gas-sensing composite (MWNT and $SiO_2$) decreases as the concentration of $CO_2$ increases.

Another effort is found in a remote sensing system using CNT. The sensor detects the presence of gases such as $NH_3$, CO, Air, $N_2$, $O_2$, etc. A microwave signal

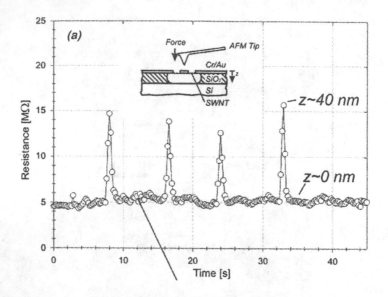

**FIGURE 33.** Electromechanical measurements: Resistance of SWNT as a function of time; each peak corresponds to continuous (~2s for up and down) deflection of the cantilever. Stampfer, Jungen et al., Nano Electromechanical Transducer Based on Single Walled Carbon Nanotubes, Transducers'05, Seoul, Korea, IEEE Electron Devices Society, (© [2005] IEEE).

interacts with the sensor resonator and produces a strong signal at its resonant frequency. This resonant frequency will be shifted away from the frequency of the transmitted signal in the presence of a test gas. The received signal at the RF receiver is compared with the transmitted microwave signal to resolve the shift in frequency as an indication of the presence of gas (Chopra and Pham). The prototype

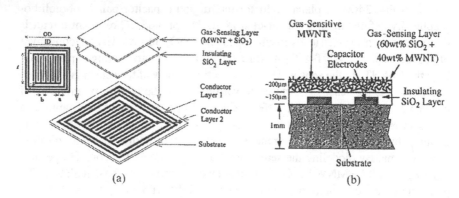

**FIGURE 34.** (a) Schematic of MWNT gas sensor. (b) Cross sectional view of interdigital capacitor. Keat Ghee, Kefeng et al., A wireless, passive carbon nanotube-based gas sensor, Sensors Journal, IEEE 2(2): 82 (© [2002] IEEE)

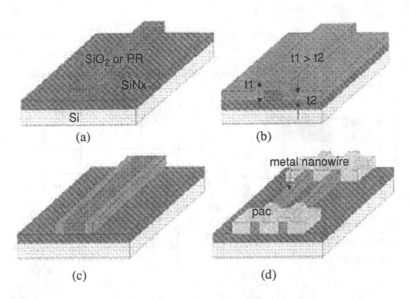

**FIGURE 35.** Fabrication Procedure of the Pt Nanowire. (a) PECVD oxide patterning on SiNx coated Si wafer. (b) Ti/Pt sputtering on the whole surface. (c) Ion-milling process to etch away the Ti/Pt layer. (d) Au lift-off process for electrical contact pad. Lee, Wang et al., 2005, Functionalized Pt Nanowire Array By Immobilizing Glucose Oxidase (Gox) In Polypyrrole (Ppy), MicroTAS 2005: 9th Intl. Conf. on Miniaturized Systems for Chemistry and Life Sciences, Boston, MA, USA, Transducer Research Foundation.

of the resonator can be developed using a milling machine that selectively etches copper conductors on a Duroid board. SWNTs or MWNTs are physically coated (thickness about 5–10 and 70–100 mm for SWNTs and MWNTs, respectively) on top of the copper disks using conductive epoxy. When the sensor is exposed to 1500 ppm of helium, the resonant frequency is shifted downward by about 0.8 MHz. The resonant frequency is recovered upon re-degassing the sensor (Grujicic et al.).

Platinum nanowire coated with polypyrrole (PPy) was used to immobilize glucose oxidase (GOx) for glucose sensing (Lee et al.). Figure 35 shows the fabrication steps of the nanowire. First, the $SiN_x$ wafer is coated with $SiO_2$ by PECVD and the oxide layer is patterned by RIE Fig. 35(a). The step height is about 100 nm. Next, Ti/Pt (2.5 nm/100 nm) is sputtered on the whole surface Fig. 35(b). Then, $SiN_x$ film is exposed by ion milling Fig. 35(c). The nanowire can be seen at the edge of the oxide step. Finally, a Au lift-off process was carried out for the electrical contacts Fig. 35(d). The nanowire is 70 nm wide, 100 nm high and 500 μm long. Then, glucose oxidase (GOx) was precisely immobilized on the Pt nanowire array by co-deposition with electro-polymerized polypyrrole (PPy). The current response to glucose of the PPy/GOx-coated nanowire at 0.7 V vs. Ag/AgCl reference electrode was measured. The current sensitivity is about 0.3 nA/mM. This verifies the GOx immobilization.

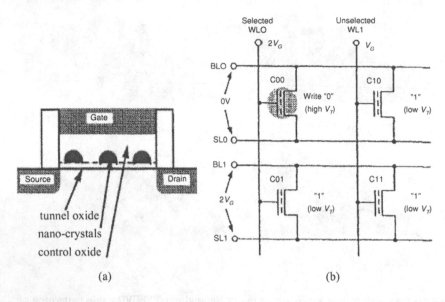

**FIGURE 36.** Single transistor nano-crystal memory. (a) Cross sectional schematic nano-crystal memory cell. (b) Memory Cell configuration for selective direct tunnel writing of "0" (high VT) or "1" (low VT). Hanafi, Tiwari et al., Fast and long retention-time nano-crystal memory, Electron Devices, IEEE Transactions on 43(9): 1553-1558 (© [1996] IEEE).

## 5.6. Nanocrystal and Nanocrescent

Nano-crystal memories achieve improved programming characteristics as a non-volatile memory, as well as simplicity of the single poly-Si gate process. The device structure is shown in Fig. 36(a) (Hanafi et al.). This memory cell with low read/write time and high retention time consists of a single field effect transistor with germanium or silicon nano-crystals, 2 to 5 nm in size, embedded in the gate oxide in close proximity to the channel. The nano-crystals are separated from each other by greater than 5 nm and from the channel by less than 5 nm. The charge in the MOSFET channel is directly tunneled and stored in the nano-crystal causing the causing a shift in the threshold voltage which can be detected by measuring the change in current. The $V_T$ window is scarcely degraded after greater than $10^9$ write/erase cycles or greater than $10^5$s retention time. The memory cell configuration is shown in Fig. 36(b). Injection of electrons occurs from the inversion layer via direct tunneling when the control gate is forward biased with respect to the source and drain. The resulting stored charge screens the gate charge and reduces the conduction in the inversion layer, i.e., it effectively shifts the threshold voltage of the device to be more positive.

The nano-crescents have been used as independent surface enhanced Raman scattering (SERS) substrates to enhance the local electromagnetic field (Lu et al.) (Love et al. 2002) as shown in Fig. 37(a). In the cross-sectional view of the nano-crescent moon shown in Fig. 37(b), we can see the nano ring and the nano tip feature

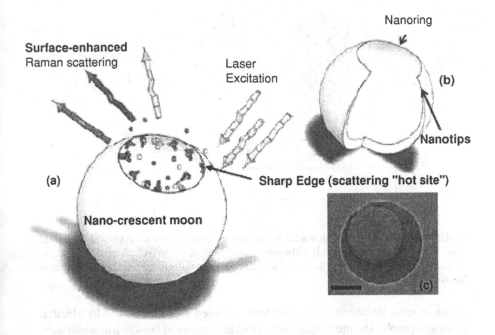

**FIGURE 37.** Gold nano-crescent moons with sharp edges. (a) Conceptual schematic of the nano-crescent moon SERS substrate. (b) Geometrical schematic of the nano-crescent moon. (c) TEM image of the nano-crescent. The scale bars are 100nm. Lu, Liu et al., 2005, Nanophotonic Crescent Structures with Sharp Edge for Ultrasensitive Biomolecule Detections by Local Electromagnetic Field Enhancement Effect, MicroTAS 2005: 9th Intl. Conf. on Miniaturized Systems for Chemistry and Life Sciences, Boston, MA, USA, Transducer Research Foundation.

of the crescent. These features enhance the scattering field and expand the SERS "hot site." A dilute solution of polystyrene colloids was drop-casted on a glass substrate to form a monolayer of sacrificial nanospheres and allowed to dry overnight. A thin gold film was then deposited by e-beam evaporation. The thickness at the bottom of the bowl was found to be around 100 nm. The gold colloids were lifted off from the glass substrate with acetone. Finally the gold-coated nanospheres were collected by centrifugation and suspended in toluene to dissolve the polystyrene. Composite multilayer nano-crescents with depositions of Au/Fe/Ag/Au with different thicknesses were performed on 50 nm sacrificial polystyrene nanospheres. The high local field enhancement factor, which is orientation dependent, can be controlled by moving or orienting a ferromagnetic nano-crescent SERS probe by an external magnetic field. The maximum local field enhancement occurs when the nano-crescent is oriented perpendicularly to the direction of the excitation laser.

## 5.7. Tools for Nanoscale Manipulation

Cell adhesion surfaces with nano structures were fabricated on microchips for use in single cell analysis, bioassay systems and bioreactor systems (Goto et al. 2005).

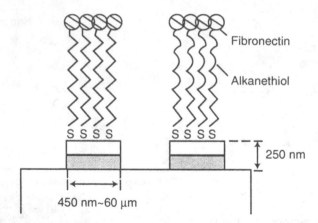

**FIGURE 38.** Side view of Substrate for cell adhesion. Goto, Sato et al., 2005, Fabrication of Nano-Patterned Surfaces for Cell Adhesion in Microchips, MicroTAS 2005: 9th Intl. Conf. on Miniaturized Systems for Chemistry and Life Sciences, Boston, MA, USA, Transducer Research Foundation.

Micro or nano structures were fabricated on fused silica substrates by electron beam lithography and metal sputtering. Dots and stripes of 60-450 μm width were fabricated by this method. Then the metal surfaces were coated with SAM of alkanethiols and fibronectin was adsorbed on the surfaces as shown in Fig. 38.

Mouse 3T3 fibroblasts were cultured on the metal patterns and in the microchannel space. Cells in the microchannels did not grow well and showed weak activity after long term static cultivation because of lack of oxygen and nutrients, accumulation of cell wastes and change of pH. It was necessary to supply fresh medium to cells by a continuous medium flow (Goto et al. 2005). Cells on stripe patterns extended along the patterns as pattern size became small; however, when the size was smaller than 1 μm, some cells seemed to have long processes. On the other hand, cells on dot patterns did not extend along the patterns. However when the size was much smaller, cells also seemed to have long processes. The cells seemed to recognize the shapes of the micro pattern but when the pattern sizes become nano scale, the effect of pattern shapes become smaller and differences between two pattern shapes becomes unclear. Nano scale (10-100 nm) adhesion surfaces are very important because these sizes are almost the same as that of focal adhesion of the cell.

Electric fields applied to a planar nanofluidic channel of a known size have been used to achieve electrokinetic trapping based pre-concentration of biomolecules (Wang et al.). The *nanofilter* shown in Fig. 39 is fabricated by sacrificial layer etching technique. Devices are made on two different substrates like silicon nitride (for nanochannels)/PDMS (for microchannels) or Si/SiO$_2$(for nanochannels)/glass (for microchannels). The nanofilters work as ion selective membranes that can generate concentration polarization or space charge extension, depending on the applied field strength. Once the extended space charge

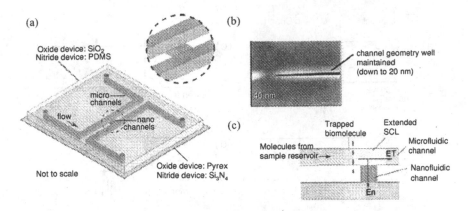

**FIGURE 39.** (a) Schematic view of the nanofluidic pre-concentrator. Both oxide and nitride devices have 40 nm deep nanochannels while the micro channels are 1.5μm deep for the oxide, 45 μm for wet-etched device and 50 μm deep for the nitride device. (b) SEM image of the nanochannels. (c) The trapping mechanism. Wang, Tsau et al. 2005, Effiecient Biomolecule Pre-Concentration by Nanofilter-Triggered Electrokinetic Trapping, Micro-TAS 2005: 9th Intl. Conf. on Miniaturized Systems for Chemistry and Life Sciences, Boston, MA, USA, Transducer Research Foundation.

layer is formed, it will work as a barrier to charged molecules, as well as a source to generate electroosmosis flow of the second kind (Wang et al. 2005). The mechanism is shown in detail in Fig. 40. The device can achieve more than a million-fold preconcentration within 30 minutes.

A novel sliding micro device with *nanoliter liquid dispensing* capability has been fabricated (Kuwata et al. 2005). The schematic of the device is shown in Fig. 41. It consists of two moving parts, B and D, while A, C and E are fixed. Surface treatment of the slide is very essential for pressure tightness. Each of the contact surfaces is treated with fluorocarbon to prevent leakage from the gaps of the slide structure. Fig. 42 shows the sequence of the dispensing procedure. The sample is introduced in the upper channel (Fig. 42I). The next part, D, is slid down to block all the flows except that in the uppermost channel (Fig. 42II). Then part B is slid down step-by-step to cut off 10 nL of the sample and store it in channels located in this part (Fig. 42III and 42IV). Finally, part D is slid up to align all the channels dispensing all the samples stored in part B into an analysis section located in the right side of the valve (Fig. 42V).

Surface acoustic waves (SAW) have been used for *nanodroplet actuation* within specifically functionalized biological liquid (Renaudin et al. 2005). The SAW device consists of interdigitated transducers (IDT) laid on a LiNbO$_3$ piezo-electric substrate. The IDT is coated with 150 angstroms SiO$_2$ followed by OTS hydrophobic surface treatment. SAW actuation produces nanoscopic scale waves on the surface of the device which have been utilized for nanodroplet manipulation. Nanodroplet displacement can be achieved by the surface acoustic waves. Fusion, mixing and splitting can be realized by using RF pulse modulation. The splitting

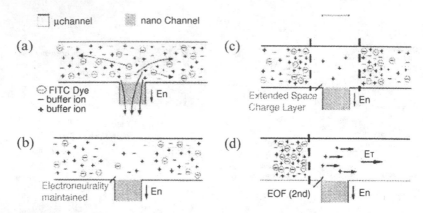

**FIGURE 40.** Mechanism of pre-concentration: (a) Ion-selective property of the nano channel Under small En; (b) Concentration polarization under diffusion-limited condition; (c) At Higher field, electroneutrality is no longer maintained, generating an extended space charge layer, (d) With proper ET and En, the trapping region and depletion region will be formed as indicated; therefore, samples will be collected in front of the virtual barrier driven by nonlinear electrokinetic flow. Reprinted with permission from (Wang, Stevens et al., Million-fold Preconcentration of Proteins and Peptides by Nanofluidic Filter, Anal. Chem. 77(14) 4293-4299.) Copyright (2005) American Chemical Society.

of a DI water drop is shown in Fig. 43. Higher power can be used to eject water from the surface as shown in Fig. 43 (Renaudin et al. 2006).

# 6. CONCLUSION

Micromachining techniques, by which MEMS devices are constructed, provide a top-down manufacturing route for nanofabrication. MEMS research combines

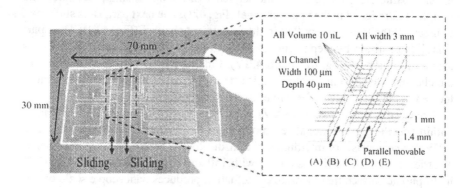

**FIGURE 41.** Structure of dispensing device. Kuwata, Sakamoto et al., 2005, Sliding Quantitative Nanoliter Dispensing Device for Multiple Anaysis, MicroTAS 2005: 9th Intl. Conf. on Miniaturized Systems for Chemistry and Life Sciences, Boston, MA, USA, Transducer Research Foundation.

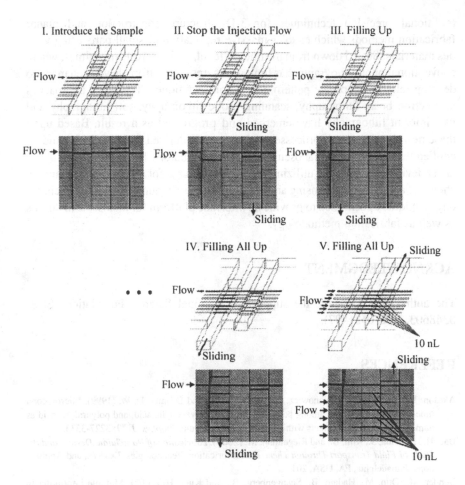

**FIGURE 42.** Dispensing procedure. Kuwata, Sakamoto et al., 2005, Sliding Quantitative Nanoliter Dispensing Device for Multiple Anaysis, MicroTAS 2005: 9th Intl. Conf. on Miniaturized Systems for Chemistry and Life Sciences, Boston, MA, USA, Transducer Research Foundation.

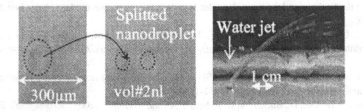

**FIGURE 43.** (Left) SAW RF pulse (power level 36 dBm) produces splitting of the DI water nanodrop. (Right) SAW RF pulse (power level 38 dBm) propagating from left to right hits the droplet and creates a water jet. Renaudin, Chuda et al., 2005, SAW Lab-on-Chip in View of Protein Affinity Purification Implemented From Nanodroplet Transport, MicroTAS 2005: 9th Intl. Conf. on Miniaturized Systems for Chemistry and Life Sciences, Boston, MA, USA, Transducer Research Foundation.

traditional precision techniques for 3-D structure construction and planar fabrication methods which enables massive replication of small parts using various materials. Scaled down from the micron level, in the 1-100 nm regimes, where conventional photolithography has limitations, research efforts have been made to develop reliable nanoscale pattern generation techniques. Top-down approaches of electron beam lithography, scanning probe lithography, soft lithography and nanoimprint lithography have emerged and progressed as a result. Based upon these new developments, processes and structures derived from MEMS could be applied to produce novel nanosensors and nanoactuators. From the review of the latest devices and systems utilizing nanoscale features for electrical, mechanical, chemical and biological sensing and manipulation, it is found that MEMS technology is uniquely positioned to provide an interface platform for nanoscale materials as well as fabrication methodology.

## ACKNOWLEDGMENT

The authors acknowledge the support of National Science Foundation (ECS-0348603, ECS-0521497).

## REFERENCES

Akeson, M., Branton, D., Kasianowicz, J. J., Brandin, E., and Deamer, D. W. (1999). Microsecond time-scale discrimination among polycytidylic acid, polyadenylic acid, and polyuridylic acid as homopolymers or as segments within single RNA molecule. *Biophys. J.* **77**: 3227-3233.

Bau, H. H., Sinha, S., Kim, B., and Riegelman, M. (2005). *Fabrication of Nanofluidic Devices and the Study of Fluid Transport Through Them.* Nanofabrication: Technologies, Devices, and Applications, Philadelphia, PA, USA, 201.

Bender, M., Otto, M., Hadam, B., Spangenberg, B., and Kurz, H. (2002). Multiple imprinting in UV-based nanoimprint lithography: relates material issues. *Microelectronic Engineering* **61-62**: 407-413.

Brinkman, W. F., Haggan, D. E., and Troutman, W. W. (1997). A history of the invention of the transistor and where it will lead us. *IEEE J. Solid-State Circuits* **32**(12): 1858-1865.

Brinkmann, M., Blossey, R., Arscott, S., Druon, C., Tabourier, P., Gac, S. L., and Rolando, C. (2004). Microfluidic design rules for capillary slot-based electrospray sources. *Appl. Phys. Lett.* **85**(11): 2140.

Bullen, D., Chung, S.-W., Wang, X., Zou, J., Mirkina, C. A., and Liu, C. (2004). Parallel dip-pen nanolithography with arrays of individually addressable cantilevers. *Appl. Phys. Lett.* **84**(5): 789-791.

Bustillo, J. M., Howe, R. T., and Muller, R. S. (1998). Surface micromachining for microelectromechanical systems. *Proc. IEEE* **86**(8): 1552-1574.

Descatoire, C., Troadec, D., Buchaillot, L., Ashcroft, A., and Arscott, S. (2005). *A Nanofluidic Electrospray Source Fabricated Using Focussed Ion Beam Etching.* Micro Total Analysis Systems 2005: 9th International Conference on Miniaturized Systems for Chemistry and Life Sciences, Boston, Massachusetts, USA, 241-243.

Cao, H., Yu, Z., Wang, J., Tegenfeldt, J. O., Austin, R. H., Chen, E., Wu, W., and Chou, S. Y. (2002). Fabrication of 10 nm enclosed nanofluidic channels. *Appl. Phys. Lett.* **81**(1): 174.

Chan, R. H. M., Fung, C. K. M., and Li, W. J. (2004). Rapid assembly of carbon nanotubes for nanosensing by dielectrophoretic force. *Nanotechnology* **15**: S672–S677.

Chang, H., Kosari, F., Andreadakis, G., Alam, M. A., Vasmatzis, G., and Bashir, R. (2004). DNA-mediated fluctuations in ionic current through silicon oxide nanopore channels. *Nano Lett.* **4**(8): 1551-1556.

Chang, T. H. P., Mankos, M., Lee, K. Y., and Muray, L. P. (2001). Multiple electron-beam lithography. *Microelectronic Engineering* **57–58**: 117–135.

Chen, P., Gu, J., Brandin, E., Kim, Y.-R., Wang, Q., and Branton, D. (2004). Probing single DNA molecule transport using fabricated nanopores. *Nano Lett.* **4**(11): 2293-2298.

Chen, X., and Guo, L.J. (2004). A combined-nanoimprint-and-photolithography patterning technique. *Microelectronic Engineering* **71**(3-4): 288-293.

Cheng, L.-J., Chang, S.-T., and Guo, L. J. (2005). *Nanoimprint of Nanofluidic Channels by Using Hydrophilic Hydrogen Silsesquioxane (HSQ)*. Micro Total Analysis Systems 2005: 9th International Conference on Miniaturized Systems for Chemistry and Life Sciences, Boston, Massachusetts, USA, 518-520.

Chiu, G. L.-T., and Shaw, J. M. (1997). Optical lithography: Introduction. *IBM J. Res. Develop.* **41**(1/2): 3-6.

Chopra, S., and Pham, A. (2002). Carbon-nanotube based resonant -circuit sensor for ammonia. *Appl. Phys. Lett.* **80**(24): 4632-4634.

Chou, S. Y. (1996). Nanoimpirnt lithography. *J. Vac. Sci. Technol.* B**14**(6): 4129-4133.

Chou, S. Y., Keimel, C., and Gu, J. (2002). Ultrafast and direct imprint of nanostructures in silicon. *Nature* **417**: 835-837.

Chou, S. Y., and Krauss, P. R. (1997). Imprint lithography with sub-10nm feature size and high throughput. *Microelectronic Engineering* **35**: 237-240.

Chou, S. Y., Krauss, P. R., and Renstrom, P. J. (1995). Imprint of sub-25 nm vias and trenches in polymers. *Appl. Phys. Lett.* **67**(21): 3114-3116.

Christopher, P. D., Chan, K. K., and Blum, J. (1992). Deep trench plasma etching of single crystal silicon using $SF_6/O_2$ gas mixture. *J. Vac. Sci. Technol.* B **10**: 1105-1110.

Chung, J., and Lee, J. (2003). Nanoscale gap fabrication and integration of carbon nanotubes by micromachining. *Sens. Actuators A* **104**: 229–235.

Chung, J., Lee, K.-H., and Lee, J. (2003). Nanoscale gap fabrication by carbon nanotube-extracted lithography (CEL). *Nanoletters* **3**(8): 1029-1031.

Cui, Y., Wei, Q., Park, H., and Lieber, C.M. (2001). Nanowire nanosensors for highly sensitive and selective detection of biological and chemical species. *Science* **293**: 1289-1292.

Davis, R. C., Williams, C. C., and Neuzil, P. (1995). Micromachined submicrometer photodiode for scanning probe microscopy. *Appl. Phys. Lett.* **66**(18): 2309-2311.

Davis, Z. J. and Boisen, A. (2005). Aluminum nanocantilevers for high sensitivity mass sensors. *Appl. Phys. Lett.* **87**(1): 013102-1 - 13012-3.

Dhaliwal, R. S., Enichen, W. A., Golladay, S. D., Gordon, M. S., Kendall, R. A., Lieberman, J. E., Pfeiffer, H. C., Pinckney, D. J., Robinson, C. F., Rockrohr, J. D., Stickel, W., and Tressler, E. V. (2001). PREVAIL—electron projection technology approach for next-generation lithography. *IBM J. Res. Develop.* **45**(5): 615-638.

Dong-Weon, L., Ono, T., Abe, T., and Esashi, M. (2002). Microprobe array with electrical interconnection for thermal imaging and data storage. *Microelectromechanical Systems, Journal of* **11**(3): 215.

Emmelkamp, J., Gardeniers, J. G. E., Anderson, H., and Berg, A. v. d. *Planar Nanoneedles On-Chip for Intracelluler Measurements*. Micro Total Analysis Systems 2005: 9th International Conference on Miniaturized Systems for Chemistry and Life Sciences, Boston, Massachusetts, USA, 400-402.

Fasching, R. J., Tao, Y., and Prinz, F. B. (2005). Cantilever tip probe arrays for simultaneous SECM and AFM analysis. *Sens. Actuators B* **108**: 964-972.

Favier, F., Walter, E. C., Zach, M. P., Benter, T., and Penner, R.M. (2001). Hydrogen sensors and switches from electrodeposited palladium mesowire arrays. *Science* **293**: 2227-2231.

Feynman, R. There's Plenty of Room at the Bottom. http://www.zyvex.com/nanotech/feynman.html.

Fologea, D., Gershow, M., Ledden, B., McNabb, D. S., Golovchenko, J. A., and Li, J. (2005a). Detecting single stranded DNA with a solid state nanopore. *Nano Lett.* **5**(10): 1905-1909.

Fologea, D., Uplinger, J., Thomas, B., McNabb, D. S., and Li, J. (2005b). Slowing DNA translocation in a solid-state nanopore. *Nano Lett.* **5**(9): 1734-1737.

French, P. J. and Sarro, P. M. (1998). Surface versus bulk micromachining: the contest for suitable applications. *J. Micromech. Microeng.* **8**: 45-53.

Fujita, H. (2003). *Micromachines as Tools for Nanotechnology*, Springer.

Gates, B. D., Xu, Q., Love, J. C., Wolfe, D. B., and Whitesides, G. M. (2004). Unconventional nanofabrication. *Annu. Rev. Mater. Res.* **34**: 339–372.

Gel, M., Edura, T., Wada, Y., and Fujita, H. *Controllable Nano-Gap Mechanism For Characterization Of Nanoscale Objects.* Micro Total Analysis Systems 2005: 9th International Conference on Miniaturized Systems for Chemistry and Life Sciences, Boston, Massachusetts, USA, 739-741.

Goto, M., Sato, K., Murakami, A., Tokeshi, M., and Kitamori, T. (2005a). Development of a microchip-based bioassay system using cultured cells. *Anal. Chem.* **77**(7): 2125-2131.

Goto, M., Sato, K., Yamato, M., Hibara, A., and Kitamori, T. (2005b). *Fabrication of Nano-Patterned Surfaces for Cell Adhesion in Microchips.* Micro Total Analysis Systems 2005: 9th International Conference on Miniaturized Systems for Chemistry and Life Sciences, Boston, Massachusetts, USA, 1282-1282.

Grigaliunas, V., Tamulevicius, S., Tomasiunas, R., Kopustinkas, V., Guobiene, A., and Jucius, D. (2004). Laser pulse assisted nanoimprint lithography. *Thin Solid Films* **453-454**: 13-15.

Grujicic, M., Cao, G., and Roy, W. N. (2004). A computational analysis of the carbon-nanotube-based resonant-circuit sensors. *Appl. Surf. Sci.* **229**(1-4): 316.

Guo, L. J. (2004). Recent progress in nanoimprint technology and its applications. *J. Phys. D: Appl. Phys.* **37**: R123-141.

Guo, L. J., Cheng, X., and Chou, C.-F. (2004). Fabrication of size-controllable nanofluidic channels by nanoimprinting and its application for DNA stretching. *Nano Lett.* **4**(1): 69-73.

Hanafi, H. I., Tiwari, S., and Khan, I. (1996). Fast and long retention-time nano-crystal memory. *Electron Devices, IEEE Transactions on* **43**(9): 1553-1558.

Heins, E. A., Siwy, Z. S., Baker, L. A., and Martin, C. R. (2005). Detecting single porphyrin molecules in a conically shaped synthetic nanopore. *Nano Lett.* **5**(9): 1824-1829.

Hertel, T., Martel, R., and Avouris, P. (1998). Manipulation of individual carbon nanotubes and their interaction with surfaces. *J. Phys. Chem. B* **102**: 910-915.

Heyderman, L. J., Schift, H., David, C., Gobrecht, J., and Schweizer, T. (2000). Flow behaviour of thin polymer films used for hot embossing lithography. *Microelectronic Engineering* **54**: 229-245.

Hirai, Y., Fujiwara, M., Okuno, T., and Tanaka, Y. (2001). Study of resist deformation in nanoimprint lithography. *J. Vac. Sci. Technol. B* **19**(6): 2811-2815.

Hoummady, M., Farnault, E., Fujita, H., Kawakatsu, H., and Masuzawa, T. (1997). *New Technique for Nanocantilever Fabrication Based on Local Electrochemical Etching: Applications to Scanning Force Microscopy.* The fourth international conference on nanometer-scale science and technology, Beijing (China), 1556-1558.

Howorka, S., Movileanu, L., Braha, O., and Bayley, H. (2001). Kinetics of duplex formation for individual DNA strands within a single protein nanopore. *Proc. Natl. Acad. Sci. U. S. A.* **98**(23): 12996-13001. From http://www.technics.com.

Kameoka, J., and Craighead, H. G. (2003). Fabrication of oriented polymeric nanofibers on planar surfaces by electrospinning. *Appl. Phys. Lett.* **83**(2): 371-373.

Kasianowicz, J. J., Brandin, E., Branton, D., and Deamer, D. W. (1996). Characterization of individual polynucleotide molecules using a membrane channel. *Proc. Natl. Acad. Sci. U. S. A.* **93**: 13770-13773.

Keat Ghee, O., Kefeng, Z., and Grimes, C. A. (2002). A wireless, passive carbon nanotube-based gas sensor. *Sensors Journal, IEEE* **2**(2): 82.

Kim, B. M., and Bau, H. H. (2005). *Hybrid Fabrication of Carbon Nanotube – Based Devices and the Measurement of Ionic Current Through Them.* Micro Total Analysis Systems 2005:

9th International Conference on Miniaturized Systems for Chemistry and Life Sciences, Boston, Massachusetts, USA, 1543-545.

Kim, B. M., Sinha, S., and Bau, H. H. (2004). Optical microscope study of liquid transport in carbon nanotubes. *Nano Lett.* **4**(11): 2203-2208.

Kovacs, G., Maluf, N. I., and Peterson, K. E. (1998). Bulk micromachining of silicon. *Proc. IEEE*, **86**(8): 1536-1551.

Kumar, A., and Whitesides, G. M. (1993). Features of gold having micrometer to centimeter dimensions can be formed through a combination of stamping with an elastomeric stamp and an alkanethiol "ink" followed by chemical etching. *Appl. Phys. Lett.* **63**(14): 2002-2004.

Kuwata, M., Sakamoto, K., Murakami, Y., Morishima, K., Sudo, H., Kitaoka, M., and Kitamore, T. (2005). *Sliding Quantitative Nanoliter Dispensing Device for Multiple Anaysis*. Micro Total Analysis Systems 2005: 9th International Conference on Miniaturized Systems for Chemistry and Life Sciences, Boston, Massachusetts, USA, 602-604.

Le Gac, S., Cren-Olive, C., Rolando, C., and Arscott, S. (2004). A novel nib-like design for microfabricated nanospray tips. *J. Am. Soc. Mass Spectrom.* **15**(3): 409.

Learmer, F., Schilp, A., Funk, K., and Offenberg, M. (1999). Bosch deep silicon etching: improving uniformity and etch rae for advanced MEMS applications. *Proc. MEMS*, Orlando, FL, 211-216.

Lee, C., Wang, J., Monbuquette, H. G., and Yun, M. (2005). *Functionalized Pt Nanowire Array by Immobilizing Glucose Oxidase (Gox) in Polypyrrole (Ppy)*. Micro Total Analysis Systems 2005: 9th International Conference on Miniaturized Systems for Chemistry and Life Sciences, Boston, Massachusetts, USA, 1237-1239.

Lee, J. A., Yun, J. Y., Lee, K.-C., Park, S. I., and Lee, S. S. (2005). *Fabrication of Microcantilever with Nano-interdigitated Elecetrodes (IDEs) for DNA Binding Protein Detection*. Micro Total Analysis Systems 2005: 9th International Conference on Miniaturized Systems for Chemistry and Life Sciences, Boston, Massachusetts, USA, 217-219.

Lee, S., Park, S., and Cho, D. (1999). The surface/bulk micromachining (SBM) process: a new method for fabricating released MEMS in single crystal silicon. *J. Microelectromech. Syst.* **8**(4): 409-416.

Li, C., Lei, B., Zhang, D., Liu, X., Han, S., Tang, T., Rouhanizadeh, M., Hsiai, T., and Zhou, C. (2003a). Chemical gating of In2O3 nanowires by organic and biomolecules. *Appl. Phys. Lett.* **83**(19), 4014-4016.

Li, C. Z., He, H. X., Bogozi, A., Bunch, J. S., and Tao, N. J. (2000). Molecular detection based on conductance quantization of nanowires. *Appl. Phys. Lett.* **76**(10): 1333-1335.

Li, H., and Huck, W. T. S. (2002). Polymers in nanotechnology. *Current Opinion in Solid State and Materials Science* **6**: 3-8.

Li, J., Gershow, M., Stein, D., Brandin, E., and Golovchenko, J. A. (2003b). DNA molecules and configurations in a solid-state nanopore microscope. *Nature Materials* **2**: 611-615.

Li, J., Ng, H. T., Cassell, A., Fan, W., Chen, H., Ye, Q., Koehne, J., Han, J., and Meyyappan, M. (2003c). Carbon nanotube nanoelectrode array for ultrasensitive DNA detection. *Nano Lett.* **3**(5): 597-602.

Li, J., Stein, D., McMullan, C., Branton, D., Aziz, M. J., and Golovchenko, J.A. (2001). Ion-beam sculpting at nanometre length scales. *Nature*, **412**: 166-169.

Liddle, J. A., Harriott, L. R., Novembre, A. E., and Waskiewicz, W. K. SCALPEL: A Projection Electron-Beam Approach to Sub-Optical Lithography.

Liu, J.-F., Zhang, L.-G., Gu, N., Ren, J.-Y., Wu, Y.-P., Lu, Z.-H., Mao, P.-S., and Chen, D.-Y. (1998). Fabrication of colloidal gold micro-patterns using photolithographed self-assembled monolayers as templates. *Thin Solid Films* **327-329**: 176-179.

Love, J. C., Gates, B. D., Wolfe, D. B., Paul, K. E., and Whitesides, G. M. (2002). Fabrication and wetting properties of metallic half-shells with submicron diameters. *Nano Lett.* **2**(8): 891-894.

Lowenheim, F. A. (1974). *Modern Electroplating*, Wiley.

Lu, Y., Liu, G. L., Kim, J., Mejia, Y. X., and Lee, L. P. *Nanophotonic Crescent Structures with Sharp Edge for Ultrasensitive Biomolecule Detections by Local Electromagnetic Field Enhancement Effect*. Micro Total Analysis Systems 2005: 9th International Conference on Miniaturized Systems for Chemistry and Life Sciences, Boston, Massachusetts, USA, 1230-1233.

Lutwyche, M. I., Despont, M., Drechsler, U., Durig, U., Haberle, W., Rothuizen, H., Stutz, R., Widmer, R., Binnig, G. K., and Vettiger, P. (2000). Highly parallel data storage system based on scanning probe array. *Appl. Phys. Lett.* **77**(20): 3299-3301.

Madou, M. J. (2002). *Fundamentals of Microfabrication*, CRC Press LLC.

Maldonado, J. R., Coyle, S. T., Shamoun, B., M. Yu, Thomas, T., Holmgren, D., Chen, X., DeVore, B., Scheinfein, M. R., and Gesley, M. (2003). A raster multibeam lithography tool for sub 100nm mask fabrication utilizing a novel photocathode. *Proc. SPIE* **5220**: 46-51.

Matsuzaka, T., and Soda, Y. (1999). Electron beam lithography system for nanometer fabrication. *Hitachi Review* **48**(6): 340-343.

McEuen, P. L., Fuhrer, M. S., and Park, H. (2002). Single-walled carbon nanotube electronics. *IEEE Trans. Nanotech.* **1**(1): 78-85.

Meller, A., Nivon, L., Brandin, E., Golovchenko, J., and Branton, D. (2000). Rapid nanopore discrimination between single polynucleotide molecules. *Proc. Natl. Acad. Sci. U. S. A.* **97**(3): 1079-1084.

Meller, A., Nivon, L., and Branton, D. (2001). Voltage-driven DNA translocations through a nanopore. *Phys. Rev. Lett.* **86**(15): 3435-3438.

Michel, B., Bernard, A., Bietsch, A., Delamarche, E., Geissler, M., Juncker, D., Kind, H., Renault, J.-P., Rothuizen, H., Schmid, H., Schmidt-Winkel, P., Stutz, R., and Wolf, H. (2001a). Printing meets lithography: Soft approaches to high-resolution. *IBM J. Res. Develop.* **45**(5): 697-719.

Michel, B., Bernard, A., Bietsch, A., Delamarche, E., Geissler, M., Juncker, D., Kind, H., Renault, J.-P., Rothuizen, H., Schmid, H., Schmidt-Winkel, P., Stutz, R., and Wolf, H. (2001b). Printing meets lithography: Soft approaches to high-resolution patterning. *IBM. J. Res. Dev.* **45**(5): 697-719.

Minne, S. C., Yaralioglu, G., Manalis, S. R., Adams, J. D., Zesch, J., Atalar, A., and Quatec, C. F. (1998). Automated parallel high-speed atomic force microscopy. **72**(18): 2340-2342.

Muller, U., Kentsch, J., Nisch, W., Neugebauer, S., Schuhmann, W., Linke, S., Kaczor, M., Lohmuller, T., Spatz, J., and Stelzle, M. (2005). *Sub-um Spaced Nano-Porous Electrode Systems: Fabrication, Properties and Application to Sensitive Electrochamical Detection.* Micro Total Analysis Systems 2005: 9th International Conference on Miniaturized Systems for Chemistry and Life Sciences, Boston, Massachusetts, USA, 473-475.

Nagahara, L. A., Amlani, I., Lewenstein, J., and Tsui, R. K. (2002). Directed placement of suspended carbon nanotubes for nanometer-scale assembly. *Appl. Phys. Lett.* **80**(20): 3826-3828.

Nakayama, Y., Okazaki, S., and Saitou, N. (1990). Electron-beam cell projection lithography: A new high-throughput electron beam direct-writing technology using a specially tailored Si aperture. *J. Vac. Sci. Technol. B* **8**(6): 1836-1840.

Nathanson, H. C., Newell, W. E., Wickstrom, R. A., and J. R. Davis, J. (1967). The resonant gate sensor. *IEEE Trans. Electron Devices* **14**: 117-133.

Park, H., Lim, A. K. L., Alivisatos, A. P., Park, J., and McEuen, P. L. (1999). Fabrication of metallic electrodes with nanometer separation by electromigration. *Appl. Phys. Lett.* **75**(2): 301.

Patolsky, F., Zheng, G., Hayden, O., Lakadamyali, M., Zhuang, X., and Lieber, C. M. (2004). Electrical detection of single viruses. *Proc. Natl. Acad. Sci. U. S. Am.* **101**(39): 14017-14022.

Peterson, K. (1982). Silicon as a mechanical material. *Proc. IEEE* **70**(5): 420-457.

Pfeiffer, H. C., Dhaliwal, R. S., Golladay, S. D., Doran, S. K., Gordon, M. S., Groves, T. R., Kendall, R. A., Lieberman, J. E., Petric, P. F., Pinckney, D. J., Quickle, R. J., Robinson, C. F., Rockrohr, J. D., Senesi, J. J., Stickel, W., Tressler, E. V., Tanimoto, A., Yamaguchi, T., Okamoto, K., Suzuki, K., Okino, T., Kawat, S., K. Morita, Suziki, S. C., Shimizu, H., Kojima, S., Varnell, G., Novak, W. T., Stumbo, D. P., and Sogard, M. (1999). Projection reduction exposure with variable axis immersion lenses: Next generation lithography. *J. Vac. Sci. Technol. B* **17**(6): 2840-2846.

Pfeiffer, H. C. and Stickel, W. (1995). PREVAIL—An E-beam stepper with variable axis immersion lenses. *Microelectronic Engineering*, **27**: 143-146.

Piner, R. D., Zhu, J., Xu, F., Hong, S., and Mirkin, C. A. (1999). Dip-pen nanolithography. *Science* **283**(29): 661-663.

Plachetka, U., Bender, M., Fuchs, A., Vratzov, B., Glinsner, T., Lindner, F., and Kurz, H. (2004). Wafer scale patterning by soft UV-nanoimprint lithography. *Microelectronic Engineering*, **73-74**: 167-171.

Quake, S. R. and Scherer, A. (2000). From micro- to nanofabrication with soft materials. *Science* **290**: 1536-1540.

Rajnikant, A., Shukla, S., Ludwig, L., Anjum, M., Cho, H. J., and Seal, S. A nanoparticle-based microsensor for room temperature hydrogen detection. *Proc. IEEE Sensors* 395-398.

Rao, S. G., Huang, L., Setyawan, W., and Hong, S. (2003). Large-scale assembly of carbon nanotubes. *Nature*: **425**: 36-37.

Renaudin, A., Chuda, K., Zhang, V., Coqueret, X., Camart, J.-C., Tabourier, P., and Druon, C. *SAW Lab-on-Chip in View of Protein Affinity Purification Implemented From Nanodroplet Transport*. Micro Total Analysis Systems 2005: 9th International Conference on Miniaturized Systems for Chemistry and Life Sciences, Boston, Massachusetts, USA, 599-601.

Renaudin, A., Tabourier, P., Zhang, V., Camart, J. C., and Druon, C. (2006). SAW nanopump for handling droplets in view of biological applications. *Sens. Actuators B: Chem.* **113**(1): 389.

Roschier, L., Penttila, J., Martin, M., Hakonen, P., Paalanen, M., Tapper, U., Kauppinen, E.I., Journet, C., and Bernier, P. (1999). Single-electron transistor made of multiwalled carbon nanotube using scanning probe manipulation. **75**(5): 728-730.

Sakitani, Y., Yoda, H., Todokoro, H., Shibata, Y., Yamazaki, T., Ohbitu, K., Saitou, N., Moriyama, S., Okazaki, S., Matuoka, G., Murai, F., and Okumura, M. (1992). Electron-beam cell-projection lithography system. *J. Vac. Sci. Technol. B* **10**(6): 2759-2763.

Sanchez, J. C. (1963). Semiconductor strain-gauge pressure sensors. *Instruments and Control Systems*: 117-120.

Scheer, H. and Schulz, H. (1998). Problems of the nanoimprinting technique for nanometer scale pattern definition. *J. Vac. Sci. Technol. B* **16**(6): 3917-3921.

Scheer, H., and Schulz, H. (2001). A contribution to the flow behaviour of thin polymer films during hot embossing lithography. *Microelectronic Engineering* **56**: 311-332.

Schellenberg, F. (2003). A little light magic. *IEEE Spectrum*: 34-39.

Shim, M., Kam, N. S. S., Chen, R. J., Li, Y., and Dai, H. (2002). Functionalization of carbon nanotubes for biocompatibility and biomolecular recognition. *Nano Lett.* **2**(4): 285-288.

Shipway, A. N., Katz, E., and Willner, I. (2000). Nanoparticle arrays on surfaces for electronic, optical, and sensor applications. *Chemphyschem* **1**: 18-52.

Shukla, S., Agrawal, R., Cho, H. J., Seal, S., Ludwig, L., and Parish, C. (2005a). Effect of ultraviolet radiation exposure on room-temperature hydrogen sensitivity of nanocrystalline doped tin oxide sensor incorporated into microelectromechanical systems device. *J. Appl. Phys.* **97**(5): 054307-054319.

Shukla, S., Zhang, P., Cho, H. J., Rahman, Z., Drake, C., Seal, S., Craciun, V., and Ludwig, L. (2005b). Hydrogen-discriminating nanocrystalline doped-tin-oxide room-temperature microsensor. *J. Appl. Phys.* **98**(10): 104306-104315.

Sinha, P. M., Valco, G., Sharma, S., Liu, X., and Ferrari, M. (2004). Nanoengineered device for drug delivery application. *Nanotechnology* **15**: S585-S589.

Smith, C. S. (1954). Piezoresistance effect in germanium and silicon. *Phys. Rev.* **94**(1): 42-49.

Sohda, Y., Ohta, H., Murai, F., Yamamoto, J., Kawanob, H., Satohb, H., and Itohb, H. (2003). Recent progress in cell-projection electron-beam lithography. **67-68**: 78–86.

Staii, C. and Johnson, J. A. T. (2005). DNA-decorated carbon nanotubes for chemical sensing. *Nano Lett.* **5**(9): 1774-1778.

Stampfer, C., Jungen, A., and Hierold, C. (2005). Nano Electromechanical Transducer Based on Single Walled Carbon Nanotubes. *Transducers'05*, Seoul, Korea, 2103-2106.

Star, A., Gabriel, J. P., Bradley, K., and Gruner, G. (2003). Electronic detection of specific protein binding using nanotube FET devices. *Nano Lett.* **3**(4): 459-463.

Storm, A. J., Chen, J. H., Ling, X. S., Zandbergen, H. W., and Dekker, C. (2003). Fabrication of solid-state nanopores with single-nanometre precision. *Nature Materials* **2**: 537-540.

Storm, A. J., Storm, C., Chen, J., Zandbergen, H., Joanny, J., and Dekker, C. (2005). Fast DNA translocation through a solid-state nanopore. *Nano Lett.* **5**(7): 1193-1197.

Suehiro, J., Zhou, G., and Hara, M. (2003). Fabrication of a carbon nanotube-based gas sensor using dielectrophoresis and its application for ammonia detection by impedance spectroscopy. *J. Phys. D: Appl. Phys.* **36**: L109–L114.

Takekawa, T., Nakagawa, K., and Hashiguchi, G. The AFM Twezeers: Integration of a Twezeers Function With an AFM Probe. *Transducers'05*, Seoul, Korea, 621-624.

Tegenfeldt, J. O., Prinz, C., Cao, H., Chou, S., Reisner, W. W., Riehn, R., Wang, Y. M., Cox, E. C., Sturm, J. C., Silberzan, P., and Austin, R. H. (2004). The dynamics of genomic-length DNA molecules in 100-nm channels. *Proc. Natl. Acad. Sci. U. S. A.* **101**(30): 10979-10983.

Torres, C. M. S., Zankovych, S., Seekamp, J., Kam, A. P., Cedeno, C. C., Hoffman, T., Ahopelto, J., Reuther, F., Pfeiffer, K., Blediessel, G., Gruetzner, G., Maximov, M.V., and Heidari, B. (2003). Nanoimprint lithography: an alternative nanofabrication approach. *Mater. Sci. Eng.* **C 23**: 23-31.

Tseng, A. A., Notargiacomo, A., and Chen, T. P. (2005). Nanofabrication by scanning probe microscope lithography: A review. *J. Vac. Sci. Technol. B* **23**(3): 877-894.

Turner, S. W. P., Cabodi, M., and Craighead, H. G. (2002). Confinement-induced entropic recoil of single DNA molecules in a nanofluidic structure. *Phys. Rev. lett.* **88**(12): 1281031-1281034.

Vercoutere, W., Winters-Hilt, S., Olsen, H., Deamer, D., Haussler, D., and Akeson, M. (2001). Rapid discrimination among individual DNA hairpin molecules at single-nucleotide resolution using an ion channel. *Nature Biotechnology* **19**: 248-252.

Wang, Y.-C., Tsau, C. H., Burg, T. P., Manalis, S., and Han, J. (2005a). *Effiecient Biomolecule Pre-Concentration by Nanofilter-Triggered Electrokinetic Trapping.* Micro Total Analysis Systems 2005: 9th International Conference on Miniaturized Systems for Chemistry and Life Sciences, Boston, Massachusetts, USA, 238-240.

Wang, Y. C., Stevens, A. L., and Han, J. (2005b). Million-fold preconcentration of proteins and peptides by nanofluidic filter. *Anal. Chem.* **77**(14): 4293-4299.

Waskiewicz, W. K., Berger, S. D., Harriott, L. R., Mkrtchyan, M. M., Bowler, S. W., and Gibson, J. M. (1995). Electron-optical design for the SCALPEL proof-of-concept tool. *Proc. SPIE*: 13-22.

Wilder, K., Quate, C.F., Singh, B., and Kyser, D. F. (1998). Electron beam and scanning probe lithography: A comparison. *J. Vac. Sci. Technol. B* **16**(6): 3864-3873.

Wu, G., Datar, R. H., Hansen, K. M., Thundat, T., Cote, R. J., and Majumdar, A. (2001a). Bioassay of prostate-specific antigen (PSA) using microcantilevers. *Nat Biotech* **19**(9): 856.

Wu, G., Datar, R. H., Hansen, K. M., Thundat, T., Cote, R. J., and Majumdar, A. (2001b). Bioassay of prostate-specific antigen (PSA) using microcantilevers. *Nature Biotechnology* **19**: 856-860.

Xia, Q., Keimel, C., Ge, H., Yu, Z., Wu, W., and Chou, S. Y. (2003). Ultrafast patterning of nanostructures in polymers using laser assited nanoimprint lithography. *Appl. Phys. Lett.* **83**(21): 4417-4419.

Xia, Y., and Whitesides, G. M. (1998a). Soft lithography. *Annu. Rev. Mater. Sci.* **28**: 153-184.

Xia, Y., and Whitesides, G. M. (1998b). Soft lithography. *Angrew. Chem. Int. Ed.* **375**: 550-575.

Yamamoto, K., Akita, S., and Nakayama, Y. (1996). Orientation of carbon nanotubes using electrophoresis. *Jpn. J. Appl. Phys.* **35** (Part 2, No. 7B): L917-L918.

Yamamoto, K., Akita, S., and Nakayama, Y. (1998). Orientation and purification of carbon nanotubes using ac electrophoresis. *J. Phys. D: Appl. Phys.* **31**: L34–L36.

Yazdi, N., and Najafi, K. (2000). An all-silicon sigle-wafer micro-g accelerometer with a combined surface and bulk micromachining process. *J. Microelectromech. Syst.* **9**(4): 544-550.

Zankovych, S., Hoffman, T., Seekamp, J., Bruch, J.-U., and Torres, C. M. S. (2001). Nanoimprint lithography: challenges ad prospects. *Nanotechnology* **12**: 91-95.

Zheng, G., Patolsky, F., Cui, Y., Wang, W. U., and Lieber, C.M. (2005). Multiplexed electrical detection of cancer markers with nanowire sensor arrays. *Nature Biotechnology* **23**(10): 1294-1301.

Zurbel, I. (1998). Silicon anisotropic etching in alkaline solutions II On the influence of anisotropy on the smoothness of etched surfaces. *Sens. Actuators A* **70**: 260-268.

Zurbel, I. and Barycka, I. (1998). Silicn anisotropic etching in alkaline solutions I. The geometric description of figures developed under etching Si(100) in various solutions. *Sens. Actuators A* **70**: 250-259.

# QUESTIONS

1. Compare advantages and disadvantages of the following top-down nanofabrication techniques.
   (a) extreme ultraviolet (EUV) lithography
   (b) X-ray lithography
   (c) electron beam lithography (EBL)
   (d) scanning probe lithography (SPL)
   (e) dip pen lithography (DPN)
   (f) soft lithography
   (g) nanoimprint lithography (NIL)
2. Discuss the resolution of features in contact printing, proximity printing and projection printing.
3. Discuss the following soft lithography techniques:
   (a) microcontact printing (CP)
   (b) micromolding in capillaries (MIMIC)
   (c) microtransfer molding (TM)
   (d) replica molding (REM)
4. Discuss the following micromachining techniques:
   (a) Bulk micromachining
   (b) Surface micromachining
   (c) Surface bulk micromachining (SBM)
5. Identify the design criteria for a highly sensitive nanobeam sensor and discuss technical challenges.

# 4

# Nanostructured Biomaterials

## Samar J. Kalita

Department of Mechanical, Materials and Aerospace Engineering, University of Central Florida, Orlando, FL 32816-2450, USA. E-mail: samar@mail.ucf.edu

## 1. INTRODUCTION

A biomaterial is a natural or man-made material used to replace, augment or aid the functions/performance of a part of an organ or the whole organ system and that remains in intimate contact with living tissues without eliciting a toxic, response. The term *biomaterial* has been defined in different words/phrases, over time. Formerly, *biomaterial* was defined as "a systematically and pharmaceutically inert substance designed for implantation within or incorporation with living systems." In 1992, Jonathan Black (1999) defined biomaterial as "a nonviable material used in a medical device, intended to interact with biological systems." Stephen D. Bruck (1900) defined biomaterials as "materials of synthetic as well as of natural origin in contact with tissue, blood, and biological fluids, and intended for use for prosthetic, diagnostic, therapeutic, and storage applications that do not adversely affect the living organism and its components". Yet, another definition was provided by D. F. Williams: "any substance (other than drug) or combination of part of a system which treats, augments, or replaces any tissue, organ, or function of the body" (William 1987) which added to the different perceptions and expressions of the same thing.

The expression *nanostructured biomaterial*, as used in this book, defines those biomaterials that consist of nanograins and/or nanostructures, or possess features of at least one dimension in the nano range (1-100 nm), developed as a result of natural processes or by laboratory/industrial experimentation. This is in line with the generic definition of nanostructured materials/system which requires at least one of the dimensions to be in the range of 1-100 nm. For example, nanoparticles are nano-sized in three dimensions; nanotubes are nano-sized in two dimensions; and nanofilms are nano-sized in one dimension. The building blocks of nanotechnology are nanomaterials, which is also true to many biological systems.

The fascination and great technical promises associated with nanoscale/ nanostructured materials are based on the significant differences in their fundamental physical, mechanical, electrical, chemical and biological properties. For eventual bioengineering applications of nanoscale/nanostructured materials, an ability to control their intrinsic structures and properties, as well as their basic features in terms of diameter, length, alignment, periodicity and spacing is essential to create the next-generation biomaterials with enhanced properties to meet the tissue engineering needs and to overcome the existing limitations. Synthesis/processing of complex nanostructures, nano-motors, and designs deviating from simplistic regular geometry can be of great clinical importance in applications ranging from drug delivery to bone scaffolds.

Challenges in tissue engineering have always motivated scientists and engineers to develop new biomaterials, designs and scaffolds that can restore and/or mimic the structural features and physiological functions of natural tissues. A wide variety of biomaterials have been developed during the past six decades and tested for applications ranging from hip replacements and vascular reconstruction to major organ replacements. Many specialty polymers, metals, ceramics and composites have been developed for numerous applications. These materials in various forms and phases are made to perform different functions in repair and reconstruction of diseased and damaged parts of the human body. Applications of these novel materials for biomedical practices have greatly revolutionized and improved the quality of modern human health care and have alleviated pains and suffering of millions of people. Examples of these novel biomaterials include metals such as austenitic stainless steels, titanium alloys, cobalt-chrome alloys, gold and nitinol; ceramics such as alumina, zirconia and calcium phosphates; polymers such as nylon, silicon rubber, polyester, polylactic acids and polytetrafluroethylene; and composites such as carbon-carbon and wire or fiber reinforced bone cement.

Biomaterials are, by definition, materials that are intended to assume the functions of tissue in natural organs or organ parts. Therefore, it is essential for a biomaterial to imitate, as closely as possible, the properties of the tissue to be replaced (Dumitriu *et al.* 1994). However, development of a biomaterial to imitate properties of a natural tissue is not an easy task. The successful application of a biomaterial *in vivo* depends on many different factors such as the material's properties, design, cytotoxicity, biocompatibility, bioactivity and biodegradability of the material used, as well as several other factors that are not under the control of engineers/manufacturers, including the surgical techniques used by the surgeons, the health and condition of the patient, and the postsurgery activities of the patient. All these factors make the field of biomaterials science and engineering truly interdisciplinary and highly challenging.

Hundreds of millions of people are affected by musculoskeletal conditions alone across the world, which costs a huge fortune on world society. According to reports prepared by Datamonitor, musculoskeletal conditions are the major causes of severe long-term pain and physical disability across the world. In the United

States alone, these conditions cost an estimated $254 billion each year (Annan 1999). It has also been reported that one out of every seven Americans suffers from a musculoskeletal injury/disease each year, and more than 80% of the adult population suffer from some form of back problem, at some point in their lives (The Bone and Joint Decade, 1998). The need for tissue grafts for surgical restoration is escalating because of the aging society combined with a precipitous increase in road accidents and sports-related injuries. The orthopedic market in the United States is growing at a steady rate of 14-16%, annually. In order to restore the lost functions in patients, tissue grafting has become a routine procedure for surgeons.

The possible tissue grafting options for a patient can be divided into five major categories based on the genetic relationship between the donor and the recipient. These options include (i) autogenous tissue graft or autograft (a tissue graft from one site to another within the same individual); (ii) isogenous tissue graft or isograft (a tissue graft between two genetically identical individuals of the same species, e.g., monozygotic twins, or animals of highly inbred strains); (iii) allogenous tissue graft or allograft (a tissue graft between individuals of the same species); (iv) xenogenous tissue graft or xenograft (a tissue graft where donor and recipient are individuals from different species); and (v) artificial tissue grafts.

Autograft is considered to be the "gold standard" in the industry because of the best immunocological responses. However, there are some serious disadvantages associated with autografting procedure such as donor site morbidity associated with infection, pain and hematoma, limitations in mechanical strength, requirements of two invasive surgical operations and limited availability of tissue for grafting. More serious problems such as transmission of diseases, and rejection by the patient's immune system are associated with other types of tissue grafting options such as allograft, isograft and xenograft. For these reasons, there is a continued interest in the development of new and improved artificial substitute materials for tissue engineering applications. The ability to fabricate complex structures using various modern manufacturing techniques to mimic physical attributes of natural tissue, and in sufficient amounts to meet the growing need, has boosted the demand for artificial implants.

Figure 1 shows a human pelvis model developed from computer tomography (CT) scan data description using a rapid prototyping technique called fused deposition modeling. It shows the structural complexity of our body and the ability of modern manufacturing techniques to mimic such complex and intricate skeletal geometries. Additionally, the need of advanced techniques to deliver drugs locally or consistently over a long period of time has spurred the need to develop new biomaterials, and more recently, nanostructured biomaterials.

The need for nanostructured biomaterials or a future in nanoscale biomaterials to meet current and future needs of the biomedical industry is obvious. Like many other disciplines of science and engineering, nanostructured/nanoscale materials are believed to have potential to overcome some of the major limitations associated with metallic, ceramic, polymeric and composite biomaterials. It has been established that in certain materials systems novel nanostructures or

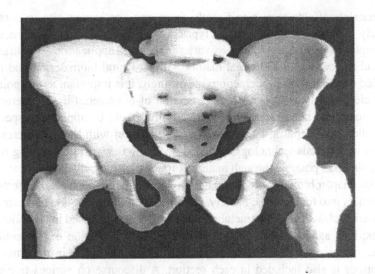

**FIGURE 1.** Prototype of human pelvis developed from actual computer tomography (CT) scan data description using Rapid Prototyping technology.

nanosystems can be designed to significantly improve the physical, mechanical, chemical and biological properties, and processes as a result of the miniature size of their constituents in the nano regime.

Nanoscale structures—for example, nanofilms and nanoparticles—possess a very high surface area to volume ratio and potentially some difference in crystallographic structures, which may radically alter the biochemical and biological activities *in vitro* and *in vivo*. Creating and using structures, devices and systems that have novel properties and functions because of their small and/or intermediate size created a new era in biomedical materials research. Research at the atomic scale has been the norm for the pharmaceutical industry in developing new drugs and medicines. However, drugs and medicine do not come under the definition of biomaterials. Nanostructured biomaterials research is relatively a new field of Biomaterials Science and Engineering, with many promising results as reported in recent literature. Such research has relevance as most of the biological materials and structures *in vivo* have dimensions at the nanorange. For instance, DNA, which is our genetic material, is of 2.5 nm, white red blood cells (WBC) are approximately 2.5 μm. Currently, medical researchers are working at the nanoscale level to develop new drug delivery methods, biomaterials and systems to enhance the quality of modern human health care.

Although the generic definitions of biomaterials provides a basic understanding of these novel materials, considering the diversity of living tissues and their associated functions, it is important to look at an application-specific description to understand the complexity, challenges and requirements involved in developing biomaterial/s to replace a natural tissue/s. The requirements of properties in biomaterials/scaffolds vary significantly from one application site to another. Similarly, the classification of various types of cell-biomaterials interactions, seen upon implantation of a device *in vivo*, varies in different applications. While many texts and reviews in the field deal mainly with structural biomaterials and have classified cell-biomaterial interactions accordingly, it is important at this point to have a clear understanding that a classification of biomaterial-tissue interaction beyond toxic and nontoxic and biocompatibility should be application specific. The challenge of defining these terms becomes harder with the ever-increasing progress in materials technology, which is changing the world by making many impossible tasks possible through materials innovation and advances.

This chapter broadly classifies emerging nanostructured and/or nanofeatured biomaterials into four groups; namely, metallic, ceramic, polymeric and composite. Different classes of nanostructured biomaterials are discussed in the following sections; then again, under separate subheadings. A discussion of conventional metallic, ceramic, polymeric and composite biomaterials and their properties and applications is also included in each section. A discourse on various tissue responses is presented in the next section to provide the readers with the basics of established concepts of cell-biomaterials interactions and the term *biocompatibility*. Improved cell responses observed in nanostructured biomaterials and some of the notable current advances in the field are presented towards the end of the chapter. The chapter ends with a summary and glossary of terms, to recapitulate the important points.

## 1.1. Biocompatibility and Types of Tissue Responses

It has always been a challenge for scientists and surgeons to define the term biocompatibility or biocompatible as both these expressions are qualitative, and a material that is compatible at a particular application site *in vivo* may not exhibit a satisfactory behavior/function/performance at another location. It has been rightly stated by William Black (1999) that the label *biocompatible* suggests that the material described exhibits universally "good" or harmonious behavior in contact with living tissue and body fluids. He also provided a generic definition of *biocompatible(-ity)* as "biological performance in a specific application that is judged suitable to that situation." Today, it is universally accepted that biocompatibility is application/situation specific and that the expected cell-biomaterials interactions is different from one application to another.

It is evident that all implant materials elicit some kind of response/s from the host tissue, upon implantation. Based on these responses, we can broadly classify the implant materials into *toxic* and *nontoxic*. It is critical that any implant material avoid a toxic response, which would kill cells in the surrounding tissues. In the case of structural biomaterials that are used as bone-grafts or in fabrication of bone implants, for example, total hip replacement (THR), spinal fusion, and knee replacements, the nontoxic responses can be further divided into three categories, namely, bioinert, bioactive and bioresorbable. The most common response

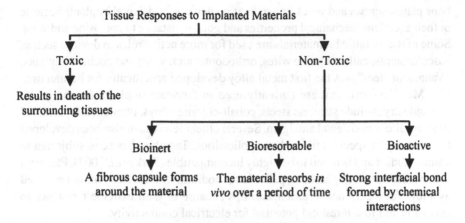

FIGURE 2. Possible tissue responses of materials tested for bone graft applications.

of tissues to an implant is formation of a nonadherent fibrous capsule that "walls-off" or isolates the implants from the host. This leads to complete encapsulation of an implant within the fibrous layer.

Biologically inactive, nearly inert materials such as alumina, zirconia, stainless steel and cobalt-chrome steel develop fibrous capsules at their interface. This is a typical response of bioinert materials. The second type of response is when a strong bond is formed across the interface between the implant and the tissue, and is termed bioactive. The interfacial bond prevents motion between the two materials and mimics the type of interface that is formed when natural tissues repair themselves. An important characteristic of a bioactive interface is that it changes with time, as do natural tissues that are in a state of dynamic equilibrium. The last type of implant is bioresorbable in which the implant material dissolves in the body fluid, with time. A material that dissolves in the body and produces by-products that can be absorbed/digested by the body through various metabolic activities is considered to be resorbable and an ideal material for many tissue engineering applications (Hench 1998). Possible tissue responses observed when an artificial material comes in contact with natural tissues are presented in Fig. 2.

# 2. CLASSIFICATION

## 2.1. Metallic Biomaterials

Metallic materials are used as biomaterials in human health care particularly because of their excellent mechanical properties and extensive knowledge-base of mankind with regard to their processing, properties and structures. They are mostly used as passive substitute materials for hard tissue replacements in load-bearing applications such as total hip and knee joint replacements, for fracture healing aids as bone plates, screws and wires, spinal fixation devices, and dental implants because of their excellent mechanical properties and good resistance to corrosion and wear. Some of the metallic biomaterials are used for more active roles in devices such as vascular stents, catheter guide wires, orthodontic arch wires and cochlea implants. "Vanadium steel" was the first metal alloy developed specifically for human use.

Metal systems that are currently used to fabricate implants for the health care industry include stainless steels, cobalt-chrome alloys, titanium alloys, nickel titanium alloys and dental amalgam. Several other metals have also been developed for a variety of specialized implant applications. Tantalum has been subjected to animal studies and is found to be highly biocompatible (Park *et al.* 2003). Platinum group metals such as platinum, palladium, rhodium, iridium and osmium are used as alloys for electrodes in pacemaker tips because of their extreme resistance to corrosion and low-threshold potential for electrical conductivity.

Table 1 presents four major groups of metallic biomaterials and their mechanical properties. Though 18-8 (type 302) stainless steels were initially used

**TABLE 1.** Classification of metallic biomaterials and their properties [Source: American Society for Testing and Materials, F139-86, F75-87, F90-87, F562-84, F67-89, F136-84, F560-86, 1992; and Materials Property Data (www.matweb.com)]

| | Metallic Biomaterials | | | | |
|---|---|---|---|---|---|
| Types | Sub-class | Density $(g/cm^3)$ | Tensile Strength (MPa) | 0.2 % Offset Yield Strength (MPa) | % Elongation |
| 316L Stainless Steel | Annealed | 7.9 | 485 | 172 | 40 |
| | Cold-worked | 7.9 | 860 | 690 | 12 |
| Cobalt-Chrome Alloys | CoCrMo (F75) | 8.3 | 655 | 450 | 8 |
| | CoNiCrMo (F562, Annealed) | 9.2 | 793-1000 | 240-655 | 50 |
| | CoCrWNi (F90) | 10 | 860 | 310 | 10 |
| Titanium Alloys | Ti6Al4V | 4.5 | 860 | 795 | 10 |
| | Grade 1 | 4.5 | 240 | 170 | 24 |
| | Grade 2 | 4.5 | 345 | 275 | 20 |
| | Grade 3 | 4.5 | 450 | 380 | 18 |
| | Grade 4 | 4.5 | 550 | 485 | 15 |
| Nitinol | High T. Phase | 6.45 | 754-960 | 560 | 15.5 |
| | Low T. Phase | 6.45 | 754-960 | 100 | 15.5 |
| Tantalum | Annealed | 16.6 | 207 | 138 | 20-30 |
| | Cold-worked | 16.6 | 517 | 345 | 2 |
| Amalgam | — | 11 | 40 | 275–345 (compressive) | — |

in implant fabrication, the current applications of stainless steel in implant fabrication include 316 and 316L. These austenitic stainless steels are nonmagnetic and possess very high corrosion resistance. The American Society for Testing and Materials (ASTM) recommends type 316L for use in implant fabrication which contains a maximum of 0.03% carbon, 2.00% manganese, 0.03% phosphorous, 0.03% sulfur, 0.75% silicon, 17.00-20.00% chromium, 12.00-14.00% nickel and 2.00-4.00% molybdenum. The presence of molybdenum improves the corrosion resistance.

Mechanical properties of 316L vary depending on whether it is cold worked or annealed. The corrosion resistance of stainless steel is inferior to other groups of metallic biomaterials and is thus used mostly in temporary implant devices such as screws, bone plates and nails. The ASTM recommends four different types of cobalt-chrome alloys for biomedical implant applications which include castable cobalt-chrome-molybdenum (CoCrMo, F75) alloy, wrought cobalt-chrome-tungsten-nickel (CoCrWNi, F90) alloy, wrought cobalt-nickel-chrome-molybdenum (CoNiCrMo, F562) alloy and wrought cobalt-nickel-chrome-molybdenum-tungsten (CoNiCrMo, F562) alloy.

Wear and corrosion resistance of cobalt-chrome alloys are superior to other groups of metallic biomaterials. Cobalt-chrome alloys also exhibit better fatigue properties compared to other groups of metallic biomaterials. However, Young's modulus (220-234 GPa) and density (8.3 g/cc) of cobalt-chrome alloys are higher than that of other metallic biomaterials. This can definitely have an impact on long-term survivability of implants, particularly due to higher Young's modulus difference with respect to bone (15-30 GPa).

Total hip prostheses (THP) is one of the major and important applications of metallic biomaterials. Figure 3 shows the schematic of a THP detailing its important components. In spite of shortcomings, metallic THP will continue to dominate the orthopedic implant market in the years to come. Titanium (Ti) is used in the biomedical industry either as commercially pure titanium (cp-Ti) or as Ti-alloys. There are four grades of cp-Ti according to ASTM classification. Ti6Al4V and Ti13Nb13Zr are the two titanium alloys mostly used in implant fabrication. Ti6Al4V contains 5.5-6.6% of aluminum and 3.5-4.5% of vanadium as alloying elements and is most widely used in the industry because its excellent corrosion resistance and fatigue strength (550 MPa) which is very close to that of cobalt-chrome alloys. One of the major advantages of Ti-alloys is their low density (4.5 g/cc) compared to other metallic biomaterials. Ti-alloys also exhibit superior biological properties due to the formation of an oxide layer that reacts with biological fluid and thus influence bonding at the interface. Limitations of Ti and Ti-alloys when used in orthopedic industry are low shear strength and low wear resistance (Long et al. 1998). The poor shear strength of Ti-alloys makes them undesirable for applications such as screws, nails and fracture plates.

Nickel-titanium (Ni-Ti) alloys have also been widely investigated for application as biomaterials. Ni-Ti exhibit shape memory effects which was first observed by Buehler et al. (1963) at the U.S. Naval Ordnance Laboratory. Ni-Ti alloys

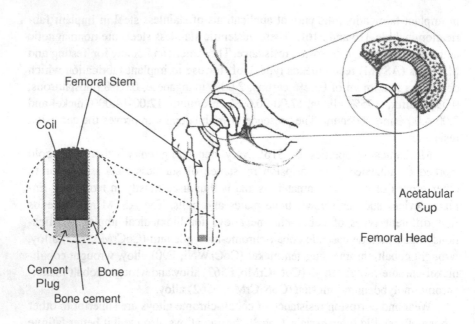

Femoral Stem

Coil

Cement
Plug

Bone cement

Bone

Acetabular
Cup

Femoral Head

**FIGURE 3.** Schematic of a Total Hip Prosthesis and its important components.

exhibits good corrosion resistance and good biocompatibility *in vivo*. According to Duerig, some of the possible applications of Ni-Ti shape memory alloys are orthodontic dental archwire, vena cava filter, contractile artificial muscles for artificial hearts, vascular stents, catheter guide wire and orthopedic staple (Duerig 1995).

For more details on various types of traditional metallic biomaterials and their properties and applications, readers are referred to well-summarized discussions in the texts *Biomaterials: Principles and Applications* edited by J. B. Park and J. D. Bronzino, and *Biomaterials Science: An introduction to Materials in Medicine*, edited by B. D. Rutner, A. S. Hoffman, F. J. Schoen and J. E. Lemenons.

Replacement of diseased or damaged bones, teeth, as well as joints, by metallic biomaterials is often needed in orthopedic surgery (Long *et al.* 1998). To meet the needs imposed by increased human life and implantation in younger patients, the artificial metallic biomaterials used in load-bearing orthopedic surgery should not only avoid short-term rejections and infections, but also provide long-term biocompatibility and long-term material properties such as compatible strength, longer lifespan, high wear and corrosion resistance *in vivo*, and absence of long-term toxic response. Long-term biocompatibility of metallic implants has been a concern because of the release of metals ions, wear and harmful effects of corrosion products and metal ions on surrounding tissues and organs. Amongst four major groups of metallic biomaterials, the stainless steels and cobalt chrome alloys have been thoroughly investigated during the late twentieth century, and their properties, applications and limitations are well established. Most of the current thrust on metallic biomaterials research is on Ti and Ti-alloys, which exhibit superior biocompatibility, possess low density, excellent corrosion resistance and minimum stiffness difference with that of bone, compared to other metallic biomaterials (Long *et al.* 1998; Noort 1987; Niinomi 2002). High stiffness difference causes *stress shielding*, which had been a major concern with metallic load-bearing prostheses. Stress shielding is a condition wherein an implant withstands the entire load, thereby limiting transfer of the load to the bone which, in the long run, leads to bone fragility and fracture, as load or stress is vital in bone formation and remodeling. To obtain the optimum combination of high strength with low stiffness, one of the limitations of metallic biomaterials and a requirement for bone tissue engineering, nanostructured materials have been explored recently and have shown potential. However, the current available literature on nanostructured biomaterials shows that most of the research has been focused on nanoscale bioceramics.

Materials scientists, orthopedic surgeons and biomedical engineers have been working for over seven decades to find a way of eliminating or somehow reducing the failures and limitations of metallic implants *in vivo*. Orthopedic experts partly blame the failure of metallic implants on incomplete osseointegration (lack of interfacial bonding) between the prostheses and surrounding bone tissues. Others hold stress shielding to be responsible and attribute the stress-shielding effect to the significant differences in the mechanical properties of surrounding bone to that of an implant. There are still others who associate these failures to the generation of wear debris at articulating surfaces between bone and orthopedic implants that leads to death of bone cells, which can lead to bone necrosis. Failures of metallic implants occur for one or more of the above-mentioned reasons limiting the life of these implants *in vivo*, which precipitate the search for new and/or improved metallic biomaterials to address these concerns.

### 2.1.1. Progress in Nanostructured Metallic Biomedical Materials

The growing list of nanocrystalline materials that facilitate and enhance activities of bone-forming cells (osteoblasts) is missing one material classification: metals. However, the design of the next-generation orthopedic metallic implants is focused on matching implant surfaces with the unique nanometer topography made by natural extracellular matrix proteins found in bone tissue. Nanoscale structures and molecules found in bone tissues show that bone-forming cells are used to interact with surfaces that have roughness on a nanometer scale. For example, immature (or woven) bone has an average inorganic mineral grain size of 10-15 nm. This woven bone is actively replaced by lamellar bone, which has an average inorganic grain size of 20-50 nm and a diameter of 2-5 nm. Conventional synthetic metals currently used in biomedical applications display surfaces with microroughness, which are smooth at the nanoscale. These smooth surfaces have been shown to encourage "fibrointegration" or the formation of calluses. These calluses can eventually encapsulate the implant placed inside our body with undesirable layered connective tissue (Webster *et al.* 2004).

*In vitro* studies to compare osteoblast adhesion between nanostructured and conventional metallic biomaterials such as Ti, Ti6Al4V, and CoCrMo alloys, was performed recently by Webster *et al.* (2004). The nanostructured and conventional metals used in their study had similar chemistry and were different only in the degree of nanometer roughness of the surface. Webster *et al.* showed that there was increased osteoblast adhesion measured on specimens having nanometer surface roughness. These may be because the nanostructured materials have increased particle boundaries at the surface, since they have smaller particle sizes.

Osteoblast adhesion occurs specifically at nanostructured particle boundaries of metals (Webster *et al.* 2004). Since nanostructured metals are made of fewer and smaller particles of the same atoms than conventional form, nanostructured metals have distinctive surface properties. Nanostructured metals have a greater number of atoms at the surface compared to their bulk counterparts, larger areas of increased surface defects (like particle boundaries and edge corner sites), and greater proportions of surface electron delocalization. These altered surface properties influence the initial protein interactions that control subsequent adhesion of osteoblast cells (Webster *et al.* 2004). Their results showed that there were greater numbers of osteoblasts on surfaces of nanostructured materials compared to conventional metals. This may be attributed to the higher percentages of particle boundaries at the surface of nanostructured metals. Because adhesion is a necessary requirement for subsequent functions of osteoblasts (i.e., the deposition of Ca-containing mineral), the results obtained by Webster *et al.* indicate the promise of nanostructured metals in orthopedic applications for the first time.

Considering a similar line of research performed on nanostructured ceramics, polymers and composites performed to study the behavior of osteoblast functions to date, the results achieved by Webster *et al.* present strong evidence that the attachment of osteoblasts may be facilitated despite materials chemistry, as long as a large degree of nanometer surface roughness is created. Nonetheless, they are not the first researchers to study nanostructured metals. This idea of creating metallic implants with decreased (nanometer-sized) surface feature dimensions that mimic the roughness of the extracellular matrices in bone has been used by others as well. However, those studies did not isolate the degree of nanometer surface roughness as the only differing material property, so these modified synthetic materials may have had other properties that influenced osteoblast functions.

As mentioned earlier, the frequently used Ti and Ti-alloys in orthopedic surgery are cp-Ti and Ti6Al4V (Niinomi 2002). Limitations of cp-Ti and Ti6Al4V include low shear strength, low wear resistance (Long *et al.* 1998) and the mismatch of the elastic modulus between the Ti-alloys (100-120 GPa) and bone (10-30 GPa) though the difference is lower than other metallic biomaterials (Niinomi 2003). For solving these problems and further improvising the biological and mechanical properties, many new Ti alloys have been developed during recent years for biomedical uses (Niinomi 2002, 2003). Niinomi (2002) designed single *h*-phase type Ti alloys to reduce the elastic modulus. However, a reduction in elastic modulus also reduced other mechanical properties in those Ti-alloys.

When the elastic modulus is reduced, the strength of those Ti-alloys decreased, and, inversely, the strength increased with the increase in the elastic modulus. He *et al.* (2005) proposed a new design strategy for Ti-alloys according to empirical rules for forming bulk metallic glasses to address this problem. They introduced a nano/ultrafine-structure into the Ti-alloys. By the composition-modification and the microstructure-control, He *et al.* reduced the elastic modulus of Ti-alloys while maintaining their high strength. Such a combination of mechanical properties and elastic modulus is attributed to the novel bimodal microstructure that consists of a micrometer-sized dendritic *h*-phase and a nano/ultrafine-structured matrix. The large volume fraction of boundaries in the nanostructured matrix may also contribute to the low elastic modulus. In nickel-free Ti-alloys, a similar bimodal microstructure was obtained that also exhibits high strength and low elastic modulus, revealing that these bimodal Ti-alloys have a potential for biomedical applications (He *et al.* 2005).

Though Ti-alloys exhibit superior biocompatibility compared to other groups of metallic biomaterials, they are still bioinert; and for prolonged clinical use it is important that the biocompatibility Ti-alloys be improved. Various surface modification techniques are mostly used to improve the biological properties of metallic prostheses. Titanium with a bioactive ceramic coating such as of hydroxyapatite ceramic has been successfully studied (Rønold *et al.* 2002; Wang *et al.* 1993; Klein *et al.* 1994].

Ti-alloys are used for dental and orthopedic implants for its superior compatibility, which is attributed to the formation of an oxide film on its surface (Rønold *et al.* 2002). The surface modification of metallic implants is an active field that involves coating of metallic prostheses mostly with a bioactive ceramic or bioactive glass by plasma spraying or electrophoretic deposition (Tsui *et al.* 1998; Klein *et al.* 1994), electrochemical deposition, basification treatment and sol-gel techniques (Korkusuz *et al.* 1999; Haman *et al.* 1995; Kyeck *et al.* 1999; Singh *et al.* 1996; Ong *et al.* 1991; Dasarathy *et al.* 1996).

Coating of metallic prostheses with nanomaterials, to enhance properties, would possibly be one of the significant contributions of nanoscience and technology to the modern human health care industry, particularly for orthopedic implants. Some of the recent work addressing the applications of nano-bioceramics as coating materials for metallic prostheses is discussed separately in this chapter. Here, I would like to discuss some of the recent works related to surface modification of Ti-implants using nanomaterials/nanotechnology.

Cui *et al.* (2005) fabricated nano-titania/titanium alloys for biomedical applications using a process wherein titanium alloys were embedded in nanometer titanium dioxide powder and sintered at elevated temperature. The particle size of $TiO_2$ particles on the surface of Ti-alloy was mainly 50-90 nm. They showed that the films of nanocrystalline $TiO_2$ powders on the surface of Ti-alloy possessed excellent biocompatibility. Specimens cultured in the simulated body fluid (SBF) showed deposition of calcium phosphate crystals having $n(Ca)/n(P)$ atom ratio of about 1.6:1 on the specimen surface, which is similar to that of hydroxyapatite ceramic and the apatite found in human bone mineral (Cui *et al.* 2005).

Hydroxyapatite, a bioactive ceramic, has been widely investigated as a coating material for various metallic orthopedic implants. A cogent understanding of the microstructure, and indeed nano-structure of hydroxyapatite ceramic and the strength at the interface with the base metal are crucial to its suitability and usefulness as a coating material for orthopedic implants. Donga *et al.* (2003) investigated the microstructure/nanostructure of plasma-sprayed hydroxyapatite/Ti6Al4V composite using analytical transmission electron microscopy (TEM) and scanning transmission electron microscopy (STEM). They observed that the composite developed was a dense-layered structure, containing hydroxyapatite, finely divided Ti6Al4V and amorphous calcium phosphate. Heat treatment at 600°C for 6 h transformed the amorphous calcium phosphate into fine hydroxyapatite crystals with minimal phase decomposition. They also reported that no grain coarsening was observed after heat treatment. The high bond strength of hydroxyapatite/Ti6Al4V was attributed to the good adhesion among the interfaces between hydroxyapatite and Ti6Al4V. However, no apparent changes in the Ti6Al4V/ hydroxyapatite interface microstructure were observed after ten weeks of immersion in the SBF (Dong *et al.* 2003).

Shih *et al.* (2003) studied the effect of surface oxide properties on corrosion resistance of 316L stainless steel for biomedical applications and have found that the stainless steel wire passivated with amorphous oxide film exhibits the best *in vitro* corrosion resistance, and this distinct advantage allows amorphous oxide to be the best surface protective film for cardiovascular devices. This improvement is attributed to the removal of plastically deformed native air-formed oxide layer and the replacement of a newly grown, more uniform and compact one, which is composed of nanoscale oxide particles with higher oxygen and chromium concentrations (Shih *et al.* 2003).

Bren *et al.* (2004) studied the effects of surface characteristics of metallic biomaterials on interaction with osteoblasts (i.e., multinucleated cells responsible for the formation of bone). Their objectives were to characterize the roughness, surface potential, surface-free energy, electron-donor and electron-acceptor surface parameters, and composition of the oxide film formed on stainless steel and Ti6Al4V alloys subjected to different surface treatments, and to investigate how these properties influence the attachment and growth of osteoblast cells. It was found that surfaces with nanoscale roughness rather than microscale roughness had a more profound impact on the growth of osteoblasts (bone-forming cells). Surfaces with nanoroughness at a level of $Ra = 7\text{-}10$ nm were found to encourage cell differentiation.

Advances in materials synthesis and processing have been pivotal in the emergence of advanced tissue engineering technologies through innovation and discovery of mechanically well-defined environment (scaffold or extracellular matrix) that is rich in biomolecular signals to achieve its objectives, namely, the growth of functional neo-tissue. The evolution of neo-tissue is governed by cell-biomaterial interactions at nanoscale, which are in turn dictated by the characteristics of the surface such as its texture, roughness, wettability and chemical functionality.

Shastri *et al.* (2003) developed and studied functionalized nanoparticles as versatile tools for the introduction of biomimetics on metallic and polymeric surfaces. Their work demonstrated that metal and polymer surfaces can be modified with functionalized nanoparticles to bear well-defined chemical functionality and topography. Functionalized nanoparticles are excellent vehicles for the modification of chemical and topological characteristics of surfaces. Roughness, texture of the surface, as well as, spatial resolution of the chemical functionality can be controlled simply by varying the chemistry and size of the nanoparticles (colloids).

Both surface structure and chemistry influence the adherence of cells to surfaces of biomaterials, although there is not much data available about cell behavior on surfaces with nanometer-sized features. There are many methods available to fabricate metal nanostructures. These include colloidal lithography, chemical dissolution, electrodeposition, and cathodic electrosynthesis (Haranp *et al.* 1999; Chu *et al.* 2003). Colloidal lithography is a quick and simple method to make uniform nanostructures over large surface areas. Particle size and coverage may be varied to produce the required structure of the surface. This method may be used to produce 3-D nanoporous particle films where repeated particle adsorption and repeated substrate charging may result in multilayer formation (Haranp *et al.* 1999).

Nanostructured materials have unique properties that are different from common materials. With the development of Al anodizing technology and commercial anodisk Al membranes, various nanostructures (metals and oxides) with a variety of morphology (tubules, fibers or wire or rods) have been made utilizing the porous membrane as templates in Sol-Gel process, chemical vapor deposition and, electro- and electroless deposition. However, there are limitations of the areas and mechanical strength makes the nanostructures difficult to handle or use in practical devices (Chu *et al.* 2003).

Metallic biomaterials will continue to be major group implant materials despite all the shortcomings until new nonmetallic biomaterials, having mechanical properties equivalent to metals, are developed in the near future. Possibly, new nanostructured metals will be developed to overcome the shortcomings of conventional metallic biomaterials. Particularly, metallic implants with nano-features at the surface will see real-time applications soon. The contribution of nanotechnology and nanoscience in improving quality and life of metallic biomaterials will be a great boon to modern human health care in the near future.

## 2.2. Ceramic Biomaterials

During the past six decades, development of many advanced ceramic biomaterials such as alumina, zirconia, hydroxyapatite, β-tricalcium phosphate, coralline, ALCAP (aluminum calcium phosphate ceramics), ZCAP (zinc calcium phosphate oxide ceramics), ZSCAP (zinc sulphate calcium phosphate ceramics) and FECAP (ferric calcium phosphate oxide ceramics) and bioactive glasses have made significant contribution to the development of the modern human health care industry and have improved the quality of human life. These are the ceramics, which can be used inside the body without rejection, to augment or replace various diseased or damaged parts of the musculoskeletal system, and which are known as *bioceramics* (Hench, 1998).

Bioceramics are primarily used as bone substitutes in the biomedical industry for their biocompatibility, low density, chemical stability, high wear resistance and compositional similarity with the mineral phase of the bone. Bioceramics have found applications in the dental industry, for example as dental crown, because of their high compressive strength, high resistance to wear, good compatibility with the body fluid and aesthetically pleasing appearance. They are also used as reinforcing material in composite implants such as artificial tendons and ligaments owing to their high specific strength as fibers and good biocompatibility (Billotte, 2003).

However, the potential of any ceramic material to be used as an implant material *in vivo* depends not only on its compatibility with the biological environment but also on its ability to withstand complex stresses at the site of application. Mechanical weakness of bioceramics associated with their inherent brittleness limits their applications in many orthopedic applications where strength under complex stress is a major concern. In a very short span of time, bioceramics have come a long way and have found applications in numerous ways, as in replacements of hips, knees, teeth, tendons and ligaments and repair for periodontal disease, maxillofacial reconstruction, augmentation and stabilization of the jawbone and spinal fusion. Table 2 lists various bioceramics used in the biomedical industry with their current/proposed applications.

### 2.2.1. Classifications of Bioceramics

Ceramic biomaterials developed and tested for tissue engineering applications are broadly classified into four major categories in terms of mechanism of tissue attachment, which is directly related to the type of tissue response at the implant interface. They are:

- Bioinert—e.g., alumina, zirconia and titania
- Porous—e.g., porous hydroxyapatite, porous alumina and porous zirconia
- Bioresorbable—e.g., tricalcium phosphate and biphasic calcium phosphates
- Bioactive—e.g., hydroxyapatite, bioactive glasses and glass ceramics

A bioinert ceramic does not form a bond with bone. A porous ceramic structure helps in the formation of a mechanical bond via the ingrowth of bone tissues into pores. A bioactive ceramic forms a strong interfacial bond with bone via chemical reactions at the interface. A resorbable bioceramic is replaced by natural bone tissues over a period of time.

**TABLE 2.** List of various bioceramics and their current/proposed applications in tissue engineering.

| | Ceramic Biomaterials | |
| --- | --- | --- |
| Names | Chemical Formulae | Applications |
| Alumina | $Al_2O_3$ | Joint replacements, total hip prostheses, reconstruction of acetabular cavities and dental implants |
| Zirconia | $ZrO_2$ | Total joint prostheses and femoral heads |
| Pyrolytic Carbon | — | Blood interfacing implants, blood vessels, repair of diseased heart valves |
| Hydroxyapatite | HAp | Bone filler, coating material for metallic prosthesis, dental composites, spinal fusion and repair of lumbo-sacral vertebrae and, drug delivery devices. |
| Tricalcium Phosphate | TCP | Resorbable bone grafts, for filling space vacated by bone screws, donor bone, excised tumors, and diseased bone loss, drug delivery devices, |
| Zinc-Calcium-Phosphorous Oxide | ZCAP | Insulin delivery devices, drug delivery. |
| Coralline | — | Repair of traumatized bone, replace diseased bone and correction of bone defects. |
| Calcium Sulfate | — | Fill defects in bone, biodegradable scaffolds in orthopedic, oral and maxillofacial surgery. |

Relatively bioinert ceramics are mostly used as structural support implants. Some of these are bone plates, bone screws and femoral heads. Examples of nonstructural support uses are ventilation tubes, sterilization devices, and drug delivery devices. These bioceramics don't interact with the physiological systems *in vivo* and the host always treats them as foreign materials. High-density, high-purity (>99.5%) $Al_2O_3$ (α-alumina) was the first bioinert bioceramic to be widely used clinically. It is used in total hip prostheses and dental implants because of its combination of excellent corrosion resistance, acceptable biocompatibility, low friction, high wear resistance and high compressive strength. Strength, fatigue resistance, and fracture toughness of polycrystalline α-alumina, are a function of grain size and purity.

Zirconia ($ZrO_2$), in tetragonal form, stabilized by either magnesium or yttrium, has also been developed for use in total joint prostheses. Zirconia ceramics have better mechanical features than alumina ceramics for flexural strength, Young's modulus, and toughness; nevertheless a specific experiment (Sudanese *et al.*, 1989) suggests that it cannot be used for prostheses with a ceramic-ceramic coupling, owing to its low wear resistance.

Titania is a biocompatible ceramic, but at present no prosthesis in bulk form has been experimented with because of its poor mechanical strength. Composite aluminous ceramics were widely experimented on for their use as implant materials for tissue engineering applications.

Calcium aluminates were the first of this kind to be tested for hard tissue engineering applications, even though their current use is practically negligible. Porous forms of calcium aluminates at one time emerged as a material suitable for application in the field of both orthopedic and odontological applications (Hentrich *et al.*, 1969; Hulbert *et al.*, 1972; Kalita *et al.*, 2002).

Among the composites of alumina, those worth mentioning are the alumina-titania ones. As an example, Metco 131 (60 wt.% alumina and 40 wt% titania) is proposed for biomedical application. Researchers have also used alumina-zirconia composites as an alternative to alumina ceramics which can prevent the possibility of delayed fracture and at the same time slows down the advancement of microcracks.

Sialon, another bioinert bioceramic, is particularly promising for the replacement of joints by virtue of its high mechanical strength, excellent biocompatibility, and resistance to biodegradation. Like any other material, Sialon is evaluated by a complete series of laboratory trials including corrosion and wear tests, tests as simulated joints, studies on biocompatibility, and tests on animals to assess bone growth on surfaces of porous Sialon.

Resorbable bioceramics are those that degrade gradually with time *in vivo* and can be replaced with natural tissues. The rate of degradation varies from one ceramic to another. This approach is considered to be the optimal solution to the problems of interfacial stability. However, development of such a ceramic material and corresponding implant fabrication possess many challenges. Complications in the development of resorbable bioceramics are (i) maintenance of strength and the stability of the interface during the degradation period and replacement by the natural host tissue, (ii) matching resorption rates to the repair rates of the body tissues.

Although Plaster of Paris, which is a resorbable ceramic, was used as a bone substitute as early as 1892 (Peltier 1961), the concept of using synthetic resorbable ceramics as bone substitutes was introduced in 1969 by Hentrich and his co-workers (Hentrich *et al.* 1969; Graves *et al.* 1972). Except Biocoral and Plaster of Paris (calcium sulphate dehydrate), almost all bioresorbable ceramics developed for biomedical use are some variations of calcium phosphate. Other of resorbable bioceramics include aluminum calcium phosphate, coralline, and tricalcium phosphate.

Bioactive ceramics offer another approach to achieve interfacial attachment. A bioactive material undergoes chemical reaction at the interface. The surface reactions lead to bonding of tissues at the interface. The bioactive concept has been expanded to include many bioactive materials with a wide range of bonding rates and thickness of interfacial bonding layers. They include bioactive glasses such as Bioglass®, bioactive glass ceramics and synthetic hydroxyapatite. The time dependence of bonding and mechanical strength of the bond, the mechanism of bonding, the thickness of bonding zone and fracture toughness differ for various materials.

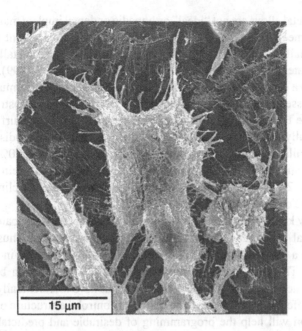

15 μm

**FIGURE 4.** SEM image of bone cells grown on bioactive hydroxyapatite substrate under standard μ cell culture conditions.

A common characteristic of all bioactive implants is the formation of a hydroxy-carbonate apatite (HCA) layer on their surface when implanted. The HCA phase is equivalent in composition and structure to the mineral phase of bone. The interface between a bioactive implant and bone is nearly identical to the naturally occurring interfaces bone and tendons and ligaments. Figure 4 shows SEM image of bone cells (osteoprecursor cells from a modified human osteoblast cell line) grown on bioactive hydroxyapatite ceramic substrate under standard cell culture conditions.

## 2.2.2. Nanostructured Bioceramics

Conventional materials that were originally developed for industrial applications such as aeronautical engineering are being used as orthopedic and dental implant materials as long as they are tolerated by the human body and meet the strength requirements. Skeletal tissues experience simple to complex stress conditions during our day-to-day activities. Examples include jaws withstanding 0.25 N during chewing and hip joint sustaining 3-5 KN during walking. Materials such as commercially pure titanium, Ti6AL4V and CoCrMo alloys are chosen for orthopedic and dental applications based on these physiological force requirements. In contrast, bioceramics find limited use in biomedical applications because of their brittleness, despite their excellent biocompatibility with the bone tissues. Unfortunately, metals and metal alloys are not "ideal" orthopedic or dental implant materials, even though they meet the mechanical requirements for bone replacement under loading. Additionally, under long-term physiological loading many metallic implants fail due to material failures that require removal (Webster et al. 1999).

There is a constant need for orthopedic and dental implant formulations that have better osseointegrative properties (Webster et al. 2001). Nanostructured ceramics can be tailored to obtain desired chemical compositions, surface properties (specifically topography), mechanical properties (ductility) and distribution of grain size similar to those of physiological bone (which contains 70% by weight of hydroxyapatite ceramic with grain sizes less than 100 nm) (Gutwein et al. 2004; Webster et al. 2001). To date, nanostructured bioceramics are not clinically used for orthopedic or dental applications.

It is now known that materials organized on multiple-length scales are more like biological structures than those with single-scale features. Thus, materials organized on a multiple-length scale should be of more advantage in biomedical applications. There has been an increasing interest in fabricating biomaterials for bone regeneration with complex structures that have hierarchically organized features of different length scales. The ability to control the structure of a material with precision will help the programming of desirable and predictable cellular responses into three-dimensional biomaterial scaffolds.

In a new hierarchical fabrication method, microfabrication technology was used by Tan et al. (2004) to create micrometer-style architecture on silicon. Then nanostructured hydroxyapatite minerals were formed on the preset structure using natural mineralization techniques (Tan et al. 2004). Based on several studies, the chemical modifications of the foundation layer were determined to be essential for achieving control over subsequent nanoscaled structure formation. Also, the rate and pattern formation of nanostructured hydroxyapatite on the micropatterned surfaces had a strong dependence on solution conditions (Tan et al. 2004).

The creation of nanomaterials has made possible the synthesis of materials that simulate the mechanical properties and chemical characteristics of normal and healthy bone for orthopedic and dental applications. For example, nanomaterials with better mechanical properties can replace conventional materials that failed because of crack imitation and propagation during loading. Because ductility is improved as grain size is reduced, nanostructured ceramics can be made (by selecting proper synthesizing parameters that can control grain size) to possess a wide range of mechanical properties. Most importantly, future nanostructured ceramic biomaterials can be designed to meet clinical requirements corresponding to anatomical differences or patient's age. These requirements may surface for many reasons, like the differences in modulus of elasticity (10% in human hip and 20% in tibia bone) between a 20- and a 90-year-old human. Also, decreasing grain size (<100 nm) can control the chemical characteristics (such as surface properties) of ceramics. Surface properties that aid in quick deposition of new bone on implant surfaces and support bonding of juxtaposed bone using nanomaterials can be of great clinical significance. This should also minimize motion-induced damage to surrounding tissues (Webster *et al.* 1999).

Studies have been done to correlate adhesion and functions of bone cells with nanoscale features of potential implant surfaces. Webster *et al.* (2004) studied and compared osteoblast functions on conventional micron-sized ceramic formulations and nanostructured substrates made separately from spherical particles of alumina, titania and hydroxyapatite. These nanostructured substrates promoted adhesion of osteoblasts and decreased adhesion of fibroblasts (i.e., cells that aid in fibrous encaptulation and formation of calluses that may lead to the loosening of the implant and failure). Nanostructured substrates also decreased adhesion of endothelial cells, which line the vasculature of the human body. An interesting outcome of this study was that compared to conventional alumina, titania and hydroxyapatite formulations; calcium deposition by osteoblast was four, three and two times greater, respectively, on the same nanostructured material after 28 days of culture (Webster *et al.* 2004; Gutwein *et al.* 2004; Webster *et al.* 2001). Other research by the same group showed that the nanostructured formulations improved and controlled the mechanical properties of alumina (Webster *et al.* 1999). Other researchers have also validated the fact that increased calcium deposition is observed on alumina nanofibers, carbon nanofibers, polylactic glycolic acid, polyurethane and other composites. The natural increase in surface area of nanostructured materials was accounted for in each of those studies, which showed that the chemistry between nanostructured and conventional materials was similar (Webster *et al.* 2004).

During recent years, there have been attempts in exploring nanotechnology to create nanoscale bioceramics, to improve properties and to overcome limitations of bioinert ceramics. Kong *et al.* (2005) of Seoul National University developed zirconia-alumina nano-composite using the Pechini process and improved the biocompatibility of the composite by addition of hydroxyapatite. The Pechini process is based on the formation of a network composed of a polymeric precursor and a cationic network modifier (Brinker *et al.* 1990). Zirconia-alumina nano-composites can be used in load-bearing applications such as dental/orthopedic implants. Because the nano-composite powder effectively decreased the contact area between hydroxyapatite and zirconia for their reaction during the sintering process, the hydroxyapatite added zirconia-alumina nano-composites contained biphasic calcium phosphates (hydroxyapatite and tricalcium phosphate) and had higher flexural strength than conventionally mixed zirconia–alumina composites. The composites showed increased osteoblast activities (as expressed by the alkaline phosphatase activity) with the increase in the amount of hydroxyapatite content. Kong *et al.* (2005) concluded that, nano-composite containing 30 vol.% hydroxyapatite and 70 vol.% zirconia-alumina is the optimal composition, based on the mechanical and biological evaluations, for load-bearing biological applications.

It has been established that the merits of nanotechnology offers a unique approach to overcome shortcomings of many conventional materials. From nanomedicine to nanofabrics, this promising technology has encompassed almost all disciplines of human life. Nanostructured materials offer much improved performances than their large-particle-sized counterparts due to their large surface-to-volume ratio and unusual chemical/electronic synergistic effects. Nanoscale ceramics can exhibit significant ductility before failure contributed by the grain-boundary phase.

In 1987, Karch *et al.* (1987) reported that, with nanograin size, a brittle ceramic could permit a large plastic strain up to 100%. Nanostructured biomaterials promote osteoblast adhesion and proliferation, osteointegration, and the deposition of calcium-containing minerals on the surface of these materials (Xu *et al.* 2004). Also, nanostructured ceramics can be sintered at a lower temperature; thereby, problems associated with high-temperature sintering processes are eliminated. It is possible to enhance both the mechanical and biological performance of calcium phosphates by controlling characteristic features of powders such as particle size and shape, particle distribution and agglomeration (Best *et al.* 1994). Nanostructured bioceramics clearly represent a promising class of orthopedic and dental implant formulations with improved biological and biomechanical properties.

## 2.2.3. Calcium Phosphate Bioceramics

Recent years have seen more emphasis on bioactive and bioresorbable ceramic research; however, the materials of choice for load-bearing applications in the implant industry are still bioinert ceramics like $\alpha$-alumina and tetragonal zirconia. In all the prospective applications of bioactive and bioresorbable bioceramics, calcium phosphates have made a significant contribution. Today, calcium phosphate–based bioceramics are the materials of choice in both dentistry and medicine.

In 1920, Albee reported the first successful medical application of calcium phosphate bioceramics in humans (LeGeros 2002), and in 1975 Nery *et al.* reported the first dental application of these ceramics in animals. They have been used in the field of biomedical engineering owing to the range of properties they offer: from tricalcium phosphates being resorbable to hydroxyapatite being bioactive, they are undeniably the current rage for clinical usage (Hench 1998; Ducheyne *et al.* 1999).

The superior biocompatibility of calcium phosphates contributed by their compositional resemblance to bone mineral has allowed them to be used for copious applications inside the body. Calcium phosphate bioceramics exhibit considerably improved biological affinity and activity compared to other bioceramics. However, unlike alumina and zirconia, these ceramics are mechanically weak and exhibit poor crack growth resistance, which limits their uses to nonload-bearing applications such as osteoconductive coatings on metallic prosthesis and as nano-powders in spinal fusion. Among different forms of calcium phosphates, particular attention has been paid to tricalcium phosphate ($Ca_3(PO_4)_2$) and hydroxyapatite ($Ca_{10}(PO_4)_6(OH)_2$) due to their outstanding biological responses to the physiological environment. The contemporary health care industry uses calcium phosphate ceramics in various applications, depending upon whether or not a resorbable or bioactive material is ideal. The recent trend in bioceramic research has been focused in overcoming the limitations of calcium phosphates, specifically hydrox-yapatite ceramics, and in improving their biological properties by exploring the unique advantages of nanotechnology.

The trend is shifting towards nanotechnology to improve the biological re-sponses of hydroxyapatite because nano-hydroxyapatite is a constituent of bone, which is a natural composite of nano-hydroxyapatite with collagen fibers. The main constituents of bone are collagen (20 wt.%), hydroxyapatite (69 wt.%), and water (9 wt.%). Additionally, other organic materials, such as proteins, polysac-charides, and lipids are also present in small quantities. Collagen, which can be considered as the matrix, is in the form of small micro fibers. It is difficult to ob-serve distinct collagen fibers because of its netlike mass appearance. The diameter of the collagen microfibers varies from 100 to 2000 nm. Calcium phosphate in the form of crystallized hydroxyapatite and/or amorphous calcium phosphate (ACP) provide stiffness to the bone. The nano-hydroxyapatite crystals, present in the form of plates or needles, are about 40-60 nm long, 20 nm wide, and 1.5-5.0 nm thick. They are deposited parallel to the collagen fibers, so that the larger dimension of crystals is along the long axis of the fiber. It is worth mentioning that the mineral phase present in bone is not a discrete aggregation of the nano-hydroxyapatite crystals. It is rather made of a continuous phase, which is evidenced by very good strength to the bone after a complete removal of the organic phase. The use of synthetic nanoscale hydroxyapatite in orthopedic applications is therefore con-sidered to be very promising, owing to its dimensional similarity with the bone crystals.

Calcium phosphates being light in weight, chemically stable and compo-sitionally similar to the mineral phase of the bone, are preferred as bone-graft materials in hard tissue engineering. They are composed of ions commonly found in physiological environment, which make them highly biocompatible. In addition, these bioceramics are also resistant to microbial attack, pH changes and solvent conditions. They exist in different forms and phases depending on temperature, partial pressure of water and the presence of impurities (Groot et al. 1990; Hench 1998). Hydroxyapatite, β-TCP, α-TCP, biphasic calcium phosphate (BCP) (Wong et al. 2002), monocalcium phosphate monohydrate (MCPM) and unsintered apatite (AP) are different forms of commercially available calcium phosphates currently used in the biomedical industry. Different phases are used in different applications depending upon whether a resorbable or bioactive material is desired (Billottee 2002). Hydroxyapatite is the ideal phase for applications inside human body be-cause of its excellent stability above pH 4.3, human blood pH being 7.3.

One of the major shortcomings of calcium phosphate bioceramics is their poor mechanical strength under complex stress states. Further, it has been proved that the bioactivity of synthetic calcium phosphates (micron-size powder) is inferior to natural apatite, the bone mineral (Driessens et al. 1978; LeGeros et al. 1978). Like other ceramic materials, the tensile and compressive strengths of calcium phosphates are governed by the presence of voids, pores or interstices, which results during the process of densification while sintering. However, unlike most advanced ceramics, calcium phosphates are difficult to sinter and thus are mechanically weak. The resistance to fatigue is another essential factor for load-bearing implants. In terms of the Weibull factor, $n$, values of $n = 50\text{-}100$ usually signify good resistance, but values of $n = 10\text{-}20$ are insufficient and may fail in several months of usage (Putter *et al.* 1983). For conventional hydroxyapatite, n = 50 in dry environment and n = 12 in a wet physiological implant bed, which is below the satisfactory limit as reported by Putter *et al.* Nanotechnology is one of the approaches, which has been explored recently to improve both the strength and toughness of this novel group of bioceramics to make them useful in load-bearing applications.

Conventional calcium phosphate based ceramic powders suffer from poor sinterability, possibly due to their low surface area (typical 2-5 $m^2$/gm). In addition, it has been also recorded that the resorption process of synthetic calcium phosphates (conventional forms) is quite different from that of bone mineral. Bone mineral crystals are nano-sized with a very large surface area. They are grown in an organic matrix and have very loose crystal-to-crystal bonds; therefore, the resorption by osteoclasts is quite homogeneous. Calcium phosphates (micron size), on the contrary, present a low surface area and have strong crystal-to-crystal bond. Resorption takes place in two steps: (i) disintegration of particles and (ii) dissolution of the crystals [Heughebaert *et al.* 1988]. Nanoscale bioceramics is one of the emerging approaches that have been extensively studied recently by various researchers to find a solution to these long standing problems associated with calcium phosphates. In a recent publication, Kim *et al.* reported that biomineralization of calcium phosphate nano-crystals on ceramics with specific compositions and structures is a core mechanism of bioactivity, that inspires acellular and protein-free biomimetic strategies for bio-interactive materials with new physical, chemical and biological functions, e.g., bioactive surface functionalizations on tough metallic and ceramic materials, sol-gel derivation of bioactive ceramic-polymer nano-hybrids and textured biomimetic depositions of nano-calcium phosphate on polymer templates [Kim 2003].

Crystallization of various salts of calcium phosphates like hydroxyapatite and $\beta$-TCP depends on the Ca/P ratio, presence of water and impurities, and temperature. For instance, in a wet environment and at a lower temperature ($<900°C$), the formation of hydroxyapatite is most likely to happen, but in a dry atmosphere and at a higher temperature, $\beta$-TCP is more likely to form. Two important phases of calcium phosphates—hydroxyapatite and $\beta$-tricalcium phosphate, which have been extensively studied at the nano-range during recent years—are discussed in detail in the next two sections.

### 2.2.3.1. Nanocrystalline Hydroxyapatite.

Hydroxyapatite ($Ca_{10}(PO_4)_6(OH)_2$) is a bioactive ceramic widely used as powders or in particulate forms for various bone repairs and as coatings for metallic prostheses to improve their biological properties (Liu *et al.* 2001). Hydroxyapatite is thermodynamically the most stable calcium phosphate ceramic compound at the pH, temperature and composition of the physiological fluid (Correia *et al.* 1995).

Recently, hydroxyapatite has been used for a variety of biomedical applications, including matrices for drug release control (Itokazu *et al.* 1998). Due to the chemical similarity between hydroxyapatite and mineralized bone of human tissue, synthetic hydroxyapatite exhibits a strong affinity to host hard tissues. Formation of a chemical bond with the host tissue offers hydroxyapatite a greater advantage in clinical applications over most other bone substitutes such as allografts or metallic implants (Bauer *et al.* 1991).

Hydroxyapatite has a hexagonal structure with a $P6_3/m$ space group and cell dimensions $a = b = 9.42$ Å, and $c = 6.88$Å, where $P6_3/m$ refers to a space group with a sixfold symmetry axis with a threefold helix and a microplane (Groot *et al.* 1990). It has an exact stoichiometric Ca/P ratio of 1.67 and is chemically very similar to the mineralized human bone (Bow *et al.* 2004). However, in spite of chemical similarities, the mechanical performance of synthetic hydroxyapatite is very poor compared to bone. In addition, the bone mineral presents a higher bioactivity compared to synthetic hydroxyapatite.

Many researchers have observed that the mechanical strength and fracture toughness of hydroxyapatite ceramics can be improved by the use of different sintering techniques; which include addition of a low melting secondary phase to achieve liquid phase sintering for better densification (Lopes *et al.* 1998; Santos *et al.* 1996), incorporation of sintering additives to enhance densification through grain boundary strengthening (Georgiou *et al.* 2001; Kalita *et al.* 2004; Kalita *et al.* 2004a), and use of nanoscale ceramic powders for better densification contributed by large surface area to volume ratios of nano-size powders (Bhatt *et al.* 2005).

It is believed that nanoscale hydroxyapatite has the potential to revolutionize the field of biomedical science from bone regeneration to drug delivery. During the past ten years, much attention has been given to nanostructured hydroxyapatite calcium phosphate ceramics, with major research emphasis on production of nanoscale powders to improve the mechanical as well as biological properties.

The importance and advantages of nanocrystalline hydroxyapatite were highlighted by Sarig and Kahana in a 2002 issue of the *Journal Crystal Growth, vol. 237*. In their work, Sarig *et al.* (2002) could synthesize hydroxyapatite with 300 nm edges, which were loosely aggregated into spherulites of 2-4 mm dimensions.

Nanocrystalline hydroxyapatite powders exhibit improved sinterability and enhanced densification due to greater surface area, which could improve the fracture toughness, as well as other mechanical properties (Yeong *et al.* 1999). Moreover, nano-hydroxyapatite is also expected to have better bioactivity than coarser crystals (Stupp *et al.* 1992; Webster *et al.* 2001).

Li *et al.* (2002) developed hydroxyapatite nano-wires in anodic oxide aluminum with similar biological *in vivo* orientation using template-assisted technology. Yingchao *et al.* (2004) synthesized nano-hydroxyapatite of 8-10 nm sizes via a nontemplate-mediated sol-gel technique. In most literature on the synthesis of hydroxyapatite by the sol–gel process, phosphorous alkoxide has been used as precursor for phosphorous. Liu *et al.* (2002) used tri-ethyl phosphate and calcium nitrate as the precursors respectively for phosphorous and calcium for hydroxyapatite synthesis. Kuriakose *et al.* (2004) suggested that agars can also be used to synthesize hydroxyapatite using similar precursors at low temperatures. These processes require high-temperature operation and produces multiphase powder. A relatively simpler sol-gel process using ethanol and/or water as a solvent has also been reported to obtain stoichiometric, nanocrystalline single-phase hydroxyapatite.

Han *et al.* (2004), synthesized nano-crystalline hydroxyapatite powder by the citric acid sol-gel combustion method. The attractive features of this method were to synthesize materials with high purity, better homogeneity and high surface area in a single step.

Shih *et al.* (2004), synthesized nano-hydroxyapatite powder (Ca/P = 1.67) of 20 nm particle size by the hydrolysis method. They also observed that the hydroxyapatite particle size increases with the annealing temperature. An annealing temperature of 1000°C resulted in an average increase in particle diameter up to 50 nm. Xu *et al.* (2005) used a radio frequency (r.f.) plasma spray process to synthesize nano-sized hydroxyapatite powder with particle size in the range of 10-100 nm.

Synthesis of stoichiometric nano-hydroxyapatite powders by the sol-gel method is relatively easy. It gives higher product purity, more homogeneous composition and a comparatively low synthesis temperature than other methods. Sol–gel derived hydroxyapatite is always accompanied by the secondary phase of calcium oxide. Phosphorous, alkoxides, gels and ethanol can be used as solvents in this method. Kuriakose *et al.* (2004) synthesized nano-crystalline hydroxyapatite of size 1.3 nm radius that is stable until 1200°C, without any by-products in the samples synthesized with pores in the crystal planes, using the latter as solvent.

Synthesis of pure hydroxyapatite crystals of 8-10 nm size can also be done by a novel sol-gel technique using agars.

Han *et al.* (2004) synthesized nanocrystalline hydroxyapatite powder at low calcination temperature of 750°C by a citric acid sol-gel combustion method. The grain size of the resulting powder was found to be between 80 to 150 nm and the open porosity to be 19% (Yingchao *et al.* 2004).

Sarig *et al.* (2002) synthesized nanocrystalline plate-shaped particles of hydroxyapatite, directly precipitated from dilute calcium chloride and sodium phosphate solutions at ambient temperature. The solution was introduced to microwave irradiation immediately after mixing. The pH of the solution was kept at 7.4 to make the hydroxyapatite suitable for medical applications. This method is a relatively fast method to synthesize nano-hydroxyapatite. Bhatt *et al.* (2005) synthesized nanocrystalline hydroxyapatite powder of size 5-10 nm in diameter using water-based sol-gel techniques. Powder X-ray diffraction results showed that the apatite phase first appeared at 400°C and the hydroxyapatite content increased with increase in the calcination temperatures (Bhatt *et al.*, 2005). **Figure 5** shows TEM micrograph of nanocrystalline hydroxyapatite powder produced via a sol-gel technique.

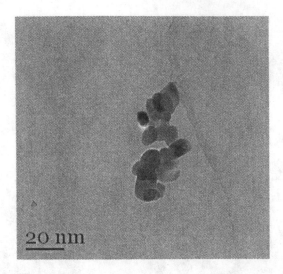

**FIGURE 5.** TEM micrograph of nanocrystalline hydroxyapatite powder produced via sol-gel technique.

Nano-sized hydroxyapatite particles can also be prepared by chemical precipitation, through aqueous solutions of calcium chloride and ammonium hydrogen phosphate. Pang *et al.* (2003) observed that the crystallinity and crystallite size of hydroxyapatite increases with the increase of synthetic temperature and ripening time when the solution is prepared by this method. Also, the morphology change of hydroxyapatite nanoparticles is related to their crystallinity. The regular shape and smooth surface of the nanoparticles can be obtained by the higher crystallinity of hydroxyapatite. They observed needlelike nanoparticles with a rough surface and blurred contour and higher combined water content for lesser crystalline hydroxyapatite, whereas a barlike shape with smooth surface, clear contour and lower water content was observed for nanoparticles with higher crystallinity.

Mechano-chemical processing (MCP) is another compelling method to produce nanostructured hydroxyapatite in the solid state. Yeong *et al.* (2001) used appropriate amounts of calcium hydrogen phosphate ($CaHPO_4.2H_2O$) and calcium oxide (CaO). Hong *et al.* (2005) synthesized nanocrystalline hydroxyapatite ($Ca_{10}(PO_4)_6(OH)_2$ powder of $\sim 50$ nm by mechanical solid-state reaction of $Ca(OH)_2$ and $P_2O_5$ mixtures in a high-energy shaker mill, using hardened steel vial and balls. SEM micrograph of nanocrystalline hydroxyapatite powder produced by mechanical solid-state reaction is shown in Fig. 6. Suchanek *et al.* (2002) performed mechanochemical-hydrothermal synthesis of carbonated apatite powders at room temperature.

### 2.2.3.2. Nanocrystalline Tricalcium Phosphate. 
Tricalcium phosphate bioceramic has been proved to be resorbable *in vivo* with new bone growth

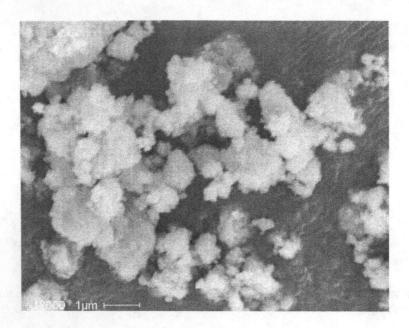

**FIGURE 6.** SEM micrograph of hydroxyapatite nano powder produced by mechanical solid-state reaction.

replacing the implanted ceramic (Gibson *et al.* 2000). Tricalcium phosphate bioceramic exists in three polymorphs, namely, β-tricalcium phosphate, α-tricalcium phosphate and α'-tricalcium phosphate (Elloitt 1994; Zhou *et al.* 1993). β-tricalcium phosphate transforms to α-tricalcium phosphate at around 1180°C. α-Tricalcium phosphate is stable between 1180°C and 1430°C and, α'-tricalcium phosphate is stable above 1430°C (Elloitt 1994; Zhou *et al.* 1993). α-Tricalcium phosphate and α/-tricalcium phosphate, however, have received very little interest and attention in biomedical materials research. The disadvantage of using α-tricalcium phosphate in this area is its fast resorption rate (Suchanek *et al.* 2002). On the contrary, β-tricalcium phosphate also known as β-whitlockite, is a slow-degrading resorbable phase (Driessens *et al.* 1978) and is thus a promising material in biomedical applications.

β-tricalcium phosphate is known to have significant biological affinity and activity and hence, responds well to the physiological environments (Kivrak *et al.* 1998). Because of these positive characteristics, porous β-tricalcium phosphate is regarded as an ideal bone substitute, which would degrade *in vivo* with time, allowing bone tissue to grow inside the scaffold. These factors give β-tricalcium phosphate an edge over other biomedical materials when it comes to resorbability and replacement of the implant *in vivo* by new bone tissues (Gibson *et al.* 2000). It is reported that the resorbability of β-tricalcium phosphate *in vivo* may be strongly related to its structure (Okazaki *et al.* 1990). β-Tricalcium phosphate belongs to

R3CH space group and has a hexagonal crystal structure. Its unit cell dimensions are: $a = b = 10.439$ Å, $c = 37.375$ Å, and $\alpha = \beta = 90°$, $\gamma = 120°$. In spite of favorable biodegradation behavior of $\beta$-tricalcium phosphate over $\alpha$-tricalcium phosphate and hydroxyapatite phases, its rate of degradation is uncontrolled, which limits its use in most bone tissue engineering applications.

A number of synthesis methods have been used to produce $\beta$-tricalcium phosphate powders. The conventional methods include solid-state process (Bigi et al. 1997) and wet-chemical method (Liou et al. 2002). The wet-chemical method gives Ca-deficient apatite [$Ca_9(HPO_4)(PO_4)_5(OH)$, CDHA] with the same molar ratio of Ca/P as that of tricalcium phosphate. It needs to be calcined above 700-800°C to transform into $\beta$-tricalcium phosphate as shown by the following reaction:

$$Ca_9(HPO_4)(PO_4)_5(OH) \rightarrow 3Ca_3(PO_4)_2 + H_2O$$

Synthesis of nano-$\beta$-tricalcium phosphate has been formulated by many researchers using starting materials as $(CH_3COO)_2Ca.xH_2O$ as the Ca sources and $H_3PO_4$ as the P sources. Bow et al. (2004) synthesized nano-sized $\beta$-TCP powder of ~50 nm particle diameter at room temperature in methanol solvent. They found that phase transformation was taking place from $CaHPO_4$, intermediate amorphous calcium phosphate (ACP) phases (including ACP1 and ACP2 with different structures) to final $\beta$-tricalcium phosphate with increase in aging time. They observed that the incorporation of carbonate helps in suppressing the transformation of ACP1 phase with hydroxyapatite-like structure into poorly crystalline CDHA and favors the formation of $\beta$-tricalcium phosphate phase.

It was observed that the presence of micropores in the sintered material affects the bioresorption of tricalcium phosphate-based ceramic implants; therefore, the size and amount of macro and micro porosity must be controlled closely during manufacturing processes (Ioku et al. 1996). SEM studies revealed that the appearance of needles or petal-like plates is characteristic of nano $\beta$-tricalcium phosphate-based cements (Genibra et al. 1998). Nano-tricalcium phosphate powders can be compacted into cylindrical pallets and then sintered for mechanical and biodegradation studies to achieve appropriate mechanical strength, to be used as drug delivery devices. Metsger et al. (1999) reported a 21 GPa value for the Young's modulus of nano-$\beta$-TCP ceramic. The mechanical strength of the cement was enhanced when the nano-structures were immersed for 24 h and 7 days in SBF (Metsger et al. 1999).

Nano $\beta$-tricalcium phosphate cements can serve as drug delivery systems for a variety of remedies such as antibiotics, antitumor and antiinflammatory drugs, etc., which can easily be added to them (Yu et al. 1992). Also, $\beta$-TCP prepared by wet precipitation procedure from an aqueous solution of $Ca(NO_3)_2$ and $NaH_2PO_4$ and calcined at 1150°C can be used as bone substitutes after grinding and sieving to obtain the desired particle size.

A subsidiary of Tredegar Corporation has recently received FDA clearance to market a new resorbable $\beta$-tricalcium phosphate bone void filler device used to treat osseous defects of the skeletal system. Future directions are aimed at creating a therapeutic nano-TCP coating that has a dual beneficial effect: osteoconductive properties combined with the ability to deliver therapeutic agents, proteins, and growth factors directly into the coating. These new coatings may offer the ability to stimulate bone growth, combat infection, and ultimately increase an implant's lifetime.

Nano-phase calcium phosphate bioceramics have gained regard in the biomedical field due to their superior biological and biomechanical properties. Hydroxyapatite and β-TCP are essentially the main calcium phosphates used in clinics at present. Several ways of synthesis of both these forms of CPCs on the nanoscale have evolved in the past few decades. Nano-hydroxyapatite is used primarily as bioactive coatings on metallic prosthesis of bioinert materials like titanium and its alloys, in bone tissue repairs and implants and also for drug delivery. Nanoscale β-TCP exhibits significant biological affinity and activity and responds very well to the physiological environments. Also, owing to its slow degradation characteristic, the porous β-TCP is regarded as an ideal material for bone substitutes that should degrade by advancing bone growth and as a good candidate material for drug delivery devices.

## 2.3. Polymeric Biomaterials

Tissue engineering has provided an alternate medical therapy using implants of polymeric biomaterials with or without living precursor cells against various transplants. A number of different polymers can be used as scaffold to promote cell adhesion, maintenance of differentiated cell functions without hindering proliferation. Synthetic and natural polymers can also be used as templates to organize and direct the growth of cells and to help in the function of extracellular matrices (ECM) (Patrick Jr. *et al.* 1988; Thomson *et al.* 2000).

Both natural and synthetic polymers have become a part of day-to-day as well as advanced medical care in the biomedical industry. The most commonly investigated natural polymers include alginate, collagen, hyaluronic acid and chitosan. Much of the interest in these natural polymers is because of their biocompatibility, relative abundance, ease of processing and/or their possible ability to imitate the microenvironment found within cartilage. Alginates are linear unbranched polymers that are produced by seaweeds. Alginate gels can easily encapsulate various pharmaceuticals, growth factors or cultured cells (Seal *et al.* 2001). They are stable in a range of organic solvents and, in contrast to other hydrogels, are possibly useful in applications that involve enzyme entrapment in nonaqueous systems. Alginate applications include encapsulated pancreatic islets, vesicoureteral reflux, topical wound care, sustained drug delivery and dental applications. A disadvantage of using alginate beads is that they become unstable upon contact with complexes such as phosphates, citrates, EDTA and lactates.

Collagen is a gelatinous protein found in most of the multicellular organisms to which it gives strength and flexibility. There are more than a dozen types of collagen. Each type is differentiated by a particular sequence of amino acids in its molecules. The collagen in the human body is called type I collagen. Each collagen microfibril has a characteristic banding pattern with a repeat period of 68 nm. Type I collagen is also found in skin, tendons, and ligaments (Martin 1999). The hydroxyapatite, or bone mineral, crystals are embedded in a collagenous matrix to form bone tissue (Baron 1996). The collagen matrix provides bone its tensile strength, while the impregnated ceramic crystals provide the compressive strength and rigidity. It forms molecular cables that strengthen tendons and many resilient sheets that support the skin and internal organs (Seal *et al.* 2001). Functions of collagen range from serving crucial biomechanical functions in bone, skin, tendon and ligaments to controlling cellular gene expression in development.

Hyaluronic acid or hyaluronan is a natural glucosaminoglycan polymer that is plentiful within cartilaginous extracellular matrix. It is used to melt hardened scar tissue, shorten injury recovery period, and to treat muscle spasm and fibromyalga. It is also used to restore circulation and relieve pain. Chitosan is another natural biopolymer which is a polysaccharide derived from chitin and is composed of a simple glucosamine monomer. Chitosen is a biocompatible and biodegradable polymer and does not evoke a strong immune response. It binds aggressively to mammalian and microbial cells, accelerates formation of osteoblasts, and has a regenerative effect on connective gum tissue. It is also helostatic, fungistatic, spermicidal and a central nervous system depressant. Chitosen is used for wound healing and burn treatment and has also been shown to reduce serum cholesterol levels.

Synthetic polymeric materials have been widely used in medical disposable supply, prosthetic materials, dental material, implants, dressings, extracorporeal devices, encapsulants, polymeric drug delivery systems, tissue engineered products, and in orthopedics. The main advantages of the polymeric biomaterial compared to metal and ceramic materials are ease of manufacturability to produce various shapes, ease of secondary processibility, reasonable cost, and availability with desired mechanical and physical properties. **Table 3** list various synthetic polymeric biomaterials developed so far with examples of their applications in the health care industry. Several excellent reviews and articles on polymeric biomaterials exist (Seal *et al.* 2001). Readers are recommended to refer to these articles for more details on generic polymeric biomaterials. This section will discuss some of the recent advances on nanostructured polymeric biomaterials, which have shown promising results.

Cells in our body live in a complex mixture of pores, ridges and fibers of ECM at nanometer scales (Desai *et al.* 2000; Flamming *et al.* 1999; Lee *et al.* 1998). Therefore, a nanostructured porous scaffold with interconnective pores and large surface area is presumed to be an ideal alternative/substitute to natural ECM for better cell in-growth in three dimensions. Such scaffolds will enable us to mimic the structure and functions of natural tissues. Nanostructured polymeric biomaterials are a new avenue in tissue engineering applications. Polymeric scaffolds or extracellular matrices with nano-features have been shown to exhibit improved biological properties as demonstrated by some of the recent findings in this area.

**TABLE 3.** List of synthetic polymeric biomaterials and their applications in the biomedical industry.

| Synthetic Polymeric Biomaterials | | | |
| --- | --- | --- | --- |
| Chemical Name | Abbr. | Tissue-response | Applications |
| Polymethylmethacrylate | PMMA | Bioinert | Bone cement, blood pump and reservoirs, membrane for blood dialyzer, implantable ocular lens, contact lens. |
| Polyethylene | PE | Bioinert | Pharmaceutical bottles, catheter, pouch, flexible container, and orthopedic implants. |
| Polyvinylchloride | PVC | Bioinert | Blood and solution bag, surgical packaging, IV sets, dialysis devices and catheter bottles. |
| Polylactic acid | PLA | Bioresorbable | Fracture fixation plates and screws, drug control-release devices, coating materials for sutures and vascular grafts. |
| Polyglycolic acid | PGA | Bioresorbable | Sutures, pins, screws, ligament reconstruction and nerve regeneration. |
| Poly($\varepsilon$-caprolactone) | PCL | Bioresorbable | Biodegradable sutures. |
| Poly(amino acid) | | Bioresorbable | Drug delivery systems, vaccines delivery devices and in gene therapy. |
| Polypropylene | PP | Bioinert | Disposable syringes, blood oxygenetor membrane, suture, nonwoven fabric, and artificial vascular grafts. |
| Poly-$p$-dioxanone | PDS | Boresorbable | Monofilament synthetic sutures. |
| Polyethyneletetrafluorate | PET | Bioinert | Implantable suture, mesh, artificial vascular grafts and heart valve. |
| Polytetrafluoroethylene | PTFE | Bioinert | Catheter and artificial vascular grafts (excellent thermal insulating properties). |
| Polyurethane | PU | Bioinert | Film, tubing, and components |
| Polyamide | Nylon | Bioinert | Packaging film, catheters, sutures, and mold parts |

Yang *et al.* (2004) fabricated and studied nanostructured porous poly-lactic acid (PLLA) scaffold intended for nerve tissue engineering (NTE). Restoring the functions of the adult central nervous system (CNS), which does not heal through self-regeneration, has been a challenge for neurobiologists and neurologists (Tresco *et al.* 2000; Woerly *et al.* 1996). Traditionally, tissue transplantation or peripheral nerve grafting are used to repair damaged or diseased regions at the CNS, which encounters some of the common problems—e.g., shortage of donor and immunological problems associated with transmittance of infectious disease (Hudson *et al.* 2000).

Beyond a doubt, NTE is one of the most promising methods to repair and restore the central nervous system in humans. Three-dimensional distribution, growth and proliferation of cells within the porous scaffold are of clinical significance in NTE. Results of Yanga *et al.'s* work suggested that the nano-structured porous PLLA scaffold showed better cell adhesion and differentiation *in vitro* and has proved to be a better potential carrier matrix for cells in nerve tissue engineering applications (Yang et *al.* 2004). Several methods have also been reported to fabricate polymeric nano-fibers by using various techniques including phase separation (Ma *et al.* 1999), electro-spinning (Bognitzki *et al.* 2001) and self-assembly (Hartgerink *et al.* 2001).

Chemical and topographical cues are of importance in guiding cell response, which can be exploited to our benefits to design and develop biomaterials with improved biological properties. Dalby *et al.* (2004) studied the changes in fibroblast morphology in response to nano-columns produced by colloidal lithography. Filopodia are actin-driven structures produced by cells, which are speculated to be involved in cell sensing of the three-dimensional environment. Dalby *et al.* (2004) quantified filopodia response to cylindrical nano-columns of 100 nm diameter and 160 nm height produced by colloidal lithography. Their results showed that the fibroblasts produced more filopodia per millimeter of cell perimeter and that filopodia often could be seen to interact with the cells' nano-environment. By understanding which features evoke spatial reactions in cells, it may be possible to design improved biomaterials that possess nano-features. The desired response may be to align the cells, to increase motility, to increase or reduce proliferation and to alter differentiation. Their results add to the theory that filopodia are involved in gathering special information from the cell's environment. When designing new-generation tissue engineering materials, it will be important to present cells with cues that will elicit responses (Dalby *et al.* 2004).

Micron fibers of dacron and polytetrafluoroethylene (PTFE) have had success in replacing large arteries in treating atherosclerotic vascular disease, which is a major cause of death in the western world. This appears in the form of coronary artery and peripheral vascular disease. Current treatment options include passing occluded arteries with either autologous veins or biocompatible synthetic materials.

To date, the material of choice for small-diameter vascular grafts is an autogeneous vein (Darling *et al.* 1972). However, some 30% of patients do not have healthy and available autologous veins due to disease or previous use (Darling *et al.* 1972; Clayson *et al.* 1976), which necessitates an alternate synthetic vascular graft. The biomaterials of choice for such uses are obviously polymeric materials. Synthetic polymers still cannot be used in smaller-diameter arteries, namely, those under 6 mm (Abbott *et al.* 1993). Thrombosis and intimal hyperplasia have plagued small-diameter vascular grafts, causing a substantial drop in potency rate to 30% within five years (Abbott *et al.* 1993), which has created a need for development of new materials. One promising solution has recently been reported by Miller *et al.* (2004).

A novel class of polymers that is under investigation to become the "next-generation" of small-diameter vascular grafts with increased efficacy is that of biodegradable polymers such as poly(lactic acid), poly(glycolic acid), and their co-polymers (Mikos *et al.* 2000). Tubular scaffolds of poly(glycolic acid) can support vascular smooth muscle and endothelial cell growth *in vitro* (Gao *et al.* 1998; Salon *et al.* 2001; Nikolovski *et al.* 2000). However, these materials are composed of micron-sized fibers and micron surface topographies that are dissimilar to the natural nanodimensional vascular tissues (Chu *et al.* 1999; Flemming *et al.* 1999).

Cellular response that is important for vascular tissue regeneration can be enhanced on PLGA through the creation of a biologically inspired nanometer surface topography. Since the arterial wall tissue has a high degree of nanometer surface roughness, it is important to design optimal tissue-engineering PLGA scaffolds having nanoscale features. Miller *et al.* (2004) studied the endothelial and vascular smooth muscle cell function on poly(lactic-co-glycolic acid) with nanostructured surface features. Their goal was to synthesize biodegradable synthetic vascular materials that mimic the natural nanostructural characteristics of the arterial wall in an effort to improve vascular cell adhesion and growth (Miller *et al.* 2004). Synthetic microstructured materials have failed to prove their efficacy in replacing small-diameter vascular grafts, possibly because of their inability to mimic the natural nanometer surface architecture of the vessel wall (Miller *et al.* 2004).

Miller *et al.* (2004) showed for the first time that by mimicking the natural surface characteristics of native vascular tissue, vascular cell adhesion and growth on PLGA films can be enhanced. Specifically, vascular smooth muscle and endothelial cell densities were improved on PLGA films that have increased nanometer surface roughness without changes in chemistry.

Morimoto *et al.* (2004) examined the nanoscale surface modification of a semi- interpenetrating polymer networks (IPN) film composed of segmented polyurethane (SPU) and cross-linked poly (2-methacryloyloxyethyl phosphoryl-choline (MPC)-co-2-ethylhexyl methacrylate (EHMA)) polymer prepared with a mica interface (Morimoto *et al.* 2004). By varying the monomer composition and the diffusion time, they controlled MPC condensation on the surface, the depth of the graded layer, and the size of MPC-enriched domains at a nanometer level. With increase in the MPC unit on the surface, the modified film showed remarkably reduced platelet adhesion and activation. Their work showed that nanoscale surface modification may be an effective method for preparing elastic polymer biomaterials (Morimoto *et al.* 2004).

Kim *et al.* (2002, 2003) prepared polyethylene oxide (PEO) grafted polyurethane/ polystyrene IPNs with controlling the sea-island nanoscale domain structure by varying the PEO chain length and evaluated the biocompatibility (Kim JH *et al.* 2003; Kim JH *et al.* 2002).

Polymers are major candidates as materials for soft tissue repairs. In order to effectively replace the resected bladders, it is necessary to design a bladder tissue replacement construct with increased efficacy. The use of synthetic biodegradable polymers with micron feature dimensions (such as poly(lactic acid), poly(glycolic acid), poly(lactic-co-glycolic acid), in such constructs has been a major focus of work, to date. These polymers are promising due to the ease of manufacturing, reproducibility, designability, and manipulation of degradation (Nikolovski *et al.* 2000).

In addition, resorbable polymers have been preferable because polymers that do not degrade carry the permanent risk of infection, calcification and unfavorable connective tissue response (Atala *et al.*1993). It has been shown that a biocompatible material that mimics the nanometer topography of native bladder tissue will enhance cellular responses and lead to better tissue integration *in vivo* (Ayad *et al.* 1994). Reduction in the surface feature dimensions of poly(lactic-co-glycolic acid) (PLGA) and poly(ether urethane) (PU) films into the nanometer regime can be achieved via chemical etching procedures.

Bladder smooth muscle cell adhesion can be enhanced on chemically treated nanostructured polymeric surfaces compared to their conventional counterparts. Since the topography of natural soft tissues (including the bladder) results from constituent extracellular matrix proteins (having nanometer length and width) (Ayad *et al.* 1994), the next generation of polymeric bladder construct materials should incorporate nanodimensional surface features. It is expected that a biocompatible material that mimics this nanometer topography of native bladder tissue will enhance cellular responses, thereby leading to better tissue integration *in vivo*.

Thapa *et al.* (2003) investigated the effects of nanostructured polymers on cell functions of bladder smooth muscles. Results of Thapa *et al.'s* research suggested that bladder smooth muscle cell function can be enhanced by synthesizing nanodimensional polymeric surfaces that mimic the topography of native bladder tissue. These findings present strong evidence that nanodimensional polymers have great potential as novel implant materials for use in the design of the next generation of tissue-engineered bladder replacements. Ramanujam (2005) developed an optical fabrication process to develop nanostructured biopolymer surfaces through a maskless nano-patterning of the surface of a biocompatible polymer that can be employed for tissue engineering and cell growth (Ramanujam 2005). The technique is based on holographic diffraction grating recording using an ultraviolet laser in a biodegradable polymer containing various amino acids.

Improvement of the control of surface structures (on a micro- and nanoscale) could lead to better performance of the degradable biomedical devices such as surgical dressings, vascular grafts, tissue engineering scaffolds, sutures and structures for guided tissue regenerations. Flow-limited field-injection electrostatic spraying (FFESS) was developed to control the deposition of polymeric materials. Reducing the tension in the solvent surface and increasing the vapor pressure of the solvent will yield distinct surface features that include uniform nanoparticles. FFESS is a simple, powerful technique used for creating biomedical devices with a precisely defined nanostructure that has the potential to use a wide range of biocompatible polymeric materials (Berkland *et al.* 2004).

In the years to come, nanostructured polymeric biomaterials will find more applications in drug delivery and tissue regeneration applications. Recently published literature shows a very promising trend with regard to nanostructured polymeric biomaterials. There is no doubt that new and advanced polymers will be developed in the future to make them useful in various tissue engineering applications.

## 2.4. Composite Biomaterials

Composite materials offer a variety of advantages over metals, ceramics and polymers, which incorporate the desirable properties of each of the constituent materials, while mitigating the more limited characteristics of each component. Generally, a composite can be defined as a materials system composed of a mixture or combination of two or more constituents that differ in form and chemical composition and are essentially insoluble in each other. The properties of composite materials depend upon the shape of the heterogeneities, upon the volume fraction occupied by them, and upon the interfaces among the constituents. Most of the biological materials found in nature are composites. Examples of natural composites include bone, skin, wood and cartilages. The benefits of using composite biomaterials are the ability to tailor their properties as per need and thus provide significant advantage over homogeneous biomaterials. However, the development of composite biomaterials is not an easy task, as each constituent phase in the composite must be proved biocompatible. The properties at the interfaces also pose significant challenges ahead of biomaterial scientists.

The composite biomaterials can be classified as either bioinert, bioactive or bioresorbable. Some of the current applications of composite biomaterials in the biomedical industry include dental composites used as filler materials, coated metallic implants and reinforced polymethyl methacryalate bone cements used in many joint replacement surgeries. All kinds of composite biomaterials with different matrices—metallic, ceramic and polymeric—were developed and tested for biomedical uses. Amongst them, the most emphasis has been given to polymer-ceramic composite biomaterials in recent years. Examples of bioceramic composites that have been fabricated and analyzed include ceramic reinforced /bioactive ceramic composites (e.g., zirconia reinforced/AW glass), and polymer ceramic composites (e.g., calcium phosphate reinforced polyethylene). Among different polymer-ceramic composites so far developed for clinical applications, bio-inert polymer with bioactive ceramics (Bonfield *et al.* 1998; Tanner *et al.* 1994), bio-inert polymer with bioresorbable ceramics (Kalita *et al.* 2003) and bioactive polymer with bioactive ceramics (Kikuchi *et al.* 1997) have found specific uses in biomedical fields. Composite materials are a relatively recent addition to the class of materials used in structural applications. In the biomaterials field, the emergence of composites has been even more recent. In view of their potential for high performance, nano-composite materials are likely to find increasing use as biomaterials in various biomedical applications. Table 4 presents some of the current research in the filed of nano-composite biomaterials.

**TABLE 4.** Present and proposed applications of nanocomposite biomaterials.

| Literature | Present and proposed applications |
|---|---|
| Chen *et al.* 2002 | Hydroxyapatite/chitosan nanocomposites of homogeneous microstructure. Proposed to be helpful in producing uniform nano materials with best properties for biomedical applications. |
| Murugan *et al.* 2004 | Bioresorbable nano-hydroxyapatite composite bone paste with natural polysaccharide and chitosan. Anticipated to act as a bioresorbable bone substitute with superior bioactivity and osteoconductivity *in vivo*. |
| Li *et al.* 2004 | Hydroxyapatite/polyanhydride nanocomposite was formed. If the Hydroxyapatite content in the polyanhydrides was appropriate and compositions in the cross linking network are suitable, it meets the rehabilitation need of different fracture bones in human body, both in mechanical properties and in the biodegradable rate. |
| Rauschmann *et al.* 2004 | Nanocrystalline hydroxyapatite and calcium sulphate as biodegradable composite carrier material for local delivery of antibiotics in bone infections offers a new treatment option in osteomyelitis. |
| Guo *et al.* 2004 | Nano-Hydroxyapatite coatings on surfaces of titanium prosthesis to get improved biocompatibility and mechanical performance of the prosthesis. |

Pure hydroxyapatite ceramics have been widely used for biomedical implants and bone regeneration applications, as was discussed previously in an earlier section in this chapter. However, its applications in periodontal and alveolar ridge augmentation are limited due to its particles mobilization and slow resorbable nature. To overcome these limitations, hydroxyapatite is widely used in combination with some polymers and other compounds as composites. To augment its usage in this area, Murugan *et al.* (2004) prepared and characterized hydroxyapatite composite bone paste with a natural polysaccharide, chitosan, using wet chemical method at low temperature. Their findings suggest that the hydroxyapatite/chitosan composite paste would be highly beneficial for the particle immobilization upon implantation and may be a candidate bioresorbable material as a bone substitute (Murugan *et al.* 2004).

Chen *et al.* (2002) from Xiamen University, China, prepared and characterized nano-sized hydroxyapatite particles and hydroxyapatite/chitosan nano-composite for use in biomedical materials. They were able to produce nano-hydroxyapatite particles of 20-30 nm width and 50-60 nm length and particles of almost homogeneous microstructure so that they can be useful in producing uniform nano materials. The nanostructured hydroxyapatite/chitosan composite promises to have excellent biomedical properties for use in the clinical applications (Chen *et al.* 2002).

Recently, Rauschmann *et al.* (2004) assessed the material properties of a calcium sulphate nanoparticulate hydroxyapatite composite material and analyzed its *in vitro* uptake and release of vancomycin (an antibacterial used for treating infections in different parts of the body, usually administered with other antibiotics) and gentamicin (an antibacterial used for treating infections of the skin) antibiotics (Rauschmann *et al.* 2004). Their results suggest this composite to be a new treatment option in osteomyelitis (an acute or chronic bone infection usually caused by bacteria) owing to its good biocompatibility and sufficient antibiotic release.

Zhang *et al.* (2002) of Tsinghua University (Beijing) described their use of conventional and high-resolution transmission electron spectroscopy (HRTEM) to study nanofibrils of mineralized collagen. They have found a key mechanism behind how these fibrils self-assemble. They have also demonstrated for the first time that nano-hydroxyapatite crystals associate specifically with the surfaces of collagen fibrils. They observed that the hydroxyapatite crystals align themselves with the long axis of the collagen fibrils. Previously, other researchers had found that anions on the collagen molecules act as nucleation sites for hydroxyapatite crystals and that the positions of the hydroxyl groups in hydroxyapatite crystals lie along the same axis as the carbonyl groups in collagen.

Nanocomposite coatings of hydroxyapatite and zirconia ceramics were prepared using high-velocity oxy-fuel spray process by Li *et al.* (2004). In recent years, nanostructured coatings attracted great deal of interest from the scientific community due to their enhanced mechanical properties (Zhu *et al.* 2001). The improvement can be related to the classical Hall-Petch relationship, like most other nanostructured materials, where a decrease in particle size enhances the strength and toughness of ceramics (Wang *et al.* 1998; Kim *et al.* 1998). Elastic modulus of the coatings, determined using a nano-indentation technique, revealed that the composite coating had a Young's modulus value of 130 GPa.

Li *et al.*'s (2004) results also showed that the nano-sized zirconia particles (−90 nm) did not coarsen drastically after high velocity oxy-fuel deposition, and were uniformly distributed throughout the coating. The crystallite size of tetragonal zirconia in the coating was found to be less than 13 nm. Furthermore, decomposition of hydroxyapatite phase and chemical interaction between hydroxyapatite and zirconia was virtually undetected, which is beneficial towards the biological performance of the coatings.

## 3. CELL RESPONSE TO NANOBIOMATERIALS AND CURRENT ADVANCES

Cytotoxicity and biocompatibility of nanocrystalline and nanostructured biomaterials have been a concern for the scientist community. Some of the nanomaterials are found to be nontoxic and biocompatible (Webster *et al.* 2000; Webster *et al.*

2004; Webster *et al.* 2001), while some others have been shown to exhibit toxic and inflammatory responses. During recent years, nanomaterials such as nanotubes, nanowires, fullerene derivatives, and quantum dots have received enormous national attention in the creation of new types of analytical tools for biotechnology and the life sciences. Despite the wide application of nanomaterials, at present there is a lack of information concerning their impact on human health and the environment. Limited studies are available on toxicity of nanoparticles for risk assessment of nanomaterials. Further, emerging technology of using nanomaterials in medicine is still in its infant stage and therefore, a generic statement on toxicity of nanobiomaterials is premature at this point of time. It is interesting to note that cells may behave differently to a biomaterial in its powder form than in its bulk forms. *In vitro* test results conducted on certain nanopowders clearly showed that toxicity depends on powder concentration in the cell culture. Some other parameters that control toxicity of biomaterials including nanostructured biomaterials are synthesis/processing parameters as well as type or nature of the precursors used.

Jia *et al.* (2005) studied cytotoxicity response of three different carbon nanomaterials, i.e., singlewall nanotube (SWNT), multiwall nanotube (MWNT), and fullerene. They observed that profound cytotoxicity of SWNTs in alveolar macrophage after a 6-h exposure in vitro. No significant toxicity was observed for C60 up to a dose of 226 μg/cm$^2$. The cytotoxicity apparently followed a sequence order on a mass basis: SWNT > MWNT> quartz > C60. SWNTs significantly impaired phagocytosis of alveolar macrophage. They concluded that carbon nanomaterials with different geometric structures exhibit quite different cytotoxicity and bioactivity *in vitro* (Jia *et al.* 2005).

Tian *et al.* (2006) of Max Planck Institute for Metals Research studied the cytotoxicity response of five carbon nanomaterials—single-wall carbon nanotubes (SWCNTs), active carbon, carbon black, multiwall carbon nanotubes, and carbon graphite—on human fibroblast cells *in vitro*. They observed twofold results (Tian *et al.* 2006). Firstly, the surface area is the variable that best predicts the potential toxicity of these refined carbon nanomaterials, in which SWCNTs induced the strongest cellular apoptosis/necrosis. Secondly, refined SWCNTs are more toxic than their unrefined counterpart. For comparable small surface areas, dispersed carbon nanomaterials due to a change in surface chemistry are seen to pose morphological changes and cell detachment, and thereupon apoptosis/necrosis (Tian *et al.* 2006).

Sayes *et al.* (2006) conducted a cytotoxicity and inflammatory response study on titania nanoparticles using human dermal fibroblasts and human lung epithelial cells. They characterize the toxicity of this important class of nanomaterials under ambient (e.g., no significant light illumination) conditions in cell culture. Only at relatively high concentrations (100 mug/ml) of nanoscale titania cytotoxicity and inflammation responses were observed. Finer particle size didn't seem to have an effect on toxicity. However, phase compositions showed a strong correlation with toxicity.

Anatase TiO$_2$ was found to be 100 times more toxic than an equivalent sample of the rutile polymorph. Nano-TiO$_2$ samples optimized for reactive species (RS) production in photocatalysis are also more likely to generate damaging RS species in cell culture. Their result highlights the important role that *ex vivo* measures of reactive species (RS) production can play in developing screens for cytotoxicity (Sayes *et al.* 2006). Braydich-Stolle *et al.* (2005) studied *in vitro* cytotoxicity of several nanoparticles in mammalian germline stem cells. They assessed the suitability of a mouse spermatogonial stem cell line as a model to assess nanotoxicity in the male germline *in vitro*. Their results demonstrated a concentration-dependent toxicity for all types of particles tested, whereas the corresponding soluble salts had no significant effect. Silver nanoparticles were the most toxic while molybdenum trioxide MoO$_3$ nanoparticles were the least toxic Similar adverse effects of other nano-particles, important to engineering, have been reported (Limbech *et al.* 2005; Murr *et al.* 2005; Brunner *et al.* 2005).

Design of biomaterials surfaces at the nanoscale level is critical for control of cell-biomaterials interactions which can be used to our benefit, to develop structures or extracellular matrices (ECM) with desired characteristic properties. Biological cells have dimensions within the range of 1-10 μm and contain many examples of extremely complex nano-assemblies, including molecular motors, which are complexes embedded within membranes that are powered by natural biochemical processes.

Various examples of nanoscale structures, nano-motors and nano-systems are present in abundance *in vivo* in different biological systems. Biomineralization of nanocrystals in a protein matrix is highly important for the formation of bone and teeth, and is also used for chemical storage and transport mechanisms within various organs. Biomineralization involves delicate biological control mechanisms to produce nanomaterials with well-defined characteristics such as particle size, crystallographic structure, morphology and architecture. Biomineralization encompasses all mineral-containing tissues formed by living organisms to accomplish a variety of diverse functions, as in shells, skeleton and teeth. The formation of these nano-crystals is often controlled in all its aspects, from incept to their orientation, dimension and assembly. Such delicate control is carried out through specialized proteins that recognize specific crystal surfaces. Recognition is based on molecular complementarities between the protein and the crystal structure on defined planes of the nano-crystals. The understanding of these nano-processes is also relevant to research in advanced biomaterials engineering. Looking at nanoscale architecture of various biological systems, the need and importance of synthetic nanostructured materials and systems to mimic nature is imperative.

Kikuchi *et al.* (2005) discussed three such nanostructured designs of biomedical materials, from the perspective of applications of cell sheet engineering to functional regenerative tissues and organs. Controlled surface chemistry is essential for rational design of materials surfaces for modulating the cell adhesion and culture behavior. Three different categories of surfaces distinguished by Kikuchi *et al.* were: (i) nonfouling surfaces such as PEG (polyethylene glycol) grafted surfaces, to which generally cells do not adhere in short-term culture, (ii) surfaces that interact with cells primarily through passive adhesion forces, on which cells adhere without cell receptor engagement or signal transduction and (iii) active surfaces where cell signaling and ATP (adenosine tri-phosphate)-dependent metabolic processes are active, using transmembrane signal processes to modulate cell adhesion/detachment.

Nanometer-thick thermoresponsive polymer grafts on polystyrene cell culture surfaces can be fabricated through polymer chemistry route. They reported that surface co-existence of two or more different chemistries, spatially controlled on the micron and submicron scales, have been shown to affect both cell and platelet adhesion processes, signaling and metabolic activation. New biomaterials surfaces that can be reliably and rationally designed to interact with cells and tissues in a responsive manner are of significant importance to obtain and maintain appropriate cellular functions in culture (Kikuchi *et al.* 2005).

Nanoparticles have been developed to treat breast cancer at the Feinberg School of Medicine of Northwestern University which has received approval from the Food and Drug Administration. Research at Northwestern University played a significant role in the approval of Abraxane indicated for the treatment of metastatic breast cancer. "Women with metastatic breast cancer no longer need to endure the toxicities associated with solvents and will no longer need steroid premedication when they receive this albumin-bound form of paclitaxel," said by the principal investigator William J. Gradishar, M.D.. Abraxane is engineered using a proprietary process (protein-bound nanoparticle technology) to create nanoparticles in which the active chemotherapeutic drug, Paclitaxel, is bound to a naturally occurring protein called albumin.

Scientists are hoping to target cancer at the cellular level, bypassing the need for radiation and chemotherapy. That dream may be getting closer through the use of nanotechnology. The emergence of nanotechnology, more precisely nanoscale biomaterials and nanomedicine, is likely to have a significant impact on drug delivery sector, affecting just about every route of administration from oral to injectable. And the payoff for doctors and patients should be lower drug toxicity, reduced cost of treatments, improved bioavailability and an extension of the economic life of proprietary drugs.

In medicine and physiology, potential uses of nano-biomaterials are nearly endless, including structures that can serve as artificial scaffolds for bone growth, forms of artificial skin and joints, implantable drug-delivery systems, and engineered materials that can carry out the function of organs. Samuel Stupp and his co-workers at Northwestern University have developed a series of molecular scaffolds resembling the structures of bone, nerves and other tissues. Each type of scaffold starts out as a gel of nanoscale molecular fibers that can be injected into the site of a broken bone or a severed nerve. Once in place, the fibers will spontaneously assemble themselves into a biocompatible matrix that will speed and guide the body's natural healing process.

## 4. SUMMARY

Biomaterials are natural or man-made materials used to replace, augment or aid the functions/performance of a part of an organ or the whole organ system and which remain in intimate contact with the living tissues without eliciting a toxic response. In earlier times, the term *biomaterial* was defined as "a systematically and pharmaceutically inert substance designed for implantation within or incorporation with living systems."

The expression *nanostructured biomaterial*, as used in this chapter, defines those biomaterials that consist of nano-grains and/or nanostructures, or possess features of at least one dimension in the nano-range (1-100 nm). The fascination and great technical promises associated with nanoscale/nanostructured materials are based on the significant difference in their fundamental physical, mechanical, electrical, chemical and biological properties.

Some of the biomaterials currently used in the industry include metals (e.g., austenitic stainless steels, titanium alloys, cobalt-chrome alloys, gold and nitinol); ceramics (e.g., hydroxyapatite, alumina, zirconia and calcium phosphates); glass-ceramics (e.g., bioactive glasses and Bioglass®); polymers (e.g., PMMA, HEMA, WHMWPE, PE, PEO, nylon, silicon rubber, polyester, PLLA, PLGA and polytetrafluroethylene); and composites (e.g., carbon-carbon and wire or fiber-reinforced bone cement).

Based on the genetic relationship between the donor and the recipient, the available tissue grafting options for a patient can be divided into five major categories. These options include (i) autogenous tissue graft or autograft, (ii) isogenous tissue graft or isograft, (iii) allogenous tissue graft or allograft, (iv) xenogenous tissue graft or xenograft, and (v) artificial tissue grafts.

*Biocompatibility* is defined as the ability of a biomaterial to perform with appropriate host response(s) in a specific application that is judged suitable to that situation.

Metallic biomaterials are used in human health care, particularly because of their excellent mechanical properties and extensive knowledge-base of mankind with regard to their processing, properties and structures. They are mostly used as passive substitute materials for hard tissue replacements in load-bearing applications such as total hip and knee joint replacements, for fracture healing aids as bone plates, screws and wires, spinal fixation devices, and dental implants because of their excellent mechanical properties and good resistance to corrosion and wear. The design of the next generation of orthopedic metallic implants is focused on matching implant surfaces with the unique nanometer topography made by natural extracellular matrix proteins found in bone tissue.

Nanostructured metals have a greater number of atoms at the surface compared to their bulk counterparts, larger areas of increased surface defects (like particle boundaries and edge corner sites), and greater proportions of surface electron delocalization. These altered surface properties influence the initial protein interactions that control subsequent adhesion of osteoblast cells.

Coating of metallic prostheses with nanostructured biomaterials to enhance properties would possibly be one of the significant contributions of nanoscience and technology towards modern health care industry, particularly for the orthopedic implants. Cui *et al.* (2005) fabricated nano-titania/titanium alloys for biomedical applications using a process, wherein titanium alloys were embedded in nanometer titanium dioxide powder and sintered at elevated temperature. Donga *et al.* (2003) investigated the nanostructure of plasma-sprayed hydroxyapatite/Ti6Al4V composite using SEM and STEM techniques.

Development of many advanced ceramic biomaterials during the last six decades has made a significant contribution to the development of the modern human health care industry. Mechanical weakness of bioceramics associated with their inherent brittleness limits their applications in many orthopedic applications, where strength under complex stress state is a major concern.

Nanostructured ceramics can be tailored to obtain desired chemical compositions, surface properties (specifically topography), mechanical properties (ductility) and distribution of grain size similar to those of physiological bone (which contains 70% by weight of hydroxyapatite ceramic with grain sizes less than 100 nm). Because ductility is improved as grain size is reduced, nanostructured ceramics can be made to possess a wide range of mechanical properties. Most importantly, future nanostructured ceramic biomaterials can be designed to meet clinical requirements corresponding to anatomical differences or a patient's age.

Nanostructured bioceramic substrates promote adhesion of osteoblasts and decreased adhesion of fibroblasts. Nanostructured biomaterials promote osteoblast adhesion and proliferation, osteointegration, and the deposition of calcium-containing minerals on the surface of these materials. Also, nanostructured ceramics can be sintered at a lower temperature, thereby problems associated with high-temperature sintering processes are eliminated. Nanostructured bioceramics clearly represent a promising class of orthopedic and dental implant formulations with improved biological and biomechanical properties.

Tissue engineering has provided an alternate medical therapy using implants made of polymeric biomaterials with or without living precursor cells against various transplants. A number of different polymeric biomaterials can be used as a scaffold to promote cell adhesion, maintenance of differentiated cell functions without hindering proliferation.

Cellular response important for vascular tissue regeneration can be enhanced on PLGA, through the creation of a biologically inspired nanometer surface topography. Since the arterial wall tissue possesses a high degree of nanometer surface roughness, it is important to design optimal tissue-engineering scaffolds having nanoscale features. Miller *et al.* (2004) studied the endothelial and vascular smooth muscle cell function on poly(lactic-co-glycolic acid) with nano-structured surface features.

Morimoto *et al.* (2004) examined the nano-scale surface modification of a semi- interpenetrating polymer networks (IPN) film composed of segmented polyurethane (SPU) and cross-linked poly (2-methacryloyloxyethyl phosphorylcholine (MPC)-co-2-ethylhexyl methacrylate (EHMA)) polymer prepared with a mica interface.

It is expected that a biocompatible material that mimics this nanometer topography of native bladder tissue will enhance cellular responses, thereby leading to better tissue integration *in vivo*. Thapa *et al.* (2003) investigated the effects of nanostructured polymers on cell functions of bladder smooth muscles. Ramanujam (2005) developed an optical fabrication process to develop nanostructured biopolymer surfaces through a maskless nano-patterning of the surface of a biocompatible polymer that can be employed for tissue engineering and cell growth.

Composite materials offer a variety of advantages over metals, ceramics and polymers, which incorporate the desirable properties of each of the constituent materials, while mitigating the more limited characteristics of each component. Murugan *et al.* (2004) prepared and characterized hydroxyapatite composite bone paste with a natural polysaccharide, chitosan, using wet chemical method at low temperature. Their findings suggest that the hydroxyapatite/chitosan composite paste would be highly beneficial for the particle immobilization upon implantation, and may be a candidate bioresorbable material for bone substitutes. Rauschmann *et al.* (2004) assessed the material properties of a calcium sulphate nanoparticulate hydroxyapatite composite material and analyzed its *in vitro* uptake and release of vancomycin.

Nanostructured coatings have attracted a great deal of interest from the scientific community in recent years due to their enhanced mechanical properties. Nanostructured composite coatings of hydroxyapatite and zirconia ceramics were prepared using high-velocity oxy-fuel spray process by Li *et al.* (2004). Design of biomaterials surfaces at the nanoscale level is critical for control of cell-biomaterials interactions, which can be used to our benefit to develop structures or extracellular matrices (ECM) with desired characteristic properties. Biological cells have dimensions within the range of 1-10 $\mu$m and contain many examples of extremely complex nano-assemblies, including molecular motors, which are complexes embedded within membranes that are powered by natural biochemical processes.

Considering the presence of the abundant and various nanoscale architectures and systems in biological systems *in vivo*, the need and the importance of synthetic nanostructured materials and systems to mimic natural systems, organs or tissues; and to replace, augment or assist during the recovery process is imperative.

# REFERENCES

The Bone and Joint Decade 2000-2010 for Prevention and Treatment of Musculoskeletal Disorders, 1988. *Acta Orthopaedica Scandinavica.* suppl. 281 69(3).

Report of the National Heart, Lung, and Blood Institute Workshop on Lipoprotein(a) and Cardiovascular Disease: *Recent Advances and Future Directions*; Washington DC: Department of Health and Human Services: 2003; p 2.

Abbott, W. M., Callow, A., Moore, W., Rutherford, R., Veith, F. and Weinberg, S. (1993), Evaluation and performance standards for arterial prosthesis, *J Vasc Surg* **17**:746-756.

Adhikary, K., Takahashi, M. and Kikkawa, S., 2004, Synthesis and sintering of nanocrystalline hydroxyapatite powders by citric acid sol-gel combustion method, *Mater. Res. Bull.* **39**:25-32.

Annan, K., *The Secretary General Message to Launch the Bone and Joint Decade 2000-2010 for the Prevention and Treatment of Musculoskeletal Disorders*; November 30, 1999.

Atala, A., Freeman, M. R., Vacanti, J. P., Shepard, J. and Retik, A. B., 1993, Implantation in vivo and retrieval of artificial structures consisting of rabbit and human urothelium and human bladder muscle., *J. Urol.* **150**:608-612.

Ayad, S., Boot-Handford, R., Humpries, M. J., Kadler, K. E. and Shuttleworth, A., 1994, *The Extracellular Matrix Facts Book*, Academic Press, San Diego.

Baron, R. E., 1996, Anatomy and ultrastructure of bone, In *Primer on Metabolic Bone Diseases and Disorders of Mineral Metabolism*, 3rd ed.; Favus, M. J., Ed. Lippincott-Raven: Philadelphia, pp. 3-10.

Bauer, T. W., Geesink, R. C., Zimmerman, R. and McMahon, J. T., 1991, Hydroxyapatite-coated femoral stems. Histological analysis of components retrieved at autopsy, *J Bone Joint Surg Am* **73**(10):1439-1452.

Berkland, C., Pack, D. W. and Kim, K. K., 2004, Controlling surface nano-structure using flow-limited field-injection electrostatic spraying (ffess) of poly(D,L-lactide-co-glycolide), *Biomaterials* **25**(25):5649-5658.

Best, S. and Bonfield, W., 1994, Processing behaviour of hydroxyapatite powders with contrasting morphology, *J. Mater. Sci., Mater. Med.* **5**:516-521.

Bhatt, H. and Kalita, S. J., 2005, In *Synthesis and Sintering Studies of Nanocrystalline Hydroxyapatite Powders Doped with Magnesium and Zinc.*, 29th International Conference on Advanced Ceramics and Composite, S8: Bioceramics, Cocoa Beach, FL, 2005; Zhu, D.andKriven, W. M., Eds. American Ceramic Society, Cocoa Beach, FL, pp. 17-23.

Bigi, A., Boanini, E., Gazzano, M., Rubini, K. and Torricelli, P., 2004, Nanocrystalline hydroxyapatite-polyaspartate composites, *Biomed Mater Eng* **14**(4):573-579.

Bigi, A., Foresti, E., Gandolfi, M., Gazzano, M. and Roveri, N., 1997, Isomorphous substitutions in beta-tricalcium phosphate: the different effects of zinc and strontium, *Jour. Inorg. Biochem.* **66**:259-265.

Bigi, A., Torricelli, P., Fini, M., Bracci, B., Panzavolta, S., Sturba, L. and Giardino, R., 2004, A biomimetic gelatin-calcium phosphate bone cement, *Int J Artif Organs* **27**(8):664-673.

Billottee, W. G., 2003, Ceramic biomaterials, In *Biomaterials: Principles and Applications*, Park, J. B. and Bronzino, J. D., Eds. CRC Press: Boca Raton, FL, pp. 21-54.

Black, J., 1999, *Biological Performance of Materials: Fundamentals of Biocompatibility*, 3rd ed.; Marcel Dekker, New York, p. 463.

Bognitzki, M., Czado, W., Frese, T., Schaper, A., Hellwig, M., Steinhart, M., Greiner, A. and Wendorff, J. H., 2001, Nanostructured fibers via electrospinning, *Adv Mater* **13**:70-72.

Bonfield, W., Wang, M. and Tanner, K. E., 1998, Interfaces in analogous biomaterials, *Acta Materialia* **46**(7):2509-2518.

Braydich-Stolle, L., Hussain, S., Schlager, J. J. and Hofmann, M. C., 2005, In vitro cytotoxicity of nanoparticles in mammalian germline stem cells, *Toxicol Sci* **88**(2):412-419.

Brunner, T. J., Wick, P., Manser, P., Spohn, P., Grass, R. N., Limbach, L. K., Bruinink, A. and Stark, W. J., 2006, In vitro cytotoxicity of oxide nanoparticles: Comparison to asbestos, silica, and the effect of particle solubility, *Environ. Sci. Technol.*, ASAP Article 10.1021/es052069i S0013-936X(05)02069-9.

Bow, J. S., Liou, S. C. and Chen, S. Y., 2004, Structural characterization of room-temperature synthesized nano-sized beta-tricalcium phosphate, *Biomaterials* **25**(16):3155-3161.

Bren, L., English, L., Fogarty, J., Policoro, R., Zsidi, A., Vance, J., Drelich, J., White, C., Donahue, S., Istephanous, N. and Rohly, K., 2004, In *Effect of Surface Characteristics of Metallic Biomaterials on Interaction with Osteoblast Cells*, 7th World Biomaterials Congress 2004, p. 1121.

Brinker, C. J. and Scherer, G. W., 1990, *Sol-Gel Science: The Physics and Chemistry of Sol-Gel Process*, Academic Press, Boston, p. 291.

Bruck, S. D., 1980, *Properties of Biomaterials in the Physiological Environment*, CRC Press, Boca Raton, FL, p. 142.

Buehler, W. J., Gilfrick, J. V. and Wiley, R. C., 1963, Effects of low-temperature phase changes on the mechanical properties of alloys near composition niti., *J Appl Phys.* 34:1475-1484.

Chen, F., Wang, Z. C. and Lin, C. J., 2002, Preparation and characterization of nano-sizes hydroxyapatite articles and hydroxyapatite/chitosan nano-composite for use in biomedical materials, *Materials Letters* 57 (4):858-861.

Choi, K., Kuhn, J. L., Ciarelli, M. J. and Goldstein, S. A., 1990, The elastic moduli of human sub-chondral, trabecular, and cortical bone tissue and the size-dependency of cortical bone modulus, *J Biomech* 23(11):1103-1113.

Chu, C. F., Lu, A., Liszkowski, M. and Sipehia, R., 1999, Enhanced growth of animal and human endothelial cells on biodegradable polymers, *Biochim Biophys Acta* 1472(3):479-485.

Chu, S. Z. e. a., 2003, Fabrication and characteristics of nanostructures on glass by al anodization and electrodeposition, *Electrochemica Acta* 48:3147-3153.

Clayson, K. R., Edwards, W. H., Allen, T. R. and Dale, A., 1976, Arm veins for peripheral arterial reconstruction, *Arch Surg* 111(11):1276-1280.

Correia, R. N., Magalhaes, M. C. F., Marques, P. A. A. P. and Senos, A. M. R., 1995, Wet synthesis and characterization of modified hydroxyapatite powders, *J. Mater. Sci. Mater. Med.* 7:501-505.

Cui, C., Liu, H., Li, Y., Sun, J., Wang, R., Liu, S. and Greer, L., 2005, Fabrication and biocompatibility of nano-tio$_2$/titanium alloys biomaterials, *Materials Letters* 59:3144-3148.

Dalby, M. J., Riehle, M. O., Sutherland, D. S., Agheli, H. and Curtis, A. S., 2004, Changes in fibrob-last morphology in response to nano-columns produced by colloidal lithography, *Biomaterials* 25(23):5415-5422.

Darling, R. C. and Linton, R. R., 1972, Durability of femoropopliteal reconstructions. Endarterectomy versus vein bypass grafts, *Am J Surg* 123(4):472-479.

Dasarathy, H., Riley, C., Coble, H. D., Lacefield, W. R. and Maybee, G., 1996, Hydroxyapatite/metal composite coatings formed by electrocodeposition, *J Biomed Mater Res* 31(1):81-89.

Desai, T. A., 2000, Micro- and nanoscale structures for tissue engineering constructs., *Med. Eng. Phys.* 22:595-606.

Dong, Z. L., Khor, K. A., Quek, C. H., White, T. J. and Cheang, P., 2003, Tem and stem analysis on heat-treated and in vitro plasma-sprayed hydroxyapatite/ti-6al-4v composite coatings, *Biomaterials* 24:97-105.

Doremus, R. H., Teich, S. and Silvis, P. X., 1978, Crystallization of calcium oxalate from synthetic urine, *Invest Urol* 15(6):469-472.

Driessens, F. C., Van Dijk, J. W. and Borggreven, J. M., 1978, Biological calcium phosphates and their role in the physiology of bone and dental tissues i. Composition and solubility of calcium phosphates, *Calcif Tissue Res* 26(2):127-137.

Ducheyne, P., 1999, Effect of bioactive glass particle size on osseous regeneration, *J Biomed Mater Res* 46(2):301-304.

Ducheyne, P. and Qiu, Q., 1999, Bioactive ceramics: The effect of surface reactivity on bone formation and bone cell function, *Biomaterials* 20(23-24):2287-2303.

Duerig, T. W., 1995, In *Present and Future Applications of Shape Memory and Superelastic Materials*, Proc. Mater. Res. Soc. Symp., Pittsburgh, PA, 1995; Materials Research Society, Pittsburgh, PA, pp. 497-506.

Dumitriu, S. D. and Dumitriu, D. D., 1994, Biocompatibility of polymers, In *Polymeric Biomaterials*, Dumitriu, S. D., Ed. Marcel Dekker: New York, p. 99.

Elliott, J. C., 1994, *Structure and Chemistry of the Apatites and Other Calcium Orthophosphates*, Elsevier, New York, pp. xiii, 389.

Evans, F. G., Lissner, H. R. and Patrick, L. M., 1962, Acceleration-induced strains in the intact vertebral column, *J Appl Physiol* 17:405-409.

Flemming, R. G., Murphy, C. J., Abrams, G. A., Goodman, S. L. and Nealey, P. F., 1999, Effects of synthetic micro- and nano-structured surfaces on cell behavior, *Biomaterials* 20(6):573-588.

Fumo, D. A., Morelli, M. R. and Segadaes, A. M., 1996, Combustion synthesis of calcium aluminates, *Mater. Res. Bull.* 31 (10):1243-1255.

Gao, J., Niklason, L. and Langer, R., 1998, Surface hydrolysis of poly(glycolic acid) meshes increases the seeding density of vascular smooth muscle cells, *J Biomed Mater Res* 42(3):417-424.

Georgiou, G. and Knowles, J. C., 2001, Glass-reinforced hydroxyapatite for hard tissue surgery. Part 1: Mechanical properties, *Biomaterials* 22(20):2811-2815.

Gibson, I. R. and Bonfield, W., 2002, Preparation and characterization of magnesium/carbonate co-substituted hydroxyapatites, *J Mater Sci Mater Med* 13(7):685-693.

Gibson, I. R., Rehman, I., Best, S. M. and Bonfield, W., 2000, Characterization of the transformation from calcium-deficient apatite to beta-tricalcium phosphate, *J Mater Sci. Mater in Med.* 11(9):533-539.

Ginebra, M. P., Fernandez, E., Driessens, F. C. M. and Planell, J. A., 1998, *Biomaterials* 11:243-246.

Graves, G. A. J., Hentrich, R. L. J., Stein, H. G. and Bajpai, P. K., 1972, *Resorbable Ceramic Implants in Bioceramics*, Wiley Interscience, New York, pp. 91-115.

Groot, K. d., Klein, C. P. A. T., Wolke, J. G. C. and Blieck-Hogervorst, J. M. A., 1990, Chemistry of calcium phosphate bioceramics, In *Handbook of Bioactive Ceramics, vol. 2, Calcium Phosphate and Hydroxylapatite Ceramics*, Yamamuro, T.,Hench, L. L. and Wilson, J., Eds. CRC Press: Boca Raton, FL, pp. 3-16.

Gutwein, L. G. and Webster, T. J., 2004, Increased viable osteoblast density in the presence of nanophase compared to conventional alumina and titania particles, *Biomaterials* 25:4175-4183.

Haman, J. D., Boulware, A. A., Lucas, L. C. and Crawmer, D. E., 1995, High-velocity oxyfuel thermal spray coatings for biomedical applications, *J Therm Spray Technol* 4:179-184.

Han, Y., Li, S., Wang, X. and Chen, X., 2004, Synthesis and sintering of nanocrystalline hydroxyapatite powders by citric acid sol-gel combustion method, *Materials Research Bulletin* 39:25-32.

Haranp, P., Sutherland D., Gold J., Kasemo B. 1999, Nanostructured model biomaterials surfaces prepared by colloidal lithography, *Nanostructured Materials* 12:429-432.

Hartgerink, J. D., Beniash, E. and Stupp, S. I., 2001, Self-assembly and mineralization of peptide-amphiphile nanofibers, *Science* 294(5547):1684-1688.

He, G. and Hagiwara, M., 2005, Ti alloy design strategy for biomedical applications, *Mater. Sci. Eng., C* 26(1):14-19.

Hench, L. L., 1998, Bioceramics, *J. Am. Ceram. Soc.* 81(7):1705-1728.

Hentrich, R. L., Graves, G. A., Stein, H. G. and Bajpai, P. K., 1969, In *An Evolution of Inert and Resorbable Ceramics For future Clinical Applications.*, Fall meeting, Ceramics Metal System, Division of the American Ceramic Society, Cleveland, Ohio, 1969; Cleveland, OH.

Hentrich, R. L., Graves, G. A., Stein, H. G. and Bajpai, P. K., 1971, An evaluation of inert and resorbable ceramics for future clinical orthopedic applications, *J Biomed Mater Res* 5(1):25-51.

Heughebaert, M., LeGeros, R. Z., Gineste, M., Guilhelm, A. and Bonel, G., 1988. Physicochemical characterization of deposits associated with HA ceramics implanted in nonosseous sites, *J. Biomed. Mater. Res.* 22(S14):257-268.

Hong, S. J., Bhatt, H., Suryanarayana, C. and Kalita, S. J., 2005. In: *Synthesis of Nano-Size Hydroxyapatite (Hap) Powders by Mechanical Alloying,* 29th International Conference on Advanced Ceramics & Composite, Cocoa Beach, FL, S8: *Bioceramics,* 2005; Zhu, D. and Kriven, W. M., Eds. American Ceramic Society, pp. 33-39.

Hudson, T. W., Evans, G. R. and Schmidt, C. E., 2000. Engineering strategies for peripheral nerve repair, *Orthop. Clin. North. Am.* 31(3):485-498.

Hulbert, S. F., Morrison, S. J. and Klawitter, J. J., 1972, Tissue reaction to three ceramics of porous and nonporous structures, *J. Biomed. Mater. Res.,* 6:347-374.

Ioku, K., Goto, S., Kurosawa, H., Shibuya, K., Yokozeki, H., Hayash, T. and Nakagawa, T., 1996, *Bioceramics* 9:201-204.

Itokazu, M., Yang, W., Aoki, T., Ohara, A. and Kato, N., 1998. Synthesis of antibiotic-loaded interporous hydroxyapatite blocks by vacuum method and in vitro drug release testing, *Biomaterials* 19(7-9):817-819.

Jarcho, M., Kay, J. F., Gumaer, K. I., Doremus, R. H. and Drobeck, H. P., 1977. Tissue, cellular and subcellular events at a bone-ceramic hydroxylapatite interface, *J. Bioeng.* 1(2):79-92.

Jia, G., Wang, H., Yan, L., Wang, X., Pei, R., Yan, T., Zhao, Y., and Guo, X., 2005. Cytotoxicity of Carbon Nanomaterials Sigle-Wall Nanotube, Multi-Wall Nanotube, and Fullerence, *Environ. Sci. Technol.* 39:1378-1383.

Kalita, S. J., Bose, S., Hosick, H. L. and Bandyopadhyay, A., 2002. Porous calcium aluminate ceramics for bone-graft applications, *J. Mater. Res.* 17:3042-3049.

Kalita, S. J., Bose, S., Hosick, H. L. and Bandyopadhyay, A., 2004. Cao-$P_2O_5$-$Na_2O$-based sintering additives for hydroxyapatite (hap) ceramics, *Biomaterials* 25(12):2331-2339.

Kalita, S. J., Finley, J., Bose, S., Hosick, H. L. and Bandyopadhyay, A., 2002. In: *Development of Porous Polymer-Ceramic Composites as Bone Grafts*, Materials Research Society Sumposium Proceeding San Francisco, San Francisco, pp. Q5.8.1-Q5.8.6.

Kalita, S. J., Rokusek, D., Bose, S., Hosick, H. L. and Bandyopadhyay, A., 2004. Effects of MgO-CaO-$P_2O_5$-$Na_2O$-based additives on mechanical and biological properties of hydroxyapatite, *J. Biomed. Mater. Res. A* 71(1):35-44.

Karch, J., Birringer, R. and Gleiter, H., 1987. Ceramics ductile at low-temperature, *Nature* 330:556-558.

Kikuchi, A. and Okano, T., 2005. Nanostructured designs of biomedical materials: applications of cell sheet engineering to functional regenerative tissues and organs, *J. Controlled Release* 101:69-84.

Kikuchi, M., Suetsugu, Y., Tanaka, J. and Akao, M., 1997. Preparation and mechanical properties of calcium phosphate/copoly-l-lactide composites, *J. Mater. Sci. Mater. Med.* 8(6):361-364.

Kim, H. M., 2003. Current opinion in solid state and materials science, 7:289-299.

Kim, J. H. and Kim, S. C., 2003. Effect of peo grafts on the surface properties of peo-grafted pu/ps ipns: Afm study, *Macromolecules* 36:2867-2872.

Kim, J. H., Kim, S.C., 2002. Peo-grafting on pu/ps ipns for enhanced blood compatibility-effect of pendant length and grafting density, *Biomaterials* 23:2015-2025.

Kim, S. K. and Yoo, H. J., 1998. Formation of bilayer ni-sic composite coatings by electrodeposition, *Surf. Coat. Technol.* 108:564-569.

Kim, S. R., Lee, J. H., Kim, Y. T., Riu, D. H., Jung, S. J., Lee, Y. J., Chung, S. C. and Kim, Y. H., 2003. Synthesis of Si, Mg substituted hydroxyapatites and their sintering behaviors, *Biomaterials* 24(8):1389-1398.

Klein, C. P. A. T., Wolke, J. G. C., de Blieck-Hogervorst, J. M. A. and de Groot, K., 1994. Calcium phosphate plasma-sprayed coatings and their stability: an in-vivo study, *J. Biomed. Mater. Res.* (28):909-917.

Kong, Y., Bae, C., Lee, S., Kim, H. and Kim, H. E., 2005. Improvement in biocompatibility of $ZrO_2$-$Al_2O_3$ nano-composite by addition of HAp, *Biomaterials* 26:509-517.

Korkusuz, F. U., O., 1999, Nonspecific inflammation and bone marrow depletion due to intramedullary porous hydroxyapatite application, *Bull. Hosp. Joint Dis. Orthop. Inst.* 58:86-91.

Kuriakose, T. A., Kalkura, S. N., Palanichamy, M., Arivuoli, D., Dierks, K., Bocelli, G. and Betzel, C., 2004. Synthesis of stoichiometric nanocrystalline hydroxyapatite by ethanol-based sol-gel technique at low temperature, *J. Crystal Growth* 263:517-523.

Kyeck, S. and Remer, P., 1999. Realisation of graded coatings for biomedical use, *Mat. Sci. Forum* 308(311):368-373.

Lee, S. C., 1998. Biotechnology for nanotechnology, *Trends. Biotechnol.* 16:239-240.

LeGeros, R. Z., 1991. Calcium phosphates in oral biology and medicine, *Monogr. Oral Sci.* 15:1-201.

LeGeros, R. Z., 1993. Biodegradation and bioresorption of calcium phosphate ceramics, *Clin. Mater.* 14(1):65-88.

LeGeros, R. Z., 2002. In: *Calcium Phosphate Ceramics in Dentistry and Medicine.* International Symposium on the New Wave of Ceramics for the 21st Century at the 40th Symposium on the Basic Science of Ceramics, Convention Center, Osaka University, January 22-23, 2002, Convention Center, Osaka University.

LeGeros, R. Z., Bonel, G. and Legros, R., 1978. Types of "$H_2O$" in human enamel and in precipitated apatites, *Calcif. Tissue Res.* 26(2):111-118.

Li, H., Khor, K. A., Kumar, R. and Cheang, P., 2004. Characterization of hydroxyapatiteynano-zirconia composite coatings deposited by high velocity oxy-fuel (hvof) spray process, *Surf. Coat. Technol.* 182:227-236.

Li, W. and Gao, L., 2003. Fabrication of hap-zro2 (3y) nano-composite by sps, *Biomaterials* 24(6):937-940.

Limbach, L. K., Li, Y., Grass, R. N., Brunner, T. J., Hintermann, M. A., Muller, M., Gunther, D. and Stark, W. J., 2005. Oxide nanoparticle uptake in human lung fibroblasts: effects of particle size, agglomeration, and diffusion at low concentrations, *Environ. Sci. Technol.* 39(23):9370-9706.

Liou, S. C. and Chen, S. Y., 2002. Transformation mechanism of different chemically precipitated apatitic precursors into beta-tricalcium phosphate upon calcination, *Biomaterials* 23(23):4541-4547.

Liou, S. C., Chen, S. Y., Lee, H. Y. and Bow, J. S., 2004. Structural characterization of nano-sized calcium deficient apatite powders, *Biomaterials* 25(2):189-196.

Liu, D. M., Troczynski, T. and Tseng, W. J., 2001. Water-based sol-gel synthesis of hydroxyapatite: process development, *Biomaterials* 22(13):1721-1730.

Liu, D. M., Yang, Q., Troczynski, T. and Tseng, W. J., 2002. Structural evolution of sol-gel-derived hydroxyapatite, *Biomaterials* 23(7):1679-1687.

Long, M. and Rack, H. J., 1998. Titanium alloys in total joint replacement: a materials science perspective, *Biomaterials* 19(18):1621-1639.

Lopes, M. A., Santos, J. D., Monteiro, F. J. and Knowles, J. C., 1998. Glass-reinforced hydroxyapatite: a comprehensive study of the effect of glass composition on the crystallography of the composite, *J. Biomed. Mater. Res.* 39(2):244-251.

Ma, P. X. and Zhang, R., 1999. Synthetic nano-scale fibrous extracellular matrix, *J. Biomed. Mater. Res.* 46:60-72.

Malik, M. A., Puleo, D. A., Bizios, R. and Doremus, R. H., 1992. Osteoblasts on hydroxyapatite, alumina and bone surfaces in vitro: morphology during the first 2 h of attachment, *Biomaterials* 13(2):123-128.

Martin, R. B., 1999. Bone as a ceramic composite material. *Mater. Sci. Forum.* 293:5-16.

Metsger, D. S., Rieger, M. R. and Foreman, D. W., 1999. Mechanical properties of sintered hydroxyapatite and tricalcium phosphate ceramic, *J. Mater. Sci. Mater. Med.* 10(1):9-17.

Mikos, A. G. and Temenoff, J. S., 2000. Formation of highly porous biodegradable scaffolds for tissue engineering, *J. Biotechnol.* 3:1-6.

Miller, D. C., Thapa, A., Haberstroh, K. M. and Webster, T. J., 2004, Endothelial and vascular smooth muscle cell function on poly(lactic-co-glycolic acid) with nano-structured surface features, *Biomaterials* (25):53-61.

Mitterhauser, M., Togel, S., Wadsak, W., Mien, L. K., Eidherr, H., Wiesner, K., Viernstein, H., Kletter, K. and Dudczak, R., 2004. Binding studies of [(18)f]-fluoride and polyphosphonates radiolabelled with [(111)in], [(99m)tc], [(153)sm], and [(188)re] on bone compartments: a new model for the pre vivo evaluation of bone seekers? *Bone* 34(5):835-844.

Morimoto, N., Watanabe, A., Iwasaki, Y., Akiyoshi, K. and Ishihar, K., 2004. Nano-scale surface modification of a segmented polyurethane with a phospholipid polymer, *Biomaterials* 25:5353-5361.

Murr, L. E., Garza, K. M., Soto, K. F., Carrasco, A., Powell, T. G., Ramirez, D. A., Guerrero, P. A., Lopez, D. A. and Venzor, J., 3rd, 2005. Cytotoxicity assessment of some carbon nanotubes and related carbon nanoparticle aggregates and the implications for anthropogenic carbon nanotube aggregates in the environment, *Int. J. Environ. Res. Public Health* 2(1):31-42.

Murugan, R. and Ramakrishna, S., 2004. Bioresorbable composite bone paste using polysaccharide based nano-hydroxyapatite, *Biomaterials* 25(17):3829-3835.

Salih, V., Georgiou, G., Knowles, J. C. and Olsen, I., 2001. Glass-reinforced hydroxyapatite for hard tissue surgery: part ii: in vitro evaluation of bone cell growth and function, *Biomaterials* 22(20): 2817-2824.

Santos, J. D., Silva, P. L., Knowles, J. C., Talal, S. and Monteiro, F. J., 1996. Reinforcement of hydroxyapatite by adding $P_2O_5$-CaO glasses with $Na_2O$, $K_2O$ and mgo, *J. Mat. Sci. Mat. Med.* 7(3):187-189.

Sarig, S. and Kahana, F., 2002. Rapid formation of nanocrystalline apatite, *J. Crystal Growth* 237:55-59.

Sayes, C. M., Wahi, R., Kurian, P. A., Liu, Y., West, J. L., Ausman, K. D., Warheit, D. B. and Colvin, V. L., 2006. Correlating nanoscale titania structure with toxicity: a cytotoxicity and inflammatory response study with human dermal fibroblasts and human lung epithelial cells, *Toxicol. Sci.* 92(1):174-185.

Seal, B. L., Otero, T. C. and Panitch, A., 2001. Polymeric biomaterials for tissue and organ regeneration, *Mater. Sci. Eng.* B 34:147-230.

Shastri, V. P., Lipski, A. M., Sy, J. C., Znidarsic, W., H. Choi, H. and Chen, I.-W., 2003. Functionalized nanoparticles as versatile tools for the introduction of biomimetics on surfaces, *Nanoengineered Nanofibrous Materials*, pp. 255-262.

Shih, W. J., Chen, Y. F., Wang, M. C. and Hon, M. H., 2004. Crystal growth and morphology of the nano-sized hydroxyapatite powders synthesized from $CaHPO_4$. $2H_2O$ and $CaCO_3$ by hydrolysis method, *J. Crystal Growth* 270:211-218.

Singh, R. K., Qian, F., Damodaran, R. and Moudgil, S., 1996. Laser deposition of hydroxyapatite coatings, *Mater. Manuf. Process.* 11:481-490.

Solan, A., Prabhakar, V. and Niklason, L., 2001. Engineered vessels: importance of the extracellular matrix, *Transplant Proc.* 33(1-2):66-68.

Su, L., Berndt, C. C. and Gross, K. A., 2002. Hydroxyapatite/polymer composite flame-sprayed coatings for orthopedic applications, *J. Biomater. Sci. Polym. Ed.* 13(9):977-990.

Suchanek, W. L., Shuk, P., Byrappa, K., Riman, R. E., TenHuisen, K. S. and Janas, V. F., 2002. Mechanochemical-hydrothermal synthesis of carbonated apatite powders at room temperature, *Biomaterials* 23(3):699-710.

Sudanese, A., Toni, A., Cattaneo, G. L., Ciaroni, D., Greggi, T., Dallart, D., Galli, G. and Giunti, A., 1989. In: *First International Symposium on Ceramics in Medicine, Tokyo*, 1989; Oonishi, H., Aoki, H. and Sawai, K., Eds. Ishuyaku Euroamerica, Inc, Tokyo, pp. 237-240.

Tan, J. and Saltzman, W. M., 2004. Biomaterials with hierchically defined micro- and nanoscale structure, *Biomaterials* 25:3593-3601.

Tan, K. H., Chua, C. K., Leong, K. F., Cheah, C. M., Cheang, P., Abu Bakar, M. S. and Cha, S. W., 2003. Scaffold development using selective laser sintering of polyetheretherketone-hydroxyapatite biocomposite blends, *Biomaterials* 24(18):3115-3123.

Tanner, K. E., Dowens, R. N. and Bonfield, W., 1994. Clinical applications of hydroxyapatite reinforced materials, *Br. Ceram Trans.* 93(3):104-110.

Tas, A. C., 2000. Synthesis of biomimetic ca-hydroxyapatite powders at 37°C in synthetic body fluids, *Biomaterials* 21(14):1429-1438.

Thapa, A., Miller, D. C., Webster, T. J. and Haberstroh, K. M., 2003. Nanostructured polymers enhance bladder smooth muscle cell function, *Biomaterials* 24(17):2915-2926.

Thomson, R. C., Shung, A. K., Yaszemski, M. J. and Mikos, A. G., 2000. Polymer scaffold processing, In Principles of tissue engineering, Lanza, R. P., Langer, R. and Vacanti, J. P., Eds. Academic Press.: San Diego, pp. p. 251-61.

Tian, F., Cui, D., Schwarz, H., Estrada, G. G. and Kobayashi, H., 2006. Cytotoxicity of single-wall carbon nanotubes on human fibroblasts, *Toxicol. In Vitro* 20:1202-1212.

Tresco, P. A., 2000. Tissue engineering strategies for nervous system repair, *Prog. Brain. Res.* 128:349-363.

Tsui, Y. C., Doyle, C. and Clyne, T. W., 1998. Plasma-sprayed hydroxyapatite coatings on titanium substrates: part 2: optimisation of coating properties, *Biomaterials* 19:2031-2043.

Wang, B. C., Chang, E., Yang, C. Y., Tu, D. and Tsai, C. H., 1993. Characteristics and osteoconductivity of three different plasma-sprayed hydroxyapatite coatings, *Surf. Coat. Technol.* (58):107-117.

Wang, F., Li, M., Lu, Y., Qi, Y. and Liu, Y., 2006. Synthesis and microstructure of hydroxyapatite nanofibers synthesized at 37°C, *Mater. Chem. Phys.* 95:145-149.

Wang, M., Joseph, R. and Bonfield, W., 1998. Hydroxyapatite-polyethylene composites for bone substitution: effects of ceramic particle size and morphology, *Biomaterials* 19:2357.

Wang, Z., Chen, X., Cai, Y. and Lu, B., 2003. Influences of $R_2O$-$Al_2O_3$-$B_2O_3$-$SiO_2$ system glass and superfine alpha-$Al_2O_3$ on the sintering and phase transition of hydroxyapatite ceramics, *Sheng Wu Yi Xue Gong Cheng Xue Za Zhi* 20(2):205-208.

Webster, T. J. and Ejiofor, J. U., 2004. Increased osteoblast adhesion on nanophase metals: Ti, Ti6Al4V and CoCrMo, *Biomaterials* 25(19):4731-4739.

Webster, T. J., Ergun, C., Doremus, R. H. and Bizios, R., 2002. Hydroxylapatite with substituted magnesium, zinc, cadmium, and yttrium. II. Mechanisms of osteoblast adhesion, *J. Biomed. Mater. Res.* 59(2):312-317.

Webster, T. J., Ergun, C., Doremus, R. H. and Lanford, W. A., 2003. Increased osteoblast adhesion on titanium-coated hydroxylapatite that forms $CaTiO_3$, *J. Biomed. Mater. Res. A* 67(3): 975-980.

Webster, T. J., Ergun, C., Doremus, R. H., Siegel, R. W. and Bizios, R., 2000, Enhanced functions of osteoblasts on nanophase ceramics, *Biomaterials* 21(17):1803-1810.

Webster, T. J., Ergun, C., Doremus, R. H., Siegel, R. W. and Bizios, R., 2000. Specific proteins mediate enhanced osteoblast adhesion on nanophase ceramics, *J. Biomed. Mater. Res.* 51(3):475-483.

Webster, T. J., Ergun, C., Doremus, R. H., Siegel, R. W. and Bizios, R., 2001. Enhanced osteoclast-like cell functions on nanophase ceramics, *Biomaterials* 22(11):1327-1333.

Webster, T. J., Siegel, R. W. and Bizios, R., 1999. Design and evaluation of nanophase alumina for orthopedic/dental applications, *NanoStructured Mater.* 12:983-986.

Williams, D. F., 1982. *Biocompatibility in clinical practice*, CRC Press, Boca Raton, FL.

Woerly, S., Plant, G. W. and Harvery, A. R., 1996. Neural tissue engineering: from polymer to biohybrid organs, *Biomaterials* (17):301-310.

Wong, L. H., Tio, B. and Miao, X., 2002. Functionally graded tricalcium phosphate/fluoroapatite composites, *Mater. Sci. Eng. C* 20:111-115.

Wu, Y. and Bose, S., 2005. Nanocrystalline hydroxyapatite: micelle-templated synthesis and characterization, *Langmuir* 21(8):3232-3234.

Xu, J. L., Khor, K. A., Dong, Z. L., Gu, Y. W., Kumar, R. and Cheang, P., 2004. *Mater. Sci. Eng. A* 374:101-108.

Xu, J. L., Khor, K. A., Gu, Y. W., Kumar, R. and Cheang, P., 2005. Radio frequency (rf) plasma spheroidized HA powders: powder characterization and spark plasma sintering behavior, *Biomaterials* 26(15):2197-2207.

Yang, F., Murugan, R., Ramakrishna, S., Wang, X., Ma, Y. X. and Wang, S., 2004. Fabrication of nanostructured porous PLLA scaffold intended for nerve tissue engineering, *Biomaterials* 25:1891-1900.

Yao, X., Tan, S. and Jiang, D., 2005. Fabrication of hydroxyapatite ceramics with controlled pore characteristics by slip casting, *J Mater Sci Mater Med* 16(2):161-165.

Yeong, K. C. B., Wang, J. and Ng, S. C., 1999. Fabricating densified hydroxyapatite ceramics from a precipitated precursor, *Mater. Lett.* 38(3):208-213.

Yeong, K. C. B., Wang, J. and Ng, S. C., 2001. Mechanochemical synthesis of nanocrystalline hydroxyapatite from CaO and $CaHPO_4$, *Biomater.* 22:2705-2712.

Yingchao, H., Shipu, L., Xinyu, W. and Xiaoming, C., 2004. *Mater. Res. Bull.* 39:25-32.

Yu, D., Wong, J., Matsuda, Y., Fox, F. L., Higuchi, W. I. and Otsuka, M., 1992, *J Pharm Sci.* 81(6):529-531.

Yu, L. G., Khor, K. A., Li, H. and Cheang, P., 2003. Effect of spark plasma sintering on the microstructure and in vitro behavior of plasma-sprayed HA coatings, *Biomaterials* 24(16):2695-2705.

Zhang, S., Marini, D. M., Hwang, W. and Santoso, S., 2002. Design of nanostructured biological materials through self-assembly of peptides and proteins, *Curr. Opin. in Chem. Biol.* 6:865-871.

Zhou, J., Zhang, X., Chen, J., Zeng, S. and Groot, K. D., 1993. High-temperature characteristics of synthetic hydroxyapatite, *J. Mater. Sci. Mater. Med.* 4:83-85.

Zhu, Y. C., Yukimura, K., Ding, C. X. and Zhang, P. Y., 2001. Tribological properties of nanostructured and conventional WC–Co coatings deposited by plasma spraying. *Thin Solid Films* 388:277.

## QUESTIONS

1. Define the following terms:
   (a) Biomaterial
   (b) Biocompatibility
   (c) Bioactive
   (d) Bioresorbable
   (e) Bioinert
   (f) Nanostructured Biomaterials
   (g) Autograft
   (h) Allograft
   (i) Xenograft
2. Explain the various types of cell-tissue responses observed, when a synthetic material intended for orthopedic application, is placed *in vivo*.
3. Summarize the advantages and disadvantages of different types of metallic biomaterials?
4. State the different ways in which an orthopedic device may be fixed in the human skeleton. Indicate with reasoning, the preferred method of fixation for prosthesis of the: *i)* hip (total hip replacement) and *ii)* severely fractured long bone (e.g. the femur).
5. Draw a schematic to show the different types of bioceramics used in biomaterial applications; classified according to their chemical activity *in vivo*. Give examples for each type.
6. List some polymers that are used as biomaterials, along with their properties. Describe some of the Ceramic/polymer composites used for hard tissue replacements and explain how the resorption process can be designed to change with the life-time of the implant. Also mention how, the use of nanofeatured polymers affect their biological and biomechanical properties.
7. Discuss the advantages of nanostructured ceramic and ceramic-polymer composites as hard tissue replacements, over their conventional counter-parts.
8. In spite of the shortcomings associated with metallic materials, they are still used in implants technology. Why? How can we use nanotechnology to reduce some of those shortcomings?
9. Discuss the unique cell-biomaterials interactions, observed in various nanostructured biomaterials.
10. Discuss the importance of surfaces in developing synthetic tissue grafts. How use of nanoscale/ nanostructured biomaterials can improve the surface properties of synthetic grafts?
11. List five non-resorbable and five resorbable synthetic polymers used in the biomedical industry. Discuss their properties and applications. Supposing these polymers are investigated at nano scale, discuss their properties, advantages and proposed application as biomaterials.

# 5

# Self-Assembly and Supramolecular Assembly in Nanophase Separated Polymers and Thin Films

## Naba K. Dutta* and Namita Roy Choudhury*

*Ian Wark Research Institute, University of South Australia, Mawson Lakes Campus, Adelaide, South Australia 5095, Australia. E-mail: naba.dutta@unisa.edu.au; Namita.choudhury@unisa.edu.au

## 1. INTRODUCTION

### 1.1. Self-assembly

Self-assembly (SA) is the process by which pre-existing molecular components, objects and devices spontaneously organize themselves into more ordered state in a discrete noncovalently bound patterns/structures without any external intervention. It is a fundamental process in nature that generates structural organization across all the scales, from molecules to galaxies. SA involves recognition or binding processes and the individual component contains enough information to recognize and interact with other appropriate accessible units to build a template for a structure composed of multiple units. The parameters that dictate the assembly are programmed in the components in the form of either topography, or functionality, or shape or electrical potential, etc.

The most widely studied aspect of self-assembly is molecular self-assembly and the molecular structure determines the structure of the assembly. The concept of SA originates from nature [1] and indeed, a wide variety of biological processes, such as folding of protein in native structure, substrate binding to a receptor protein, protein-protein complexes, antigen-antibody association, enzymatic reaction, transduction of signal, cellular recognition, translation and transcription of the genetic code are excellent examples of the ingenuity of biological processes governed by such interaction [1-3].

Molecular self-assembly is also ubiquitous in chemistry and material science: the formation of molecular crystals, liquid crystals, semicrystalline polymers, colloids, phase-separated polymers, and self-assembled monolayers are unique examples. The association of a ligand with a receptor may also be considered as a form of self-assembly. The semantic boundary between the different processes that form more ordered from less organized assemblies of molecules is not yet very precise [4, 5]. Lehn elaborated on the processes in great length and proposed to use the terms *self-assembly* and *self-organization* to categorize different processes involved and to resolve the semantic disputes [6, 7]. However, often in literature self-assembly is taken as being synonymous with self-organization and they are used interchangeably to encompass all the self-ordering processes. There are two types of molecular self-assembly: intramolecular self-assembly and intermolecular self-assembly. Intramolecular self-assembly is an interesting aspect in complex polymers that enables them with the ability to assemble from random coil conformation to a well-defined structure (secondary and tertiary structures), such as folding of proteins. The intermolecular self-assembly is the ability of the molecules to form supramolecular assemblies (quaternary structures) such as the formation of micelles by surfactant molecules in solution.

The classical molecular chemistry is the chemistry of kinetically stable covalent bonds and is concerned with an understanding of the rules that govern the structures, properties and transformation of molecular species [7]. In molecular self-assembly the interactions involved are usually much weaker and kinetically labile noncovalent intermolecular forces. Lehn coined the term *supramolecular chemistry* for such field of complex intermolecular noncovalent bonding interactions.

SA is a fundamental process that derives many important structural organization and function in nature; for example, the cell contains an astonishing range of complex structures such as lipid membranes, folded proteins, structured nucleic acids, protein aggregates and many more that are formed by SA [4, 8, 9]. Synthetically engineered self-assembled peptides and other biomaterials have also been shown to have superior handling, biocompatibility and functionality. SA is scientifically interesting and technologically important; and the development of functional materials based on tailor-made molecular building blocks has gained significant importance and success in last two decades [4-11]. In many disciplines, control and placement of molecules via self-assembly represent a facile and radically different approach towards the fabrication of functional nanostructures and macroscopic components. Biological systems serve as an idealized model for many current trends in supramolecular science; indeed, functional supramolecular systems may be considered as the connecting link between life science and materials science (Fig. 1). This is due to the fact that the creation of a complex functional nanodevice; using the classical approach, in discrete steps, is highly inefficient.

The potential advantages of construction of such a device by self-assembly include: (i) facile formation of the complex end-product with unprecedented levels of structural control; (ii) reduction in structural error—the reversible nature of interaction induces a "self-checking" process that ensures the high integrity of the

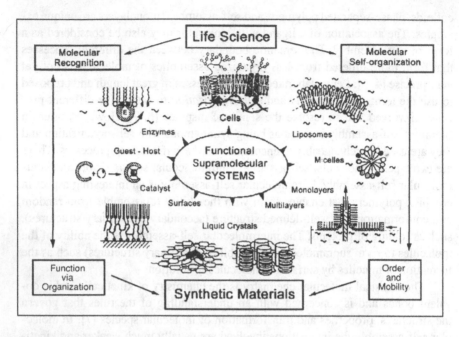

**FIGURE 1.** Diversity of functional supramolecular systems in a variety of biological, and synthetic materials. (This article was published in Molecular architecture and function based on molecular recognition and self-organization, Vol 9, Reichert, Anke; Ringsdorf, Helmut; Schuhmacher, Peter; Baumeister, Wolfgang; Scheybani, Tschangiz; Comprehensive Supramolecular Chemistry, pp. 313-350, Copyright Elsevier, 1996.)

complete structure; (iii) highly convergent synthetic protocol—fewer steps to convert starting material to final product; (iv) economy in fabrication. The unique self-assembled morphologies of very well defined structures of nano-scale dimension are potentially useful in many applications such as; nano-reactors, synthesis of nanoporous materials, formulation of coatings, pharmaceuticals, controlled drug delivery systems, microelectronics, detergents, paints, oil recovery and personal care products. It has also generated widespread interest in the field of photovoltaic devices, miniaturized sensors, nanotube-based electronic components, thin membranes with interesting and well-defined properties, self-assembling nanotubes, nanopipelines for liquid transport, tailored nanostructure, development of new catalysts, genomics and medicine, shape memory polymers and alloys, etc.

The precise control over chemical structure, their fabrication and interconnection using supramolecular assembly have the potential to take molecular nanostructure into realm beyond current technology based on a "top-down" approach. While the overall field of self-assembly is very broad and vast, and beyond the scope of this chapter, in this chapter we will focus on polymer-based supramolecular assemblies. This article will cover the basic understanding of self-assembly in

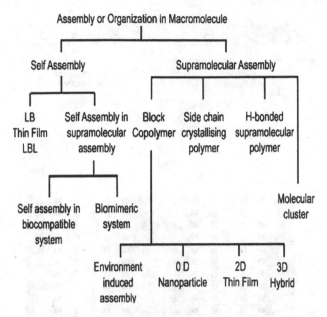

SCHEME 1. Outline of the layout of the chapter.

various phase-separated supramolecular systems; however, special emphasis will be placed on block copolymer based systems due to their enormous future potential. In the first part, we will discuss in some details the guiding principle of nano-phase separation in polymer materials using diblock copolymer as an example. The effects of composition and environment on their spontaneous and directed supramolecular assembly will also be discussed. How this unique assembly can be utilized to construct nanostructured complex materials, ultrathin films, hybrids, crystals, etc., with a special emphasis to the results obtained in our laboratory, will be addressed. A layout of the chapter is outlined in Scheme 1. In the end, some advanced characterization techniques that can be employed for an in-depth understanding of self-assembled structure are illustrated.

## 1.2. Strategies of Self-assembling Supramolecular Complexes

A variety of strategies can be employed to design and fabricate zero-, one-, two- and three-dimensional (0D, 1D, 2D, 3D) ordered structures ranging from molecular to macroscopic sizes. The foundation of the self-assembly is built upon a detailed understanding of the way in which noncovalent forces interact, compete, and eventually guide generating specific functionality to the supramolecular assemblies [12]. The interaction of the components with their environment can also strongly influence the dynamics of the organization process. Such ordered association can be reversible or may allow the components to adjust their positions within an aggregate, once it has formed. Indeed, in nanoscale, mesoscopic, and

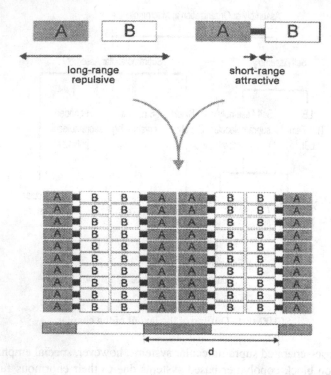

**FIGURE 2.** Illustration of long-range repulsive and short-range attractive forces leading to an ordered structures by self-organization; $d$ is the periodic length of the structure. (Reproduced by permission of Wiley-VCH from S. Foster and T. Plantenberg, Angew. Chem. Int. (41) 688, 2002.)

macroscopic self-assembled systems, the challenge often is to maintain the mobility of the components. As the size grows and becomes larger than molecules, Brownian motion becomes irrelevant, and gravity and friction become dominant [4]. Thus, the self-assembled structure is the result of a delicate balance of competing short-range attractive and long-range repulsive interactions in diverse physical and chemical systems (Fig. 2) [13]. The classical noncovalent forces that facilitate SA, are columbic interactions (permanent charge and dipoles), van der Waals interactions (induced dipoles, charge transfer, dispersive attraction), hydrogen and solvophobic bonds, water-mediated hydrogen bond, ionic bond, $\pi$ stacking, charge transfer complexation, etc. The shape effects also have an important role to play in determining the best fit of complementary sites and in favoring growth of supramolecular assembly. Table 1 presents the most important secondary interactions and their characteristics.

## 1.2.1. Self-assembly by Ionic Interaction

The ionic self-assembly represents the association due to the interaction between two ionic species and involves the interaction between ion-ion, ion-dipole or ion

**TABLE 1.** The most important secondary interaction forces and their characteristics

| Types of interaction | Strength kJ mol$^{-1}$ | Range | Character |
|---|---|---|---|
| Ion-ion interactions | 50-250 | long | Nonselective |
| Ion-dipole interactions | 50-200 | long | Nonselective |
| Dipole-dipole interactions | 5-50 | | Selective |
| Hydrogen bonding | 4-20 | short | Selective, directional |
| Cation-π interactions | 5-80 | | Directional |
| π − π stacking | 0-50 | | Directional |
| van der Waals force | <5 | short | Nonselective, nondirectional |
| Coordination binding | 50-200 | short | Directional |
| Fit interaction* | 10-100 | short | Very selective |
| Amphiphilic** | 5-50 | short | Nonselective |
| Covalent | 350 | short | irreversible |

*Key-lock principle: the precise fitting of interacting surfaces with respect to geometry and cohesion energy density.
**Amphiphilic association as within surfactant micelles.

quadruple, and π-π interactions. Interaction takes place between fixed and complementary ionizable groups, which is modulated by co- and counter ions. The energy $U$ associated in such an ion pair of charge $q_1$ and $q_2$ is expressed by Coloumb's law, $U = kq_1q_2/Dr$, where $k$ is a universal constant ($k = 9.0 \times 10^9$ Nm2C$^{-2}$), $D$ is the dielectric constant of the medium ($D = 1$ for vacuum and increases with polarity of the medium), and $r$ is the distance between ion pairs. The ion-ion interaction is the strongest force of all the noncovalent interactions, with binding free energy >50 kJ (depends on the polarity of the medium) and increases with decrease in solvent polarity. This interaction enhances the stability of host-guest complexes, and influences self-organization in ionic polymers, polyelectrolytes and in biological supramolecular polymers.

π-π **interaction**: π-π interactions are partial charge interactions and play a role in the association of π-conjugated molecules. For example benzene and other condensed aromatic rings display a partial negative charge in the plane of the ring within the π system, and partial positive charge on the hydrogen atoms. Consequently, the parallel stacking of such rings is hindered by the repulsive interaction, resulting an edge-to-face arrangement for successive rings [14, 15]. However, in solution the energy difference between the edge-to-face and face-to-face arrangement appears to be negligible. By proper electrostatic compensation, the interaction may also be surmounted [16]. It is also possible to program π-conjugated molecules such as oligo(p-phenylenevinylene)s to self-assemble into cylindrical aggregate in solution and in solid, which may be of significant promise in supramolecular nanosize optoelectronic devices [17].

**Cation-π interaction:** Cation-π interaction is a strong and specific interaction between a cation and the π face of an aromatic ring [18]. These interactions play a significant role in many self-assembly process particularly in biomolecular structures [19, 20]. In proteins, cation-$\pi$ interactions can occur between the cationic side chains of lysine (Lys), arginine(Arg), or histidine (His) on one hand and the aromatic side chains of phenylalanine (Phe), tyrosine (Tyr), or tryptophan (Trp) on the other and are increasingly being recognized as important structural and functional features of proteins and other biomolecules. This interaction plays role both in intermolecular recognition at the protein-protein interface and intramolecular process such as protein folding [21].

## 1.2.2. Self-assembly Using van der Waals Force

The noncovalent interactions between electrically neutral molecules are collectively known as van der Waals force that arises from electrostatic interactions among permanent and/or induced dipoles. These interactions are generally much weaker than the charge-charge interaction of ion pairs. These energies vary with $r^{-3}$ so they rapidly decrease with distance.

**Dipole-induced dipole interaction:** A permanent dipole also induces a dipole moment on a neighboring group and consequently forms an attractive interaction force. Such interaction is generally much weaker than dipole-dipole interaction.

**London dispersion force:** At any instant, even nonpolar molecules exhibit a dipole moment resulting from the rapid fluctuation of their electron cloud density. The resulting transient dipole moment polarizes the electrons in a neighboring group. These forces are very weak attractive force and only significant for contacting groups (near their van der Waals contact distance) as their association energy is proportional to $r^{-6}$. However, the London dispersion force significantly influences the conformation of polymers and proteins due to the large number of interatomic contacts.

## 1.2.3. Self-assembly by Hydrogen Bonding

The hydrogen bonding force is a special class of dipole-dipole interaction and is predominantly an electrostatic interaction between a weakly acidic donor group (C-H, O-H, N-H, F-H) and an acceptor atom that has a lone pair of electron with distance of separation of $\sim$3 Å. The strong directionality of the hydrogen bond reflects the anisotropy of charge distribution (lone pair) of the acceptor atom. It plays a major role in the self-assembling of synthetic polymeric materials. The reversible hydrogen-bond formation is a critical feature in biological system for creation of complex functional materials such as proteins, carbohydrates and in particular, nucleic acids. It demonstrates the ability of hydrogen bonding to control molecular recognition and association to construct large supramolecules by simple aggregation of relatively simple building blocks. Hydrogen bonding can be purposefully designed and used as a general strategy for controlling molecular association to create novel material with predictable architecture and properties [22, 23].

### 1.2.4. Self-assembly by Solvophobic and Incompatibility Effects

The solvophobic effect is the influence that causes exclusion of the solvophobic segments of a molecule to minimize their contact with solvent and derives from the special property of the solvent. The formation of micelles by amphiphilic molecule-like soap is a well-known example. The unfavorable free energy of hydration of the nonpolar segment in soap causes its ordering and has the net effect of exclusion from the aqueous phase. By doing so it minimizes the overall entropy loss of the entire system. The free energy change to remove a $-CH_2$-group from an aqueous solution is about 3 kJ by thermodynamic measurements. This is a relatively small quantity of free energy; however, considering the large number of contacts involved in a supramolecular assembly, the cumulative solvophobic interaction is significant. In fact, the hydrophobic force is a key driver in folding proteins into native conformations.

### 1.2.5. Self-assembly by Geometrical Fit

In self-assembly, good geometric fit of the interacting molecules is as important as the functionality and microenvironment inside the binding cavity. Geometric size may impose selectivity, efficiency and stereo specificity in the resulting interaction. Recently, design and development of molecules that bind substrate according to the geometric requirements has been used to develop highly specific zeolite and cyclodextrin-based enzyme mimic catalysts. Indeed, this success is rather rudimentary in relation to the elegant use of chemistry by nature for many remarkable manifestations such as mechanism of enzyme action.

### 1.2.6. Metal-ion Directed Self-assembly

Transition metal centers and their coordination chemistry are a widely used strategy for the self-assembly in organometallic chemistry. Through the directional bonding, they provide highly predictable corners or side units that result in various geometric shapes such as triangles, squares, rectangles, cubes, etc., and is a widely used strategy to develop novel functional materials. Considering the many coordination environment the metal centers can adopt and a vast range of geometries available from coordinating ligands, a variety of complexes are accessible through metal-ion-directed self-assembly. The catalytic performance of the encapsulated transition metal complexes is significantly different from a nonencapsulated analogue, they are more active and highly selective in nature.

In general, all the secondary interaction forces are weak; however, they may lead molecules from some random state to a highly complex final ordered state. Rule of additivity is effective for various free-energy contributions of a system. In addition, the occurrence of several sites for complementary interactions can magnify the effect of single pairwise interaction. Cooperative effects among different secondary interactions are also common. In a specific case a single interaction may play major role; however, in general the supramolecular structures are established by mutual interplay of combination of several of the above forces and quantitative assessment of their relative contribution is very difficult. Nevertheless, the knowledge and understanding to control the mechanism of the process is the key to developing new functional materials of novel performance by SA.

The self-assembly is primarily driven by enthalpy in the case of Coulombic and van der Waals interactions as translational entropy being lost due to the establishment of aggregation. However, entropy effects may play a fundamental role when the formation of assembly is driven by solvophobic effects, or by the formation of a mesophase. Detailed quantitative analyses and simulation, based on molecular mechanics, molecular dynamics and Monte Carlo methods for host-guest and self-assembly, are available [24-26]. However, precise predictions of morphology, structural and dynamic properties of complex mesoscopic or macroscopic system, based on local interaction of the constituent molecular units, require the development of more detailed theories of supramolecular assembly. Tailoring molecular building blocks with specific functionality to develop a specific supramolecular assembly is still based on experimental knowledge, experience and intuition. To develop designer polymer material using supramolecular assembly, development of polymer chains having appropriate functional sites, site distribution, and shape are crucial.

## 2. MESOPHASE SEPARATION IN BLOCK CO-POLYMER SYSTEM

### 2.1. Evolution of Supramolecular Assembly in Block Copolymers

In solid systems, supramolecular assembly and ordered mesophase separation are observed in liquid crystal, block copolymers, hydrogen and pi-bonded complexes, amphiphilic polymers, biomacromolecules, and natural polymers. Amongst many self-assembling systems, block copolymers represent a unique and important class of self-assembling materials because of their unique molecular architecture, exquisite control over self-assembly and emerging applications of the resulting nanostructures [27]. They are also extensively investigated to understand the general principle of supramolecular assembly and ordering in macromolecular systems. A polymer chain sequence with different repeating units of thermodynamic incompatibility between the segments can be chemically linked together through covalent bonds to form a variety of block copolymers (BCs) of unique ordered microstructure. Such BCs form mesophase separated (size ca. several tens of nanometers) morphology by spontaneous, thermodynamically driven aggregation of the amphiphiles (complete separation is impossible since the components are chemically joined). They form well-defined mesoscopic aggregates, both in melt and in solution in selective solvents (solubilizes one component). By controlling

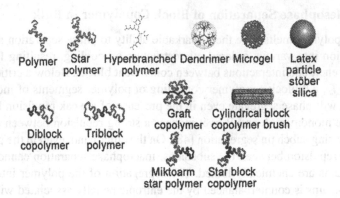

FIGURE 3. Different polymer topologies arranged in order of increase segment density. (Reproduced by permission of The Royal Society of Chemistry from S. Foster, M. Konrad, J. Mater Chem, (13) 2671, 2003.)

appropriately the segment nature, length and architecture of each constituent of the block copolymers, a wide variety of meso-domain structure of high degree of richness and complexity in bulk as well in solution phase is possible.

Diblock copolymer represents the simplest polymeric amphiphiles, and proteins the most complex. Intermediate cases include, amongst others, block copolymers of various composition, architectures, topology, graft and comb copolymers. The block copolymer architecture that comprises diblock, triblock, and multiblock copolymers may be arranged linearly, or a graft, star, or H-shaped blocks (Fig. 3). It is also possible to synthesize more exotic structures such as four- or six-arm star block copolymers and comblike block copolymers [10, 28]. Some other polymer architectures are also shown in the figure for comparison. The architectural diversity of the polymeric amphiphiles is enormous due to the "combinatorial effect" of the possible architectural contrast. The composition, and topology (segment length, copolymer architecture) of BCs can be programmed, controlled and manipulated such that the interfaces between materials with very different chemical nature, polarity, and cohesive energy can be created to a much broader extent than currently possible with low molecular weight surfactants.

BCs have unique, diverse and complex microdomain morphologies [29-33] microdomain sizes [34, 35] interface of the microdomains, [36-40] order-disorder transitions, [40, 41] order-order transitions, [42-44] and deformation and orientation of microdomains [44-47]. The kinetic stability of the BC aggregate structures is sensitive to the macromolecular architecture and the interaction it entails with the environment; and the lifetime of the aggregates can be adjusted to be in the order of second, minute, hour or almost permanent. This kinetics of aggregation has a significant effect on static, dynamic and other functional properties. The control of functionality of block copolymer is an important issue that has been specifically motivated, recently, by the necessity to stabilize metallic, semiconductor, ceramic, or biological interfaces in nanotechnology.

### 2.1.1. Mesophase Separation of Block Copolymer in Bulk

Block copolymer melts have the remarkable ability to phase separation and self-organization into various ordered microstructure on cooling, resulting from the repulsive energetic interactions between constituent blocks. Below a critical temperature, $T_c$, a diblock copolymer consisting of polymer segments of monomers A and B, will phase segregate even in the presence of a weak repulsion between the unlike monomers A and B as it induces a strong repulsion between the subchains causing subchain segregation [48]. On the other hand, even in the presence of strong repulsion between the subchains, macrophase separation cannot occur, as the chains are chemically linked. The segregation of the polymer into A and B rich domains is counterbalanced by the entropic penalty associated with chain stretching. Due to this energetic competition block copolymers form a highly regular periodic structure. The state and geometry of the phase-separated state in the bulk of a block copolymer are principally governed by the parameter $f\chi N$, where $f$ is the polymer volume fraction, $\chi$, is the Flory-Huggins interaction parameter measuring the incompatibility between the individual components (inversely proportional to the temperature) and $N$ is the total degree of polymerization. Typical dimensions of micro domains in a BC are $\sim$10-200 nm, depending on the molecular weight of the blocks.

Significant theoretical and experimental efforts have been directed at characterizing the phase behavior of the bulk materials. The factors that govern the selection, location and morphological organization of various ordered phases in A-B diblock copolymers are relatively well understood [49]. The selective interaction and self-organization characteristics of block copolymers have been exploited to develop designer soft materials and interfacial adhesive [50-52], improve polymer blend compatibility and properties [53, 54]. The phase behavior of a block copolymer of even a simple architecture such as linear diblock copolymer can be quite complex. Most of the theoretical approaches used are primarily based on the standard mean-field theory (SFT) [55] and differ only in relation to the approximations added to the SFT.

Self-consistent mean field theory (SCFT) has been the most successful in describing the phase behavior of high molecular weight polymer [56-59]. Using SCFT approximation, where thermal fluctuations are ignored; the phase diagram for conformationally symmetric diblock may be parameterized using only two parameters: $\chi N$, the segregation parameter, and $f$, the block length ratio (Fig. 4A). The experimentally observed phase diagram of a typical diblock copolymer, poly(styrene-$b$-isoprene) (PS-PI) is also shown in the same figure (Fig. 4B) [60, 61].

The phase diagram exhibits several regimes of the phase segregation. For any fixed value of $f$, with increase in $\chi N$, one passes through a disordered regime where the melt exhibits a disordered state to a weak segregation regime (WSR), to an intermediate segregation regime, to finally the strong segregation regime (SSR). The sizes of the A- and B-rich domains in WSR are of the same order

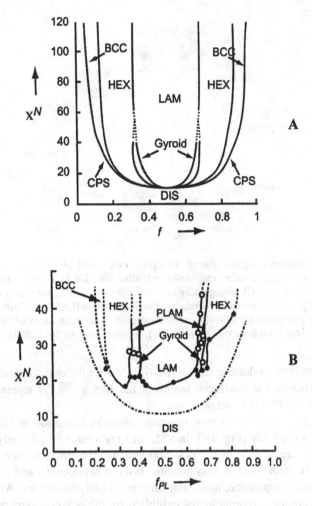

**FIGURE 4.** (A) Theoretical and (B) experimental phase diagram for block copolymers. Various self-organized morphologies can be obtained through variation of the block length, $f$, and the segregation parameter, $\chi N$. The symbols are explained in the text and in Fig. 5. A: (Reprinted with permission from S.Foster, A. K. Khandpur, J. Zhao, F.S. Bates, I.W. Hamley, A.J. Ryan and W. Bras, Macromolecules 27, 6922, 1994. Copyright (1994) American Chemical Society) B: (Reprinted with permission from A.K. Khandpur, S. Foster, F.S. Bates, I.W. Hamley, A.J. Ryan and W. Bras, K. Almdal, and K. Mortensen, Macromolecules 28, 8796, 1995. Copyright (1995) American Chemical Society)

as the interfacial regions around the bonding points and in SSR the domain size is much larger than the interfacial length. In the mesophases, dissimilar blocks exist in distinct "microdomains", which are highly enriched in blocks of the same type, sometimes to the point of being pure. The covalent bonds linking the dissimilar blocks are localized to the vicinity of the microdomain interfaces. With an increase in $\chi$, the interfacial energy increases and chain-stretching leads to

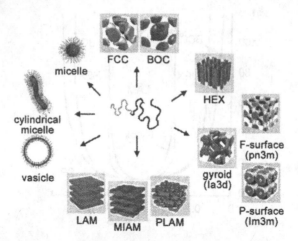

**FIGURE 5.** Common morphologies of mesophase-separated block copolymers in bulk and solution: spherical micelles, cylindrical micelles, vesicles, face centered cubic (FCC), body centered cubic (BCC), hexagonally ordered cylinders (HEX), various complex minimal surfaces (gyroid, $F$-surface, $P$-surface), simple lamellae (LAM), modulated lamellae (MLAM), and perforated lamellae (PLAM). (Reproduced by permission of The Royal Society of Chemistry from S. Foster, M. Konrad, J. Mater Chem, 13 2671, 2003.)

increased equilibrium domain size, $d$ [62-64]. In SSR the domain size scale like, $\chi^{1/6}N^{2/3}$, whereas the interfacial length scales like $\chi^{-1/2}$ for styrene-isoprene block copolymer (SI) [65] are observed.

The rich polymorphism near the order-disorder transition in WSL, where $\chi N \sim 10$ is remarkable [Fig. 4B]. In SSL, at large values of $\chi N$, only a limited morphologies are stable. In diblock copolymers, the microdomain structures formed can be controlled in a systematic fashion between 10 and 200 nm depending on the composition, molecular dimension and architecture. An overview of the most common morphologies exhibited by diblock copolymers is shown in Fig. 5 [28]. It is clearly observed from the phase diagram (Fig. 4A) that for a particular system by adjusting the block length, $f$, it is possible to tailor the morphology. Spherical micelles with cubic packing (FCC, BCC), hexagonally packed cylindrical micelles (HEX), and lamellar phases (LAM) are the most common in solid (Fig. 5). Besides these phases, modulated lamellar [MLAM] and perforated lamellar [PLAM] as well as cubic bicontinuous structures (like gyroid) are also observed in experimental phase diagram of different diblock copolymers [66-69].

The ABA triblock copolymers self-assemble and form physical network and gels due to exchange interactions. In ABA rigid/soft block copolymer, the glassy microdomains serve to anchor soft segments of the polymer and provide them with the characteristics of rubber (thermoplastic elastomer) or melt-processable characteristics. In triblock copolymers, the presence of a third block of different chemical identity creates an ABC triblock copolymer that leads to a much richer phase behavior, with distinctive feature in the mesophase. They can produce astonishingly complex nanostructures [70]. In such a case, three Flory-Huggins interaction parameters $\chi_{AB}$, $\chi_{BC}$, $\chi_{CA}$ are responsible for the phase transitions [71], and this unique situation creates many unfamiliar and much richer repertoires of motifs and morphologies, which has not been exploited in details. ABC triblock copolymer PS-$b$-EB-$b$-PMMA (PS = polystrene, EB = ethylene butylene, PMMA = polymethyl methacrylate) demonstrates a morphology with unique knitting pattern [72]. A complex helical morphology is reported for PS-$b$-PB-$b$-PMMA [73] (PB = polybutadiene). (PS-$b$-Poly (4-vinylbenzyl dimethyl amine)-$b$-PI, (PI = poly isoprene) shows a scheme of threefold symmetry [74]. PI-$b$-PS-$b$-PVP (PVP = poly vinyl pyrrolidone) with similar amount of all the components in PI-PS-PVP exhibits a lamellar morphology [75]. On the other hand, similar amount of all components in PS-PI-PVP exhibits a hexagonal packing arrangement [76].

Petschek and Wiefling [77] reported a strategy to design ferroelectric liquid crystal from ABC block polymer, where the incompatibility between terminal A and C blocks can result in microphase separation of the coronal block and align the electrical dipole residing on the middle rodlike B block [78]. PI-$b$-PS-$b$-PMMA star copolymer exhibits novel two-dimensionally periodic nonconstant mean curvature morphologies [79]. Multiblock (AB)$_n$ block copolymers are capable of undergoing purely intrachain self-assembly. Fig. 6 illustrates a brief overview of the possibilities to tune the phase behavior, morphologies and properties of block copolymers by varying the number of component polymers, the number of blocks, the chemical composition, and architecture of the blocks [80]. Figure 6A indicates the lamellar morphology of a model symmetric SI diblock (I stained with osmium tetraoxide OsO$_4$ and appears black). Figure 6B displays morphology of polybutyl acrylate (PBUA)/polymethylmethacrylate three-arm star copolymer (PBUA-PMMA)3 (composition 65 wt% PMMA), PBUA stained with liquid ruthenium tetraoxide, RuO$_4$ and appears black) prepared by nitroxide-mediated controlled radical polymerization (CRP); and Figure 6C represents a semicrystalline diblock of syndiotactic polypropylene (sPP) and PE, (partly crystalline sPP lamellae appears white, whereas amorphous PE, stained with RuO$_4$ is dark, image) [81]. Figure 6D illustrates complex "helices on cylinders" morphology for an PS-$b$-PB-$b$-PMMA ter block copolymer (PB stained with OsO$_4$ appears black, PS forms grey cylinders hexagonally packed in the PMMA matrix) [73] and indicates the highly complex morphologies that can be accessible for ABC terpolymers. Blending of block copolymer with homopolymer or other block copolymer can provide access to a wide variety of complex morphologies, which has not yet been fully understood and properly exploited [80, 82]. Figure 6E demonstrates that blends of suitably chosen ABC triblock and AC diblock copolymer (here PS-$b$-PB-$b$-PT and ps-$b$-pt, (where PT or pt is poly($t$-butyl methacrylate)) can form non-centrosymmetric periodic arrangements of lamellae (ABCcaABCca)m, where corresponding diblock and triblocks form the usual (ABCCBA)m and (acca)m centrosymmetric lamellar morphologies. In the Fig. 6E, B is block stained with OsO$_4$ and appears black, S is grey and T is white [80]. Remarkable recent advancements in controlled

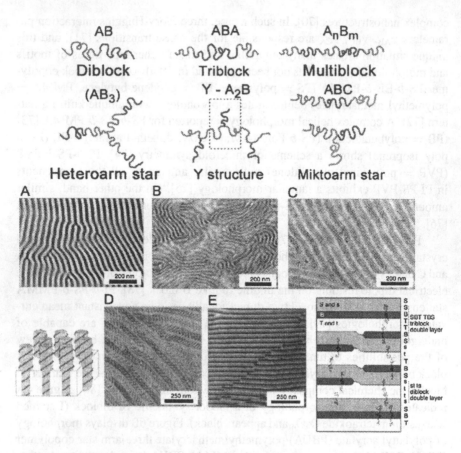

**FIGURE 6.** Effect of block copolymer architecture on ordered mesophase organization:
(A) A model symmetric SI diblock. (B) Polybutyl acrylate/polymethylmethacrylate three-
arm stat copolymer (BUAM)3. (C) A semicrystalline diblock of syndiotactic polypropylene (sPP) and PE. (Reprinted with permission from J. Ruokolainen, R. Mezzenga, G.H.
Fredrickson, E.J. Kramer, P.D. Hustad, G.W. Coats, Macromolecules 38, 851, 2005. Copyright (2005) American Chemical Society). (D) SBM ter blockcopolymer. (Reprinted with
permission from U. Krappe, R. Stadler, L. Voigt-Martin, Macromolecules 28, 4558, 1995.
Copyright (1995) American Chemical Society). (E) Blends of suitably chosen ABC triblock
and ac diblock copolymer. (Reproduced by permission of Nature publishing group from
T. Goldacker, V. Abetz, R. Stadler, I. Erukhimovich, L .L. Leibler, Nature, 398, 137, 1999;
A.-V. Ruzette, L. Leibler, Nature Materials, 4, 23, 2005.)

polymer synthesis have generated unparalleled opportunities to control molecular-
scale morphology in this class of materials.

## 2.1.2. Solvent Regulated Ordering in Block Copolymer: Micelles and Mesophases

In block copolymer solutions, the interaction between the polymer and the solvent
adds an extra dimension to the mesophase behavior. The conformation that any

macromolecule assumes in solution is directed by the interaction strengths of the polymer segments among themselves and with the solvent molecules. The dissolution process of any polymer is dictated by the chemical structure, molecular weight of the polymer, thermodynamic compatibility of the solvent with the polymer, molecular size of the solvent and temperature. The situation is versatile and complicated. For a given block copolymer system, a solvent may be a neutral (a good solvent for all the blocks) or selective (a good solvent for one block but a nonsolvent for the other). Therefore, the dissolution and phase behavior of block copolymers can vary greatly depending on the selectivity of the solvent. Even a solvent of slight selectivity can lead to a measurable preferential swelling or segregation of one component.

This balance becomes more complex and spectacular when the solvent used is selective—a poor solvent for at least one of the blocks and a good solvent for the others. This self-assembling behavior produces micelles of variety of shapes and mesophases, involving different geometries and arrangements. In block copolymer solutions, a wealth of micellar, lyotropic and thermotropic order-order transition (OOT) and order-disorder transition (ODT) are possible. Such copolymer self-assembling behavior has been observed for a variety of BCs in water, polar and nonpolar organic solvents, and more recently, in supercritical fluids. For a complete assessment, exploration is indispensable with block copolymers of a wide range of compositions, $N$ (total degree of polymerization) and $f$ (block volume fraction) and different types of solvents to access all morphological states of different degrees of separation. Tailoring block copolymers with three or more distinct types of blocks creates more exciting possibilities of exquisite self-assembly. The possible combination of block sequence, composition and block molecular weight provides an enormous space for the creation of new morphologies. In multiblock copolymer with selective solvents, the dramatic expansion of parameter space, poses both experimental and theoretical challenges.

### 2.1.2.1. *Micellization of Nonionic Block Copolymer in Organic Solvent.*

The classical self-assembly mechanism for amphiphilic AB-diblock copolymer in apolar solvent is the formation of well-defined micelles with core consisting of the insoluble block and a shell or corona of the soluble block. The thermodynamics of micelle formation and structure in diblock copolymer was examined thoroughly and is well understood [83-87]. The solvent-polymer interactions control the phase behavior and morphology at lower polymer concentrations, whereas polymer-polymer interactions are dominated at higher concentration regime [86-88]. A rich phase variation has also been reported in solvents in which the selectivity varies with temperatures [87, 89]. These micelles are known to exhibit a transition between a liquidlike state and an ordered solid-like state at higher copolymer concentration [90-92]. When the concentration of the micelles is high enough to pack them into ordered array; the systems exhibit elastic gel properties.

The structure and dynamics of ABA triblock copolymer solutions and gels in selective solvents have been investigated extensively using different scattering

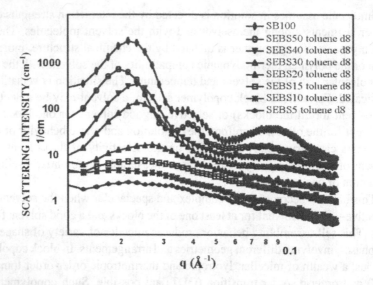

**FIGURE 7.** SANS intensity, $I(q)$ profile of polymer gels of ABA block copolymer (SEBS), of different polymer concentration. In SEBS$n$, $n$ represents the wt% of SEBS in the gel. SB100; represents pure partially deuterated SB diblock copolymer. (Reproduced by permission of Elsevier Science Ltd. from N. Dutta, S. Thompson, N. Roychoudhury, R. Knott, Physica B: Condensed matter 773, 385-386, 2006.)

techniques [93-110]. The swelling of this block copolymer differs significantly from that of the corresponding AB diblock analogue. Figure 7 shows the small angle neutron scattering (SANS) intensity profile of ABA triblock copolymer gels (SEBS) (EB = ethylene butylene) over a wide range of concentrations in deuterated toluene (only slightly selective to PS phase) at 28°C [110]. The SANS profile for the partially deuterated neat dPS-$b$-polybutadiene is also shown in the figure for comparison. The characteristic features of the scattering patterns are one or two well-pronounced structure factor maxima at low $q$ region near 0.019 Å$^{-1}$ and broad form factor maxima at higher $q$ value near 0.09 Å$^{-1}$. The effect of polymer concentration on the scattering pattern and trends of the interference maxima in the low $q$ regime (structure factor) are clearly observed from the figure. The highly concentrated gel (>20 wt% gels) exhibits at least two orders of reflections in low $q$ range. The significant sharp interference peak in the lower $q$ regime may be attributed to interdomain interference and indicates relatively higher level of ordering of the micelle cores. A second maximum at >50 wt% gels demonstrate a further increase in degree of segregation with a higher level of order. The interference maximum associated with the interdomain interaction is relatively larger and confined to the narrow small-angle region.

Small angle neutron and X-ray scattering (SANS and SAXS) measurements on ABA triblock copolymer indicate that the morphology of the micelles depends on whether the solvent selective to the outer (A) block or to the inner (B) block leading either to the formation of isolated core-shell micelles (A block remains in the same core domain, the B block forms a loop) or bridged micelles (A blocks reside on different domains and the B forms a bridge). ABA block copolymer aggregates into the microdomain when mixed with A-block selective solvent. These microdomains consist of a dense core of B blocks from which the corona of the flexible A block chain emerges and reveals either spherical, rod-, thread-, or disklike shape.

When the experimental temperature is higher than glass transition temperature ($T > T_g$), the system morphology will change with polymer concentration and yield a gel when polymer concentration is sufficiently high. If the volume fraction of the microphase-separated microdomains increases significantly, intermicellar correlations become more pronounced and may be so strong that micelles order in a crystalline structure on the mesoscopic level. A rather different situation is expected in solution of a midblock selective solvent—the polymer midblock emerging from the core forms either loops or micellar bridges, depending on whether polymer endblocks are located in the same or in different microdomains. Dutta et al. [107-110] examined triblock copolymer SEBS in both neutral and midblock selective solvents using SANS, transmission electron microscopy (TEM) and rheological methods over the entire range of concentrations including bulk polymer.

In neutral solvent, a micellar structure was confirmed from TEM images; however, in midblock selective solvent a loose bridged structure is predominant (Fig. 8). The intermicellar bridging results in a cluster of highly connected micelles. Above a certain concentration they may extend over the whole sample volume, and provide a macroscopically isotropic physical gel or soft solid. The loose bridged structure contributes significantly to the high elasticity of such system. The crosslink dy-

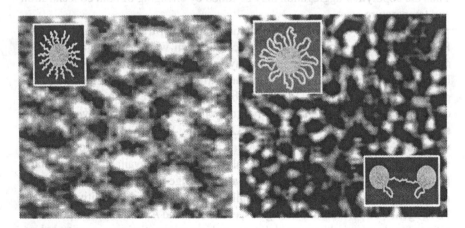

**FIGURE 8.** TEM image of amphiphilic ABA triblock copolymer (SEBS) thin film cast from its dilute solution in selective solvent: (A) solvent slightly selective to end block, the characteristic spherical micelles in solution are shown in the inset; (B) solvent selective to middle block, the corresponding complex micelles in solution are shown in the inset.

namics in such systems is governed by the $T_g$ of the polystyrene terminal block. If $T < T_g$ the residence time of the terminal block within the micelle core is very long and the bridge is effectively permanent. In such case, the gel will not dissolve in the presence of excess solvent. The entropy penalty is stricter for solvents, which are selective to the middle block. Kleppinger *et al.* [89] demonstrated a cubic structure for SEBS copolymer in a selective solvent good for the middle block. Raspaud *et al.* [111] have employed SANS to investigate polystyrene-block-polyisoprene-block-polystyrene in a selective solvent good for the middle block and observed that the network formed by the outer blocks has a cubic ordered structure at high concentrations. In tetra block and higher multiblock copolymers, bridged systems with more complex structure are formed [112]. Lodge *et al.* [113] demonstrated a method using, SANS to quantify the small degree of selectivity for a block co-polymer in selective solvent.

### 2.1.2.2. *Block Copolymer Micelles in Water.*
Structure and dynamics of block copolymer colloids and micellization in aqueous media have been the subject of considerable interest and investigated in details [114-116]. Due to their pharmacological application, extensive investigations have been carried out on poly(ethylene glycol) (PEG)-containing di- and triblock copolymers [117-124]. PEO (polyethylene oxide) is highly soluble in water due to its strong hydration (very favorable fitting of ethylene oxide monomer into water structure). However, the hydration is strongly temperature dependent and PEO exhibits a upper critical solution temperature (UCST) behavior and theta ($\Theta$) temperature of PEO in water is close to 100°C. Similar temperature effects on solubility are also displayed by aqueous solution of other nonionic block copolymers based on poly($N$-isopropyl acrylamide)(NIPA), poly(lactic acid), hydrophobically modified celluloses, etc. Thus the copolymer aggregation may be tuned by changing of both concentration and temperature.

In the family of PEO containing block copolymer triblock copolymer PEO-PPO-PEO (PPO = polypropylene oxide) has been widely investigated as they are low in cost, biocompatible and display varied micellar behavior depending on PEO/PPO ratio and molecular weight. They exhibit a very rich structural polymorphism. Alexandridid *et al.* [125, 126] observed nine different mesophases (four cubic, two hexagonal, and lamellar lyotropic liquid crystalline and two micellar) structure in one ternary diagram of PEO-PPO-PEO copolymer in selective solvents (water and oil). Shearing of concentrated solution may also produce novel mesophase structure. These block copolymers are also available commercially [Poloxamers (ICI), Pluronic, BASF)] in a wide range of compositions, PEO/PPO ratio and molecular weight and industrially used as antifoaming additives, emulsifying agents, dispersants and detergents, etc. Booth and Atwood [127] reported the effect of molecular architecture of these copolymers upon their aggregation behavior and noted that in terms of block composition PEO units have little effect and aggregation is controlled by the hydrophobic segment. It has also been demonstrated that micellization of PPO-PEO-PPO is less favorable than that of PEO-PPO-PEO due to constrains related to the curving of the macromolecular chain as discussed in Sec. 2.1.2.1.

These water soluble nanostructured block copolymers find numerous applications in different areas. Many of the uses are related to their ability to dissolve otherwise water-insoluble substances and find applications in the remediation of contaminated soil or wastewater treatment. Another important area of application involves solubilization and controlled release of pharmaceutical substances (neither renal filtration nor uptake occurs for micelle of diameter ~50-100 nm and it circulates in the blood stream for long) and carrier of hydrophobic drugs [128-131]. In this respect besides micelles solubilization capacity, their biocompatibility and interaction/lack of interaction with biostructures and biomolecules are important. Owing to the low CMC (critical micelle concentration), block copolymers associate into micelles even when highly diluted in blood. For such applications PEO is the most commonly used hydrophilic segment not only due to their efficiency and biocompatibility, but also its capacity to prevent protein adsorption, and consequently, phagocytosis, increase *in vivo* circulation lifetime, etc. The most widely reported systems include PEO-PPO-PEO [132, 133], polyaminoacid-PEO copolymer with hydrophobic block aspartic acid [134] and aspartate derivatives, [135] polylysine [136], polycaprolactone, [137] and polylactide [138].

**2.1.2.3. Ionic Block Copolymers.** The presence of electrical charge in the block copolymer molecule adds a novel features to the already complex aggregation process in block copolymer, and the micellar behavior of ionic block copolymer has been the subject of significant interest [139]. Due to enormous interfacial tension these systems are in a thermodynamic state close to the super-strong segregation limit (see Fig. 4) and under these conditions, a sequence of shape transitions from spherical to cylindrical to lamellar is possible. Such transition can be induced by increasing the ionic strength of the solution or by increasing the length of the core block. A salt-induced sphere to cylinder transition has also been demonstrated for such block copolymers. Depending on the solvent selectivity, these aggregates may be classified as block ionomers or block polyelectrolytes. In the former the ionic moieties form the core of the aggregate, and apolar blocks form the corona in organic solvents selective to apolar phase. In polar solvent the nonionic block forms the core and the corona containing the ionic block. Due to high incompatibility between the ionic and nonionic (hydrophobic) they aggregate at very low concentration (low CMC $\sim 10^{-8}$ mol cm$^{-3}$).

Block copolyelectrolytes display interesting rheology often forming highly elastic gel at lower concentration than their neutral counterparts. However, strong influence of pH and electrolytes on CMC and properties of the aggregates are observed. The most widely studied diblock polyelectrolyte solution and gels are those based on polyacrylic acid (PAA) and its derivatives as the ionic block. PAA and its derivatives are weak polyelectrolyte; therefore, the chain conformation and the solution properties of the block copolymer depend on solution pH as well as ionic strength [140]. For example, micelles of PS-poly(methacrylic acid) in water increase their hydrodynamic radii from 40-50 nm to ca. 80 nm, as pH increases from 4 to 8, which is ascribed to the dissociation of the acid PMA block.

Polymer-surfactant complexes formed from ionic block copolymers with single and multi-tail surfactant form unique self-assembled structured nanoparticles that may have potential application in dug delivery [141]. Environmentally responsive nanoparticle from block ionomer complexes (BIC) synthesized by reacting block ionomer (PEO-*b*-PMA) and oppositely charged surfactant (hexadecyltrimethylammonium bromide) exhibit excellent pH and salt sensitivity, which are reversible [142]. They may be used as smart gels for a wide range of applications, that react to environment conditions such as pH, temperature, salt concentration. By specific variation of these parameters it is possible to shrink, to swell, or to dissolve the gels. Hydrophobically modified block polyelectrolyte such as PEO-PPO-PEO-PAA exhibits sensitivity of their associative property in aqueous solution to pH, ionic strength and temperature. They also exhibit "anti-polyelectrolyte effects." PEO-PPO-PEO-/polyacrylic acid/concanavalin gel for glucose-responsive insulin release has been developed [143]. Block copolymer based on linear polyelectrolyte block and hydrophobic block carrying pendant dendritic moiety has been made to develop structure selective dye uptake into aggregates of copolymer in water [144]. In proteins, the presence of ionic, hydrogen-bonding and hydrophobic groups makes them prone to all interactions and governs the properties of proteins [145].

## 2.2. Synthetic Strategy of Multiblock Copolymers

Remarkable recent advancements in controlled polymer synthesis have generated unparallel opportunities to control molecular-scale morphology of block copolymers. There are different methods available to synthesize block copolymer which include: radical, living anionic, living cationic, living radical and polycondensation. The method of choice depends on the type of monomer used and dictates the final architecture and dimension of the polymer.

### 2.2.1. Synthesis of Block Copolymer by Living Ionic Polymerization

In the design of self-assembling polymeric materials, it is extremely important that control over all aspects of polymer structure, such as molecular weight and architecture, polydispersity, block volume fraction, number of functional groups, their nature and position, etc., is achievable precisely. To synthesize well-defined block copolymers of various controlled architectures, solubility, and functionality, precise control over the polymerization processes is crucial. Living polymerization provides the maximum control over the synthesis of block copolymers. In this process, all polymer chains start growing simultaneously, and during chain growth no termination or chain transfer takes place, resulting in a very narrow molecular weight distribution polymer.

When all the monomers have been consumed, the active center persists, and upon further addition of new monomer, polymerization continues to the final block copolymer. The concept of block copolymer of controlled architecture was first introduced with the invention of living anionic polymerization by Szwarc in 1956 [146-148]. Many block copolymers have been successfully synthesized using this technique that includes polystyrene (PS), polybutadiene (PB), polyisoprene (PI), polymethyl methacrylate (PMMA), poly(methacrylate) (PMA), polyethylene oxide (PEO), poly(propylene oxide)(PPO) and poly(dimethylsiloxane) (PDMS) as block segments. Many of such copolymers such as PS-PB-PS triblock copolymer (Kraton) (thermoplastic elastomers), PEO-PPO-PEO (Pluronics) (surfactants and steric stabilizers) are commercially important polymers. Amphiphilic block copolymers based on PS-PEO, PS-PVP-PEO(PVP = poly(2-vinylpyridine) and PEP-PEO (PEP = poly(ethylene propylene)) have also been successfully synthesized [149].

A series of well-defined functional groups with amphiphilic block copolymers containing styrene and poly[oligo(ethylene glycol)methacrylate] (POEGMA) were synthesized by sequential anionic copolymerization of styrene and trialkylsilyl-protected oligo(ethylene glycol) methacrylates followed by deprotection [150]. Living anionic polymerization is the most important polymerization technique of synthesizing well-defined block copolymers [151-157]. Block copolymer synthesis has also been successfully achieved by other living polymerization techniques such as cationic polymerization, [158] ring opening polymerization [159] or by pseudo-living polymerization techniques using Ziegler-Natta catalyst [160].

A number of new amphiphilic block copolymers based on isobutylene and derivatives of vinyl ethers including poly(isobutylene-b-methyl vinyl ether) [161], poly(styrene-b-hydroxyethyl vinyl ether) [162], poly(styrene-b-ionic acetylene), [163], poly(methyl tri(ethylene glycol)vinyl ether-b-isobutyl vinyl ether) [164], poly(α-methylstyrene-b-2-2-hydroxyethyl vinyl ether), [165], poly(2-(1-pyrrolidonyl)ethyl vinyl ether-b-isobutyl vinyl ether) [165] were synthesized using living cationic polymerization. The key to this living polymerizations is the strong interaction between the nucleophilic counter anions and the cationic active sites. The process may produce copolymer with well-defined architecture and size but requires stringent control of the reaction conditions and purity of the monomers.

Group-transfer polymerization (GTP) [166] is a valuable method for controlled polymerization of $\alpha,\beta$ unsaturated esters, ketones, nitriles and carboxamides. GTP is not only useful to synthesize block copolymers based on methacrylate and acrylate polymer surfactants that stabilizes hydrophilic/hydrophobic interfaces but also biological interfaces. The technique can also produce living polymers at very mild conditions. Okano et al. [167] prepared a highly blood compatible polymer surface with a poly(styrene-b-2-(hydroxyethyl)methacrylate) (PS-PHEMA) block copolymer. PBM-b-PHEMA (PBM = polybutyl methacrylate) that exhibits high oxygen permeability has also been successfully synthesized by Ito et al. [168].

A ring-opening metathesis polymerization [ROMP] technique has been employed to synthesize block copolymers and functionalized block copolymers based on norbornene, norbornene derivatives, functional norbornenes, methyl tetracyclodecene (MTD), etc. [169-173]. They are useful to synthesize functional block copolymer with functional groups including ~COOH, amine and cyclopentadiene, contained on one block, which may be employed to form stable complex with a variety of metals to form semiconductor nanoclusteres. However, all the above polymerization techniques suffer from serious disadvantages, such as: (i) most of them do not tolerate even extremely low levels of impurities, (ii) special reaction conditions are needed to perform the reactions, (iii) they are not suitable for synthesizing block copolymer from many dissimilar vinyl monomers, (iv) processes are very expensive.

## 2.2.2. Synthesis of Block Copolymer by Controlled Radical Polymerization

The concept of living free radical polymerization was advanced to expand the scope of synthesis of complex block copolymer architecture using diverse range of monomers. The reason was its high tolerance against a large number of monomers, as well for copolymerization, a convenient temperature range and requirements of minimal purification of monomers, solvents, etc. Since early 1980s, attempt has been made to synthesize block copolymers using regular free-radical polymerizations [174-176]. The main disadvantages of such radical polymerization techniques are the lack of macromolecular structure and dimension control (poor molecular weight control, high polydispersities, and low blocking efficiencies. This is attributed to the slow initiation, slow exchange, direct reaction of counter radicals with monomers, and thermal decomposition of the initiators, transfer and terminating agents. The possibility of polymerization can be greatly extended if radical polymerization can be manipulated to lead to well-defined materials; however, the crucial problem is to avoid the presence of unavoidable termination between growing chains.

In the last two decades, the aim of synthesis of well-defined polymers by controlled free radical polymerization technique has been accomplished that combine the versatility of free radical polymerization with the control of anionic polymerization. Living radical polymerization (LRP) is currently a very rapidly growing field of research in polymer synthesis [177-200]. In general, living polymerization is a chain growth polymerization without any termination or disproportion. With these conditions it is possible to control the molecular weight of the resulting polymer by varying the monomer-initiator ratio, while the molecular weight distribution (MWD) stays narrow. It is also possible to functionalize the polymer through the initiator and the terminating end group. Indeed, in LRP the termination can only be suppressed and the reactions are called controlled free radical polymerization (CFP) rather than LRP and are generally based on two principles; (i) reversible termination and (ii) reversible transfer.

Recently, the most successful controlled free radical techniques, are respectively, nitroxide-mediated polymerization (NMP) [178-182], atom transfer radical polymerization [ATRP] [183, 184] and reversible addition-fragmentation chain transfer (RAFT) polymerization [185]. NMP and ATRP are reversible termination type CFP where the chain end is end-capped with a moiety that can undergo reversibly homolytic cleavage. In NMP, this moiety is a nitroxide, while in ATRP, a halide is reversibly transferred to a transition-metal complex. RAFT is a reversible transfer-based process, where fast exchange of growing radicals via a transfer agent takes place. In the RAFT process, dithiocarboxylates are responsible for this exchange that proceeds via intermediate radicals. These techniques have the potential to synthesize block copolymers under less stringent conditions than those necessary for living ionic polymerization and enable the synthesis of block copolymer of controlled $M_n$ with narrow molecular weight distribution at elevated temperatures over prolonged period of reaction time.

ATRP is a versatile method to control radical polymerization of a wide range of monomers, including styrenes, (meth)acrylates, (meth)acrylonitrile, and dienes. Matyjaszewski and coworkers accomplished the synthesis of linear, comb, star and hyperbranched/dendritic architectures, all with the additional facility of controlling functionality [186-189]. ATRP has been used since 1995 as a promising new CRP method and has become rapidly one of the preferred methods in polymer synthesis. ATRP minimizes the termination by reducing the stationary radical concentration. In a typical ATRP, an alkyl halide is used as an initiator, with a transition metal in a lower oxidation state, and a ligand that complexes with the metal. Various metals and ligands have been used for ATRP chemistry, with the most used metals being nickel and copper. It has been reported that $Cu(n)Xn$/ligand (X = Cl or Br, ligand = dipyridyl) catalyzed ATRP can be accomplished in the presence of air, when Cu(0) metal is added to the polymerization mixture.

A wide range of well-defined block copolymers from dissimilar vinyl monomers has been reported using NMP [190]. Yousi *et al.* [191] demonstrated the polymerization of methyl methacrylate (MMA), *n*-butyl methacrylate (BMA), ethyl methacrylate (EMA), octyl methacrylate (OMA), vinyl acetate (VAc), *N,N*-dimethylacrylamide (DMA), 2-(dimethylamino) ethyl acrylate (DAEA) using polystyrene macro initiators with terminal TEMPO (2,2,6,6-tetramethyl-L-piperidinyloxy) groups to produce a wide range of hard-soft, hydrophilic-hydrophobic, and other block copolymers with various functionality using NMP. RAFT is a very robust polymerization technique and is compatible with the broadest range of monomers and reactions conditions [192-199]. In RAFT polymerization, contrary to some other CFP, a conventional free radial initiator is employed. The RAFT process is also capable of controlling polymerization in aqueous dispersions. NMP and ATRP are somewhat less suitable for such process [200].

### 2.2.3. Polycondensation

The synthesis of block copolymer by polycondensation of, $\alpha,\omega$ difunctional telechelic oligomers with another bifunctional block or precursor of end block to yield block/multiblock copolymer is an important industrial process. Novel segmental block copolymer containing alternating soft block and crosslinkable block [201], oligosaccharides blocked by (condensation with amino-group-ended) oligomer [202] have been successfully synthesized by this method. Metal-catalyzed polycondensation, particularly palladium and nickel derivatives, have been employed successfully for polymers that could not be prepared other ways. Takagi et al. [203] synthesized block copolymer of alkoxyallenes with phenylallene, of narrow molecular weight distribution using [(p-allyl) NiOOCCF$_3$]$_2$ as catalyst and it was observed to be eniantiomer specific.

Block copolymer synthesis by enzyme-catalyzed polycondensation and polyadditions has also developed as they exhibit higher selectivity, efficient control of reactions and structure than the classical catalyst, and can be carried out in mild conditions. Wallace and Morrow [204] reported the enzyme-catalyzed polycondensation of bis(2,2,2trichloroethyl)-3,4-epoxyadipate with 1,4-butanediol optically active, epoxy-substituted polyester in the presence of porcine pancreas lipase and it was observed to be enantiomer specific. Blinkovsky and Dordick [205] has demonstrated that the peroxidase-catalysed synthesis, of lignin phenol block copolymers. Loos and Stadler [206] prepared amylose-block-polystyrene rod coil block copolymer using potato phosphorylase as the catalyst. Though there are only few examples of synthesis of block copolymers by enzyme-catalyzed synthesis; however, they are interesting development with significant future potential.

## 2.2.4. Sequential Polymerization

A sequential reaction, where several associated techniques are used in a clever fashion, is a powerful new tool in block copolymerization and promises fantastic developments. An entirely different concept towards block copolymer synthesis with a wide range of structure and architecture makes the use of active-center transformation approach. In this approach, first an active block is prepared, and then its active center is modified to initiate another polymerization. This transformation involving anions, cations, and radicals is possible. Goldschmidt [207] used this unique technique to synthesize PS-PEO and PMMA-PEO diblock copolymers. In this reaction, CRP is employed in presence of mercaptoethanol as chain transfer agent to make hydroxy functionalized PS, which was then used as the starting point for the ring-opening polymerization of ethylene oxide. PS-PEO graft copolymer with PEO as a side block that is water soluble and micelle forming, is useful as support for solid-phase peptide synthesis. Amphiphilic graft copolymers have become the focus of significant investigation because of their ability to undergo unique microphase separation and micellization [208-213]. Similarly associating radical-to-cationic polymerization, [214, 215] cationic to-radical polymerization [216-218] anionic-to-radical polymerization, [219] cationic-to-anionic polymerization, [220] anionic-to-cationic polymerization [221] have also been employed successfully to synthesize many novel A-B and ABA-type block copolymers.

Coca et al. [222] successfully carried out ROMP and ATRP to prepare diblock copolymer based on polynorbornene such as polynorbornene-b-polystyrene, polynorbornene-b-poly(methyl acrylate). Recently, Grubbs dedicated

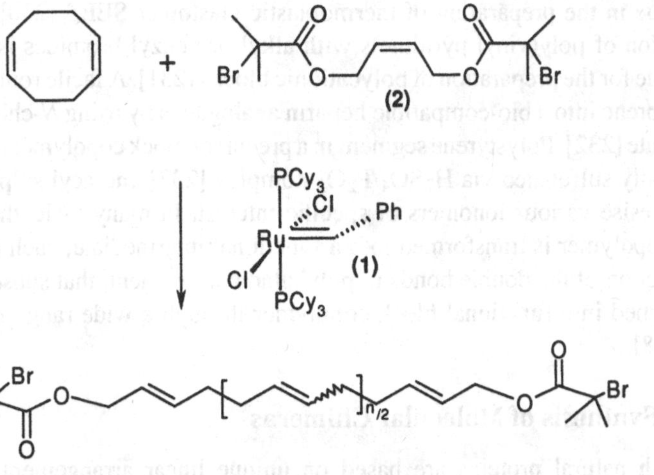

SCHEME 2. Ru-catalyzed preparation of macroinitiator.

research efforts to synthesize low molecular weight poly methyl methacrylate-b-polybutadiene (PB-b-PMMA). Grubbs et al. [223, 224] had synthesized telechelic polymers like hydroxyl-, carboxyl-, amino-, methylmethacrylate- and epoxide-terminated telechelic PBDs by ROMP using a Ruthenium-based catalysts, such as $(PCy_3)_2Cl_2Ru = CHPh$ (1).

Synthesis of triblock copolymers, containing poly(n-alkyl methacrylate)s as the outer blocks is difficult to achieve, as carbonyl groups and the $\alpha$-hydrogen atom of acrylates are sensitive toward nucleophilic attacks. However, telechelic polymers prepared by ROMP can be used as macro initiators for the following polymerization of a monomer belonging to a different class (Scheme 2) to prepare the outer block. Dutta et al. [225] reported a facile route towards the synthesis of a series of triblock polymers with alkyl methacrylate as the outer block by ATRP with telechelic poly(butadiene)s as a macro-initiator. The telechelic poly(butadiene)s were synthesized by ring-opening metathesis polymerization of 1,5-cyclooctadiene in the presence of the chain transfer agent (2) using a ruthenium-based metathesis catalyst (1) (Scheme 2). A series of triblock copolymers was reported by the authors with several long-chain n-alkyl methacrylates (such as lauryl and stearyl methacrylate) as outer blocks.

## 2.2.5. Polymer-Analogue Reaction (PAR)

In this process, a desirable functional block copolymer of desirable functionality is formed by chemical modification of the precursor block copolymers, which is accessible in large scale in a variety of compositions, architectures and molecular lengths. It is a very useful method to broaden the diversity of the block copolymer systems [226-228]. A very commonly used PAR is the hydrolysis of poly{(meth)acrylic ester}s to prepare poly((meth)acrylic acid) [229]. Hydrogenation of polydienes containing block copolymer such as SBS, is a standard industrial technique in the preparation of thermoplastic elastomer SEBS [230]. The quarternization of poly(vinyl pyridine)s with alkyl or benzyl bromides is a standard technique for the preparation of polycationic blocks [231]. A facile route to convert polyisoprene into a bio-compatible heparin analogue is by using N-chlorosulfonyl isocyanate [232]. Polystyrene segment in a precursor block copolymer may be fully or partially sulfonated via $H_2SO_4/P_2O_5$ complex [233] and acyl sulphates [234] to synthesise various ionomers of specific interest. In many PAR, the precursor block copolymer is transformed into a functional intermediate; such as epoxidation reaction of the double bonds in polybutadiene segment, that subsequently be transformed into functional block copolymer through a wide range of reactions [235-238].

## 2.2.6. Synthesis of Molecular Chimeras

Although natural proteins are based on unique linear arrangement of just 20 different monomers, folded proteins achieve a tremendous breadth of physical and chemical activities, ranging from exquisitely specific room temperature catalysis to the formation of unusually strong and tough biomaterials, such as collagen and spider silk. However, attempts to synthesize polypeptides with well-defined amino-acid sequences, as biopolymer have been plagued by unwanted side reactions, polydispersity and poor organization. Synthetic polymer, on the other hand, can be manufactured in bulk at low cost, but due to the lack of precise control over composition and dimension, complex folding architectures cannot be achieved.

There are varieties of synthetic building blocks in hydrocarbon polymers, which hold tremendous potential to create an improved class of non-natural bioinspired polymers of hybrid molecular structure with peptide and nonpeptide sequences. These polymers can not only mimic protein, structure, activity and properties with novel backbone and side chain chemistry (Scheme 3), but also can

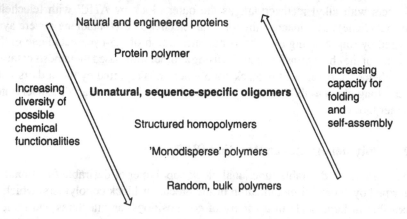

**SCHEME 3.** The spectrum of polymeric materials and the wide range of possibilities for non-natural sequence specific block copolymers (Reprinted by permission of Elsevier from Ref. 239.)

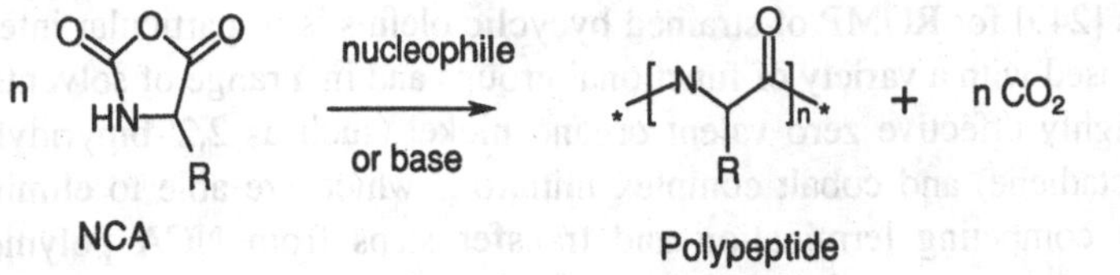

SCHEME 4. Ring-opening polymerization of α-amino acid-N-carboxyanhydrides (NCAs). R is a functional group that can be replaced.

be obtained at low cost, chemical diversity and biological stability [239-241]. It also holds the promise for the development of biomimetic polymers that cannot only closely match the functionality of those of biological molecules in one aspect, but which are also enhanced in other novel functionalities. A potential way of combining all the advantages of synthetic block copolymer and peptide structures lies in the creation of "molecular chimeras" or "hybrid," which are block copolymers with synthetic segments and amino acid sequences [241, 242]. The most economical and rapid process for the synthesis of long polypeptide chains is the ring-opening polymerization (ROP) of α-amino acid-N-carboxyanhydrides (NCAs). This method involves the simplest reagent, and high-molecular-weight polymers can be prepared in both good yield and large quantity with no detectable racemization at the chiral centers. NCA polymerizations have been initiated using many different nucleophiles and bases, the most common being primary amines and alkoxide anions (Scheme 4).

Recently, synthesis of many compositionally and topologically different block copolypeptides have been reported [242-244]. In most cases, the copolypeptide block is composed of γ-benzyl L-glutamate, β-benzyl L-aspartate, or Nε-benzyloxycarbonyl L-lysine as the polymerization of these NCAs are the best controlled of all. The chemical nature of the synthetic block segment has been varied to a much greater extent using both hydrophobic (such as polystyrene, polybutadiene, polyisoprene, poly(methyl methacrylate), poly(propylene oxide), poly(dimethyl siloxane), etc., and hydrophilic segment [poly(ethylene oxide), poly(vinyl alcohol), poly(2-methyl oxazoline), etc]. However, in all the cases the block copolymers formed are chemically dispersed and are often contaminated with homopolymer. The problem in conventional NCA polymerization is that there is no control over the reactivity of the growing polymer chain-end during the course of the polymerization and the monomer can undergo a number of side reactions and exhibits a very broad molecular weight, $M_w$ distribution (PDI~7) [245, 246].

Recent advances in polymer synthesis are now beginning not only to allow control of polymer backbone and side chain architectures but also to enable the incorporation of a very broad range of functional groups appended to the polymer chains. ROMP has proved to be very useful in the synthesis of oligopeptide functional polymers [247, 248]. The ROMP technique with a well-defined catalyst is able to control polymer structure and to prepare multifunctional materials with designed architectures. Ruthenium carbene catalyst developed by Fraser and

Grubbs [249] for ROMP of strained bycyclic olefins is of particular interest as it can be used with a variety of functional groups and in a range of solvents

Highly effective zero valent organo nickel (such as 2,2'-bipyridyl Ni(1,5-cyclooctadiene) and cobalt complex initiators, which are able to eliminate significant competing termination and transfer steps from NCA polymerization to allow preparation of well-defined peptides has recently been demonstrated by Deming et al. [245]. This polymerization yields narrow $M_w$ distribution ($M_w/M_n = 1.05$-$1.15$) and is obtained in excellent yield (95-99% isolated). Recently, Seidel et al. [250] also reported use of chiral ruthenium and iridium amido-sulfonamidate complex for controlled enantioselective polypeptide synthesis. These new methods can now be employed successfully to realize stereocontrol in NCA polymerization that has been difficult to achieve before. Coupling of natural or engineered polypeptides to synthetic polymers has also been attempted to develop biomimetic macroamphiphiles [251, 252]. Recently, unique vesicle-forming (spherical bilayers that offer a hydrophilic reservoir, suitable for incorporation of water-soluble molecules, as well as hydrophobic wall that protects the loaded molecules from the external solution) self-assembling peptide-based amphiphilic block copolymers, that mimic biological membranes, have attracted great interest as polymersomes or functional polymersomes due to their new and promising applications in drug delivery and artificial cells [253].

## 2.3. Nanophase Separation in Side Chain Crystalline Polymers

Flexible polymer chains bearing linear long alkyl side chains exhibit unique morphological arrangement in solid-state that consists of biphasic arrangement of alternating layer of crystalline side chain and flexible main chain. Side-chain crystallizing comblike polymers are unique in their special architecture and intimate contact of crystalline order and amorphous disorder. Side chain crystallinity is observed in poly-$n$-alkylmethacrylate (PnMA) and poly-n-acrylates (PnAA) with flexible $n$-alkyl side chain from 12 to 18 carbon atoms in length (Fig. 9A). In such a polymer, every monomer unit of the chain is involved in both crystalline (in branch) and amorphous (in backbone) regions. These crystalline polymers are different from semicrystalline polymers such as polyethylene in that the crystallites are made up 'of $n$-alkyl groups, which extend from the backbone of the molecule rather than the segment of the backbone itself. They are characterized by the small domain size of the order of few nanometers, and a large number of topological constrains due to microstructure with short incompatible sequences. This squeezed-in situation considerably affects the segmental mobility of the main chain. This unique nanophase separation has been reported in several homologous series of polymers with long alkyl groups in the side chain, [254-258]; however, it is less investigated compared to mesophase-separated block copolymers. The effects of (i) side chain length [259], (ii) influence of a functional group situated along the side chain, [260], (iii) result of interrupting the long ordered side chains by randomly interspersed amorphous chains [261, 262] and stereo regularity [263] on structurisation

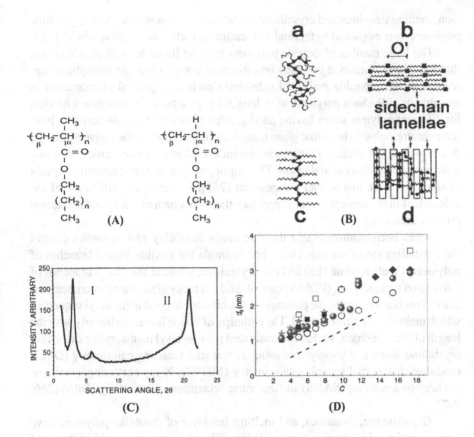

**FIGURE 9.** Morphological self-organization in side-chain crystalline polymer with long alkyl group in the side chain: (A) structure, (i) polynalkylmethacrylate (PnMA) and (ii) polynalkylacrylates (PnAA). (B) Quasi one-dimensional picture to describe the nanophase-separated side-chain crystal polymer. (a) coil conformation in the melt, (b-d) layer structures, ordered due to side-chain crystallization, (b) lamellar structure (main chains, perpendicular to the paper; d long period), (c) double coil conformation, (d) top view of the structure of b (x, crystalline; a, amorphous). (Reprinted with permission from Ref. 262. Copyright 1998 American Chemical Society.) (C) SAXS curve for poly(stearyl methacrylate), a low angle peak (I) and van der Waals peak are indicated on the figure. (D) Equivalent Bragg spacing $d$ as a function of carbon in poly($n$-alkyl acrylates) (■), poly($n$-alkyl methacrylate), (●), poly(di-$n$-alkyl itaconates), (★), hairy rod polyimides (♦), poly(alkylbenzimidazol-alt-thiophene) (○) and poly-1-olefines (□). (Reproduced by permission of Nature publishing group from M Beiner; H. Huth, Nature Materials 2(9), 595, 2003.)

and crystallization have been widely examined. Plate and Shibaev [264] employed dielectric spectroscopy, mechanical methods, and NMR spectroscopy as tools to investigate the molecular mobility in comblike polymers with polymethacrylate, polyacrylate, poly(vinyl ether), and poly (vinyl)ester main chains. The heats of fusion, melting transition and crystallization behavior of many side-chain crystalline polymers were evaluated by thermal and scattering techniques such as DSC, SAXS.

The large number of flexible polymers bearing linear long alkyl side chains that have been examined by many investigators confirm that the nanophase separation of incompatible main and side-chain parts is a general phenomenon in amorphous side-chain polymers with long alkyl groups. A comparison with data for different polymer series having alkyl groups of different lengths reveals a low-temperature polyethylene-like glass transition, in addition to the conventional $T_g$ due to main chain motion. This implies the independent cooperative motion occurs within the small alkyl nanodomain. The important structural and dynamic aspects of side chains are main-chain independent [265]. However, crystallization of the side chains in the homopolymer suppresses the main chain mobility and relaxation process drastically.

It has been confirmed that the side chain flexibility and dynamics control the crystallization of the side chain. For example the flexible stearyl branches of polystearyl methacrylate (PSMA) can crystallize, whereas the oleoyl branches of polyoleoyl methacrylate (POMA) cannot, and is a tacky elastomer at room temperature. This is attributed to the presence of the cis double bond in the alkyl sequence, which makes the chain very rigid. The enthalpy of fusion increases linearly with the length of the side chain. We have investigated poly-$n$- alkyl methacrylate side chain crystalline homo and copolymers using differential scanning calorimetry (DSC), modulate differential scanning calorimetry (MDSC), X-ray scattering, small angle neutron scattering (SANS) and dielectric relaxation spectroscopy (DRS) [266, 267].

The structure, dynamics, and melting behavior of comblike polymers have been investigated in details. A polyethylene-like glass transition within the alkyl nanodomains is clearly observed by DRS for the experimental polymers and attributed to the hindered glass transition in self-assembled confinements. We employed thermal AFM ($\mu$-TA) and pulsed-force mode (PFM) scanning force microscopy for visualization and characterization of nanostructured side-chain crystalline polymeric materials. Pulsed force microscopy (PFM) coupled with local thermal analysis [266, 267] for poly stearyl methacrylate (PSMA) confirms the growth of the crystals and formation of self-organized nanodomains below melting with a typical size of 20-30 nm (Fig. 10A). The local thermal analysis performed on nine different lateral positions on the topographic image (Fig. 10B). The melting point ($T_m$) was determined from the onset of the slope change of the sensor signal with temperature. The $T_m$ observed in the range of 20-25°C. Most of the curves retrace each other indicating the uniform distribution of the nanocrystals in the amorphous matrix. The MDSC data on (PSMA) reveals the complex melting behavior of the polymer and indicates that alkyl groups of monomeric units aggregate in the melt. In comblike polymer the regular spacing imposed by the main chain sequence contributes to retain some ordering in the molten state and the fusion process is complex, and close to above melting and directs the side

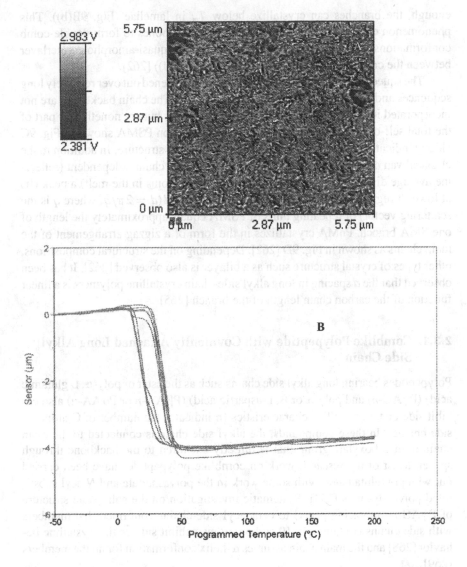

**FIGURE 10.** Pulsed-force-mode (PFM) scanning force microscopic image of the stearyl methacrylate homopolymer. The unique phase separation nature of the side-chain crystallizing comblike polymers are clearly identified. (B) Local thermal analysis (LTA) of the sample in nine different lateral positions. (Reproduced with permission from World Scientific publishing group from S. Thompson, N. Dutta, N. Choudhury, International Journal of Nanoscience, 6, 839, 2004.)

chain to organize in a quasihexagonal arrangement and acts as a nucleating agent for rapid crystallization upon cooling.

At higher temperatures, above the melting point, $T_m$ such comblike chains form random coil (Fig. 9B(a)); however, if the branches are long and regular enough, the branches can crystallize below $T_m$ in lamellae (Fig. 9B(b)). This phenomenon can force the amorphous polymer backbones to form double-comb conformations (Fig. 9B(c)) that finally arrange in a quasi-amorphous interlayer between the crystalline side-chain lamellae (Fig. 9B(d)) [262].

The squeezed-in polymer main chains are straightened out over relatively long sequences and cross sometime through the lamellae. The chain backbones are not incorporated in the crystalline order of the side chains but are nonetheless part of the total self-organized structure. Our SAXS results on PSMA shown in Fig. 9C clearly indicate the presence of two different levels of structure. In addition to the classical van der Waals peak (I), which is nearly side chain independent (reflects the average distance between nonbonded neighbor atoms in the melt) a peak (II) at lower $\theta$ regime is also observed. The long period $d$ ($d = 2\pi/q$, where $q$ is the scattering vector) of repeating units of PSMA equals approximately the length of one SMA branch. PnMA crystallizes in the form of a zigzag arrangement of the main chains as shown in Fig. 9D [262]. Depending on the structural compositions, other types of crystal structure such as a bilayer is also observed [262]. It has been observed that the $d$ spacing in long alkyl side-chain crystalline polymers is a linear function of the carbon chain length of the branch [265].

### 2.3.1. Comblike Polypeptide with Covalently Attached Long Alkyl Side Chain

Polypeptides bearing long alkyl side chains such as the ester of poly($\alpha$, L-glutamic acid) (P$\gamma$AG-$n$) and poly($\alpha$ or $\beta$, L-aspartic acid) (P$\beta$AA-$n$ or P$\alpha$AA-$n$) also exhibit side-chain crystalline characteristics ($n$ indicates the number of C-atoms in side chains). In these compounds, the alkyl side chain is connected to the main chain by a caboxylate group that is directly anchored to the backbone through spacer. Most of the research work on comblike polypeptides have been carried out with polyglutamates, with some work in the polyaspartate and $N$-acyl substituted poly(L-lysine)s [254]. Systematic investigation on the solid-state structure of P$\gamma$AG-$n$ embracing pentyl to octadecyl side-chains confirmed that members with side chains containing $> 10$ carbon atoms exhibit side-chain crystalline behavior [268] and the main chain assumes $\alpha$-helix conformation for all the members [269].

In such cases, a crystallized phase induces a layered arrangement with the $\alpha$-helices aligned in sheets and the paraffin crystallites placed in between. The melting temperature of the paraffin crystallites increases from $-24°C$ to $+62°C$ as the number of carbon atoms in the alkyl side chain increases from 10 to 18. The polypeptides with sufficient long polymethylene side chains can display liquid-crystalline phases upon melting of the side chain. P$\gamma$AG-$n$ is the most extensively investigated comblike polypeptide, and the phases exhibited observed to be largely dependent on polymer size. In P$\gamma$AG-18 sample with $M_w$ above 100 $k$, a cholesteric phase is formed immediately after melting at about 50°C that is converted into a columnar liquid-crystal phase upon heating at higher temperature

[270]. Theoretical prediction indicates that this structure appears as a result of excluded volume effects associated with the lateral packing of the rods [271]. Stable liquid crystalline phases in bulk may also be attained by combining short and long alkyl groups in the same polypeptide [272, 273].

Similar to P$\gamma$AG-$n$, in P$\alpha$AA-$n$ the degree of assembly, ordering and packing also depend on the length of alkyl side chain present and the supramolecular properties exhibited in bulk and solution may be different. P$\alpha$AA-$n$ with $n > 12$ reveals the occurrence of two first order transitions at temperatures at $T_1$ and $T_2$ separating three structurally distinct phases. $T_1$ was attributed to the melting of the paraffinic phase composed of alkyl side chain. A color change from red to blue was displayed by uniaxially oriented film upon heating from $T_1$ up to $T_2$ (caused by the selective reflection of circularly polarized light with wavelength similar to half-pitch of the supramolecular helical structure that varies with temperature). The phase between $T_1$ and $T_2$ is considered to be cholesteric with the ability to crystallize upon cooling below $T_1$ and to convert into a nematic phase upon heating above $T_2$, and the structural change appears to follow a smectic $\leftrightarrow$ cholesteric $\leftrightarrow$ nematic sequence [254, 274].

# 3. SELF-ASSEMBLED NANOPARTICLE SYSTEM

## 3.1. Zero-dimensional Self-assembly

**Block copolymer nanoparticle:** Nanoparticles are key components in the field of nanotechnology for their potential applications in drug delivery systems, photonic crystals, and as templates for other periodic structures. Supramolecular assemblies of nanoscale dimension can also be extremely useful in the design of such novel functional materials. Block copolymers may be optimum nanomaterial, either for their intrinsic properties as self-organized assemblies or for their ability to template other organic, inorganic, semiconductor, metallic or biologically relevant materials. Heterogeneity in such systems results within the system due to the presence of a microdomain leading to anisotropic nature of the polymer, which can be harnessed to template the organization of the desired particles into nanoplanes, nano-wires, or spheres within the polymer matrix. The new chemical bottom-up approach includes a large variety of chemical processes with the common technique of using reactions-in-solution to produce particles of different materials. In order to control the shape and size of the nanoparticles, it is important to control the parameters that influence the nucleation and growth. Also stabilization and prevention of agglomeration of the particles once they are formed is of crucial importance. For this reason many modern techniques have been developed to make use of confined geometries.

In particular, polymeric micelles are developed in pharmaceutics as drug, gene delivery and also in diagonostic imaging techniques as carrier or contrasting agents. Core-shell architecture of polymeric micelles is essential for their utility

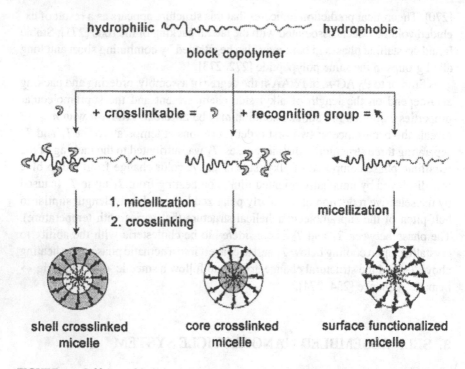

**FIGURE 11.** Self-assembled drug delivery system: core-shell architecture of polymeric micelles is essential for their utility in drug delivery applications. Simplified representation of different types of functional block copolymer core-shell particle. (Reproduced by permission Elsevier from A. Rosler, G.W.M. Vandermeulen, H.-A. Klok, Advanced Drug Delivery Review, 53, 95, 2001.)

in those applications (Fig. 11) [275]. Amphiphilic block and graft copolymers self-assemble in an aqueous medium to polymeric micelles. The aggregation number, size, and shape of the micelle vary depending on the relative and overall lengths of the soluble and nonsoluble blocks and the nature of the surrounding medium. Spherical micelles with core-shell morphology have significant potential as nanocontainer for hydrophobic drug delivery or other release agents. In such cases, proteins and synthetic polymers can play a complementary role to each other. Functionalization of block copolymer with cross-linkable groups can increase the stability of the micelle's control over release and distribution. Temporal and distribution controls are two important issues in drug therapy. Temporal control offers the ability to adjust the drug release time or to trigger the release process, while distribution control, precisely, directs the delivery system to the desired site. Use of block copolymer, where one of the blocks possesses lower critical solution temperature (LCST), offers the most elegant approach to improve temporal control. Depending on the type of functionalization, either shell cross-linked or core cross-linked or surface functionalized micelles can be obtained.

Cross-linked micelles are stable at concentrations below critical micelle concentrations (CMC) of the block copolymer, and are less likely to collapse in the blood stream. It also enhances circulation time, hence allows temporal control. Kataoka et al. [276, 277] reported on the block copolymer with core cross-linkable micelles, where the polymerizable group was present at their hydrophobic chain ends. In poly (D, L-lactide)-b-poly(ethylene glycol) (PLA-PEG) copolymers containing methcryloyl group at the PLA chain end, the methacryloyl group can be polymerized after micellization either thermally by adding initiator or by UV using photo initiator.

The concept of chemically cross-linking the corona of diblock copolymer micelles (shell cross-linkable micelle) was reported by Wooley et al. [278]. Various hydrophobic core-forming blocks from styrene, isoprene, butadiene, caprolactone and methyl methacrylate have been used, while monomeric block or hydrophilic corona based on 4-vinyl pyridine, methacrylic acid, 2-dimethyl aminoethyl methacrylate have been employed to chemically link in aqueous solution after the self-assembly process. Various types of cross-linking can be designed with ionic or covalent bonds and a combination of the two. Cross-linked poly(styrene)-b-poly(2-cinnnamoyl ethyl methacrylate) shells have been prepared by UV irradiation, 4-vinyl pyridine based micellar corona was cross-linked by quarternization with p-chloromethyl styrene, followed by photo polymerization of the styrene double bonds [278].

The performance of a block copolymer micelle based drug delivery system can be enhanced using auxiliary agents, e.g., magnetic nanoparticles can bind to drugs, proteins, enzymes, antibodies. If properly encapsulated they can also be directed to an organ, tissue, or tumor using an external magnetic field, or can be heated in alternating magnetic fields for use in hyperthermia. The applications of some particles in vivo diagnostics have been practiced, such as magnetic resonance imaging enhancement, and immunoassay, detoxification of biological fluids and cell separations. However, these materials in bulk are very dense, heavy, and can be prepared only in restricted size and shape. The stabilization of such a metal oxide particle is a critical issue.

The use of microphase-separated morphology of BC to harbor metal particles on the surfaces, the use of functional polymers to mediate the growth of gold, magnetic nanoparticle or the use of dendritic polymer as container for nanoparticle represent unique routes and well-defined polymeric templates for controlling nanoparticle growth. Core-shell hybrid nanoparticles offer a potential route to stabilize them. They can be prepared from BC and inorganic oxide or metals particles. There are two ways to introduce inorganic particles into BC: (i) to precipitate in a cross-linked polymer matrix or network; (ii) to prepare as core from colloidal reagent with a biodegradable shell of gel, which often prevents coagulation of the particles.

Incorporation of nanoparticles into the dendrimer templates has also drawn significant interest due to the potential applications in sensors, catalysts, and delivery systems. Magnetite particle formation in dendrimers has been reported in solution process. Such nanoparticle-cored dendrimer (NCD) represents an organic-inorganic hybrid nanostructure with a nanoparticle core and well-defined dendritic wedges. Such NCD was synthesized by the divergent approach (based on multistep reactions) and reported by Lundin *et al.* [279]. The synthesis involves the iterative reactions of ester-functionalized nanoparticles with 1,2-diaminoethane followed by the reaction with methacrylate. The multivalency and highly branched nature of dendrimers make them ideal scaffolds for supramolecular architectures. Recently, supramolecular architectures based on dendrimers, peptides and proteins were reported by Meijer [280]. The synthesis and properties of oligo-peptide and protein modified dendrimers, as well as their applications in dynamic combinatorial libraries have been documented by this author. Furthermore, he designed a general guest-host system to assemble a well-defined number of guests around a dendrimer in a reversible way.

Other BCs such as polypeptide-PEO block copolymer are highly suitable for drug delivery. Anticancer drugs have been covalently attached to the micellar polypeptide block or physically included into the core polypeptides such as polyaspartic acid (PAsp may solubilize large quantity of drugs). The PEO-shell reduces the antigenic effect of the block copolymer/drug conjugate. Ligands with targeted function may also bind to the end of the PEO segment [281-286]. PEO-PPO-PEO micelles conjugated with anthracyclin antibiotics have shown significant promise in chemotherapy and inhibition of tumor There is promising evidence that they also could be useful to overcome multidrug resistance (MDR) which represents a considerable problem in the successful treatment of cancer by chemotherapy [287, 288]. Recently, Bae et al. [289] reported a specially designed self-assembling conjugate micelle of amphiilic block copolymer folate-PEG-poly(aspartate hydrazones adriamycin) [Fol-PEG-P(Asp-Hyd-ADR)] to develop multifunctional polymeric micelles with folate-mediated cancer cell targeting and pH-triggered drug release properties. Folate was installed at the end of the shell forming PEG segment to enhance intracellular transport, and the anticancer drug adriamycin (ADR) was attached to the core-forming PAsp segment through an acid-sensitive hydrazone bond (Fig. 12). Because folate-bound proteins are selectively over-expressed on the cancer cell membranes, the folate-bound micelles (FMA) can be guided to the cancer cells in the body, and after the micelles enter the cells, hydrazone bonds are cleaved by the intracellular acid environment.

Recently, block copolymer vesicles (spherical bilayers that offer a hydrophilic reservoir, suitable for incorporation of water-soluble molecules, as well as the hydrophobic wall that protects the loaded molecules from the external solution) that mimic lipid amphiphilicity have attracted great interest as polymersomes or functional polymersomes due to their new and promising application in drug delivery and artificial cells [290, 291]. The polymersomes exhibit increased size, strength, plasticity, and reduced permeability, while having most of the properties of natural living cells. The surface can be provided with "homing devices", such as antibodies, which impart crucial function such as latching onto the cells they are targeting. Many different amphiphilic block copolymers such as polystyrene

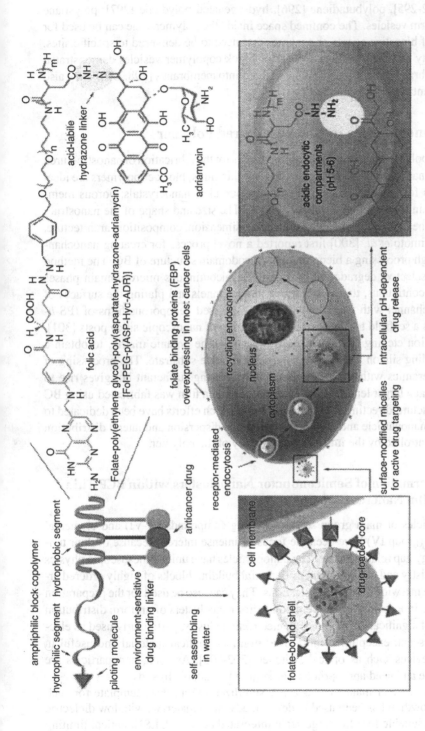

**FIGURE 12.** Multifunctional polymeric micelles with tumor selectivity for active drug targeting and pH sensitivity for intercellular site-specific drug transport. Folic acid with high tumor affinity due to the over expression of its receptors was conjugated onto the surface of the micelle. (Reproduced by permission of Royal Society of Chemistry from Y. Bae, W-D. Jang, N. Nishiyama, S. Fukushima, K. Kataoka, Molecular Biosystem, 1(3), 242, 2005.)

[253, 292-295], polybutadiene [296], hydrogenated polydiene [297], polysilane blocks form vesicles. The confined space inside the polymersome can be used for storage of bioactive agents or enzymes that need to be delivered to specific sites. Bioactivity and membrane properties of block copolymer vesicles demonstrated that membrane proteins can be reconstituted into membranes [298,299]. They also have potential use in gene therapy.

## 3.2. Nanoparticles in Nanostructured Polymer

Block copolymers can be utilized as precursors for the fabrication of nanostructured network, nanoparticles, membranes, etc. As thin films, block copolymers are ideal templates for creating quantum dots, nanoparticles, nanocrystals, porous membranes, thin films, patterned films, hybrids. The size and shape of the nanostructures can be tailored by changing molecular dimension, composition, architecture, etc. Hashimoto *et al.* [300] first reported a novel process for creating nanochannel through processing a bicontinuous microdomain structure of BC. The method involves selective degradation of one of the bicontinuous micro domain phases to create continuous, tortuous holes or nanochannels and plating the surfaces of the nanochannel with nickel metal. Kim *et al.* used nanoporous films of (PS-*b*-MMA) as a scaffold to prepare an ordered array of nanoscopic silica posts [301]. Reactive ion etching (RIE) was used to remove the organic matrix, to obtain a free-standing silicon oxide posts on a silicon oxide substrate. The processing of sol-gel ceramics with BC is similar to that of using surfactant but gives rise to structure at a larger length scale. Mesoporous silica film was fabricated using BC as the structure-directing agent. Significant research efforts have been dedicated to combine nanoparticle and polymer, wherein the dispersion and lateral distribution can be controlled by the intrinsic morphology of the polymer.

### 3.2.1. Formation of Semiconductor Nanoclusters within BCP Thin Film Nanoreactor

Nanoparticles of main group semiconducting compounds (II-VI) and elemental materials (group IV) have been the focus of intense interest because of their tunable energy gap related to size. The nanoparticles have unique mesoscopic physics and chemistry and are important as potential building blocks of highly ordered architecture in a wide range of applications. They can also be used for the preparation of organic-inorganic hybrids. Semiconductor nanoclusters of uniform distribution are also of significant importance in electronic industry, particularly used as electroluminescent, exceptional third-order nonlinear optical material and useful in optical devices such as optical switches [302-305]. As such, nanoparticles are metastable and need appropriate stabilization by suitable ligands.

As a polymer matrix can act as a stabilizer and a robust template for such hybrid growth, it has been used to develop advanced materials with low dielectric property (suitable for ultra large scale integrated circuits, ULSI), optical limiting characteristics (suitable for eye and sensor protection) and electroactive property (suitable for a light-emitting diode). Specifically, functionalized block copolymers offer a wide range of potential for their successful use as templates for the synthesis of ordered structure or nanoparticles with subtle control of particle growth, particle size distribution and particle surface interactions. Such nanoparticle formation can be carried out either in solid nanostructured polymer, or in pores or cavities of nanometer-size present in the nanostructured polymer or in solution and thin films of nanostructured polymers. Semiconductor clusters have been synthesized within microphase-separated diblock copolymer films [306-308]. In a particular method, metal complexes are initially attached to one of the blocks before microdomain formation. Sankaran *et al.* [309] and Tassoni *et al.* [310] have used this approach to synthesize lead sulfide (PbS) nanocluster using lead-containing norbornene derivatives as one block of the diblock copolymer.

Yue *et al.* [307] developed a new approach to prepare uniformly distributed semiconductor cluster, which involves the selective sequestering of metal into the preformed mesodomain of a block copolymer film. Kane *et al.* [311] reported the synthesis of PbS nanoclusters of controllable sizes, narrow size distribution and surface characteristics within the microphase-separated films of diblock copolymers containing carboxylic acid units as one of the blocks. The block copolymer $(MTD)_n(NORCOOH)_y$ [MTD = methyl tetra cyclodecene; NORCOOH = 2-norbornene-5,6-dicarboxylic acid) was synthesized using ring-opening metathesis polymerization (ROMP) using $Mo(CHCMe2Ph)(NAr)(o-t-Bu)2$ (Ar = 2, 6-diisopropylphenyl). The technique involved selectively sequestering of metals into the acid containing block copolymer domain treating the block copolymer film with tetraethyl lead, and PbS was formed by subsequent treatment of the film with hydrogen sulphide ($H_2S$). They also demonstrated that the cluster size can be controlled by varying the processing parameters such as temperature, presence of coordinating bases, and the $H_2S$ exposure time. This universal cluster synthesis process has the flexibility of multiple passes through the process and may be used to increase the cluster size, passivation or to produce mixed semiconductor clusters.

### 3.2.2. Self-assembly in Biomimetic System

Nanostructured materials are widespread in biological systems. The naturally occurring nanostructured materials are primarily hierarchically organized, where organization occurs in discrete steps from atomic to macroscopic scale. These are formed in nature by a process called bio-mineralization (*in vivo* formation of inorganic crystal or amorphous component in biological system). The formation of mineral phases in organic matrices is a key feature of natural bio-mineralization processes, e.g., in multicellular organisms, enamel and bones, gel-like extracellular network with remarkable properties are formed. Recent advances in synthetic methods allow us to create such well-defined structures capable of mimicking many of the features of biological and other natural materials. Such materials have the potential to function at a molecular level in bulk, thin films, and in solution.

The drive came from biological systems, where an intimate association between ordered organic self-assembly and deposition of inorganic is very common. The self-assembly in functionalized block copolymers provides an environmentally benign, effective method towards the synthesis of novel ceramic and nanocomposite materials. Thus various approaches such as self-assembly of colloidal crystal, self-assembled monolayer and three-dimensional (3D) SA are widely used in the synthesis of designer materials and composites. Self-assembly of colloidal crystal allows the synthesis of periodic optical devices and quantum dots from simple monodispersed particles. This process is used in nature to create opal with brilliant iridescent color.

A monodispersed colloid can be influenced to self-assemble into an ordered crystal or cluster, if the particles experience sufficient interaction energies. Three major interaction forces that dictate the behavior of such particles are van der Waals interaction, electrostatic repulsion and polymeric interactions. The van der Waals attraction between two particles can be derived from all intermolecular dipole-dipole interactions. The electrostatic double layer repulsive force, results from the surface charge of the particle in the medium. This surface charge in conjunction with diffused ionic double layer surrounding the particle gives rise to columbic repulsion between the particles.

Apart from these two interaction forces, other forces such as steric interactions also come into play when an organic or polymeric species is used to control the particle growth. Depending on the molecular weight, its chemical nature, the force can be attractive or repulsive. A long chain polymer induces repulsion as the interparticle distance becomes close to the dimension of the polymer. Thermodynamic parameters such as temperature, concentration, particle number density, confinement of the particle in a viscous medium etc., can induce solidification of such colloidal suspension. However, the critical step in a self-assembled crystal is the accurate control of particle dimension and the thermodynamic parameters. Depending on that, one can obtain an ordered array of colloidal particles with repulsive interactions or an disordered fractal aggregates with attractive interactions. The process is sensitive to the nature of the substrate. The most studied biomimetic system is the nacre of abalone shell, an orientated coating composed of alternating layers of aragonite ($CaCO_3$) and an acidic macromolecule (1vol%) [312]. Its laminated structure simultaneously provides strength, hardness and toughness: nacre is twice as hard and 1,000 times as tough as its constituent phases. Calcite crystals were grown in a collageneous matrix using counter diffusion arrangement and reported by Lobmann et al. [313]. The analogy of the gelatin grown particles to some biomineral, however, suggests that biological crystallization may take place under comparable conditions. The use of synthetic block copolypeptide of cysteine-lysine block allows us to mimic the properties of silicatein, a protein that directs silica growth in certain sponges [314].

### 3.2.2.1. Self-assembly in DNA Crystal. Molecular self-assembly represents a "bottom-up" approach towards the fabrication of nanometer-scale structure.

In biological systems, nature uses biomacromolecular assemblies for a wide rang of purposes. Self-assembly is the common route for the formation of the cell and its components. Chains of RNA to functional *t*-RNA, proteins can self-assemble into reversibly assembled supramolecular structures. Such structures are extremely small and complex. DNA molecular structures and intermolecular interactions are highly suitable to the design and development of complex molecular architectures.

Self-assembled core shell particles comprised of calcium phosphate, oligonucleotide, and block copolymers of poly(ethylene glycol)-block-poly(aspartic acid) (PEG-PAA) were reported by Kakizawa *et al.* [315]. Although a simple mixing of calcium/DNA and phosphate/PEG-PAA solution leads to the formation of such monodispersed particles, however, the block copolymer of PEG and PAA segments is essential to prevent precipitation of calcium phosphate crystals and to allow nanoparticles to form. Dynamic light scattering measurements showed the diameters of the particles to be around 100 nm with a significantly narrow size distribution. Gel permeation chromatography (GPC) and fluorescence spectroscopy showed the particles have the ability to incorporate DNA in the core with sufficient efficiency. Furthermore, the cytotoxicity of the particles, assessed by MTT (3-(4,5-dimethyl-2-thiazolyl)-2,5-diphenyl-2H-tetrazolium bromide) assay, was significantly low [316]. The organic-inorganic hybrid nanoparticles containing DNA molecules, thus prepared. can be utilized in the DNA delivery systems for gene and antisepses therapy. Recently, self-assembly of cross-linked DNA-gold nanoparticle layers was visualized by Zou *et al.* using in situ scanning force microscopy [317]. DNA-directed assembly of cross-linked gold-nanoparticle aggregates was prepared on two-dimensional streptavidin crystals substrate. In situ scanning force microscopy showed the interparticle distance could be controlled by the length of the DNA linker connecting individual particle. Such nanostructured materials with programmable functionalities have potential for use in nanobiotechnology.

## 3.3. Two-dimensional Thin Film

The behavior of a small quantity of self-assembling molecules on a surface is strongly influenced by the way they are attached to the surface. Surface forces predominate in small or ultra thin assemblies and thus in practice the behavior of nano-assemblies can be more readily manipulated. The SIM (surface induced mineralization) process uses the idea of nature's template-mediated mineralization by chemically modifying the substrate to produce a surface that induces heterogeneous nucleation. SIM-produced bioactive coatings provide greater control of the thickness of the mineral phase, also a better way to coat porous metals, complex shapes and large objects. An emerging class of application of block copolymers is as template for porous materials, e.g., membranes, catalyst, drug delivery devices, etc. A priori knowledge of various templates or domain size in block copolymers, ionomer (Fig. 13) can also be applied to the synthesis of nanoparticles or nanoclusters. The sol-gel process allows formation of inorganic network in a specific block

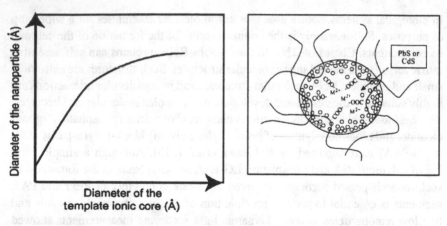

**FIGURE 13.** Degree of molecular sizing as a function of Ionic aggregate.

of a block copolymer. Although the concept appears simple, often it poses problem in a macroscopic scale. While the formation of nanostructure using block copolymer is really facile, the degradation step (depolymerization by plasma, ozonolysis, digestion, etc.) in the bulk samples such as in thin films often creates difficulty due to many reasons. The size control of quantum-confined nanoparticles within such template or the degree of molecular sizing is a function of template size in the system or in a thin film [Fig. 13].

Thin film coatings are used for a wide variety of purposes from molecular electronic based devices to protection from environmental attack, abrasive wear, erosion, impact and overheating etc. The fabrication of molecular electronic based devices relies heavily on generating defect-free supramolecular assemblies with carefully optimized molecular ordering and properties. The commonly used methods are the Langmuir-Blodgett (LB) deposition method, spin coating, molecular beam epitaxy and vacuum evaporation techniques [318-321]. Micro contact printing, lithography, mixed SAM and replaced SAM are other different routes to produce such functionalized surfaces. Employing self-assembly, one can fabricate inhomogeneously functionalized surface, which is subsequently used as template for selective deposition of nanoparticle, polymer and protein adsorption. Thin film of incompatible block copolymer can self-assemble to highly ordered supramolecular structures with dimensions 10-100nm. Domains of block copolymers in a coated thin film can be tailored to arbitrary structures by patterning the surface of the substrate with monolayers of different interfacial energies. Additionally, as different monomers respond in a different manner to etching either by plasma or reactive ion etc, a spin-coated film of block copolymer is often used as a mask to transfer the domain pattern of the film to an underlying substrate. Thus, posts or arrays of holes of less than 20 nm diameter have been prepared on substrate by this technique [301]. Such phase separated morphology has also been used to guide the growth or deposition of various colloids such as gold, silver, Pt, Pd, etc.

The layer-by-layer (LBL) self-assembly technique is widely used to build multilayer film on a nanometre length scale and is particularly attractive for its exceptional simplicity [322, 323]. It is a viable alternative to the Langmuir-Blodgett technique, spin coating, in situ polymerization and other methods of preparation of organic and inorganic nanostructures. Using electrostatic interaction between alternately deposited charged species, it allows deposition of inorganic nanoparticle, biological macromolecules, dyes, photochromic molecules, conducting polymers, semiconductor quantum dots, self-assembly of biological compounds, etc. In a typical process, it consists of four consecutive steps, adsorption of a positive component, washing, adsorption of a negative component and final washing. These steps can be repeated in a cyclic manner to produce film automatically. As a class of interesting composite material, they find use in catalysis, optical, magnetic and electronic applications. The driving force for such assembly is not only the intermolecular interactions between the oppositely charged groups of the polyelectrolytes, but also other such as ion-dipole, dipole-dipole, van-der-Waals interactions. Table 2 shows various features of LB and SA thin film, while Fig. 14 shows the schematic steps of thin film preparation using LBL and block copolymer SA.

### 3.3.1. Organic-Inorganic Hybrid Thin Film and Coating

Supramolecular assemblies of self-assembled polymers represent a unique class of templates to produce functional nanostructures in a controlled way. Materials can

**TABLE 2.** The similarities and differences between LB and SA thin film

| Langmuir-Blodgett film | Self-assembled thin film |
| --- | --- |
| Developed by Agnes Pockels in 1891; demonstrated by Irving Longmuir and Katerine Blodgett in 1930, | First demonstrated by Sagiv in 1980. |
| Less versatile; limited to those molecules that can be compressed to stable monolayer, thin film or layer-by layer (LBL) | Versatile; a large variety of molecules and macromolecule can be assembled by thin film or layer-by layer (LBL) |
| Requires film balance, Better organisation of the film | Simple and rapid preparation; no need of surfactant or polyelectrolyte compression in film balance; molecular beaker epitaxy [323] prepared by substrate immersion from a very dilute solution; spontaneous self-assembly of thin film; no deposition equipment required. |
| Surfactant suitable for compression; primarily held by weak van der Waals interactions; less stable | Requires strong interaction to anchor the molecule to the clean substrate; prolonged stability. |
| Measured by thickness; a number of layers, can be formed | Measured by the diameter or size of the molecule, less number of layers can be formed ca. LB film. |

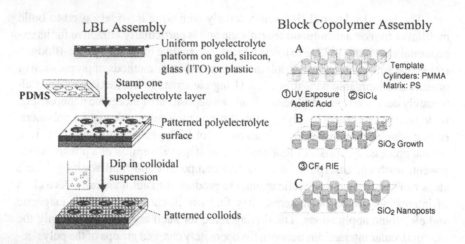

**FIGURE 14.** Schematic steps of thin film preparation using LBL. (Reproduced by permission of Wiley VCH from 322 'Colloid and Colliod Assemblies', ed. F.Caruso, Wiley VCH, NY, 2004.) Block copolymer self-assembly. (Reproduced by permission of Wiley VCH from H-C. Kim, X. Jia, C.M. Stafford, D. H. Kim, T.J. McCarthy, M. Tuominen, C. J. Hawker, T.P. Russel, Adv. Mater. 13, 795, 2001.)

behave very differently when they are nanostructured. Finer sizes (hence larger active surface per unit mass) produce more dense material with improved barrier properties. There are various routes to prepare hybrids e.g., self-assembled monolayer, layer-by-layer or thin stacked layer (with different band gap) use of mesophase material for controlling molecular orientation of rod or disk liquid crystals and subsequent polymerization, use of microporous material, hybridization of organic and inorganic material [324, 325] and so forth. The molecular building block approach to the synthesis of organic-inorganic hybrid material involves incorporation of nanosized oxometallate clusters such as polyhedral oligosilsesquioxane (POSS), oxotitanate, oxozirconate or mixed titanate/zirconate clusters into various polymeric matrices. Typically, a cluster is a discrete metal and oxygen framework bearing reactive functional groups, which can be incorporated into polymer matrices to significantly enhance their physical properties. Thus a new series of organic-inorganic hybrid materials have gained interest, in which organic polymers are efficiently cross-linked by structurally well-defined oxometallate clusters. These hybrid materials have interesting properties. In such nanoscale materials, the properties are predominantly controlled by the nature, type and size of the confined cluster. The clusters are made by functionalization of metal alkoxide with organic group, thus preventing their further growth.

Recently, Schubert *et al.* [326-329] prepared and reported a series of acrylate and methacrylate substituted oxotitanate, oxozirconate, oxotantalum based clusters and a mixture of them in different sizes and shapes by reacting the metal alkoxide with calculated excess of methacrylic acid or anhydride. Finally, the clusters were copolymerized with methacrylate at various cluster-polymer ratios.

These new materials have a similar type of structure as the POSS (polyhedral oligomeric silsesquioxane) hybrid. These cluster dopants can also act as reinforcing agent and nano-crosslinker for the matrix polymer creating nano-confined hybrid. POSS [330-333] based hybrid materials are also emerging as unique materials with characteristic architectural features and enhanced properties. Recently, from our laboratory we reported improved synthesis of some organic-inorganic hybrids [334]. Poly(methyl methacrylate) containing octakis (methacryloxypropyldimethylsiloxy) octasilsesquioxane (OMPS) was synthesized via the reaction of methacryloyl chloride [334] or methacrylic acid anhydride with octakis(3-hydroxypropyldimethylsiloxy)octasilsesquioxane (OHPS), with the latter giving improved purity. Polymerization of OMPS with methyl methacrylate using dibenzoyperoxide initiator gave a highly cross-linked polymer with enhanced thermal stability and glass transition temperature.

Recently, Zheng *et al.* [335] reported a new bottom-up approach to obtain polymer nanocomposite using controlled self-assembly of a cubic silsesquioxane scaffold. Although the mutual affinity between POSS units causes the particles to aggregate and closely pack into a crystal lattice, the organic polymer chemically attached to each POSS unit limits the crystal growth. This is further confirmed by TEM and small angle X-ray scattering data, which confirm the presence of two-dimensional lamellar-like nanostructure of assembled cubic silsesquioxane. A similar observation was also noted in our recent work on TEM and AFM of POSS-based hybrid (Fig. 15), where the aggregate structure of POSS in a POSS-PMMA hybrid was quite evident [334]. The TEM-EDAX further confirmed the presence of Si and oxygen in those segregated regions, in an expected ratio. Wu *et al.* [336] while studying a PU-based hybrid hydrogel, observed crystallization of POSS in a unique series of PEG-based multiblock PU. Wide-angle X-ray diffraction (WAXD) studies revealed that both the hydrophilic soft segments (PEG) and hydrophobic hard segments (POSS) can form crystalline structures driven by microphase separation., itself due to thermodynamic incompatibility. Both composition and thermal history are the key factors in controlling the internal network built by the POSS nanophase.

Recently, transparent linear polyurethane hybrid containing an inorganic open cage polyhedral oligomeric silsesquioxane was synthesized and reported by Oaten *et al.* [337, 338] for thin film application. Analysis by angle resolved X-ray photo electron spectroscopy (XPS) and small angle neutron scattering (SANS) have shown the interface between the substrate and the POSS-urethane coating forms a lamellar structure rather than a random 3D network (Scheme 5). The long-range ordered lamellar structure of the hybrid results from the strong hydrogen bonding interactions of the amide groups in the urethanes and the long-range hydrophobic interactions of the hexamethylene and iso-butyl groups.

The most popular and practical approach to prepare organic-inorganic hybrids is by sol-gel reaction [339-342]. The sol-gel process with its unique mild processing characteristics can be used to prepare pure and well-controlled compositions. The process involves the hydrolysis of metal alkoxides, followed by a condensation

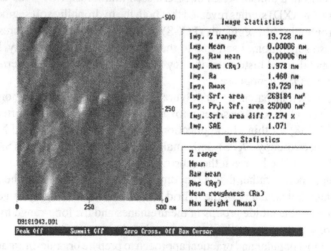

**FIGURE 15.** AFM image of of POSS-based hybrid where the aggregate structure of POSS in a POSS-PMMA hybrid was quite evident. (Reprinted with permission from O.Toepfer, D. Neumann, N. Roy Choudhury, A. Whittaker, J. Matisons, Chem. Mater. 17 (5), 1027, 2005. Copyright (2005) American Chemical Society).

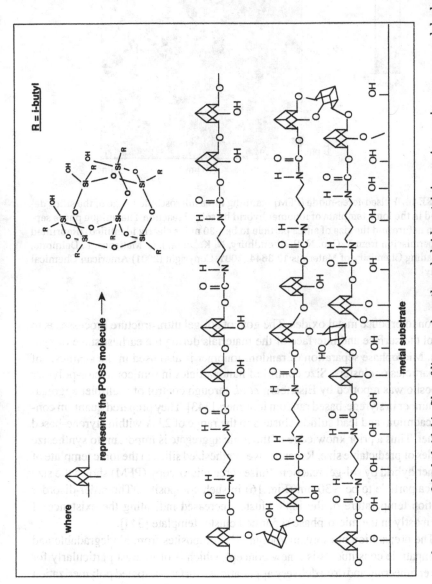

**SCHEME 5.** Proposed structure of the hybrid thin film of linear polyurethane hybrid containing an inorganic open cage polyhedral oligomeric silsesquioxane. The interface between the substrate and the POSS-urethane coating forms a lamellar structure rather than a random 3D network. (Reprinted with permission from Ref. 337. Copyright 2005 American Chemical Society.)

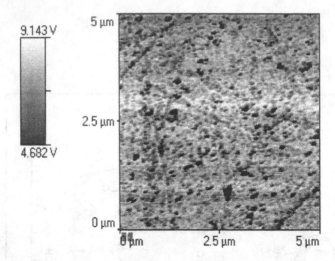

**FIGURE 16.** Pulsed-force-mode (PFM) scanning force microscopic image of the silica deposited in the ionic template of ionomer hybrid by sol-gel reaction The unique phase separation nature and the size of silica particle to be ∼30 nm is clearly identified. (Reprinted with permission from Y. Gao, N. R. Choudhury, N. K. Dutta, J. G. Matisons, L. Delmotte, M. Reading Chemistry of Materials 13 3644, 2001. Copyright (2001) American Chemical Society).

reaction to produce metal oxides. The goal of sol-gel ultrastructure processes is to control the surface and interface of the materials during the earliest stage of synthesis. Microphase separation of random ionomer is also used in the synthesis of nanoparticles or clusters. Size control of nanoparticles in semiconductor-polymer composite was reported by Eisenberg *et al.* through control of multiplet aggregation number in styrene-based random ionomers [343]. They prepared quantum confined cadmium and lead sulfide clusters in the range of 23 Å within styrene-based ionomer. Thus a prior knowledge of the ionic aggregate is important to synthesize particles of predictable size. Recently, we synthesized silica in the ionic template of ionomer hybrid by sol-gel reaction. Pulse force microscopy (PFM) shows the size of silica particle to be ∼30 nm (Fig. 16) in such composites. The order-disorder transition temperature of the ionic cluster increased indicating the existence of silica mostly in the micro-phase separated cluster template [344].

The preparation of inorganic/polymeric composites from biodegradable and biocompatible constituents is a new concept, which is of interest particularly for tissue engineering and drug delivery applications. Surface-initiated polymerization (SIP) represents an elegant approach to prepare well-defined tethered copolymers or brushes (Fig. 17). Hydrophilic Si surfaces of poly[styrene-*b*-(*t*-butyl acrylate)] were prepared by such a method followed by hydrolysis to form poly (styrene-*b*-acrylic acid) $p$[(S-*b*-AA)]. Contact angle study of the surfaces shows a decrease in water contact angle from $(86 +/- 4°)$ to $(18 +/- 2°)$. Recently, we reported the

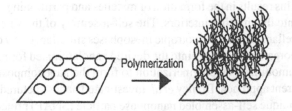

**FIGURE 17.** Schematic representation of the prepare of well-defined tethered copolymers or brushes. Hydrophilic Si surfaces of poly[styrene-*b*- (*t*-butyl acrylate) is prepared by such a method followed by hdrolysis to form poly(styrene-*b*-acrylic acid) *p*[(S-*b*-AA)].

synthesis of nanostructured porous silicon (pSi) and poly(*L*-lactide) (PLLA) composites. The composites were produced using tin(II) 2-ethylhexanoate-catalyzed surface-initiated ring opening polymerization of *L*-lactide onto self-assembled silanized porous silicon films and microparticles. The surface roughness as measured by AFM after surface-initiated polymerization increased significantly and AFM images showed the formation of PLLA nano-brushes [345]. It is evident that new types of organic/inorganic hybrid structures can be made by attaching organic molecules to ultra small particles or ultra thin layered structures or super lattices. And there are many new diverse applications in different products or processes, some of which is summarized below in Table 3.

Recently Sellinger *et al.* [312] reported an efficient, rapid self-assembly of organic-inorganic nanocomposite coatings that mimic nacre. A solution of silica, surfactant and organic monomers was used followed by evaporation during

**TABLE 3.** Applications of various organic-inorganic hyrid systems

| Application | New property or improved property |
|---|---|
| Colloids (0D) | New optical, structural, electronic and thermodynamic properties |
| Dispersions | To tailoring nanopaticle size, molecular structure |
| Nano-emulsion | To engineer viscosity, particle size, control viscosity and absorption characteristics |
| Anticorrosion coating (2D) | Enhanced surface mechanical barrier properties and environmental stability |
| Ferro fluids | Fluid magnet to exhibit new rheological properties |
| Lubricants | Tailored viscosity and thermal rheological properties |
| Drug delivery system | Chemical reactivity, distribution, release profile and critical time of drug delivery can be controlled |
| Bioreceptors | Nanocomposites designed to exhibit well-defined excitonic absorption spectra |
| Cosmetics/pharmaceutical | Nanoparticles can enhance flow properties and absorb harmful radiation. |
| Tissue engineering scaffold (3D) | Designed physico-chemical and mechanicals for cell adhesion and growth. |

dip-coating. This results in the formation of micelles and partitioning of the organic constituents into the micellar interiors. The self-assembly of the silica-surfactant-monomer micellar species into lyotropic mesophases simultaneously organizes the organic and inorganic precursors into the desired nanolaminated form. Finally, this structure is immobilized by polymerization to form the nanocomposite assembly.

In a different approach, Donley et al. investigated nanostructured hybrid coating based on unique self-assembled nanophase particle (SNAP) protective coating [346, 347]. They used organo functional silanes to engineer nanostructured coating for corrosion protection of aluminum alloys. This process allows in situ formation of functionalized silica nanoparticle in an aqueous sol-gel process followed by cross-linking the nanoparticle to form a thin film. The versatility of the approach allows the formation of a hybrid coating with tailored structure and improved adhesion characteristics. Such multifunctional coating can produce dense, defect-free, thin nanocomposite film on the substrates that possess a high barrier property and corrosion resistance. In a similar way, an organic-inorganic hybrid can be formed. Recently, Dutta et al. [348] prepared self-assembled nanophase particle (SNAP) based hybrid for barrier film application. Poly (ethylene acrylic acid) and its ionomer were chosen to make such hybrid with various organofunctional silanes.

In hybrids, the organic component itself can exhibit diverse self-assembling characteristics and can direct the formation of three-dimensional hybrid network in a self-organization process. Ben et al. obtained evidences of such anisotropic organization due to the presence of rigid aromatic segments through birefringence property investigation of synthetic silsesquioxane materials [349] Moreau et al. reported the synthesis of self-organized hybrid silica with a medium long-range ordered lamellar structure using strong supramolecular interactions of hydrogen bond association of urea groups and van der Waals interactions of long hydrocarbon chains [350]. By tailoring the host-guest interaction and using self-assembly as a tool, various ordered inorganic silicates were prepared using small surfactant molecule or a block copolymer as templates [351, 352]. Thus by selectively swelling one domain in the microphase separated block copolymer, using sol-gel approach, films of 1 mm thickness can be obtained with domain size ranging from several to 100 nm. However, creation of hydrophobic periodic organosilicates by this approach is a challenging task as the interaction between block copolymer and organosilicates at the interface, accessibility of organosilicates into polymer domain govern the formation of hydrophobic organosilicate's periodic structure. Yang et al. [353] examined the routes to self-organize periodic structure in hydrophobic organosilicates, which is not compatible with block copolymer templates.

## 3.4.  Self-assembly in Biocompatible System and Biomolecular Assembly

Self-assembly of biomolecules is emerging as a new route to fabricate a novel material that can complement existing materials and composites. Most biological systems rely on the self-assembly of polypeptides and polysaccharides into functional three-dimensional nanostructures dictated by the delicate balance of relatively weak interactions (hydrophobic, hydrogen bonding, electrostatic). The exquisite property of many proteins is directly related to their highly organized nanoscale structures. However, they have other limitations such as limited solubility, restricted temperature and pH stability, susceptible to enzymatic degradation, etc. Unlike many proteins, synthetic polymers are able to recognize and bind certain guests, and are biologically inactive. Thus, proteins and synthetic polymers are complementary to each other.

Therefore, hybrid materials containing these two moieties can synergistically combine the properties of individual components and overcome their limitations. The conjugation of biological and synthetic polymers thus represents a powerful strategy not only for mediating self-assembly of the polymer but also for modulating protein binding and recognition [354]. In protein-polymer hybrids or chimeras, a very specific type of self-assembly (folding and organization) is coded in the primary structure of the peptide segments. Thus, it is possible to obtain very complex, hierarchically organized and highly functional materials suitable for use in medicines, bioanalytical application and bioseparations. Also the self-assembling process or the ability of the peptide sequence to direct the self-assembly of the hybrid block copolymer can offer an effective and environmentally benign method of synthesizing novel magnetic nanocomposite particle.

Recently, Millicent et al. [355] reported the effect of film architecture in biomolecular assemblies. The authors presented strategies for the optimization of supramolecular architecture aimed at controlling the organization of biomolecules at solid surfaces. Myoglobin, modified by site-directed mutagenesis to include a unique cysteine residue, is selectively chemisorbed to self-assemble haloalkylsilylated silica surfaces of varying $n$-alkyl chain length ($n = 2, 3, 8, 11, 15$) to yield a series of surface-immobilized recombinant protein supramolecular assemblies. These assemblies were then probed using tapping mode atomic force microscopy and wetability measurements, While surface roughness is found to be a minor contributor in the determination of macromolecular protein ordering in these assemblies, the nature of the underlying silane self-assembled coupling layer was found to strongly influence both the spatial and functional properties of the chemisorbed protein. Silane coupling layers with short aliphatic chain lengths ($n = 2, 3$) produce highly $trans$-conformationally ordered structures upon which differential heme prosthetic group orientation can be achieved, whereas long alkyl chain ($n > 11$) silane-derivatized surfaces form ordered structures. Overall, the stability of myoglobin appended to long-chain aliphatic silylated surfaces is poor; the apparent protein instability arises due to the increased hydrophobic character of these films.

In designing rational protein-based nanostructures, it is important to understand the assembly of protein, protein-protein interactions, the role of specific secondary structure motif on such interactions and their relationship with diseases. Another important parameter is the phenomenon of order-disorder transition or coacervation of the protein. Recently, Yang *et al.* [356] studied substrate facilitated assembly of elastin-like polypeptides by variable temperature in situ AFM. From the observed propensity of ordered growth of the protein film on highly ordered graphite, the authors proposed that hydrophobic peptide-substrate interactions facilitate organization of the peptide at the graphite-peptide interface. They pointed out that such model is in line with the coacervation mechanism with the fact that specific alignment of individual domain is required for β sheet, β spiral formation. Recently Elvin *et al.* [357, 358] reported a novel bioelastomer based on resilin. Resilins are composed of naturally occurring protein polymers in which biological control of the polypeptide sequence made a material with mechanical and elastic properties that exceed those of any synthetic and nonpeptide natural polymer. Although the molecular basis of the unusual resilience characteristics of the resilin-gel is unknown but due to its immense potential, we [359] employed small angle neutron scattering (SANS) to explore the self-organization behavior of cross-linked resilin gel in equilibrium-swollen condition over a wide range of temperatures. While significant research is necessary and currently underway, a correlation between water uptake, self-organization and unique resilience behavior is already noticed.

## 3.5. Supramolecular Assembly via Hydrogen Bonding

By combining several hydrogen bonds in an array, it is possible to create strongly bound supramolecular polymer and three-dimensional structures. With this approach, first hydrogen bonded supramolecular polymer was developed by Lehn *et al.* [360] where the polymers are held together by triple hydrogen bonds with an association constant of $10^3$-$M^{-1}$. Supramolecular polymers in which the monomeric units are held together by noncovalent interactions thus provide unique opportunities for designing tunable polymeric materials [361-363]. Due to the environmental sensitivity and reversibility of such interaction, the property of such material may be manipulated by adjusting the environment using external stimuli. These systems may also provide the possibility to develop supramolecular co-polymer by mixing different monomer. It is also possible to create branched or reversibly cross-linked polymer by using monomer with higher functionalities. Functionalization of low molecular weight bifunctional telechelic polymer with strong quadrupole bonded 2-ureido-4[1H] pyrimidinone (UPy) units combines excellent mechanical properties at room temperature with good processability at elevated temperature (weakening of the hydrogen bond at higher-temperature, low-melt viscosity). Well-defined UPy-terminated polystyrene (PS), polyisoprene and PS-*b*-PI block copolymer were synthesized [364]. They exhibit increase in glass transition temperature (Tg) and melt viscosity upon functionalization. The functionalized polymers also form aggregates in the melt state.

Meijer *et al.* developed a range of new, self complementary, quadrupole hydrogen bonding units based on pyrimidine and triazine [365]. In such self-assembling structures, the components remain assembled via DADA or DDAA arrays

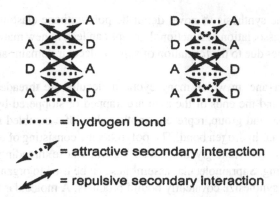

...... = hydrogen bond

...✏ = attractive secondary interaction

✏ = repulsive secondary interaction

**FIGURE 18.** Self-assembling structures through hydrogen bonding; the components remain assembled via DADA or DDAA arrays (donor-acceptor-donor-acceptor) as shown in the figure. The difference in the level of intramolecular interaction gives different stability into these arrays e.g., DDAA arrays are more stable that DADA arrays. (Reprinted with permission from F.H. Beijer, R.P. Sijbesma, H. Kooijmman, A.L. Spek, E.W. Meijer, J. Am. Chem. Soc., 120, 6761, 1998. Copyright (1998) American Chemical Society.)

(donor-acceptor-donor-acceptor) as shown in Fig. 18 of double or triple hydrogen bonds which remain stable by intramolecular hydrogen bonding. The difference in the level of intramolecular interaction gives different stability into these arrays e.g., DDAA arrays are more stable that DADA arrays (Fig. 18).

Supramolecular assembly via hydrogen bonding has thus been actively used as a means of polymer formation, or modification via a self-assembly process [366]. Hamachi *et al.* [367] used glycosylated amino acid to prepare a supramolecular unit that can act as a carrier for thermally controlled release of DNA. Thus weak, controllable hydrogen bonding leads to bonds with high energy when assembled together. A few studies report on the polymer modification using hydrogen bonds between the side chains of polymers or the side chain of a polymer and small molecule [368-370]. A significant challenge in material science is in the area of toughening of brittle polymer. Recent work by Stupp *et al.* shows that small amount of self-assembling molecules known as dendron rod coils (DRC), [371, 372] when added to PS (0.1 wt%), spontaneously assemble into a gel-forming network, can induce orientational order in drawn PS and significantly enhances its impact strength. The self-assembled gels contain supramolecular ribbons of 10 nm wide, 2 nm thick and up to 10 um long and are birefringent. Thus an extremely large interfacial area between the DRC nanoribbons and bulk PS is responsible for a mechanically interlocking network, which can provide multiple crack deflection sites. Recently, Chino *et al.* [373] investigated thermo reversible cross-linking of rubber using a triazole-based supramolecular hydrogen bonding network with an aim to obtain final mechanical properties similar to vulcanized rubber. Also, hydrogen-bond-rich dendritic structures are ideal building blocks for making high-performance functional materials. Chen *et al.* prepared a series of hydrogen bonding rich polyurea/malonamide dendrons for the synthesis of novel dendritic polyurethane elastomers [374]. In such cases, the association of functional groups can lead to new materials with enhanced properties due to the formation of supramolecular domains acting as labile cross-links.

A polyrotaxane, in which many cyclic molecules are threaded on a single polymer chain and the ends of the axle are trapped or stoppered by capping the chain with bulky end group, represent another class of assembled structure held by supramolecular hydrogen bond. The polyrotaxane consisting of α cyclodextrin (α-CD) and poly (ethylene glycol) is well known, whose threading is guided by hydrogen bonding. Supramolecular assemblies can be used to organize molecular components to synthesize covalently bound products. A molecular tube has been synthesized by Harada et al. [375] by threading several cyclodextrin molecules on a linear polymer thread (poly ethylene glycol) to form a polyrotaxane (molecular necklace), then rim to rim covalent linking of the beads and removing the thread. Such "supramolecular assistance to molecular synthesis" uses molecular threads as a mold or template around which the molecular tube is formed. Wenz and coworkers [376, 377] made polyrotaxane from CD and polymer chains containing hydrophobic and hydrophilic units. CD is expected to encompass the hydrophobic part, while the hydrophilic part makes it water soluble with 37 beads on the thread (Fig. 19). Recently oligopeptide-terminated polyrotaxane has been developed with potential to be used as drug carrier [378].

Crown ether and cryptands provide a set of versatile constructional elements such as ether chains, ring, proton binding group, nitrogen junction, etc., and hence function is based on molecular recognition (i.e., recognition of a substrate by receptor). Strong and selective binding is often the result of increasing degree of pre-organization of binding groups in the receptor.

In 1960, Pederson [379] while investigating additives for rubber found related cyclic ethers that bind metal ions and assumes a corrugated conformation similar to a crown. He named theses receptors as crown ethers. Crown ethers are cyclic molecules that bind metal ions via lone pairs of electrons (chains of oxygen atom and methylene units linked together in a ring). In 1967, Lehn and coworkers [380] modified the cyclic molecules to obtain stronger and more selective metal binding. They synthesized bicyclic molecules in which nitrogen atoms unite three ether chains. They have three-dimensional roughly spherical cavities that bind alkali metal ions several order of magnitude more tightly than single-crown ether does. These three-dimensional receptors are called by Lehn as cryptands and the bound metal complexes as the cryptates. Cryptands are very selective to metal ion binding, which depends not only on the size of the cavity but also on a delicate balance between binding energy or enthalpy and gain or loss of configurational freedom or entropy. Although such bi- and tricyclic cryptands have higher binding capacity due to their enveloping nature, the metal ions passage in or out is rather slow, hence the rate-limiting step in many cases. Later, Gokel et al. [381] developed a different type of crown ether called lariat ether which possesses the binding ability of cryptands but the release profile of monocyclic crown ether. Such a receptor

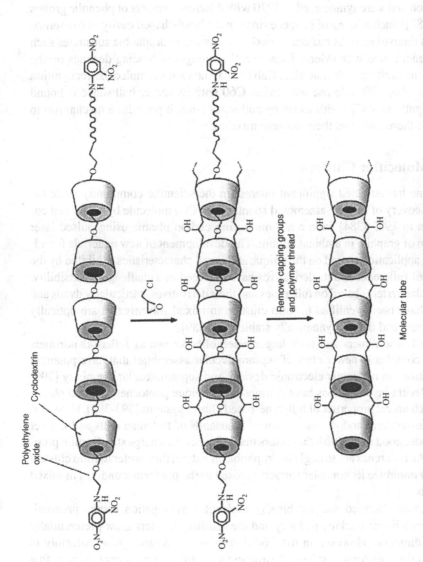

**FIGURE 19.** Polyrotaxane from cyclodextrin CD and polymer chains containing hydrophobic and hydrophilic units. CD is expected to encompass the hydrophobic part of the unit, while the hydrophilic part makes it water soluble. (Reproduced by permission of Wiley-VCH from G. Wenz, B. Keller, Angew. Chem. Int. Ed. Engl 31, 197, 1992.)

has a single ether ring with a flexible ether chain arm capable of swinging to offer additional binding ability of a captured metal ion.

Receptor molecules with relatively rigid cavities and cup shape known as calixerenes (*calix* means cup shape; *arene* denotes the presence of a benzene aryl group) and are a class of cyclooligomers formed via a phenol-formaldehyde condensation and were synthesized in 1970 with different numbers of phenolic groups [382, 383]. Such a lining of benzene rings in the bowl-shaped cavity of calixerene enables them to serve as molecular basket to contain hydrophobic substrates such as toluene, benzene or xylene. However, the strength of binding depends on the complementarity of substrate size. Calix [7] arene exhibits molecular recognition towards C60, C70 fullerene molecules. C60 with its soccer ball shape is bound more tightly than C70 with its rugby ball shape; thus it provides a mechanism to separate these two from their intimate mixture.

## 3.6. Molecular Clusters

Fullerene has attracted significant interest in the scientific community since the first discovery of the self-assembled spontaneous C60 molecule by Kroto and co-workers in 1985 [384] from a hot nucleating carbon plasma using pulsed laser ablation of graphite in ambient helium. The development of new materials for advanced applications based on the unique and novel characteristics exhibited by the family of fullerenes and its derivatives have emerged as a challenging possibility. Molecular systems based on fullerenes and their derivatives, particularly dyads and triads, have been identified to form clusters in mixed solvents that are optically transparent and thermodynamically stable [385-389].

Fullerene dimers and their larger assemblies known as fullero-dendrimers form a completely newer class of supramolecular assemblies that have potential applications in molecular electronic devices and supramolecular chemistry [390-392]. Recent investigations have demonstrated unique photochemical and photo-electrochemical properties of fullerene-based cluster systems [393-395]. However, such clusters are random association of thousands of fullerene units and not yet well understood to establish the factors that influence the charge stabilization properties. As fullerenes are strongly hydrophobic in nature, they prefer to form clusters so as to minimize its nonpolar surface exposed to the polar surroundings in mixed solvents.

It was observed that for bis-$C_{60}$ molecules aggregation occurs predominantly in a linear stacking pathway and the resulting clusters grow preferentially in one direction. However, in tris-C60 derivative, aggregates grow uniformly in all directions to form a spherical structure in a three-dimensional fashion that mimics dendrimeric-type molecular growth (intermolecular interactions between fullerene moieties become favorable). In sharp contrast, the monofunctionalized fullerene derivatives form isolated clusters. The selection of the linker is also critical in the formation of supramolecular self-assemblies. This provides the ability to design different carbon nanostructure by simple modification of fullerene functionalization. Template-driven clustering can further aid in defining and diversifying the shapes and sizes of these clusters. Biju *et al.* [396] demonstrated that the cluster network of nanorods or spherical entangled structures exhibits better charge stabilization properties than the monomeric forms.

As the fundamental knowledge is established, three main fields are emerging in the preparation of materials from the fullerene and its derivatives (family of fullerene materials, organometallic fullere-C60 derivatives, fullerenthiolate-functionalized gold nanoparticles, fullerene-based organic-inorganic nanocomposites, fullerene derivative embedded hybrid sol-gel glasses, mono- and multiple-functionalized fullerene derivatives, inclusion complex of single-C60-end-capped poly(ethylene oxide) with cyclodextrins); fullerene films, organic polymeric materials containing fullerene molecules, and fabrication of organic-inorganic nanocomposites [397]. They have opened several interesting perspectives for different applications, especially in nonlinear optics and photo electrochemistry [398, 399].

## 4. CHARACTERIZATION OF A SELF-ASSEMBLED SYSTEM

While it is clear that self-assembly and self-organization will be the tool fundamental to nanotechnology, as it bridges the gap between the top-down and bottom-up approaches, the capabilities for control of fabrication and properties at the nanoscale and precise characterization of these properties at the same length scale is essential for future growth. A complete analysis of such a system requires careful structural analysis at length scales ranging from the atomic $10^{-10}$ m to the microscopic, $10^{-6}$ m. However, only a handful techniques can be used to follow either the structural evolution at nanometer scale or morphology or phase behavior of such nanostructured systems.

The small angle scattering techniques using either X-ray or neutrons are very versatile tools to derive fundamental information on the size, shape, morphology, and dispersion of colloidal systems in complex material and to investigate various processes such as ordering, aggregation, crystallization, phase separation, self-assembly, etc. Properties such as micellar aggregation number, size and density profiles can be examined using static and dynamic light scattering (SLS, DLS), small angle neutron scattering (SANS) and high-resolution scanning and transmission electron microscopy (SEM, TEM). Advanced thermo-rheological investigations provide information on thermal transitions such as the order-disorder transition (ODT), which is the dissolution of the microdomain structure, and the lower temperature transition that involves destruction of the lattice within the domains while the nearest neighbor distance of the domains remains intact. Small angle X-ray scattering (SAXS), calorimetric techniques and spectroscopic techniques are employed for in-depth understanding of the nature of phase separation. Scanning probe microscopy (SPM) has also been developed as a key tool for the fabrication and characterization of nanostructured materials. Surface plasmaon resonance spectroscopy and quartz crystal microbalance (QCM), and secondary ion mass spectrometry (SIMS) are also important tools to investigate nanostructured materials.

## 4.1. Advanced Scattering Techniques

### 4.1.1. X-Ray/Neutron Reflectometry

X-ray (XR) and neutron reflectivity (NR) techniques are powerful methods for investigating structure of surfaces, monolithic and multilayered film structures. The technique can provide information about layer thickness and surface roughness, while in multilayered samples properties of the interface between layers can also be investigated. They allow not only extracting information on the free surface and the interface(s), but also determining the mass density and the thickness of thin layers along the direction normal to the specimen surface, precisely. Molecular structure and segmental mobility of a polymer film within a few nanometer of the surface has significant influence on its application as adhesives, lubricants, biosensors and protective coatings.

In many electronic and optical devices, the performance depends on the nature of interfaces of the multiplyer. The use of both XR and NR measurements in the investigation of thin films at interfaces is now well established. The noninvasive nature and high spatial resolution of the techniques make them ideal for studying a diverse range of systems of high complexity. The method involves guiding an X-ray/neutron beam of very low divergence onto the sample surface by a slit system and simultaneously the signal and the background reflectivities are quantified and the reflected intensity (either X-ray or neutron) as a function (typically small) of incident angle. Specular and diffuse reflectivities are the two basic types of measurements possible. In specular reflectivity, the incident beam impinges on the sample at a small angle $\theta$ and the intensity of the specularly reflected beam is detected at an angle $\theta$ from the surface and data are collected as function of $\theta$ or scattering vector, $Q[Q = (4\pi/\lambda) \sin \theta, \theta$ is the glancing angle of incidence, $\lambda$, the wavelength of the beam]. In diffuse or off-specular reflectivity the incidence angle is $(\theta - \omega)$, while the reflected beam are detected at an exit angle of $(\theta + \omega)$ from the surface. In this situation, the scattering vector has a component parallel to the surface $(Q_x = (4\pi/\lambda) \sin \theta \sin \omega)$. Measurements of the intensity as a function of $Q_x$ allow determination of the lateral correlation function of roughness or the lateral length scale of surface or interface roughness.

X-ray reflectometry has been widely used to characterize interfacial structure in a nondestructive manner [400]. While X-ray specular reflectivity provides average information in the direction perpendicular to the surface of the film, diffuse reflectivity is sensitive to the lateral structure of rough interfaces. The lateral information of an interface provides access to structural details about the morphology of the interfaces of a heterostructure. These methods have been employed to investigate interfacial structure of diverse range of multilayers such as W (tungsten)/ Si (Silicon) [401], Si(Silicon)/Mo(Molybdenum) [402], Co(Cobalt)/Cu(Copper) [403], thin lubricant films, self-asembled polymers [400] and hybrid organic-inorganic multilayer films [404]. Thin films supported on flat solid substrates are the commonly accepted sample geometry. Limited success has been obtained from free standing thin films. Samples with a thickness ranging from 5 nm to 1 μm can be measured by the X-ray technique. The lateral dimensions of samples typically to be 0.5 cm × 3.0 cm or larger but not to exceed 10 cm × 10 cm. There is another central requirement for the sample—its flatness. It must be optically flat on both along wavelength (centimeters) and short wavelength (nanometers) scales. Typical substrates used are polished silicon single crystal wafers and float glass. Due to the typically low number of atoms interacting with the beam, the high flux and very small vertical beam divergence of a synchrotron source are essential for this type of work.

Neutron reflectometry (NR) has become an increasingly popular and powerful technique in the characterization of thin-film surfaces and interfaces during the last 10-15 years. NR allows us to obtain structural and magnetic information about surfaces and interfaces—the detection of interdiffusion between neighboring layers of different isotopes, the layer formation at surfaces and interfaces of liquids or in polymers or the adsorption of biomolecules at interfaces. In NR the incoming neutron beam at low incident angle is reflected from the surface or interface depending on the neutron scattering length density of the structure. Grazing incident geometry and the wavelength range of cold neutrons provides a wave-vector transfer, $Q$ range well matched to the length scale of interest in soft condensed matter (~1-400 nm). Cold neutrons are also a penetrating probe and provide access to 'buried' interfaces; structures up to 100 nm thickness can be penetrated with a depth resolution of even below 1 nm. Neutrons being nondestructive and highly penetrating make them ideal probes to study buried interfaces as well as surfaces under a wide range of extreme environments. Recent advances in neutron source, instrumentation, experimental design, sample environments and methods of data analysis now make it possible to obtain an angstrom precision depth profile of the film composition. The specular reflectivity of neutrons provides information about the refractive index or scattering length density distribution normal to the surface, which is directly related to the composition or concentration profile in the interfacial region. The soft matter research benefits from the large difference between the neutron scattering length density of a hydrogen nucleus and a deuterium, which provides the opportunity to manipulate the refractive index distribution by H/D isotope substitution, without substantially altering the chemistry. The refractive index of neutrons is defined as $n = \frac{Nb}{2\pi}\lambda^2$, where $N$ is the atomic number density and $b$ is the neutron scattering lengh density ($0.6674 \times 10^{-2}$ cm for deuterium and $-0.374 \times 10^{-12}$ cm for hydrogen). It is now routine to enhance the contrast of adjacent layers by deuterating one specific layer or deuterating certain parts of a molecule. The emerging patterns of application of NR involve the investigatopn of comples interfaces, including biomembranes, in situ electrochemistry, adsorption at the liquid-solid and liquid-liquid interfaces; and more complex environments such as complex mixture, development of nano-structure and surfaces under shear or confinement [405]. Off-specular NR is used to achieve detail information of the in-plane structure of multilayer or lamellar systems where correlations between layers may influence physical/chemical behaviour. This information is not accessible using specular reflectivity, and rquires the study of the distribution of the off-specular reflectivity. However, the theoretical treatment of off-specular reflectivity is significantly complex compared to the specular signal, and makes the interpretation of data less direct. However, recent development reflects that quantitative analysis is possible and is now frequently used, especially in the investigation of correlated roughness between layers [406]

### 4.1.2. Small Angle X-Ray/Neutron Scattering

Small angle scattering of X-ray (SAXS) and neutron (SANS) are widely used in structural studies of noncrystalline materials at relatively low resolution and provide access to structural information of the order of approximately a nanometer to submicrometers [407, 408]. A great deal of similarity exists between X-ray and neutron scattering methods as applied to the study of the structure of matter. Information about size, shape, dispersity, periodicity, interphase boundary area and solution properites can be obtained. Specimens may be solids, films, powders, liquids, gels, crystalline or amorphous. Applications of SAXS and SANS in polymers and biological samples are abundant and well known. In structural biology this technique complements crystallographic structural analysis, which requires hard-to-get high-quality crystals of macromolecules, and is one of a few structural techniques for studying proteins in solution such as those in unfolded states or simply those whose crystallization conditions have not been determined. Biological fibers such as skeletal muscles are quasi-crystalline but not as well ordered as crystals, thus giving relatively broad diffraction peaks mostly at small angles. Many biological lipids exist as vesicles or liquid crystals, physical states that are rather poorly ordered. Micellar structures and synthetic polymer materials are often studied with this technique. Small angle scattering has been used in studying crystallization processes within polymers or alloys, and is also useful in studying a certain component within an amorphous material using anomalous scattering effects. In this experiment solids and films must be thin enough to be X-ray/neutron transparent and thick enough to present adequate sample volume in transmission. Estimates of nominal thickness are 1-2 mm for polymers and biological samples, 0.3-0.8 mm for ceramics, 0.05-0.2 mm for metals and alloys, 1 mm for water solutions and 1.5-2 mm for organic solvent solutions. The area required is as small as 1 mm$^2$.

### 4.1.2.1. Small Angle X-Ray Scattering (SAXS).
Small angle X-ray scattering (SAXS) is a well-known standard analytical X-ray application technique for

the structural characterization of solid and fluid materials in the nanometer range. [409, 410, 411]. In SAXS experiments, the sample is irradiated with a well-defined, monochromatic X-ray beam and the coherent scattering pattern that arises from electron density inhomogeneities within the sample is monitored. When a non-homogeneous medium—for example proteins in water—is irradiated, structural information of the scattering particles can be derived from the intensity distribution of the scattered beam at very low scattering angles. With SAXS it is possible to study both monodisperse and polydisperse systems. In the case of monodisperse systems one can determine size, shape and internal structure of the particles. For polydisperse systems, a size distribution can be calculated under the assumption that all particles have the same shape. Since the dimensions typically analyzed are much larger than the wavelength of the typical X-ray used (1.54 Å, for Cu), dimensions from tens to thousands of angstroms can be analyzed within a narrow angular scattering range. SAXS is applied to investigate structural details in the 0.5-50 nm size range in materials such as particle sizing of suspended nanopowders, life science and biotechnology (proteins, viruses, DNA complexes), polymer films and fibers catalyst surface per volume, micro-emulsions, liquid crystals, etc.

The availability of synchrotron radiation has made small angle X-scattering studies much more effective. It provides very small beam divergence, high beam flux, and energy tunability. It is crucial to have small beam divergence in order to isolate weak scattering at very small angles from the direct beam, which is orders of magnitude stronger. The flux level at a synchrotron source is usually several orders of magnitude higher than those from conventional X-ray sources; thus studies of weak scatterers have become much more practical. The high flux beam also makes it possible to conduct time-resolved measurements of small angle scattering to investigate fast structural transitions in the submillisecond time region in solutions and partly ordered systems. It is possible to conduct anomalous small angle X-ray scattering (ASAXS) only when beam energy tunability of synchrotron radiation is used. ASAXS offers contrast variation that enables the separation of species-specific structures.

*4.1.2.2. Small angle neutron scattering (SANS).* SANS is a relatively new technique for the investigation of polymer structure and supramolecular organization and has been able to produce definitive data for the critical evaluation of molecular theories of polymer structure [407, 412]. The scattering of neutrons by polymer or its solution or gel yields information not only on the structure and distribution of different components in the system, but also on their dynamics. During the last decade, SANS technique has been widely used to investigate the structure and thermodynamics issues in hybrid material, porous material, micelle, microemulsion etc. The technique is very versatile because of the highest sensitivity in the length scale of 1-100 nm and allows also bulk material investigation. SANS techniques play a unique role in the study of both the structural and dynamic properties of a wide range of "soft matter" due to the suitability of the length and time scales accessed by neutrons. The most interesting feature of SANS arises from the large neutron scattering cross-section of hydrogen atoms and the considerable difference in scattering cross section for hydrogen and deuterium. It is this difference that generates a contrast enabling individual molecules to be observed and their dimensions measured. Neutron scattering is the only tool for unraveling the molecular morphology and motion in a soft matter system at different relevant length scales. Solvent-regulated ordering in soft matter, particularly block copolymer, has been the subject of many recent investigations by SANS as it can provide in-depth information of phase behavior, microstructure of aggregates of different morphologies and information on hierarchical structure [413]. Neutron scattering in combination with computer simulations will have a very large impact on the future scientific endeavors in multicomponent soft condensed matter of increasing complexity.

## 4.2. Advanced Surface Analysis Techniques

Raman, IR, UV-vis spectroscopy, X-ray reflection, X-ray photoelectron spectroscopy (XPS), Scanning tunneling and atomic force microscopy (TEM, STM, AFM) are extensively used to analyze macromolecular assemblies. Various techniques such as ellipsometry, surface plasmon resonance spectroscopy, as quartz crystal microbalance (QCM), secondary ion mass spectrometry (SIMS), transmission electron microscopy, voltammetry, etc., are widely used to study replaced SAMS. XPS, IR spectroscopy, ellipsometry, and coulometry are used to investigate chemical composition and properties of mixed SAMS.

Surface plasmon resonance spectroscopy (SPRS) has been widely used to study self-assembled polymer, nanoparticle and protein systems [414, 415]. It is primarily used to measure layer thickness and deposition kinetics. The method produces a transverse magnetic electromagnetic wave (surface plasmon) travelling along the interface between the layers. The surface plasmon mode can be excited by the evanescent field generated from total internal reflection inside a prism, on which nanometer-thick gold film is deposited. A resonance angle is finally obtained measuring the intensity of the reflected light as a function of incident angle into the prism. When a thin film is deposited on a gold layer, resonance angle varies. From the reflectivities, refractive index and film thickness can be obtained.

In QCM, a thin quartz crystal is located between two electrodes. On application of voltage across these electrodes, the crystal is oscillated near its resonant frequency. The frequency of oscillation is related to the mass of the crystal. QCM is suitable for measuring adsorption of polymer in a thin layer or SAM. On adsorption, the frequency of crystal oscillation decreases. The change in frequency is related to mass by the following equation:

$$\Delta f = \frac{-2f_0^2 \Delta m}{A\sqrt{\mu_i \rho_i}} = K \Delta m$$

where $f_0$ = resonant frequency of the crystal
$A$ = piezoelectric area
$\rho_i$ = the quartz density
$K$ = experimental mass coefficient
$\mu_i$ = shear modulus of the quartz.

### 4.2.1. Elliposometry

Ellipsometry can be used to measure the thickness of thin film. A beam of monochromatic laser light is plane polarized and directed towards the substrate film. The reflected light is elliptically polarized due to the contributions of the parallel and perpendicular components of the incident lights ($s$ and $p$ polarized), which are reflected differently. With a compensator, this elliptically polarized light can be plane polarized and the angle of polarization is analyzed and detected by analyzer and detector. The thickness of a film can be obtained by comparison of the angles of polarization of the incident and reflected light and the refractive index of the film.

### 4.3. MALDI-MS, TOF-SIMS

The advances in the field of self-assembly and thin film require a sensitive tool to measure molecular mass information of the topmost monolayer with high mass and spatial resolution. In recent years, mass spectrometry techniques have gained significant interest for the analysis of synthetic polymers, in particular thin polymer film. Among many mass spectrometry (MS) techniques, the matrix-assisted laser desorption ionization technique MALDI-MS, gained importance to study self-assembled behavior of various polymeric materials [416]. Detailed information about the molecular weight and structure of the polymer can be obtained using this technique. Developed initially for biomaterials characterization, MALDI-MS permits fast identification of the polymer molecular weight without time consuming chromatographic process.

Primarily it is a soft ionization process that produces (quasi)molecular ions from large nonvolatile molecules such as proteins, oligonucleotides, polysaccharides, and synthetic polymers with minimum fragmentation. On irradiation by a pulsed laser beam, the prepared sample produces pseudo molecular ions, which are separated by their mass-to-charge ratio in the mass analyzer. Compared with other mass spectrometry techniques such as direct pyrolysis-mass spectrometry spectrometry (DP-MS), or time of flight secondary ion mass spectrometry spectrometry (TOF-SIMS) where ionized low mass molecular ions (fragments) are obtained, the MALDI-MS spectra show mainly single charged quasi molecular ions with hardly any fragmentation and, especially in the low molecular range, exhibit high-mass resolution. Various mass spectrometer systems are used for MALDI, such as TOF, Fourier transform, magnetic sector, Paul trap, and quadrupole. Among them MALDI-TOFMS and MALDIFTMS are the most popular. The growth of this technique has been accelerated by recent innovative MS method for polymer analysis, including new matrices for analyzing high-mass compounds [417, 418], reproducible sample preparation procedure [419]. improved instrumentation including postsource decay [420], delayed extraction [421, 422] and enhanced ion optics and improved data processing software.

Time-of-flight secondary ion mass spectrometry (TOF-SIMS) is a powerful and versatile technique for the analysis the elemental and chemical composition of a wide range of surfaces. In this method, the sample is irradiated with a focused pulsed primary ion beam to desorb and ionize species from a sample surface. The resulting secondary ions are accelerated into a mass spectrometer, where they are mass analyzed by measuring their time-of-flights from the sample surface to the detector. TOF-SIMS can provide spectroscopy for characterization of chemical composition (surface area of ~10 nm2), imaging for determining the distribution of chemical species, and depth profiling for thin film characterization. In the spectroscopy and imaging mode, only the outermost atomic layer of the sample is analyzed. Depth profiling by TOF-SIMS allows monitoring of all species of interest simultaneously and with high mass resolution. Chemical images are generated by collecting a mass spectrum at every pixel as the primary ion beam is rastered across the sample surface [413]. TOF-SIMS has been used successfully for quantitative evaluation of polymer surfaces (top 1-3 nm) [414] and additives. Investigation of annealed PS-PMMA block copolymer and the blend shows that in the copolymer, PS with lower surface free energy tends to segregate to the surface and PMMA tends to reside at the substrate interface due to its affinity for hydrophilic silicon surface [426]. However, complete phase separation in the blend on annealing leads to the formation of PS droplet on a thin matrix layer of PMMA. TOF-SIMS spectra coupled with nano-SIMS image can be utilized to reveal not only the effect of annealing on the blend surface characteristics but also the formation of submicron domain of the most hydrophobic domain.

## 4.4. Microscopic Techniques

Microscopic techniques such as TEM, AFM, SFM and high-resolution transmission electron microscopy (HRTEM) are very powerful tools for characterizing nanomaterials and self-assembly. With the combination of energy-dispersive X-ray analysis (EDS) and electron energy loss spectroscopy (EELS), the TEM becomes a versatile and comprehensive analysis tool characterizing electronic and chemical structure of materials at a nanoscale. TEM provides information from very small volumes over short length scales, $10^{-10}$ m to $10^{-8}$ m. It is an extremely powerful technique in providing atomic/molecular level information but is not suitable for information over the entire length scale of physical features in self-assembled systems. It is also destructive in nature since it requires an extremely thin electron transparet sample for investigation.

Wang [427, 428] reviewed the most recent developments in electron microscopy for nanotechnology with a special emphasis on in situ microscopy to study dynamic shape transformation of nanocrystals, in situ nanoscale property measurements on the mechanical, electrical and field emission of properties of nanotubes, nanowires, environmental microscopy for direct visualization of surface reaction, high spatial resolution electron energy loss spectroscopy (EELS) for nanoscale electronics and chemical analyses. HRTEM is used extensively to measure the shape and size of nanoparticles, and the structure of nanotubes. The self-assembly in block-selective solvent has been shown to result in a variety of nano-scale morphologies including spheres, rods, lamellae, and cylinders. Selective staining is essential to create sufficient contrast between different blocks and phase separation of polymers. For example in the case of SEBS, the styrene block can be made distinguishable from the ethylene-butylene block by staining with ruthenium tetroxide, which selectively oxidizes the aromatic ring in the styrene block. Oxidized aromatic rings have significantly increased electron density, which renders the styrene block much darker than the ethylene-butylene block on TEM images (Fig. 8).

### 4.4.1. Scanning Probe Microscopy

In 1982, Binning et al. [429] invented scanning tunneling microscopy (STM) which allows the imaging, manipulation, and electronic characterization of metal and semiconductors on an atomic scale. The STM became the archetype of a whole class of microscopes known as scanning probe microscopes. The techniques include: atomic force microscopy, which allows imaging and manipulation on an atomic scale for insulating and biological materials; and scanning near-field optical microscopy (SNOM) (which allows all the contrast mechanism known from regular optical microscopy to be used with 10-50 nm lateral resolution, and measures mechanical properties of molecular bonds, quantifies charge distributions, and determines material properties with very high resolution and sensitivity). Many other tip-based microscopes, such as friction force, scanning capacitance, scanning ion conductance, and scanning thermal microscopes have found important specific applications [430, 431]. In SPM the topography of a surface is measured by moving a very fine probe tip mounted on a cantilever, which rasters across the surface of the sample. The sample is vertically positioned using a piezoelectric crystal. A laser beam is reflected from the top of the cantilever and goes to the sensor. Thus, the interaction between the tip and the sample is measured in terms of bending force of the cantilever. The image of a sample can be obtained to a molecular resolution by oscillating the tip vertically near or at its resonant frequency. SPM can be operated in ambient air, in a vacuum, or in liquids over a wide range of temperatures, and can be operated in a range of environments.

Depending on the configuration, it can perform imaging, manipulation or various spectroscopies; often various types of SPM techniques are combined together to achieve multi-probe capabilities to answer complete questions. Force-distance measurement can also be performed, where the cantilever is lowered slowly towards the substrate and the deflection is recorded. A curve is obtained with the bending of the cantilever with the distance. The deflection is converted to force from Hooke's law (from the spring constant of the cantilever). SPM, particularly AFM, has become important for self-assembled structures, biological systems and soft matter investigation and is able to directly measure the force of a chemical bond. Thermal AFM ($\mu$-TA) has evolved as a very useful technique for the characterization of nanostructured polymer material and thin films [432]. Recently, the possibility of obtaining SPM images in millisecond resolution has been realized, giving access to the macromolecular relaxation time. Ultrafast SPM imaging can be used to follow a process in situ, in real time [433]. Recent developments in SPM are producing new and fundamental insights across a broad spectrum of disciplines and emerging as a key tool for nano- and biotechnology. The techniques are versatile, do not require a vacuum, are generally nondestructive and are capable of providing information over the complete length scale. However, they are normally sensitive to the surface alone and do not provide information about the subsurface, buried interface or structural information.

## 4.5. Solid-state NMR in Characterizing Self-assembled Nanostructures

NMR is a spectroscopic technique of great diversity that relies on the magnetic dipole moment of the atomic nuclei. Compared to other techniques, NMR is sensitive to the local environment, and provides information on both structure and dynamics of the system. Additionally, the molecular conformation, relaxation behavior and mobility of functional molecules in a self-assembled structure can be obtained through different dynamic experiments. Using NMR not only a wide range of materials and systems can be investigated but also the wealth of details on a molecular level achieved is unique. High-resolution NMR is sensitive to short-range structure and hence complementary to the longer-range structural information obtained from imaging and diffraction techniques [434-438]. NMR experiments are normally performed without the need to disturb the systems with probe molecules and without the need of synthetic work to intoduce isotopic levels in the molecules (in contrast to SANS that often rely on deuterium labeling to achieve the desired contrast).

Previous theoretical and instrumental limitations have now recently been overcome to employ NMR in widespread application in the realm of colloid and interface science [439-441]. Diffusion nuclear magnetic resonance methods traditionally used within colloid and porous media science have recently been applied to the study of microstructures in biological tissue using magnetic resonance imaging. Improved experimental design utilizing multiple diffusion periods or the length of the displacement encoding gradient pulses increases the information content of the diffusion experiment, while new numerical and analytical methods allow for the accurate modeling of diffusion in complicated porous geometries [441]. Protein association is an extremely complicated colloidal phenomenon. Understanding its intricacies is affected by limited experimental approaches and simplistic theoretical models. NMR diffusion (i.e., pulsed gradient spin-echo NMR) measurements

are emerging as an extremely powerful technique for addressing protein association phenomena [443]. By proper choice of excitation scheme, it is now possible to
selectively probe a specific interaction. Pulsed-field-gradient NMR spectroscopy
(PFG-NMR) yields ensemble-averaged local information about dynamics in diverse soft systems including polymer gels. PFG-NMR can be used to probe complex structure and structural anisotropy such as diffusion, adsorption dynamics
and flow in complex anisotropic gels. The recent use of Rheo-NMR and velocity
imaging confirms how the method may be used to elucidate molecular ordering
and dynamics and their relation to complex fluid rheology [444].

## 4.6. Advanced Thermal Analysis

Advanced thermal, thermomechanical and rheological investigations are powerful
tools and can provide information on many interesting aspects of soft materials
including colloidal suspension, emulsion, polymer solutions, gels, membranes,
liquid crystals, etc. Many industries including, adhesive, agrochemical, biotechnology, construction, cosmetic, electronics, food, imaging, oil, packaging, paint,
pharmaceuticals, paints use such soft materials. The structure and molecular dynamics of such systems are strongly influenced by entropy and are relatively weak.
They are easily deformable by external stresses, thermal fluctuations, electrical and
magnetic fields. Rheological behavior of such materials is determined by mesoscopic structure and dynamics. Rheology can be critical to their performance during
manufacture or in the application of the final product.

The storage and dissipation of mechanical energy by soft materials is often
surprisingly complex and a result of the hierarchical internal structure of these
media. Viscoelastic properties and the relaxation time play an important role in
the function of soft matter. It can provide specific information not only about,
deformation and flow characters but also about molecular dynamics, creep and
relaxation behavior, phase transitions, temperature sensitivity of various order-
disorder, order-order transitions in ordered mesophase systems, etc. [445-448].
Rheological experiments on such materials can provide critical insight not only
on deformation behavior for both equilibrium and nonequilibrium systems but
also on their precise origin. Micro-rheology has recently emerged as an important
methodology for probing the viscoelastic properties of soft materials, providing
access to a minuscule sample volume over a broader range of frequencies than
conventional rheometry.

A typical rheological investigation to observe the effect of dilution and temperature in the complex phase window of a soft gel system consists of triblock
copolymer block copolymer of SEBS (linear block copolymer with two rigid end
blocks, each with a molecular weight 12,600, and a rubbery mid-block with molecular weight of 87,000) and PAO4 (poly $\alpha$ olefin oil, midblock selective solvent) is
shown in Fig. 20. The figure illustrates the temperature dependence of the shear
storage modulus, $G'$ of SEBS ($\sim$29% styrene endblock) and SEBS physical gels
with 5, 10, 20, 40, 60, 80 wt% polymer under isochronal condition (1 Hz) over

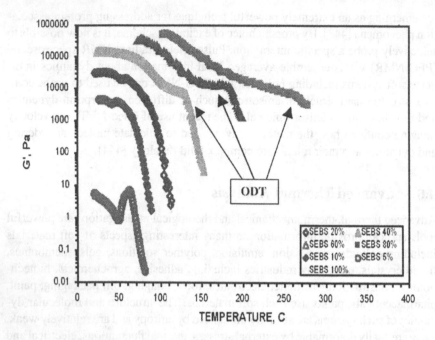

**FIGURE 20.** Effect of environment (solvent concentration and temperature) on the shear storage modulus $G'$ of neat SEBS (~29% styrene end block) and SEBS physical gels with 5, 10, 20, 40, 60, 80 wt% polymers under isochronal condition (1 Hz) over a wide range of temperature. Different transition processes including order-disorder (ODT) and order-order transitions (OOT) are clearly identified from such plots.

a wide range of temperatures. The experiment was carried out within the linear viscoelastic region of the samples. Different transitions including order-disorder (ODT) and order-order transitions (OOT) can be clearly identified from such plots.

**Order-disorder transition:** From Fig. 20, it can be clearly observed that $G'$ is decreasing significantly with an increase in both temperature and solvent concentration. From the rubbery plateau to the terminal zone of each sample within the experimental regime, multiple relaxations can be identified from the figure. The discontinuous drop in $G'$ observed beyond the OOT in the shear modulus, indicates the order-disorder transition temperature (ODT). At this point, on transformation from order to disordered state, the modulus value decreases by about three decades. The ODT occurs between long-range well-ordered microdomain structures (or lattices) and disordered micelles (short-range liquid-like order) and homogeneous phase during heating. Such a decrease in $G'$ is relevant to the disruption of the PS microdomain, and ODT is related to the homogenization for the SEBS block co-polymers and therefore, complete dissolution of the PS microdomains takes place when temperature is above the ODT. At higher temperature the number of micelles decreases, however, some may persist even at higher temperature. The composition dependence of the ODT is also clearly visible from the figures; the temperature for ODT ($T_{ODT}$) is systematically lowered with increase in oil concentration. The addition of the solvent selective to the mid-block selective solvent component appears to be an efficient way of reducing the ODT and increasing the block compatibility. This is a composite effect owing to an increase in the free energy of the ordered block and to a decrease of the disordered free energy.

**Order-order transition:** Order-order transition (OOT) is the transition where the microdomain structure transforms from one type of microdomain structure to another type of microdomain structure during heating/cooling. In rheology OOT is manifested as a drop of $G'$ followed by an increase in shear storage modulus above the rubbery plateau region The temperature at which a minimum in $G'$ occurs may be regarded as being the OOT temperature ($T_{OOT}$) of the block copolymer [450]. It is obvious that the addition of PAO4 causes significant changes in the rubbery plateau region of the shear storage modulus curve of the block copolymer. The rubbery plateau reduced to a narrow temperature range. The terminal zone of the rubbery plateau reduces due to the decrease in Todt with increase in the PAO4 level. A clear trend of decreasing $T_{oot}$ with PAO4 concentration is clearly observed for the composite containing PAO4. The $T_{OOT}$ may also be identified from the maxima in tanδ ($G''/G'$) curve. Different microdomains in block copolymers and other organized soft materials show a marked variation in viscoelastic behavior, and rheology may be used to identify OOT [451]. Sakamoto et al. [452] reported multiple OOT before ODT in asymmetric polystyrene-b-polyisoprene-b-polystyrene (SIS triblock, having 0.158 volume fraction PS) using SAXS.

# 5. APPLICATION AND FUTURE OUTLOOK

The future of nanotechnology will depend on our ability to do science in nanometer scale, just as the current microelectronic age is a consequence of harnessing materials on a micrometer scale. The use of self-assembly and organization opens the door to a host of well-defined materials suitable for use in nano- and bio-applications. Since the term "supramolecular" introduced in 1978 by Jan-Marie Lehn, the field has attracted intense interest, and an enormous research effort has been dedicated to the area of research by many pioneering scientists. Due to the interdisciplinary nature of this field, "today it pervades much of chemistry and extends to the interfaces of chemistry with a diverse array of other disciplines, including biology, physics, materials science, and engineering. Manifestations and applications include: molecular recognition, selective binding and encapsulation, receptors and sensors, drug delivery strategies, catalysts, biological mimics, and nanoscale electronic and mechanical devices" [453].

Since the pioneering research of Kresge et al. [454] a significant research effort has been also dedicated to develop mesotextured inorganic or hybrid phases, and thin films from self-assembled polymer-templates, which has potential application in the fields of catalysis, optics, photonics, sensors, separation, drug delivery,

**TABLE 4.** Summary of the nanomaterials that have been developed from phase-separated block copolymer (BCs)*

| BC tool | As nanomaterials | | As templates | |
| --- | --- | --- | --- | --- |
| | Direct use | Through processing | Direct use | Through processing |
| BCs in bulk | Photonic crystals Conducting BCs | Individual polymeric nano-objects: spheres, fibers, tubes | Nanolithographically patterned materials | Nanolithographically patterned materials; silicon, silicon nitride, gold, others |
| | | Nanoporous membranes | Hierarchically self-assembled metal nanowires | High-density arrays of metallic or semiconductor materials: cobalt, silica, chromium, others |
| | | Nanostructured carbons | | |
| Amphiphilic BCs | Drug delivery | Individual polymeric nano-objects: spheres, hollow spheres, cylinders, tubes | Metal nanoclusters Inorganic colloids | Nanolithographically patterned materials |

*Reproduced by permission of Wiley-VCH from Ref. 457.

sorption, acoustic and electronics [455, 344]. The choice of the organic template (which relies on the supramolecular arrays formed by self-organization) employed to spatially control the mineralization process in the mesoscale is a key issue in the synthesis of nanostructured and textured porous materials. For example, self-assembled block copolymers (BCs) are indeed a suitable template for the creation of highly controlled large-pore and highly stable meso-stuctured and mesoporous materials (silica, nonsilica oxide, carbons) of diverse shape such as powders, films, monoliths or aerosols. BCs are interesting as they are capable of imparting larger pores and thicker walls compared to the surfactants [456].

Table 4 summarizes the nanomaterials that have been fabricated using self-assembled block copolymer (BCs) [457]. Numerous self-assembling systems have been developed including DNA based structures and models to unravel protein folding, misfolding and conformational disease. Immense potential exists if one can combine biomacromolecular assemblies to ultra thin organic self-assembly for hybrid molecular design. The versatility and possibilities of this approach, has recently been discussed by some pioneers in the area of research [453]; however, the area is still in its "embryonic stage of development." The development is at the best experimental and the underlying design concepts are largely based on intuition and on simple theoretical concepts. It is challenging now to develop more detailed theories of supramolecular assembly to predict structural and dynamic properties of very complex microscopic or mescoscopic systems based on the microscopic interactions of their constituent molecular units. It will be ambitious to develop such predictive theories from analytical tools rather than using computer modeling. The theoretical viewpoint of polymeric amhiphiles is also incomplete.

Little attention has been given to the common features, which may be attributed in part to the multidisciplinary nature of this field. Various scientific communities (polymer synthesis, polymer science, surface science, materials science, biology, physics, chemical engineers, mechanical engineers, metallurgy) must interact with one another for the recognition of the common features and the development of the field as a whole.

# 6. ACKNOWLEDGMENT

The authors acknowledge the financial support of the Australian Research Council, Government of Australia, to carry out some of the research works presented in the chapter. It is also a pleasure to acknowledge the conscientious and patient cooperation of M. Ankit K. Dutta that made this chapter possible. We dedicate this work to the loving memory of our fathers: S. N. Roy Choudhury (deceased December 1998) and S. K. Dutta, (deceased March 2005) who have always been and will remain sources of strength and insperation for us.

# REFERENCES

1. P. Ball *Nanotechnology* 13 R15-R28 (2002); 16, R1-R8 (2005).
2. J. L. Atwood, J. Eric, D. Davies, D. D. MacNicol, and F. Vögtle, eds., *Comprehensive Supramolecular Chemistry*, Pergamon, Oxford (1996).
3. J.-M. Lehn, *Angew Chem. Int. Ed. Engl.* 29, 1304 (1990).
4. G. M. Whitesides and M. Boncheva, *PNAS* 16, 4769 (2002).
5. G. M. Whitesides and B. Grzybowski, *Science*, 293, 2418 (2002).
6. J.-M. Lehn, *Supramolecular Chemistry*, New York VCH (1995).
7. J.-M. Lehn and P. Ball, in *The New Chemistry* ed. N. Hall, Cambridge University Press, pp. 300-351.
8. K. Jakab, A. Neagu, V. Mironov, R. R. Markwald, and G. Forgacs, *PNAS*, 101, 2864 (2004).
9. J.-M. Lehn, *Angew Chem. Int. Ed. Engl.* 127, 89 (1988).
10. *Supramolecular Polymers*, ed. A. Ciferri and Marcel Dekker, New York (2000).
11. F. Oosawa and S. Asakura, *Thermodynamics of the Polymerization of Protein*. London: Academic Press, 1975, p25.
12. H.-J. Schneider and A. Yatsimirsky, *Principles and methods in supramolecular chemistry*, Wiley, New York (2000).
13. S. Foster and T. Plantenberg, *Angew. Chem. Int. Ed.* 41, 688 (2002).
14. S. L. Price and A. J. Stone, *J. Chem. Phys.* 86, 2859 (1987).
15. S. K. Burley and G. A. Petsko, *Adv. Protein Chem.*, 39, 125 (1988).
16. H. L. Anderson, C. Hunter, M. N. Meach, and J. K. M. Sanders, *J. Am. Chem. Soc.* 112, 5780 (1990).
17. A. P. H. J. Schenning, P. Jonkheijim, F. J. M. Hoeben, J. van Herrikhuyzen, S. C. J. Meskers, E. W. Meijer, L. M. Herz, C. Daniel, C. Silva, R. T. Phillips, R. H. Friend, D. Beljonne, A. Miura, S. De Feyter, M. Zdanowska, H. Uji-I, F. C. De Schryver, Z. Chen, F. Würthner, M. Mas-Torrent, D. den Boer, M. Durkut, and P. Hadley, *Synthetic Metal*, 147, 43 (2004).
18. D. A. Dougherty, *Science*, 271, 163 (1996).
19. J. C. Ma and D. A. Dougherty, *Chem. Rev.* 97, 1303 (1997).

20. N. S. Scrutton and A. R. C. Raine, *Biochem. J.*, 319, 1 (1996).
21. E. V. Pletneva, A. T. Laederach, D. B. Fulton, and N. M. Kostié, *J. Am. Chem. Soc.* 123, 6232 (2001).
22. J.-M. Lehn, *Supramolecular Chemistry: Concepts and Perspectives*, Wiley-VCH, Weinheim, Germany (1995).
23. M.-E. Perron, F. Monchamp, H. Duval, D Boils-Boissier, and J. D. Wuest, *Pure Appl. Chem.* 76, 1345 (2004).
24. H.-J. Schneider, *Angew Chem. Int. Ed. Engl.* 30, 1417 (1991).
25. P. Hobza and R. Zahradnik, *Intermolecular Complexes*, Amsterdam, Elsevier (1988).
26. W. J. Jorgensen, *Acc. Chem. Res.* 22,184, 1989. H. Bruning, D. Feil, *J. Comput. Chem.* 12, 1 (1991).
27. T. P. Lodge, *Macromol. Chem. Phys.* 204, 265 (2003).
28. S. Foster and M. Konrad, *J. Mater. Chem.*, 13 2671 (2003).
29. D. Meier, *J. Polym. Sci., Part C.* 26, 81, 1969, *J. Polym. Sci., Polym. Phys.* 34, 1821 (1996).
30. E. Helfand and Z. R. Wasserman, *Macromolecules*, 9, 879, 1976, 11, 960, 1978, 13, 994 (1980).
31. T. Hashimoto, K. Nagatoshi, A. Todo, H. Hasegawa, and H. Kawai, Macromolecules, 7, 365 (1974).
32. T. Hashimoto, M. Fujimura, and H. Kawi, *Macromolecules* 13, 1660 (1980).
33. E. L. Thomas, D. M. Anderson, C. S. Henkee, and D. Hoffman, *Nature* 334, 598 (1988).
34. T. Hashimoto, M. Shibayama, and H. Kawai, *Macromolecules* 13, 1237 (1980).
35. R. Séguéla and J. Prud'homme, *Macromolecules* 11, 1007 (1978).
36. G. Kraus and K. W. Rollmann, *J. Polym. Sci., Polym. Phys.*, 14, 1133 (1976).
37. S. Krause, Z.-H. Lu, and M. Iskandar, *Macromolecules*, 15, 1076 (1982).
38. B. Morese-Séguéla, M. St-Jacques, J. M. Renaud, and J. Prud'homme, *Macromolecules* 28, 5043 (1995).
39. J. Denault, B. Morese-Séguéla, R. Séguéla, and J. Prud'homme, *Macromolecules*, 14, 1091 (1981).
40. R.-J. Roe, M. Fishkis and J. C. Chang, *Macromolecules* 14, 1091 (1981).
41. C. D. Han, D. M. Baek, J. K. Kim, T. Ogawa, N. Sakamoto, and T. Hasimoto, *Macromolecules* 28, 5043 (1995).
42. N. Sakamoto, T. Hashimoto, C. D. Han, D. Kim, and N. Y. Vidya, *Macromolecules* 30, 1621 (1997).
43. K. Almdal, K. A. Koppi, F. S. Bates, and K. Mortensen, *Macromolecules* 25, 485 (1993).
44. T. Pakula, K. Saijo, H. Kawai, and T. Hashimoto, *Macromolecules* 18, 1294 (1985).
45. R. Séguéla and J. Prud'homme, *Macromolecules* 216, 35 (1988).
46. J. Sakamoto, S. Sakurai, K. Doi, and S. Nomura, *Polymer* 34, 4837 (1993).
47. Y. Cohen, R. J. Albalak, B. J. Dair, M. S. Capel, and E. L. Thomas, *Macromolecules* 33, 6502 (2000).
48. R. Choksi, *J. Nonlinear Sci.* 11, 223 (2001).
49. F. Balta Calleja, Z. Roslaniec, and Marcel Dekker, eds. *Block Copolymers*, New York (2000).
50. P. Alexandridis and B. Lindman, eds. *Amphiphilic Block Copolymers: Self-Assembly and Applications*, Elsevier, Amsterdam (2000).
51. F. S. Bates and G. H. Fredrickson, *Physics* Today 52, 32 (1999).
52. N. K. Dutta, N. Roy Choudhury, and A. Bhowmick, *Handbook of Thermoplastics*, Marcel Dekker, New York (1996).
53. Alexandridis, *Curr. Opin. Colloid Interface Sci.* 2, 478 (1997).
54. K. Schillen, K. Bryskheand, and Y. S. Mel'nikova, *Macromolecules* 32, 6885 (1999).
55. E. Helfand and Z. R. Wesserman, *Macromolecules* 9, 879 1976, 11, 960, 1978;13, 994 (1980).
56. M. W. Matsen and F. S. Bates, *J. Chem. Phys.* 106, 2436 (1997).
57. F. S. Bates and F. S. Fredricson, *Annu. Rev. Phys. Chm.* 41, 265 (1990).

58. L. Leibler, *Macromolecules* 13, 1602 (1980).

59. M. W. Matson and F. S. Bates, *Macromolecules* 29, 1091 (1996).

60. S. Foster, A. K. Khandpur, J. Zhao, F. S. Bates, I. W. Hamley, A. J. Ryan, and W. Bras, *Macromolecules* 27, 6922 (1994).

61. A. K. Khandpur, S. Foster, F. S. Bates, I. W. Hamley, A. J. Ryan, W. Bras, K. Almdal, and K. Mortensen, *Macromolecules* 28, 8796 (1995).

62. A. N. Semenov, *Macromolecules* 26, 6617 (1993).

63. T. Ohta and K. Kawasaki, *Macromolecules* 19, 2621 (1986).

64. E. Helfand and Z. R. Wasserman, *Macromolecules* 9, 879 (1976).

65. T. Hashimoto, M. Shibayama, and H. Kawai, *Macromolecules* 13, 1237, 1660 (1980).

66. F. S. Bates, M. F. Schultz, A. K. Khandpur, S. Föster, J. H. Rosedale, K. Almdal, and K. Mortensen, Faraday *Discuss. Chem. Soc.* 98, 7 (1994).

67. J. Zhao, B. Majumdar, M. F. Schulz, F. S. Bates, A. K. Khandpur, K. Almdal, K. Mortensen, D. A. Hajduk, and S. M. Gruner, *Macromolecules* 29, 1204 (1996).

68. M. F. Schulz, A. K. Khandpur, F. S. Bates, K. Almdal, K. Mortensen, D. A. Hajduk, and S. M. Gruner, *Macromolecules* 29, 2857 (1996).

69. H. Fischer, S. Poser, and M. Arnold, *Liq. Cryst.* 18, 503 (1995).

70. K. Jung, V. Abetz, and R. Stadler, *Macromolecules* 29, 1076 (1996).

71. X. Ren and J. Wei, *J. Nonlinear Sci.* 13, 175 (2003).

72. U. Breiner, U. Krappe, and R. Stadler, *Macromol. Rapid Commun.* 17, 567 (1996).

73. U. Krappe, R. Stadler, and L. Voigt-Martin, *Macromolecules* 28, 4558 (1995).

74. Y. Matsushita, K. Yamada, T. Hattori, T. Fujimoto, Y. Sawada, M. Nagasawa, and C. Matsui, *Macromolecules* 16, 10 (1983).

75. Y. Mogi, M. Nomura, H. Kotsuji, K. Ohnishi, Y. Matsushita, and I. Noda, *Macromolecules* 27, 6755 (1994).

76. S. P. Gido, D. W. Schwark, E. L. Thomas, and M. do Carmo Goncalves, *Macromolecules* 26, 2636 (1993).

77. R. G. Petschek and K. M. Wiefling, *Phys. Rev. Lett.* 59, 343 (1987).

78. A. Halperin, M. Tirrell, and T. P. Lodge, *Adv. Polym. Sci.* 100, 31 (1992.

79. S. Sioula, N. Hajichristidis, and E. L. Thomas, *Macromolecules* 31, 5272 (1998).

80. T. Goldacker, V. Abetz, R. Stadler, I. Erukhimovich, and L. L. Leibler, *Nature*, 398, 137, 1999, A.-V. Ruzette, L. Leibler, *Nature Materials*, 4, 23 (2005).

81. J. Ruokolainen, R. Mezzenga, G. H. Fredrickson, E. J. Kramer, P. D. Hustad, and G. W. Coats, *Macromolecules* 38, 851 (2005).

82. J. Peng, X. Gao, Y. Wei, H. Wang, L. Hanfu, H. Binyao, and Y. Han, *Journal of Chemical Physics* 122(11), 114706/1-114706/7 (2005).

83. M. Shibayama, T. Hasimoto, and H. Kawai, *Macromolecules* 16, 16 (1983).

84. E. Raspaud, D. Lairez, M. Adam, and J. P. Carton, *Macromolecules* 27, 2956 1994, 29, 1269 (1996)).

85. C. Lai, W. B. Russel, and R. A. Register, *Macromolecules* 35, 841, 202.

86. K. J. Hanley and T. P. Lodge, *Macromolecules* 33, 5918 (2000).

87. T. P. Lodge, B. Pudil, and K. J. Hanley, *Macromolecules* 35, 4707 (2002).

88. E. Raspaud, D. Lairez, M. Adam, and J. P. Carton, *Macromolecules* 29, 1269 (1996).

89. R. Kleppinger, K. Reynders, N. Mischenko, N. Overbergh, M. H. J. John, K. Mortensen, and H. Reynaers, *Macromolecules* 30, 7008 (1997).

90. C. R. Harless, M. A. Singh, S. E. Nagler, G. B. Stephenson, and J. L. Jordan-Sweet *Phys. Rev. Lett.* 64, 2285 (1990).

91. K. Mortensen and J. S. Pedersen, *Macromolecules* 26, 805 (1993).

92. C. Lai, W. B. Russel, and R. A. Register, *Macromolecules* 35, 841-849 (2002).

93. Lairez, M. Adam, J. P. Carton, and E. Raspaud, *Macromolecules* 30, 6798 (1997).

94. Z. Tuzar, C. Konak, P. Stepanek, J. Plestil, P. Kratochvil, and K. Prochazka, *Polymer*, 31, 2118 (1990).

95. J. Plestil, D. Hlavata, and Z. Hrouz, *Polymer* 31, 2112 (1990).
96. C. Konak, G. Fleischer, Z. Tuzar, and R. J. Basil *J. Polym Sci. part B: Polym. Phys.* 38, 1312 (2000).
97. N. P. Balsara, M. Tirrell, and T. P. lodge, *Macromolecules* 24, 1975 (1991).
98. S. Sakurai, T. Hashimoto, and I. Fetters, *Macromolecules* 29, 740-747 (1996).
99. R. Kleppinger, N. Mischenko, H. L. Raynaers, and M. H. J. Koch *J. Polym Sci. Part B: Polym. Phys.* 37, 1833-1840 (1999).
100. J. H. Laurer, J. F. Mulling, S. A. Khan, R. J. Spontak, and R. J. Bukovnic, *Polym Sci.* Part B: *Polym Phys.* 36, 2379 (1998).
101. R. Bansil, H. Nie, Y. Li, G. Liao, K. Ludwig, M. Steinhart, C. Konak, and J. Lai, *Macromol. Chem. Phys.* (inpress).
102. O. Glatter, G. Scherf, K. Schillen, and W. Brown, *Macromolecules* 7, 6046 (1994).
103. I. W. Hamley, J. P. A. Fairclough, A. J. Ryan, C. Y. Ryu, T. P. Lodge, A. J. Gleeson, and J. S. Pedersen, *Macromolecules* 31, 1188 (1998).
104. C. Konak, M. Helmstedt, and P. Bansil, *Polymer* 41, 9811 (2000).
105. C. Konak, G. Fleischer, Z. Tuzar, and R. J. Basil, *Polym. Sci. part B: Polym. Phys.* 38, 1312 (2000).
106. K. Mortensen, W. Brown, and B. Norden *Phys. Rev. Lett.*, 13, 2340 (1992). 106
107. S. Thompson, N. K. Dutta, N. Choudhury, and R. Knott, *International Conference on Neutron Scattering*, ICNS 2005, Sydney p. 270.
108. N. K. Dutta, Sandra Thompson, N. Roy Choudhury, and R. Knott, *Proceedings Int. Conf. on Nanomaterials: Synthesis, characterization and application*, Calcutta, India, p. 243, Nov 4-6, (2004).
109. Naba K. Dutta, Sandra Thompson, N. Roy Choudhury, and Robert Knott, *Australian Polymer Symposium*, p. D3/1 Adelaide, 28 Nov-2 Dec (2004).
110. N. Dutta, S. Thompson, N. Roychoudhury, and R. Knott, *Physica B: Condensed matter* 773, 385-386(2006).
111. E. Raspaud, D. Lairez, M. Adam, and J. P. Carton, *Macromolecules* 27, 2956, 1994, 29, 1269 (1996).
112. R. Bansal, H. Nie, C. Konak, M. Helmstedt, and J. Lal, *J. Polym. Sci., Part B: Polymer Phys.*, 40, 2807 (2002).
113. T. P. Lodge, M. W. Hamersky, K. J. Hanley, and C.-I. Huang, *Macromolecules* 30, 6139 (1997).
114. Z. Tuzar and P. Kratochvil, in *Surface and Colloid Science*, ed., E. Matijevic, Plenum New York (1993).
115. B. Chu, *Langmuir* 11, 414 (1995).
116. P. Alexandridis, Curr. Opin. Colloid. *Interface Sci.* 1, 490 (1996).
117. C. Booth and D. Attwood, *Macromol. Chem. Rapid Commun.* 21. 501 (2000).
118. J. L. M. Cornelissen, M. Fischer, N. A. J. M. Sommerdijk, and R. J. M. Nolte, *Science* 280, 1427 (1998).
119. C. Nardin, T. Hirt, J. Leukel, and W. Meier, *Langmuir* 16, 1035 (2000).
120. I. W. Seymour, K. Kataokaa, and A. V. Kabanov, in *Self-assembling Complexes for Gene Delivery: From Laboratory to Clinical Trials*, K. V. Kabanov, P. L. Felgner, L. W. Seymour, eds., John Wiley & Sons, New York, p. 219 (1998).
121. P. Dubin, J. Bock, R. M. Davis, D. N. Schul, and C. Thies, Eds, Macromolecular Complexes in Chemistry and Biology, Springer-Verlag, Berlin (1994).
122. I. Goodman'Development in Block Copolymers, Elsevier, Applied Science, New York Vols 1 and 2, 1982 (1985).
123. F. S. Bates, *Science* 251, 898 (1991).
124. B. M. Disher, Y. Y. Won, D. S. Ege, J. C. M. Lee, F. S. Bates, D. E. Disher, and D. A. Hammer, *Science* 284, 1143 (1999).
125. P. Alexandridid and R. J. Spontak, *Curr. Opin. Colloid Interface Sci.* 4(2), 130 (1999).

126. P. Alexandridid, U. Olsson, and B. Lindman, *Langmuir* 14(10), 2627 (1998).
127. C. Booth and D. A. Atwood, *Macromol. Rapid Commun.* 21(9), 501 (2000).
128. J. H. Bader, H. Ringsdorf, and B. Schmidt, *Angrew. Makomol. Chem.* 123, 457 (1984).
129. M. K. Pratten, J. B. Lioyd, G. Hurpel, and H. Ringsdorf, *Makromol. Chem.* 186, 725 (1985).
130. G. S. Kwon, *Crit. Rev. Ther. Drug. Carrier Syst.* 15, 481 (1998).
131. Y. Bae, W-D. Jang, N. Nishiyama, S. Fukushima, and K. Kataoka, *Molecular BioSystems*, 1, 242 (2005).
132. E. V. Batrakova, S. Li, D. W. Miller, and A. V. Kabanov, *Pham. Res.* 16(9), 1366 (1999).
133. D. W. Miller and A. V. Kabanov, *Colloids. Surf.* B, 16, 321 (1999).
134. A. Harada and K. Kataoka, 31, 288 (1998).
135. G. Kwon, M. Naito, M. Yokoyama, T. Okano, Y. Sakurai, and K. Kataok, *J. Controlled Release* 48, 195 (1997).
136. S. Katayose and K. Kataoka, *J. Pharm. Sci.* 87, 160 (1998).
137. C. Allen, Y. S. Yu, D. Maysinger, and A. Eisenber, *Bioconjugate Chem.* 9, 564 (1998).
138. A. S. Hagan, A. G. A. Coombes, M. C. Garnett, S. E. Dunn, M. C. Davies, L. Illum, and S. S. Davis, *Langmuir* 12, 2153 (1996).
139. M. Mofitt, K. Khougaz, and A. Eisenberg, *Acc. Chem. Chem. Res.* 29, 95 (1996).
140. A. S. Kimerling, W. E. Rochefort, and S. R. Bhatia, *Ind. Eng. Chem. Res.* (2006).
141. T. K. Bronich, A. M. Popov, A. Eisenberg, V. A. Kabanov, and A. V. Kabanov, *Langmuir* 16, 481 (2000).
142. S. V. Solomatin, T. K. Bronich, T. W. Bargar, A. Eisenberg, V. A. Kabanov, and A. V. Kabanov, *Langmuir* 19, 8069 (2003).
143. L. Bromberg and L. Salvati, *Bioconjugate Chem.* 10, 678 (1999).
144. K. Tamano, T. Imae, S.-I Yusa, and Y. Shimada, *J. Phys. Chem.* B 109, 1226 (2005).
145. L. E. Bromberg and D. P. Bar, *Macromolecules* 32, 32 (1999).
146. M. Szwarc, *Nature* 178, 1168 (1956).
147. M. Szwarc, M. Levy, and R. M. Milkovich, *J. Am. Chem. Soc.* 78, 2656 (1956).
148. J. P. Kennedy and B. Ivan, *Designed Polymers by Cabocationic Macromolecular Engineering: Theory and Practice*, Hanser, Munich (1992).
149. S. Föster and M. Antonietti, *Adv. Mater.* 10, 195 (1998).
150. T. Ishizone, S. Han, M. Hagiwara, and H. Yokoyama, *Macromolecules* 39, 962 (2006).
151. J. M. Yu, Y. Yu, and R. Jerome, *Polymer* 38, 3091 (1997).
152. G. Hild and J. P. Lamps, *Polymer* 36, 4841 (1995).
153. N. Hadjichristidis, *J. Polym. Sci. Part A, Polym. Chem.* 37, 857 (1999).
154. H. Iatrou and N. Hadjichristidis, *Macromolecules*, 26, 5812, 2479 (1993).
155. T. Fujimoto, H. Zhang, T. Kazama, Y. Isono, H. Hasegawa, and T. Hashimoto, *Polymer* 33, 2208 (1992).
156. G. Riess, M. Schlienger, and S. Marti, *J. Macromol Sci. Polym. Phys.* Ed 17, 355 (1990).
157. N. Hadjichristidis, S. Pispas, and G. A. Floudas, *Synthetic Strategies, Physical Properties and Applications*, Wiley, New York (2003).
158. S. Penczek, P. Kubisa, and K. Matyjaszewski, *Adv. Polymer Sci.*, 68/69, 1 (1985).
159. W. H. Janes and D. C. Allport, *Block Copolymers*, Applied Science Publishers, London (1973).
160. T. Otsu, M. Yoshida, and A. Kuriyama, *Macromol. Chem. Rapid. Commun.*, 3, 133, 1982. Polym. Bull 7, 45 (1982).
161. S. Hadjikyriacou and R. Faust, *Macromolecules* 29, 5261 (1996).
162. S. Beinat, M. Schappacher, A. Deffieux, *Macromolecules*, 29, 6737 (1996).
163. L. Balogh, L. Samuelson, K. S. Alva, and A. Blumstein, *Macromolecules* 29, 4180 (1996).
164. C. S. Patrickios, C. Forder, S. P. Armes, and N. C. Billingham, *J. Polym. Sci., Part A: Polymer. Chem.*, 34, 1529 (1996).
165. H. Fukui, S. Yoshihashi, M. Sawamoto, and T. Higashimura, *Macromolecules* 29, 1862, 196.
166. O. W. Webster, W. R. Hertler, and D. Y. Sogah, *J. Am. Chem. Soc.* 105, 5706 (1983).

167. T. Okano, S. Nishiyama, I. Shinobara, T. Akaike, Y. Sakuri, K. Katakoka, and T. Tsuruta, *J. Biomed. Mater. Res.* 15, 393 (1981).
168. H. Ito, A. Taenaka, Y. Nagasaki, K. Kataoka, and M. Kato, *Polymer* 37, 633 (1996).
169. C. C. Cummins, R. R. Schrock, and R. E. Cohen, *Chem. Mater.*, 4, 27 (1992).
170. J. Yue, V. Sankaran, R. E. Cohen, and R. R. Schrock, *J. Am. Chem. Soc.* 115, 4409 (1993).
171. V. Sankaran, C. C. Cummins R. R. Schrock, and R. E. Cohen,
172. R. J. Silby, *J. Am. Chem. Soc.*, 112, 6858 (1990).
173. R. S. Kane, R. E. Cohen, and R. Silby, *Chem. Mater.*, 8, 1991 (1996).
174. T. Otsu, M. Yoshida, and T. Tazaki, *Macromol. Chem. Rapid. Commun.* 3, 127 (1982).
175. D. R. Suwier, P. A. M. Steeman, M. N. Teerenstra, M. A. J. Schellekens, B. Vanhaecht, M. J. Monterio, and C. E. Koning, *Macromolecules* 35, 6210 (2002).
176. M. K. Geeorges, K. A. Moffat, R. P. N. Vereginis, P. M. Kazmaier, and G. K. Hamer, *Polym. Mater. Sci. Eng.* 69, 305 (1993).
177. K. Matyjaszewski, *Curr. Opin. Solid State Interface Sci.*, 1, 769 (1996).
178. M. K. Georges, R. P. N. Vereginis, P. M. Kazmaier, and G. K. Hamer, *Macromolecules*, 26, 2987, 5316, 1993, *Trends Polym. Sci.*, 2, 66, 1994, *Macromolecules* 28, 4391 (1995).
179. M. K. Georges, M. D. Saban, P. M. Kazmaier, R. P. N. Vereginis, G. K. Hamer, K. A. Moffat, and U. S. Pat, 5412047, May, 21995.
180. C. J. Hawker, *J. Am. Chem. Soc.* 116, 11185 (1994).
181. C. J. Hawker and J. L. Hedrick, *Macromolecules* 28, 2993 (1995).
182. J. S. Wang and K. Matyjaszewski, *Macromolecules* 34, 4416 (2001).
183. M. Kato, M. Kamigaito, M. Sawamoto, and T. Higashimura, *Macromolecules* 28, 1721 (1995).
184. J. Chiefari, Y. K. Chong, F. Ercole, J. Krstina, J. Jeffery, T. P. T. Le, R. T. A. Mayadunne, G. F. Mejis, C. L. Moad, G. Moad, E. Rizzardo, and S. H. Thang, *Macromolecules* 31, 5559 (1998).
185. Č. Koňák and M. Hellmstedt, *Macromolecules* 34, 6131 (2001).
186. Y. Zhang, I. S. Chung, J. Huang, K. Matyjaszewski, and T. Pakula, *Macromol. Chem. Phys.* 206, 33-42 (2005).
187. K. Matyjaszewski and J. Spanswick, *Materials Today*, 26-33 (2005).
188. K. Matyjaszewski (ed), *Advances in controlled/Living Radical Polymerization*, Americal Chemical Society, Washington, D.C. (2003).
189. K. Matyjaszewski and J. Spanswick, Controlled/living radical polymerization, in *Handbook of Polymer Synthesis*, H. Kricheldorf, O. Nuyken, G. Shift, eds., Marcel Dekker, New York, p. 895 (2004).
190. N. Hadjichristidis, S. Pispas, and G. A. Floudas, *Block Copolymers: Synthetic Strategies, Physical Properties, and Applications*, Wiley-VCH, New York (2003).
191. Z. Yousi, L. Jian, Z. Rongchuan, Y. Jianliang, D. Lizong, and Z. Lansun, *Macromolecules* 33, 4745 (2000).
192. E. Rizzardo, J. Chiefari, Y. K. Chong, F. Ercole, J. Kristina, J. Jeffery, T. P. T. Le, R. T. A. Mayadunne, G. F. Mejis, C. L. Moad, G. Moad, and S. H. Thang, *Macromol. Symp.* 143, 291 (1999).
193. R. T. A. Mayadunne, J. Jeffery, G. Moad, and E. Rizzardo, *Macromolecules* 36, 1505 (2003).
194. R. T. A. Mayadunne, G. Moad, and E. Rizzardo *Tetrahedron Lett.*, 43, 6811 (2002).
195. E. Rizzardo, J. Chiefari, R. T. A. Mayadunne, G. Moad, and S. H. Thang, *Macromol. Symp.* 174, 209 (2001).
196. L. P. T. Le, G. Moad, E. Rizzardo, S. H. Thang, and Dupont, *PCT Int. Appl.* (1998) WO 9801478.
197. J. Chiefari, Y. K. Chong, F. Ercole, J. Kristina, J. Jeffery, T. P. T. Le, R. T. A. Mayadunne, G. F. Mejis, C. L. Moad, G. Moad, E. Rizzardo, and S. H. Thang, *Macromolecule* 31, 5559 (1998).
198. Y. K. Chong, T. P. T. Le, G. Moad, E. Rizzardo, and S. H. Thang, *Macromolecule* 32, 2071 (1999).
199. M. S. Donovan, T. A. Sanford, A. B. Lowe, B. S. Sumerlin, Y. Mitsukami, and C. L. McCormic, *Macromolecles* 35, 4570 (2002).
200. R. Bussels *Multi block copolymer synthesis via controlled radical polymerization in aqueous dispersion*, PhD. Dissertation, Eidhoven: Technische Universiteit, Eidhoven (2004).

201. Z. Jedlinski and K. Brandt, Novel segmental polymers containing rigid and soft blocks, *Acta Polymerica*, 39, 13 (1998).
202. G. Ziegast and B. Pfannemuller, *Makromol. Chem.* 185, 1855 (1984).
203. T. Takagi, I. Tomita, and T. Endo, *Polym. Bull.* 39, 685 (1997).
204. S. Wallace and C. J. Morrow, *J. Polym. Sci. PartA: Polym. Chem. Ed.* 27, 2553 (1989).
205. A. M Blinkovsky and J. S. Dordick, *J. Polym. Sci., PartA:Polym Chem. Ed.* 31, 1839 (1993).
206. K. Loos and R. Stadler, *Macromolecules* 30, 7641 (1997).
207. T. Goldsmith and A.G., German *Patent* 4134967 (1992).
208. Y. Qui, X. Yu, L. Feng, and S. Yang, *Makromol. Chem.* 193, 1377 (1992).
209. T. Ramasami, G. Bhaska, and A. B. Mandal, *Macromolecules* 26, 4083 (1993).
210. G. Barany, S. Zalipsky, J. L. Chang, and F. Albericio, *React. Polym.* 22, 243 (1994).
211. I. V. Berlinova, A. Amzil, S. Tsvetkova, and I. M. Panayotov, *J. Polym. Sci. Part A: Polym. Chem.* 32, 1523 (1994).
212. B. Wesslen, C. Friej-Larsson, M. Kober, A. Ljungh, M. Pauisson, and P. Tengvall, *Mater. Sci. Eng.*, C1, 127 (1994).
213. B. Wesslen, and H. Derand, *J. Polym. Sci, PartA:Polym Chem.* 33, 571 (1995).
214. G. Galli, E. Chiellini, Y. Yagci, E. I. Se Rhatli, M. Laus, and A. S. Angeloni, *Macromol. Symp.* 107, 85 (1996).
215. Y. Yagci, A Önen, and W. Schnabel, *Macromolecules* 24, 4620 (1991).
216. Cationic to-Radical polymerization, S. Coca, K. Matyjaszewski, *Macromolecules* 30, 2808 (1997).
217. E. Yoshida and A. Sugita, *J. Polym. Sci. Part A: Polym. Chem.* 36, 2059 (1998).
218. O. Nguyen, H. Kröner, and S. Aechner, *Makromol. Chem. Rapid. Commun.* 9, 671 (1998).
219. S. Kobatake, H. J. Harwood, R. P. Quirk, and D. B. Priddy, *Macromolecules* 31, 3735 (1998).
220. R. Nomura, Y. Shibasaki, and T. Endo, *Polym. Bull.* 37, 597 (1996).
221. M. P. Labeau, H. Cramail, and A. Deffieux, *Macromol. Chem. Phys.* 199, 335 (1998).
222. S. Coca, H. Paik, and K. Matyjaszewski, *Macromolecules* 30, 6513 (1997).
223. T. Morita, B. R. Maughon, W. Belawski, and R. H. Grabbs, *Macromolecules* 33, 6621 (2000).
224. M. A. Hillmyer, S. T. Nguyen, and R. H. Grubbs, *Macromolecules*, 30, 718 (1997).
225. J. Trautmann and N. K. Dutta *Synthesis of block copolymer by ATRP*, Research report, University of South Australia, Jan (2002).
226. M. Antonietti, S. Föster, J. Hartmann, and S. Oestreich, *Macromolecules* 29, 3800 (1996).
227. D. R. Invergar, S. M. Perutz, C. A. Dai, C. K. Ober, and E. J. Kramer, *Macromolecules* 29, 1229 (1996).
228. M. Antonietti, S. Föster, M. A. Micha, and S. Oestreich, *Acta Polym.* 48, 262 (1997).
229. C. Ramireddy, Z. Tuzar, K. Prochazka, S. E. Webber, and P. Mun, *Macromolecules* 25, 2541 (1992).
230. L. A. Mango and R. W. Lenz, *Macromol. Chem.* 163, 13 (1973).
231. L. Danicher, M. Lambla, and F. Leising, *Bull. Soc. Fr.* 9/10, 544 (1979).
232. C. G. Gebelein *Advances in Biomedical Polymers*, Springer, Dordrecht, Netherlands (1987).
233. H. Vink *Makromol. Chem.* 131, 133 (1970).
234. W. A. Thaler, *Macromolecules*, 16, 623 (1983).
235. J. C. Brosse, J. C. Soutif, and C. Piazzi, *Macromol. Chem.* 2109 (1979).
236. A. Iraqi and D. J. Cole-Mamilton, *J. Mater. Chem.* 2, 183 (1992).
237. M. Chini, P. Chrotti, and F. Machchia, *Tetrahedron Lett.*, 31, 4661 (1990).
238. T. Nishikubo and A. Kameyama, *Prog. Polym. Sci.* 18, 963 (1993).
239. A. E. Barron and R. N. Zuckermann, *Cur. Op. Chem. Bio.* 3, 681 (1999).
240. T. J. Deming, *Adv. Drug Del. Rev.* 54, 1145 (2002).
241. A. Taubert, A. Napoli, and W. Meier, *Cur. Opin. Chem. Bio.* 8, 598 (2004).
242. H. Schlaad and M. Antonietti, *Eur. Phys. J. E10*, 17 (2003).
243. B. Gallot, *Prog. Polym. Sci.* 21, 1035 (1996).
244. H.-A. Klok and S. Lecommandoux, *Adv. Mater.* 13, 1217 (2001).

245. T. J. Deming, *Nature* 390, 386 (1997).
246. H. R. Kricheldorf, *Alpha-Aminiacid*-N-*Carboxyanhydrides and Related Heterocycles*, Springer-Verlag, Berlin (1987).
247. D. Cunliffe, S. Pennadam, and C. Alexander, *Eur. Polym. J.* 40, 5 (2004).
248. H. S. Bazzi and H. F. Sleiman, *Macromolecules* 35, 9617 (2002).
249. C. Fraser and R. H. Grubbs, *Macromolecules* 28, 7248 (1995).
250. S. W. Seidel and T. J. Deming, *Macromolecules* 36, 969 (2003).
251. M. J. Boerakker, J. M. Hannink, P. H. H. Bomans, P. M. Frederik, R. J. M. Nolte, E. M. Meijer, and N. A. J. M. Sommerdijk, *Angew Chem Int. Ed. Engl.* 41, 4239 (2002).
252. K. Velonia, A. E. Rowan, and R. J. M. Nolte, *J. Am. Chem. Soc.* 124, 4224 (2002).
253. A. Taubert, A. Napoli and W. Meier, *Current Options in Chemical Biology*, 8, 598 (2004).
254. K. Loos and S. Munoz-Guerra, Microstructure and crystallization of rigid-coil comblike polymers and block copolymer, in *Supramoleculars Polymers*, ed A. Ciferri, Marcel Dekker, New York (2000).
255. H. Ringsdorf and P. Tschirner, *Mackromol. Chem.* 188, 1431 (1987).
256. W. Chen, M. Pyda, A. Habenschuss, J. D. Londono, and B. Wunderlich, *Polym. Adv. Technol.* 8, 747 (1997).
257. M. Beiner, K. Schroter, E. Hempel, S. Reissig, and E. Donth, *Macromolecules* 32, 6278 (1999).
258. G. Floudas and P. Stepanek, *Macromolecules* 31, 6951 (1998).
259. S. A. Greenberg and T. Alfrey, *J. Am. Chem. Soc.* 74, 6280, 1954 E. F. Jordan, Jr., D. W. Feldeisen, and A. N. Wrigley, *J. Polym Sci.*, Part A-1, 9, 1835 (1971).
260. J. M. Barrales-Rienda and J. M. Mazon-Arechederra, *Macromolecules* 20, 1637 (1987).
261. E. F. Jordan, Jr., B. Artymyshyn, A. Speca, and A. N. Wrigley, *J. Polym. Sci. Part A*, 9, 3349 (1971).
262. I. Alig, M. Jarek, and G. P. Hellmann, *Macromolecules* 31, 2245 (1998).
263. N. A. Plate, V. P. Shibaev, B. S. Petrukhin, and V. A. Kargin, *J. Polym. Sci.*, Part C 23, 37 (1968).
264. N. A. Plate and V. P. Shibaev, *J. Polym. Sci., Macromol. Rev.* 8, 117 (1974).
265. M. Beiner and H. Huth, *Nature Materials* 2(9), 595 (2003).
266. S. Thompson, N. Dutta, and N. Choudhury, *Proceedings, International Conference on Materials for Advanced Technology*, 7-12 Dec. 2003, Suntec Singapore International and Exhibition Centre, Singapore, p. 734 (2003).
267. S. Thompson, N. Dutta, and N. Choudhury, *Int. J. Nanosci.*, 6, 839 (2004).
268. A. Thierry and A. Skoulios, *Mol. Cryst. Liq. Cryst.* 41, 125 (1978).
269. J. Watanabe, H. Ono, I. Uematsu, and A. Abe, *Macromolecules*, 18, 2141 (1985).
270. J. Watanable and Y. Takashina, *Macromolecules*, 24, 3423 (1991).
271. D. Frenkel, *Liq. Cryst.* 5(1989)929. M. Hasimo, H. Nakano, H. Kimura, *J. Phys. Soc. Jpn.* 46, 1709 (1990).
272. J. Watanable, M. Gotoh, and T. Nagase, *Macromolecules* 20, 298 (1987).
273. J. Watanable and T. Nagase, *Polymer J.* 19, 781 (1987).
274. A. N Semenov and S. V Vasilenko, *Sov Phys JETP* 63, 70 (1986).
275. A. Rosler, G. W. M. Vandermeulen, and H.-A. Klok, *Advanced Drug Delivery Review*, 53, 95 (2001).
276. M. Iijima, Y. Nagasaki, T. Okada, M. Kato, and K. Kataoka, *Macromolecules*, 32, 1140 (1999).
277. K. Emoto, Y. Nagasaki, and K. Kataoka, *Langmuir*, 15, 5212 (1999).
278. K. B. Thurmond, T. Kowalewski, and K. L. Wooley, *J. Am. Chem. Soc.*, 118, 7239 (1996).
279. E. J. Lundin and Y.-S. Shon, *231st ACS National Meeting*, Atlanta, United States, March 26-30 (2006).
280. E. W. Meijer, *231st ACS National Meeting*, Atlanta, United States, March 26-30 (2006), PMSE-121.
281. S. Cammas, N. Nagasaki, and K. K. Kataoka, *Bioconjugate Chem.* 6, 226 (1995).
282. S. Cammas and K. K. Kataoka, *Macromol. Chem. Phys.* 196, 1899 (1995).

283. M. Yokoyama, M. Miyauchi, N. Yamada, T. Okano, Y. Sakurai, K. Kataoka, and S. Inoue, *J. Controlled Release*, 11, 269 (1990).

284. M. Yokoyama, T. Okano, Y. Sakurai, H. Ekimoto, C. Shibazaki, and K. Kataoka *Cancer Res.*, 51, 3229 (1991).

285. K. Kataoka, G. S. Kwon, M. Yokoyama, T. Okano, and Y. Sakuri, *Macromolecules*, 267 (192).

286. G. S. Kwon, M. Naito, M. Yokoyama, T. Okano, Y. Sakurai, and K. Kataoka, *Pharm. Res.*, 12, 192 (1995).

287. V. Y. Alkhov, E. Y. Moskaleva, E. V. Batrakova, and A. V. Kabanov, *Bioconjugate Chem.*, 7, 209 (1996).

288. D. W. Miler, E. V. Batrakova, and A. V. Kabanov, *Pharm. Res.* 16, 396 (1999).

289. Y. Bae, W-D. Jang, N. Nishyama, S. Fukushima, and K. Katakoa, *Mol. Biosyst.*, 1(3), 242 (2005).

290. D. E. Discher and A. Eisenberg, *Science* 297, 967 (2002).

291. R. Savic, L. Luo, A. Eisenberg, and D. Maysinger, *Science* 300, 615 (2003).

292. A. Choucair, P. Lim Soo, and A. Eisenberg, *Langmuir* 21, 9308 (2005).

293. X. Liu, J.-S. Kim, J. Wu, and A. Eisenberg, *Macromolecules* 38, 6749 (2005).

294. L. F. Zhang and A. Eisenberg, *Science* 268, 1728 (1995).

295. L. F. Zhang, K. Yu, and A. Eisenberg, *Science* 272, 1777 (1996).

296. S. Jain and F. Bates, *Science* 33, 56180 (2003).

297. B. M. Discher, H. Bermudez, D. A. Hammer, D. E. Discher, Y. Y. Won, and F. S. Bates, *J. Phys. Chem. B*, 106, 2848 (2002).

298. C. Nardin, T. Hirt, J. Leukel and W. Meier, *Langmuir*, 16, 1035 (2002).

299. M. Valentini, A. Napoli, N. Tirelli, and J. A. Hubbell, *Langmuir*, 19, 4852 (2003).

300. T. Hashimoto, K. Tsutsumi, and Y. Funaki, *Langmuir*, 13, 6869 (1997).

301. H.-C. Kim, X. Jia, C. M. Stafford, D. H. Kim, T. J. McCarthy, M. Tuominen, C. J. Hawker, and T. P. Russel, *Adv. Mater.* 13, 795 (2001).

302. P. E. Lippens and M. Larinoo, *Phys. Rev. B* 39, 935, 1989, 41, 6079 (1990).

303. N. A. Hill and K. B. Whaley, *J. Chem. Phys.* 99, 3707 (1993).

304. Y. Wang, N. Herron, and J. Caspar, *J. Mater. Sc. Eng.* B19, 61 (1993).

305. B. O. Dabbousi, C. B. Murray, M. F. Rubner, and M. G. Bawendi, *Chem. Mater.* 6, 216 (1994).

306. C. C. Cummins, R. R. Schrock, and R. E. Cohen, *Chem. Mater.* 4, 27 (1992).

307. J. Yue, V. Shankaran, R. E. Cohen, and R. R. Schrock, *J. Am. Chem. Soc.* 115, 4409 (1993).

308. V. Shankaran, J. Yue, R. E. Cohen, R. R. Schrock, and R. J. Silbey, *Chem. Mater.* 5, 1133 (1993).

309. V. Shankaran, C. C. Cumins, R. R. Schrock, R. E. Cohen, and R. J. Silbey *J. Am. Chem. Soc.* 112, 6858 (1990).

310. R. Tassoni and R. R. Schrock, *Chem. Mater.* 6, 744 (1994).

311. R. S. Kane, R. E. Cohen, and R. Silbey, *Chem. Matter.* 8, 1919 (1996).

312. A. Sellinger, P. M. Weiss, and L. Anh Nguyen, A. Yunfeng, A. Roger, W. George, C. J. Brinker, *Nature* 394, 256 (1998).

313. O. Grassmann, G. Muller, and P. Lobmann, *Chem. Matl.* 14, 4530 (2002).

314. J. N. Cha, G. D Stucky, D. E. Morse, and T. J. Deming, *Nature* 403, 289 (2000).

315. Y. Kakizawa and K. Kataoka, *Langmuir* 18(12), 4539 (2002).

316. J. N. Cha, G. D Stucky, D. E. Morse, and T. J. Deming, *Nature* 403, 289 (2000).

317. B. Zou, B. Ceyhan, U. Simon, and C. M. Niemeyer, *Adv. Materials* 17(13), 1643 (2005).

318. K. B. Blodgett, *J. Am. Chem. Soc.* 56, 495 (1934).

319. D. Chinn and J. Janata, *Thin solid Films* 252, 145 (1994).

320. J. R. Arthur, J. J. lePore, *J. Vac. Sci. Technol.*, 6, 545 (1969).

321. H. M. Manaseuit, *Appl. Phys. Lett.* 12, 156 (1968).

322. F. Caruso, ed., *Colloid and Colliod Assemblies*, Wiley-VCH, New York (2004).

323. S. W. Keler, H. N. Kim, T E. Mallouk., *J. Am. Chem. Soc.* 116, 8817 (1994).

324. Y. Gao, N. Roy Choudhury, in *Organic-Inorganic Hybrid Materials and Nanocomposites*, ed. H. S. Nalwa, Americal Scientific Publishers, USA pp. 271-293.

325. Y. Gao, N. Roy Choudhury, J. Matisons, and U. Schubert, *Chem. Mater.* 14(11), 4522 (2002).

326. U. Schubert, N. Husing, and A. Lorenz, *Chem. Mater.* 7, 2010 (1995).
327. U. Schubert, E. Arpac, W. Glaubitt, A. Helmerich, and C. Chau, *Chem. Mater.* 4, 291 (1992).
328. G. Trimmel, P. Fratzl, and U. Schubert, *Chem. Mater.* 12, 602 (2000).
329. U. Schubert, *J. Chem. Soc.*, Dalton Trans., (1996), 3343.
330. C. Zhang and R. M. Laine, *J. Am. Chem. Soc.* 122, 6979 (2000).
331. T. S. Haddad and J. D. Lichtenhan, *Macromolecules* 29, 7302 (1996).
332. R. O. R. Costa, W. L. Vasconcelos, R. Tamaki, and R. M. Laine, *Macromolecules* 34, 5398 (2001).
333. R. M. Laine. J. Choi, and I. Lee, *Adv. Mater.* 13(11), 800 (2001).
334. O. Toepfer, D. Neumann, N. Roy Choudhury, A. Whittaker, and J. Matisons, *Chem. Mater.* 17 (5), 1027 (2005).
335. L. Zheng, S. Hong, G. Cardoen, E. Burgaz, S. Gido, and E. Bryan Coughlin, *Macromolecules* 37, 8606 (2004).
336. J. Wu, Q. Ge, K. A. Burke, P. T. Mather, Materials Research Society Symposium Proceedings (2005), Volume Date 2004, p. 847.
337. M. Oaten and N. Roy Choudhury, *Macromolecules* 38, 6392 (2005).
338. M. Oaten, N. Roy Choudhury, N, Duttta and R, Knott, *Proceedings Int. Conf. Nanomaterials Synthesis Characterization and Application*, Nov 4-6, Kolkata, India, P192.
339. J. Wen and G. L. Wilkes, *Chem. Mater.*, 8, 1667, 340 (1996).
340. B. Novak, M. W. Ellsworth and C. Verrier, *Hybrid Organic-Inorganic Composites*, Eds. J. E. Mark, C. Y-C Lee and P. A. Bianconi, ACS Symp. Ser 585, ACS, USA (1995), p. 86.
341. J. E. Mark, C. Y. Jiang and G. L. Wilkes, *Macromolecules* 17, 2613 (1984).
342. N. Roy Choudhury, *J. Sol-Gel Sci. Technol.* 31, 37 (2004).
343. M. Moffitt and A. Eisenberg, *Chem. Mater.* 7, 1178 (1995).
344. Y. Gao, N. R. Choudhury, N. K. Dutta, J. G. Matisons, and L. Delmotte, M. *Read. Chem. Mater.* 13, 3644 (2001).
345. S. McInnes, M. Thissen, N. R. Choudhury, and N. H. Volcker, *Proc. SPIE*, V 6036 (2006).
346. N. N. Voevodin, V. N. Balbyshev, M. Khobaib, and M. S. Donley, *Prog. Org. Coat.*, 47, 416 (2003).
347. A. N. Kharmov, V. N. Balbyshev, N. N. Voevodin, and M. S. Donley, *Prog. Org. Coat*, 47, 207 (2003).
348. O. Didear, N. K. Dutta, and N., Roy Choudhury, *University of South Australia, Project Report* (2003).
349. F. Ben, B. Boury, R J. P. Corriu, and V. Le Strat *Chem. Matl.*, 12, 3249 (2000).
350. J. J. E. Moreau, L. Vellutini, M. W. C. Man, C. Bied, J-L. Bantignies, P. Dieudonne, and J.-L. Sauvajol, *J. Am. Chem. Soc.*, 123, 7957 (2001).
351. C. T. Kresge, M. E. Lenowicz, W. J. Roth, J. C. Vartuli, and J. S. Beck, *Nature* 359, 710 (1992).
352. P. D. Yang, D. Y. Zhao, D. I. Margolese. B. F. Chmelka, and G. D. Stucky, *Chem. Mater.*, 11, 2813 (1999).
353. S. Yang, Y. Horibe, C-H Chen, P. Mirau, T. Tatry, P. Evans, J. Grazul, E. and M. Durfesne, *Chem. Mater.*, 14, 5173 (2002).
354. H.-A. Klok, *J. Polym. Sci., Polym. Chem.* 43, 3 (2005).
355. M. A. Firestone, M. L. Shank, S. G. Sligarand, and P. W. Bohn, *J. Am. Chem. Soc.*, 118, 9033 (1996).
356. G. Yang, K. A. Woodhouse, and C. M. Yip, *JACS*, 124, 10648 (2002).
357. C. M. Elvin, A. G. Carr, M. G. Huson, J. M. Maxwell, R. D. Pearson, T. Vuocolo, N. E. Liyou, D. C. C. Wong, D. J. Merritt and N. E. Dixon, *Nature*, 437, 999 (2005).
358. N. K. Dutta, N. D. Tran, N. Roy Choudhury, A. J. Hill, C. Elvin, and R. Knott, *International Conference on Neutron Scattering*, ICNS2005, p132.
359. N. K. Dutta, N. D. Tran, N. Roy Choudhury, A. J. Hill, and C. Elvin, *Proc. 3rd Australian Korean Rheology Conference*, SG5, 17-20 July 2005, Cairns Convension Centre, Cairns.
360. C. Fouquey, J.-M. Lehn, and A.-M. Levelut, *Adv. Mater.* 2, 254 (1990).
361. J.-M. Lenh, *Science*, 295, 2400 (2002).

362. A. Ceferrri, *Supramolecular polymers*, Marcel Dekker Inc. New York (2000).
363. L. Brunsveld, B. J. B. Folmer, E. W. Meijer, and R. P. Sijbesma, *Chem. Rev.* 101, 4071 (2001).
364. K. Yamauchi, J. R. Lizotte, D. M. Hercules, M. J. Vergne, and T. E. Long, *J. Am. Chem. Soc.* 124, 8599 (2002).
365. *Supramolecular Assembly via Hydrogen Bonding*, ed. D. M. Mingos, Springer-Verlag, New York (1998).
366. F. H. Beijer, R. P. Sijbesma, H. Kooijmman, A. L. Spek, and E. W. Meijer, *J. Am. Chem. Soc.*, 120, 6761 (1998).
367. S. Kiyonaka, K. Sugisau, K. Shinkai, and I. Hamachi, *J. Am. Chem. Soc.*, 124, 10954 (2002).
368. T. Kato and J. M Frechet, *Macromolecules*, 22, 3818 (1989).
369. U. Kumar, T. Kato, J. M Frechet, *J. Am. Chem. Soc.* 11, 6630 (1992).
370. R. F. Lange and E. W. Meijer, *Macromolecules* 28, 782 (1995).
371. J. C. l Stendahl, L. Li, E. R. Zubarev, Y. Chen, and S. I. Stupp, *Adv Matl.*, 14 (21), 1540 (2002).
372. E. R. Zubarev, M. U. Pralle, E. D Sone, and S. I. Stupp, *J. Am. Chem. Soc.*, 123, 4105 (2001).
373. K. Chino and M. Ashiura, *Macromolecules*, 34, 9201 (2001).
374. C. Chen, S. A. Dai, H.-L. Chang, W. C. Su, CT-M. Wu, and R.-J. Jeng, *Polymer*, 46, 11849 (2005).
375. Y. Hasegawa, M. Miyauchi, Y. Takashima, H. Yamaguchi, and A. Harada, *Macromolecules*, 38, 3724 (2005).
376. G. Wenz and B. Keller, *Angew. Chem. Int. Ed. Engl.* 31, 197 (1992).
377. E. Lee, J. Jungseok, and K. Kim, *Angew. Chem. Int. Ed. Engl.* 39, 1433 (2000).
378. T. Ooya, K. Arizono, and N. Yui, *Polym. Adv. Technol.* 11, 642 (2000).
379. C. J. Pederson, *J. Am. Chem. Soc.* 89, 2495, 7017 (1967).
380. B. Dietrich, J.-M. Lehn, and J.-P. Sauvage, *Tetrahedron Lett.* 2886, 2889 (1969).
381. G. W. Gokel, D. M. Dishing, and C. J. Diamond, *J. Chem. Soc. Chem. Commun.* 1053 (1980).
382. D. Diamond and M. A. McKervey, *Chem. Soc.* Rev15 (1996).
383. F. Cadogan, K. Nolan, D. Diamond, Sensor applications in *Calixarenes 2001*, eds. Z. Asrafi, V. Bohmer, J. Harrowfield, J. Vicens, Kluwer Academy Press, Dordrecht, The Netherlands (2001).
384. H. W. Kroto, J. R. Heath, S. C. O. Brien, R. F. Curl and R. E. Smalley, *Nature* 318, 162 (1985).
385. Y. M. Wang, P. V. Kamat, and L. K. Patterson, *J. Phys. Chem.* 97, 8793 (1993).
386. F. Diederich and M. Gómez-López, *Chem. Soc. Rev.* 28, 263 (1999).
387. K. G. Thomas, Interface 8(1999)30. G. M. Guldi, C. Luo, C. Da Ros, M. Prato, E. Dietel, A. Hirsch, *Chem. Commun.*, 375, 388 (2000).
388. H. Imahori, K. Tamaki, D. M. Guldi, C. Luo, M. Fujitsuka, O. Ito, Y. Sakata, S. Fukuzumi, *J. Am. Chem. Soc.* 123, 6617, 389 (2001).
389. H. Imahori, T. Hasobe, H. Yamada, P. V. Kamat, S. Barazzouk, F. Fujitsuka, O. Ito, and S. Fukuzumi, *Chem. Lett.*, 784 (2001).
390. J. F. Nierengarten, *Chem. Euro J.* 6, 3667 (2000).
391. J. L. Segura, and N. Martin, *Chem. Soc. Rev.* 29, 13 (2000).
392. V. Biju, P. K. Sudeep, K. G. Thomas, M. V. George, S. Barazzouk and P. V. Kamat, *Langmuir*, 18, 1831 (2002).
393. P. V. Kamat, S. Barazzouk, K. G. Thomas, and S. Hotchandani, *J. Phys. Chem. B*, 104, 4014 (2000).
394. M. V. George, V. Biju, S. Barazzouk, K. G. Thomas, and P. V. Kamat, *Langmuir*, 17(2001)2930, 399.
395. P. V. Kamat, S. Barazzouk, S. Hotchandani, *Electrochemical aspect of C60-ferrocene cluster films, in fullerenes-2001*, P. V. Kamat, D. Guldi, K. Kadish, eds., *The Electrochemical Society*, Pennington, New Jersy (2001).
396. V. Biju, P. K. Sudeep, K. George Thomas, M. V. George, S. Barazzouk, and P. V. Kamat, *Langmuir*, 18, 1831 (2002).
397. I. Plinio and B. Giovanna, *Chem. Mater* 13, 3126 (2001).
398. J. E. Riggs and Y-P. Sun, *J. Chem. Phys.*, 112, 4221 (2000).

399. Y. Song, G. Fang, Y. Wang, S. Liu, C. Li, L. Song, Y. Zhu, Q. Hu, R. Signorini, M. Meneghetti, R. Bozio, M. Maggini, G. Scorrano, M. Prato, G. Brusatin, P. Innocenzi, and M. Guglielmi, *Carbon*, 38, 1653 (2000).

400. J. Wang, Y. J. Park, K. -B. Lee, H. Hong, and D. Davidov, *Phy. Rev.*, 66, 161201(R) (2002).

401. M. Jergel, V. Holý, M, M. Majková, S. Luby, and R. Senderák, *J. Appl. Cyrst.*, 30, 642 (1997).

402. J. M. Freitag and B. M. Clemens, *J. Appl. Phys.*, 89, 1101 (2001).

403. A. de Bernabé, M. J. Capitán, H. E. Fiswcher, and C. Priero, *J. Appl. Phys.*, 84, 1881 (1998).

404. G. Evmenenko M. E. van der Boom J. Kmetko, S. W. Dugan, T. J. Marks and P. Dutta *J. Chem. Phys.* 115, 6722 (2001).

405. J. Penfold, *Curr. Opin. Coll. Int. Sci.*, 7, 139 (2002).

406. R. Dalgliesh, *Curr. Opin. Coll. Int. Sci.*, 7, 244 (2002).

407. Ryong-Joon Roe, *Methods of X-Ray and Neutron Scattering in Polymer Science*, Oxford University Press, New York (2000).

408. Naba K. Dutta, S. Thompson, N. Roy Choudhury, R. Knott, *Physica. B: Condensed matter*. 773, 385-386, (2006).

409. A. Gunier and G. Fournet, *Small Angle Scattering of X-rays*, John Wiley, New York (1955).

410. H. Brumberger ed., *Modern Aspects of Small-Angle Scattering*, Kluwer, Dordrecht (1994).

411. S. Vitta, *Current Science*, 79, 61 (2000).

412. R. W. Richard and J. L. Thomas, *Macromolecules*, 16, 982 (1983).

413. P. Alexandridis and R. J. Spontak, *Curr. Opin. Coll. Int. Sci.* 4, 130 (1999).

414. R. Advincula, E. Aust, W. Meyer, and W. Knoll, *Langmuir*, 12, 3536 (1996).

415. J. H. Fendler, *Chem. Mat.*, 8, 1616, 1996. F. Caruso, K. Niikura, D. N. Furlong, Y. Okahata, *Langmuir*, 13, 3422 (1997).

416. S. F. Macha and P. A. Limbach, *Curr. Opin. Sol. St. & Mat. Sci.*, 6, 213 (2002).

417. P. O. Danis, D. E. Karr, Y. Xiong, and K. G. Owens, *Rapid Commun. Mass Spectrom.* 10, 82 (1996).

418. A. M. Belu, J. M. DeSimone, R. W. Linton, G. W. Lange, and R. M. Friedman, *J. Am. Soc. Mass Spectrum.* 7, 11 (1996).

419. R. R. Hensel, R. C. King, and K. G. Owens, *Rapid Commun. Mass Spectrom.*, 11, 1785 (1997).

420. B. Spengler, F. Lutzenkirchen, and R. Kaufmann, *Org. Mass Spectrom.*, 28, 1482 (1993).

421. J. K. Olthoff, I. A. Lys, and R. J. Cotter, *Rapid. Commun. Mass Spectrom.* 2, 17 (1988).

422. R. Brown and J. J. Lennon, *Anal. Chem.* 67, 1998 (1995).

423. *Tof-SIMS: Surface Analysis by Mass Spectroscopy*, eds J. C. Vickerman and D. Briggs, IM Publication, West Sussex (2001).

424. A. A. Galuska, *Surf. Int. Anal.*, 25, 1 (1998).

425. R. W. Linton, M. P. Mawn, A. M. Belu, J. M. DeSimone, M. O. Hunt Jr., Y. Z. Menceloglu, H. G. Cramer, and A. Benninghoven, *Sur. Int. Anal.*, 20, 991 (2004).

426. L. Kailas, J.-N. Audinot, H.-N. Migeon, and P. Bertrand, *Appl. Surf. Sci.*, 231, 289 (2004).

427. Z. L. Wang and Z. C. Kang, *Functional and Smart Materials: Structural Evolution and Structure Analysis*, Plenum Press, New York (1998).

428. Z. L. Wang, *Adv. Matl.*, 15, 18, 1497 (2003).

429. G. Binning, H. Rohrer, Ch. Gerber, and E. Weibel, *Phys. Rev. Lett.* 49, 57 (1982), Surfacr science, 131, L379 (1983).

430. R. Wiesendanger, ed., *Scanning probe microscopy: Analytical methods*, Springer-Verlag, Berlin (1998).

431. Scanning Probe Microscopy: Methods and Applications, R. Wiesendanger, Cambridge University Press, London 1994.

432. N. D. Tran, N. K. Dutta, and N. Roy Choudhury, *J. Polymer Sci. Part B: Polymer Physics*, 43, 1392 (2005).

433. J. K. H. Hörber and M. J. Miles, *Science*, 302, 1005 (2003).

434. *Nuclear Magnetic Resonance: Concepts and Methods*, D. Canet, Wiley, New York (1996).

435. *NMR spectroscopy: Basic Principle, Concepts, and Applications in Chemistry*, 2nd edition, H. Günther, Wiley, New York (1997).

436. Carbon 13 NMR Spectroscopy H.-O. Kalinowski, S. Berger, S. Braun, Wiley, New York (1991).

437. *Encyclopedia of Nuclear Magnetic Resonance*, 9 volume set, D. M. Grant, R. K. Harris, (ed. in chief), Wiley, New York (2003).

438. R. deGraaf, *In Vivo NMR Spectroscopy: Principles and Techniques*, Wiley, New York (1999).

439. P. Stilbs nd István Furó, *Current Opinion in Colloid & Interface Science*, 11, 3 (2006).

440. N. K. Dutta, N. Roy Choudhury, B. Haidar, A. Vidal, J. B. Donnet, L. Delmotte and J. M. Chezeau, *Polymer*, 35, 4923 (1994).

441. Naba K. Dutta, N. Roy Choudhury, B. Haider, A. Vidal, J. B. Donnet, L. Delmotte, and J. M. Chezeau, *Rubber Chemistry and Technology* 74, 260 (2001).

442. Daniel Topgaard, *Current Opinion in Colloid & Interface Science* 11, 7 (2006).

443. William S. Price, *Current Opinion in Colloid & Interface Science* 11, 19 (2006).

444. Paul T. Callaghan, *Current Opinion in Colloid & Interface Science* 11, 13 (2006).

445. R. A. Larson, *The Structure and Rheology of Complex Fluids*, Oxford University Press, N.Y., USA (1998).

446. Rheology: Principles, *Measurements*, and Applications, Wiley VCH, New York (1994).

447. F. A. Morrison, *Understanding Rheology*, Oxford University Press, London (2001).

448. R. G. Owen and T. N. Phillips, *Computational Rheology*, Imperial College Press, London (2002).

449. J. Verhas, *Thermodynamics and Rheology*, Springer-Verlag, Berlin (1997).

450. K. Almdal, K. A. Koppi, F. S. Bates, and K. Mortinsen, *Macromolecules*, 26, 485 (1993).

451. K. J. Hanley and T. P. Lodge, *J. Polym. Sci., Part B: Polym. Phys.*, 36, 3101 (1998).

452. N. Sakamoto, T. Hashimoto, C. D. Han, D. Kim, and N. Y. Vaidya, *Macromolecules*, 30, 5321 (1997).

453. Special feature issue on *Supramolecular Chemistry and Self-assembly*, PNAS, 99, (Apri 16 2002).

454. C. T. Kresge, M. E. Leonowicz, W. J. Roth, J. C. Varuli, and J. C. Beck, *Nature*, 359 (1992), 710.

455. Special issue on *Nanostructure and Functional Materials*, *Chemitry of Materials*, 13(10) (2001).

456. G. J. de A. A. Soler-Illia, E. L. Crepaldi, D. Grosso, and C. Sanchez, *Cur. Opin. Coll. Int. Sci.*, 8, 2003, 109.

457. M. Lazzari and M. A. López-Quintela, *Adv. Mater.*, 15, 1583 (2003).

## QUESTIONS

1. What is the principal difference between classical molecular chemistry and supramolecular chemistry? What are the origins of order and resulting molecular self-organization? Describe the major noncovalent interactions that dictate the ultimate organization of the molecules. What do we mean by "program chemistry"? Explain its advantages in respect to nanofabrication.

2. In a self-organization process the entropy of the overall system decreases; still it is a spontaneous process. Explain from the thermodynamic point of view with examples.

3. Block copolymers are the best model of amphililic behavior. Explain. Narrate different types of molecular architecture possible in block copolymers with examples. Describe the major advantages of using self-organized block copolymer in nanofabrication.

4. What are the most important parameters that influence the phase morphology in block copolymer? Show different types of classical phases accessible in mesophase-separated diblock copolymer. Explain with a typical phase diagram.

5. Explain with an example the statement that in supramolecular chemistry, "information and programmability, dynamics and reversibility, constitution and structural diversity, point towards the emergence of adaptative and evolutionary chemistry."

6. Describe a suitable method for the fabrication of ordered metal, metal oxide and semiconductor nanoparticles on solid surfaces with uniform and controllable size and shape and with a high spatial density.

7. Explain the terms *top-down approach* and *bottom-up approach* as used in the fields of nanomaterials and nanotechnology with examples.

8. How do you classify hybrids and nanocomposites in terms of interfacial interactions? Give examples of each type and describe their methods of preparations.

9. How do you prepare a 2D material on a substrate? How can you make a patterned surface? Using the concept of self-organization, how will you prepare nano-patterned surfaces and decorate them with nanoparticles?

10. Which methods would you choose to characterize and quantify the organic and inorganic contents, structure and morphology of a hybrid?

11. Define the roles of "interface" and "interphase" in composite and nanocomposite/ hybrid materials.

12. Is there any interphase in a polymeric nanocomposite? How does the interface influence the mechanical performance of a polymer nanocomposite?

13. What are the applications for which only polymer hybrids are competent and why?

14. Give a list of applications of 2-D nanomaterials. How are they important in our daily lives?

# 6

# Nanostructures: Sensor and Catalytic Properties

**B. Roldan Cuenya,**[1] **A. Kolmakov**[2]

[1]Department of Physics, University of Central Florida, Orlando, FL 32816
E-mail: roldan@physics.ucf.edu

[2]Department of Physics, SIUC, Carbondale, IL 62901.
E-mail: akolmakov@physics.siu.edu

## 1. INTRODUCTION

### 1.1. Overview

It is not a coincidence that very often the materials and sensitization methodologies used in chemical sensing are basically the same as the ones practiced in the field of catalysis. In both fields, a chemical reaction on the surface of the sensor or catalyst is the origin of the desired reaction products (in catalysis) or the electrical signals (in sensing). These fields were widely explored during the past few decades but both are currently experiencing a renaissance thanks to exciting new opportunities offered by nanotechnology.[1]

Catalysis encompasses the science of enhancing chemical reaction rates and selectivities based on the use of specific materials (catalysts) that are not consumed during the course of the process. The potential benefits of a new understanding and mastery of catalysis have broad societal impacts. About 80 percent of the processes in the chemical industry (including fuels, plastics and medicines manufacture) now depend on catalysts to work efficiently. This is why it is necessary to re-examine existing methods of synthesizing all chemical feedstocks in order to ensure that environmentally benign "green" processes are adopted.[2] More specifically, in heterogeneous catalysis the gaseous reactants utilize the surface of solid catalysts to form intermediate states that lower the energetic barrier to form products.

Under "real-world" industrial conditions, a number of poorly controlled factors complicate this simplistic picture, and the understanding of the fundamental

305

processes and steps in catalysis becomes challenging. That is why, for many years, the only way to develop new or improved catalysts was by empirical testing in so-called "trial-and error" experiments. During the last decade, this time- and cost-consuming effort has been complemented by more rational methods based on combinatorial approaches and a deeper fundamental understanding of catalysis at the molecular level. The application of model surface science methods helped in opening the door to mastering the art of heterogeneous catalysis. However, the transformation of this field to a truly consistent science based on the vast amount of empirical observations available has yet to be achieved.

The advent of nanoscience and nanotechnology offered the possibility of creating new model catalysts with precisely controlled composition, structure and morphology and with previously inaccessible chemical properties.[3]

Among the prospective candidates for the next generation of catalysts are semiconductor and metal nanoparticles with dimensions less than 10 nm. These nanoparticles possess unique chemical, optical, electrical and magnetic properties that deviate drastically from their macroscopic counterparts. Moreover, these properties can be effectively modified by tuning their size. With decreasing particle size, the optical absorption edge shifts towards the blue region, the melting point decreases, metal/non-metal transitions occur and the surface reactivity changes significantly.[4-12] In particular, the appearance of two-size ranges in the reactivity of supported clusters has been reported.[13-17] For large particles (more than a few hundred atoms), the kinetics of a surface chemical reaction are determined by geometric shell-filling effects which manifest themselves at steps and kinks. For smaller clusters (1-30 atoms), atom-by-atom size-dependent catalytic properties appear, where the individual electronic structure of each cluster size plays a decisive role.

In the last decade, a great variety of highly dispersed metallic nanoparticles have been shown to be active and selective for many industrially important reactions including low-temperature CO oxidation and water shift reactions, NO reduction with hydrocarbons, partial hydrogenation and oxidation of hydrocarbons, the selective oxidation of propylene to propylene oxide, and the methanol oxidation and production reactions.[13,18-41] All of these are processes with broad technological and environmental application, including fuel-cell technology, air and automobile outgas purification and cleaning of ground soil and water.

Despite multiple research efforts, fundamental knowledge of key features that influence catalysis is still lacking. Included in these features are the nanoparticle's ability to modify its structure in the course of a chemical reaction, adsorbate mobility, selective active site blocking, catalyst surface poisoning, or promoter effects. Nevertheless, it is generally accepted that a nearly molecular control over the cluster size, structure, location and composition is required to systematically design highly active and selective systems.[7,8,13,42-46] For example, nanoparticle interactions, tunable by changing the interparticle distance, have been recently shown to affect the stability and lifetime of the catalysts.[47]

The understanding and technological evolution of gas sensors closely parallels the developments in catalysis described above. In the last few decades, novel solid-state gas sensors based on chemically induced modifications of the electrical properties of thin films have been designed. These sensors have had a profound impact on many fields: semiconductor processing, medical diagnosis, environmental sensing, and personal safety and security, with economic impact in agriculture, medicine and the automotive and aerospace industries.

Thanks to the field of catalysis, great progress has been achieved in our atomic-level understanding of metal- and semiconductor-adsorbate interactions. However, the complexity of the processes involved in manufacturing and operating sensors with predetermined characteristics under realistic conditions is far of being resolved. In addition, the further miniaturization of sensor chips requires active elements with higher levels of sensitivity and selectivity. Furthermore, the decrease in the active area poses the challenges of acquiring a sufficient amount of analyte and overcoming possible cross-talk between neighboring individual sensing elements. These limitations, together with the growing demand for ultra low power consumption and highly selective materials sensitive to more complex molecules, shows that further progress in gas sensing is required. Modern nanotechnology with its new material development, innovation in structure, architecture and sensor design principles is providing great breakthroughs in this field.

In this chapter, we intend to give an overview of several new trends in the field of chemical sensing based on a nanotechnology approach.

## 1.2. Why are Nanostructures Important for Gas Sensing and Catalysis? (Structure-Sensitivity Relationship)

One of the major goals in chemo- (bio-) sensing is the building of new active elements that will respond to the analyte selectively and sensitively. Modern nanotechnology offers a number of prospective sensing elements including nanometer-thick 2D metal/oxide/semiconductor (MOS) diodes, quasi-1D semiconducting metal oxides, and metal nanoparticles supported on nanostructured metal or semiconducting oxides. In particular, the quasi-1D metal oxide sensing elements inherit the sensitive responsiveness to surface redox reactions of the traditional nanostructured thin films, and yet posses simple electron transport mechanism of the single crystal that make these structures potential next-generation chemical sensors (see Fig. 1). Some of the unique properties of nanostructured materials that are of advantage in the field of catalysis and chemical sensing are:

- High surface-to-volume ratio of these structures implies an exposure of a large fraction of atoms (about 30% for 10 nm structure) to the ambient environment. For nano-porous, tubular and ribbon-like nanostructures this ratio is even higher.
- At the 10 nm scale and below, the mean free paths of the electronic excitations (electrons, holes and excitons) become comparable with the effective radius of the nanostructure, thus favoring the delivery of these active "reactants" toward the surface over their recombination. Activating surface

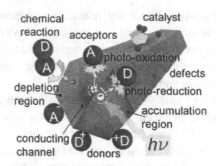

**FIGURE 1.** The basic processes relevant to gas sensing with quasi-1D $n$-type metal oxide. The adsorption of acceptor (A) or donor (D) molecules leads correspondingly to depletion or accumulation of the electrons in the nanostructure, thus decreasing or enhancing the conductance through the chemiresistor. Due to the small diameter of the nanostructure, photo-generated electrons and holes are able to reach the surface and participate in surface redox reactions before recombination. The chemical action of catalyst particles and surface defects may have a pronounced effect on the charge transport through the nanostructure due to greater ratio between the size of local depletion (accumulation) and spillover zones versus nanostructure diameter.

atoms and imperfections and localizing at them, these species can drastically enhance surface oxidation-reduction reactions with target molecules from the gas phase. The latter is particularly important for the chemisensors based on photocatalytic effects and can also lead to significant reduction of the operation temperature of the sensors.[48,49] A detailed description of chemical sensors based on the ballistic transport of hot electrons/holes through nanometer-thick metal films will be given in a later section.

- 1D nanostructures offer a well-defined (structurally and compositionally) conducting channel that is at least microns long (and therefore easily addressable) and that is a responsive "antenna" for the surface and near surface events.

- For moderately doped oxides operating within the practical temperature range, the width of the space charge layer (SPL) $W_D$ falls into the 10-100 nm range. When the effective diameter of the nanostructure $D$ becomes comparable to (or smaller than) this parameter, the tiny changes in the surface's charge state due to the presence of chemical or biological agents can result in the depletion (or accumulation) of electrons. This happens not only at the near surface, but in the entire volume of the nanostructure, with a concomitant change of the conductance. Combined with the large surface-to-volume ratio, this effective transduction provides the basis for superior chemical sensor function of the *individual* oxide nanowire device.[49-53]

- Different from granular thin films, the quasi 1D and 0D nanostructured sensors can be made from a single crystal with well-defined facets without degrading sensitivity. The latter allows for achieving better stability, reproducibility, predictability and modeling. In addition, recent progress in nanotechnology offers a new level of control over the morphology and composition of quasi-1D nanostructures (see examples of the nanostructures in Fig. 1). In fact, the individual functional nanostructured building block can be pre-engineered at the nanoscale to better serve as a gas-sensing element.

## 1.3. The Impact on the Fundamental Science

Understanding the mechanisms of the transduction of surface phenomena to electronic signals (and vice versa) in nanostructures is a frontier of the fundamental surface science of nano-objects. It is crucial for their applications as promising active elements of chemical and biological sensors, as well as photovoltaic and nanoelectronic devices (see Rfs. 54-56 and sources therein). This is particularly true when their size approaches the electron confinement regime (see as an example Ref. 57).

In spite of the fact that the influence of the confinement effects on surface reactivity of metal clusters (0D), catalyst nanoparticles[4,58] and thin films[59] (2D) is an active area of experimental studies, the manifestation of electron confinement effects on the chemical properties of 1D nanostructures (and semiconducting oxides in particular) still remains a largely unexplored field of research.♣ When the electrons and holes become confined inside nanostructures with diameters less than their de Bröglie wavelength (typically 1-10 nm), the drastic changes in the electronic structure of such objects will apparently induce changes in their surface reactivity. These changes involve: (i) blue shift, narrowing of the electronic bands and concomitant increase of the density of states that will influence the strength and probability of the formation of chemical bonds with target molecules; (ii) an increase of the redox potentials of photo-generated electrons and holes that will result in enhancement of photoredox and photocatalytic properties of the pristine nanostructures; (iii) change in the kinetics of the surface reactions due to reduced electron-phonon coupling and preferable interaction of electrons and holes with the surface species rather than their recombination.

## 2. PHENOMENA AT NANOSCALED METAL AND SEMICONDUCTING OXIDE SURFACES RELEVANT TO GAS SENSING AND CATALYSIS

## 2.1. Pristine Oxide Surfaces: Physisorption vs. Chemisorption

The interaction of molecules with the pristine surfaces of homogeneous metal and metal oxide nanostructures can be in the form of *physisorption*, which does not involve charge transfer, or *chemisorption*, which proceeds with significant alteration

---

♣ The extensive studies on this issue conducted on carbon nanotubes are beyond the scope of this review.

of the electron densities of both the target molecule and the nanostructure's surface.

In principle, the physisorption of the target molecules is able to contribute to the conductance change inside a metal oxide nanostructure via increase of surface scattering and some degree of electrostatic gating when highly polarizable molecules are physisorbed. These effects may have experimental significance predominately for very thin (1-10 nm) metal oxide nanostructures.

Despite a much stronger disturbance of the electronic structure, the chemisorption itself does not necessarily result in strong responsiveness of the chemiresistor. For an appreciable conductance change it is important that the chemisorption act involves the electrons (holes) from the conduction (valence) band. Semiconducting metal oxides having oxygen vacancies as electron donors, and surface oxygen vacancies as adsorption-dissociation sites for the target molecules, inherently possess these useful properties. Essentially, all experiments carried out to date on metal oxide nanowires (or other nanostructures) indicate that the role of oxygen vacancies dominate the electronic properties of pristine metal oxide nanowires along much the same lines as they do in bulk systems. For nanostructures with diameters on the order of 100 nm, one can safely adopt the principles of chemical sensitivity developed over many years to explain the function of polycrystalline and thin film metal oxide gas sensors.[60-68]

As an example, let us consider the intereaction of $SnO_2$ nanowires with oxygen (an electron acceptor) and CO (an electron donor in the surface redox process) as a model oxide semiconductor system. Under normal conditions, the surface of stoichiometric tin oxide is inert. However, it can be easily reduced as a result of a moderate annealing in vacuum, or under an inert or reducing atmosphere. These treatments cause a fraction of the surface oxygen atoms to desorb, leaving oxygen vacancy sites behind (Fig. 2(b)). The latter can also be realized at low tempratures during exposure of the oxide to UV light. These vacancy sites form donor states lying just below the conduction band edge (Fig. 2(c)). The majority of these energy states are shallow enough that most are ionized even at low temperatures, thus defining the material to be an $n$-type semiconductor.

At any given temperature, the nanowire's transport properties (i.e., conductance $G$) are defined by the concentrations of ionized vacancy states:

$$G = \pi R2e\mu n/L \qquad (1)$$

Here, $R$, $L$ are the radius and length of the nanostructure, respectively; $e$, $\mu$, $n$, are the charge, mobility, and density of available free electrons in the nanostructure, respectively.

The conductance of $SnO_2$ changes during the exposure to oxygen. During oxygen adsorption, a fraction of the surface vacancies becomes repopulated, resulting in ionized (ionosorbed) surface oxygen of the general form $O_{\beta S}^{-\alpha}$. Under steady state conditions, the surface oxygen coverage, $\theta$, depends on the oxygen partial pressure and the system temperature through the temperature-dependent adsorption/desorption rate constants, $k_{ads/des}$. It is important to stress that the

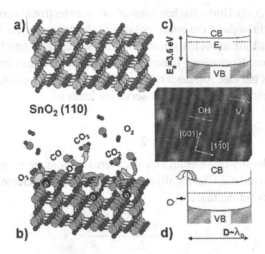

**FIGURE 2.** The model of the (a) stoichiometric rutile $SnO_2$ (110) surface; (b) partially reduced $SnO_2$ with missing bridging oxygens. Molecular oxygen (left cartoon) binds to the vacancy sites, dissociates and captures an electron as an electron acceptor. CO molecules react with preadsorbed oxygens (central cartoon). $CO_2$ is formed and electron are released back to the nanowire (c) Band diagram of nearly stoichiometric $SnO_2$. The vacancies present in the bulk of slightly reduced $SnO_2$ form shallow donor levels and define its $n$-type behavior. (d) During oxygen ionosorption, band bending occurs. When the Debye length is comparable to the radius of the nanowire, the entire structure becomes depleted. In the center: an STM image of surface vacancies (bright spots) and hydroxyl groups at similar $TiO_2$ (110) rutile.

oxygen coverage is dependent on the concentration of free electrons, $n$, and unoccupied chemisorption (vacancy) sites, $N_s$.

$$\frac{\beta}{2} O_2^{gas} + \alpha \cdot e^- + N_S \Leftrightarrow O_{\beta S}^{-\alpha} \tag{2}$$

$$k_{ads} \cdot N_S \cdot n \cdot p_{O_2}^{\beta/2} = k_{des} \cdot \theta \tag{2a}$$

(where $\alpha, \beta = 1, 2$ accounts for the charge and molecular or atomic nature of the chemisorbed oxygen[66]).

## 2.2. Band Bending and Charge Depletion

During the formation, the ionosorbed oxygens capture electrons from the bulk and create a $\sim 10-100$ nm thick SCL electron-deficient surface layer, corresponding approximately to the Debye length of $SnO_2$. For the macroscopic semiconductor oxide and mesoscopic wires, the latter results in band bending near the surface region. Thus, cylindrical nanowires loose a part of their cross-section for electron conduction. For the nanowires with diameters less than 50 nm, SCL embraces the entire nanowire. Under these so-called "flat band" conditions the relative position of the Fermi level shifts away from the conduction band edge during oxygen inosorption (Fig. 2(d)). Under flat band conditions the electrons become distributed homogeneously throughout the entire volume of the nanowire. As a result, due to facile access of bulk electrons to the nanostructure surface and their limited number ($\sim 10^5 - 10^6$) in ca. 10 mm long nanowire), any surface charge transfer processes drastically affects the electron density throughout the entire nanostructure. Under the flat band conditions, the charge conservation leads to:

$$N_s \cdot \theta = \frac{R}{2} \cdot (n - n_m) \tag{3}$$

where $n_m$ is the density of itinerant electrons remaining in the nanostructure after exposure to the adsorbate. The localizaton of $\Delta n = 2N_s \theta / R$ electrons results in a significant drop in conductance:

$$\Delta G = \frac{\pi R^2 e \mu}{L} \cdot \frac{2N_s \theta}{R} \tag{4}$$

and concomitant increase in the activation energy.[52]

Now let us consider the admission of a reducing gas such as CO. During the exposure to CO molecules, an oxidation surface reaction takes place with the ionosorbed oxygen being an oxidant:

$$\beta \cdot CO^{gas} + O_{\beta S}^{-\alpha} \rightarrow \beta \cdot CO_2^{gas} + \alpha \cdot e^- \tag{5}$$

The net results of the CO exposure is the recovery of the adsorption (defect) sites and the redonation of electrons back to the $SnO_2$ (Fig. 2(b)). This classical reaction scheme can be checked experimentally since, as was shown in Ref.[66], under flat-band conditions a monotonic increase in the electron concentration ($\Delta n$) would be:

$$\Delta n_{CO} \sim p_{CO}^{\frac{\beta}{\alpha+1}} \tag{6}$$

and therefore the change in nanowire's conductivity ($\Delta G$) will be:

$$\Delta G_{CO} \sim e \cdot \Delta n_{CO}(T) \cdot \mu(T) \tag{7}$$

thus proportional to a 0.5 power law with increasing CO partial pressure, assuming that $O^-(\alpha, \beta = 1)$ is the dominant reactive surface species (see Ref. 52). This simple reaction model accounts for the operation of tin-oxide nanowire sensors under ideal ambience consisting of dry oxygen and a simple combustible gas. In a "real-world" environment the reactive molecules such as water can react with adsorption sites, thus modifying the possible reaction's pathways.[69]

It is important to note that the simple picture described above is mainly applicable to semiconducting mesoscopic oxides, and has a limited applicability to nanowires with diameters on the order of 10 nm. For such small nanostructures, the distance between dopants becomes comparable to the radius of the nanostructures, and therefore the continuum model of the SCL loses its applicablility.[70]

## 2.3. Chemisorption and Magnetization

In recent years, the novel properties of magnetic nanostructures have been intensively investigated.[71-73] The main interest arises from the potential technological applications of these nano-designed materials as magnetic recording media, magnetic sensors and biocompatible magnetic nanodevices.

A significant breakthrough for the application of the magnetic properties of nanostructured materials in the field of sensing came with the discovery of the *giant magnetoresistance* (GMR) effect.[74] GMR is a very large change in the electrical resistance of ferromagnetic/paramagnetic (FM/PM) multilayer structures that is observed when the relative orientations of the magnetic moments in alternate FM layers are modified by an external magnetic field. Within ten years of its discovery, GMR was adapted to a new generation of high-performance magnetic read-heads in commercial hard drives. This event also initiated the spintronics era, dominated by solid-state sensors that conjugate electronic and magnetic properties in a single hybrid device.[75]

The interest in the magnetic behavior of low-dimensional systems has been expanded now to one-dimensional (1D) and nearly zero-dimensional (0D) structures. Magnetic particles are well known for their use in recording devices. For an application involving information storage, the nature, size, and shape of the particles are the relevant parameters. At present, the major challenges facing the magnetic sensing industry (GMR, spin valves) are related to the necessity of reducing the storage unit's size in the recording media. This progressive process is approaching the physical frontier imposed by the superparamagnetic limit.[76] Therefore, further increasing the storage density will require new technological approaches, for example, the use of specific tailored nanostructured materials as magnetic media.

The latest developments in biocompatible functionalized magnetic nanoparticles also show considerable promise for both enhanced and novel applications in the biomedical and diagnostic fields, ranging from targeted drug delivery to contrast enhancement in magnetic resonance imaging (MRI).[78] Magnetic nanoparticles loaded with drugs can be attracted to specific body areas by applying a magnetic field. Concentrating the particles in areas requiring treatment would enhance the therapeutic benefits while reducing side effects on other areas of the body. In addition, magnetic nanoparticles are currently routinely used in MRI medical applications to enhance the contrast between biological structures and for early tumor detection. Furthermore, magnetic nanoparticles can replace the radioactive materials that are currently used as drug tracers. Physicians can then determine the location of drugs by measuring magnetic instead of radioactivity variations, eliminating potential human harm from radiation.

The expected applications of nanomaterials in the magnetic recording industry require high-yield syntheses of narrow-size/shape-distributed nanoparticle arrays showing controlled magnetic properties. A variety of chemical synthesis method have been developed in the last decade for that purpose.[79-84] Unfortunately, the highly monodispersed magnetic nanoparticles produced by chemical procedures usually display low magnetization. This effect is generally attributed to the presence of residues of the precursor or reduction agent. The interaction between the magnetic nanoparticle surface and the matrix or coordination agents (ligands) can also introduce perturbation on the electronic properties and thus on the surface magnetism.[81] Recent results have demonstrated that chemical interactions between adsorbates and magnetic particles are expected to play an increasingly important role in determining the main properties of the magnetic devices as their dimensions are decreased to the nanometer-size range. Chemically induced changes in the intrinsic magnetic moments of interfacial atoms, surface magnetic anisotropy and structure have been reported.[81,86,87] It has been often observed that chemisorption on transition metal clusters is accompanied by a magnetization reduction due to "spin quenching" of the surface atoms. This behavior has been generally interpreted by considering that an absorbed atom or molecule bound to a surface metal atom will entirely quench the magnetic moment of that atom while leaving the moments of neighboring metal atoms unaffected.[88]

As an example, carbon monoxide chemisorption on nickel nanoparticles (2-12 nm in diameter) was found to be accompanied by a decrease in their magnetization of up to 1.1 $\mu_B$ per absorbed CO. This result was independent of the particle size, and corresponded to a "magnetic quenching" of ~2 surface atoms.[89] A similar outcome was observed for a molecular beam of Ni nanoclusters composed of 8-18 atoms per cluster (Fig. 3).[87]

First-principles calculations of the structural, electronic and magnetic properties of clean and CO-adsorbed Ni(110) surfaces also revealed that the spin-structure

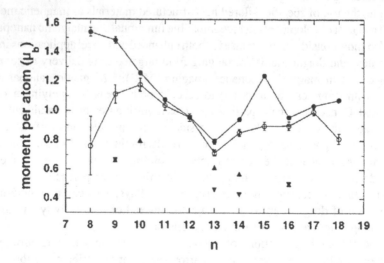

**FIGURE 3.** Experimentally determined magnetic moments per atom of $Ni_n$ (full circles) and $Ni_n$ CO (open circles). The moments per atom calculated by Raatz and Salahub for fcc model $Ni_n$ (upright triangles) and $Ni_n$ CO (downward triangles) are also shown. [Reused with permission from Mark B. Knickelbein, Journal of Chemical Physics, 116, 9703 (2002). Copyright 2002 by the American Institute of Physics].

of the surface depends on the adsorption geometry in a highly localized fashion. The enhanced magnetic moments of the top-layer atoms are attenuated after CO adsorption. Strikingly, the adsorbate-induced demagnetization is limited primarily to those surface atoms directly bonded to the molecule. The adsorbate itself was found to be only weakly magnetized in the opposite sense to the surface majority spin.[90] The influence of CO adsorption on small Co particles deposited on sapphire single crystal has also been recently investigated in situ by ferromagnetic resonance (FMR). The results confirmed a reduction of the magnetic moment of surface CO atoms due to the CO adsorption.[91]

The role of ammonia and hydrogen chemisorption in surface magnetism was theoretically investigated using Co(0001) and small Co clusters as model systems. At the Co(0001) surface, the atomic magnetization is predicted to diminish locally by 0.26 $\mu_B$ due to the adsorption of an isolated hydrogen atom; for $Co_{13}$ clusters, the change is smaller (0.1 $\mu_B$) but less localized. At H(1 × 1) - Co(0001), the magnetization of surface Co atoms drops to 0.88 $\mu_B$. The magnetic moment induced at H is very small and couples antiferromagnetically to Co atoms. Ammonia adsorption is found to locally reduce the Co atom magnetization by $\sim$0.1 $\mu_B$ or less.[92]

The superparamagnetic behavior of small iron particles (1.5 nm) with about half of their atoms at the surface can be reversibly changed by adsorption and desorption of hydrogen below the superparamagnetic transition temperature. Such change is ascribed to a modification of the crystalline shape and was not observed for larger iron particles (8 nm).[93]

Chen et al.[77] investigated chemically induced changes on the spin anisotropy due to oxygen adsorption on ultrathin Fe films (1-3 monolayers thick) deposited on Ag(100). Their experiments showed that small doses of oxygen can produce striking changes in the spin anisotropy of bilayer magnetic films. In particular, the preferred spin orientation changes from perpendicular to the surface to in-plane (Fig. 4). While some of the observed effects can be explained by assuming oxygen-induced quenching and by taking into account the structural changes of the film that accompany oxygen adsorption, the primary effect (oxygen-induced change of preferred spin direction) appears to be produced by electronic effects that govern the surface-anisotropy parameter.[86]

As can be inferred from this previous section, a promising future in the field of sensing can be envisioned for chemically modified magnetic nanostructures.

## 3. NANOSTRUCTURED GAS SENSORS: SOME EXAMPLES OF DETECTION PRINCIPLES

### 3.1. Two-dimensional Nanoscaled Metal/Oxide/Semiconductor Diodes

Selective and nonselective solid-state chemical sensors have been developed for a variety of applications in process control, environmental monitoring, and hazardous substance detection. Most devices rely on an indirect detection mechanism

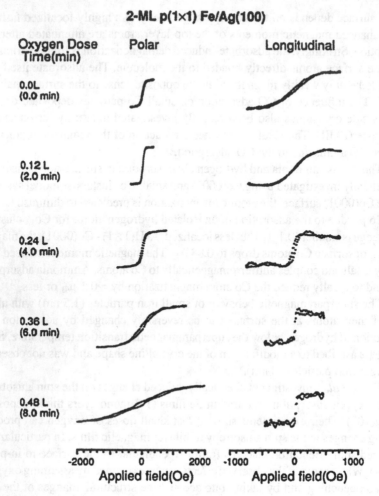

**FIGURE 4.** Left panel: polar configuration (applied magnetic field perpendicular to surface) magneto-optical Kerr- effect hysteresis loops for 2-ML p(1 × 1)Fe on Ag(100) as a function of oxygen dose in Langmuirs (1L = 1 × 10-6 Torr sec) and time in minutes. Right panel: corresponding results for longitudinal configuration (applied magnetic field parallel to surface and in the plane of incident light). All hysteresis loops have been normalized to the same height to emphasize shape differences in the loops resulting from oxygen adsorption. [Reprinted figure with permission from J. Chen, Physical Review B, 45, 3636 (1992). Copyright 1992 by the American Physical Society].

whereby their electronic or electro-optical properties are altered by the substance of interest, allowing for a measurable response. Sensors based on analyte-sensor interactions that modify capacitance, conductivity, magnetization, and refractive index operate in this manner.[94-100]

Metal-oxide semiconductor (MOS) structures have been intensively investigated in the last decades and have been used as gas-sensitive devices.[98,101-108]

Traditional MOS sensors are based on chemically induced changes in the electric field across the oxide. These changes modify the space charge layer in the semiconductor as well as the electrical properties of the device. In the praxis, the flat-band voltage, capacitance-voltage characteristics, or photocurrents are monitored as a function of the gas exposure. In most cases, accurate information on how the gas species interact with the metal gate is lacking, and a generalized model that can fit experimental observations on different metal/oxide systems has not been found.

Some examples of the most commonly reported detection mechanisms are given in the following paragraph. Hydrogen detection by Pd-MOS diodes is believed to be due to the polarization of the metal-oxide interface by hydrogen atoms that penetrate the Pd film after being created by breaking oxygen bonds at the catalytic metal.[104] A device temperature well above 400 K has been found necessary for this mechanism. The penetration of gas particles into metal films through pores, voids, or grain boundaries is also held responsible for the sensitivity of MOS devices with noble metal gates to reactive molecular species such as $NO_2$.[98,102,105] A capacitive coupling of adsorbed molecules on a discontinuous gate metal film to the space charge layer was proposed to understand the gas-sensitive field effect in which ammonia detection by $Pt-SiO_2-Si$ diodes is based.[109]

### 3.1.1. Chemically Induced Electronic Excitations: Chemicurrent Sensors

Metal-semiconductor (MS) Schottky diodes with nanometer-thick metal films can also be used as active sensors for atomic and molecular gas phase species.[110-121] On these devices, gas-surface interactions are monitored, not indirectly by a change in the device properties (passive sensing), but by direct detection of charge carriers produced from the gas-metal interaction. When an adsorbate interacts with the surface, the adsorption energy may appear as an energetic electron-hole (e-h) pair generated in the metal surface. The excited electron can travel ballistically through the thin metal film and traverse the Schottky barrier if the kinetic energy of the electron is larger than the barrier height ($\Phi_B$) and if the film thickness is comparable to the ballistic mean free path (see Fig. 5).[122] Once inside the semiconductor, the electron is detected as a *chemicurrent*, analogous to the photocurrent in a photodiode. Chemicurrent detection is very effective at temperatures between 120 and 200 K.

Hot electrons have interesting and unusual reactivity and transport properties which may be taken advantage of in Schottky-barrier MS (metal/semiconductor) and MOS diode sensors. Using ultrathin metal films on MS and MOS device structures, Roldan et al.[122-124] investigated the "chemielectronic" phenomena associated with a variety of molecular and atomic interactions with transition metal surfaces (Ag, Au, Pt, Pd). Distinct differences in the mechanism of signal production between highly energetic atomic species (H, O) and more weakly interacting species such as xenon, carbon dioxide and hydrocarbons were found. In this section, the basis for the chemical sensitivity of these diode structures will be addressed,

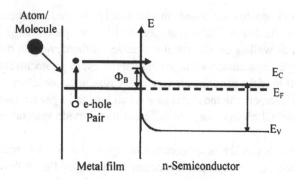

FIGURE 5. Schematic plot describing the process of electron-hole pair excitation upon chemical adsorption and subsequent generation of a ballistic electron transported over the Schottky barrier of an *n*-type semiconductor junction.

emphasizing the importance of oxide interfacial states in the detection of electronic signals from weakly interacting gases. Experimental evidence linking the surface chemistry to the electronic processes giving rise to signals in MS and MOS diode sensors will be presented.

Chemically induced electrical signals were measured in a UHV system. Mixed beams of atomic and molecular gas species were produced and delivered by a three-stage differentially pumped atomic beam source.[118] Hydrogen and oxygen molecules were dissociated within a microwave plasma cavity. Since the plasma emits intense ultraviolet light with a photon energy of 10.2 eV and the Schottky diodes are sensitive photodetectors, the plasma source was specially designed to extract any photons from the gas beam. The room temperature thermalized beams had a flux that varied between 2 and $5 \times 10^{12}$ atoms cm$^{-2}$ s$^{-1}$. To investigate the diode response to gas exposure, the active sensors were positioned in line to a shuttered collimated gas beam modulated at 4 Hz (50% duty cycle). The reaction-induced chemicurrents were detected by the diode connected in series with a current preamplifier. The preamplifier output was connected either to a lock-in or an A-D convertor (oscilloscope). All measurements were carried out at low temperature (125 K) to minimize the noise caused by the thermal drift. Other details of the experimental set-up are described elsewhere.[111]

Figure 6 shows the demodulated lock-in amplifier output from a 6 nm thick Pd-MS diode during exposure to a highly energetic atomic oxygen beam. The metal film was deposited in UHV on an HF-terminated *n*-Si(111) wafer.

The initial signal peak of ~930 pA corresponds to ~$5.6 \times 10^{-3}$ electrons detected per incident O atom on the clean polycrystalline Pd surface. The time-decay profile is consistent with formation of an oxide coverage assuming a surface site density of $1.5 \times 10^{15}$ cm$^{-2}$. Similar behavior has been observed with H atoms, NO, and NO$_2$.[121] The phase ($\phi$) of the lock-in signal, $-83°$, is essentially identical to the photocurrent phase corrected for the difference in beam transit times (~0.5 ms), indicating a reverse bias current analogous to inverse electron emission

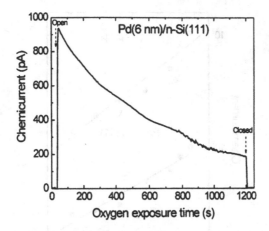

**FIGURE 6.** Lock-in detected chemicurrent signal from a Pd(6 nm)/n-Si(111) diode vs. exposure time to atomic oxygen. The measurement temperature was 125 K. The dashed arrows signalize when the gate valve to the beam chamber was opened and closed.[122] [Copyright ©2004 From Dekker Encyclopedia of Nanoscience and Nanotechnology by B. Roldan Cuenya/J.A. Schwartz, C.I. Contescu, K. Putyera. Reproduced by permission of Routledge/Taylor & Francis Group, LLC.)

(IEE) generated by a photocurrent. The MS devices are insensitive to hydrocarbons, Xe, $CO_2$ or other low adsorption energy species.

The energetic charge carriers have a ballistic mean-free path of tens of nanometers. Thus for low-dimensional metal structures, the "hot" charge carriers persist at energies well above the Fermi level. In agreement with the concept of mean free path of ballistic charge carriers in metals, the chemicurrents are attenuated exponentially with increasing metal film thickness, d. Figure 3 shows this thickness dependence for Pd-MS diode exposed to O and H atoms. The solid lines represent a single exponential fit [$\sim$exp(-d/$\lambda$)] to the experimental data. The attenuation length constants are $\lambda \approx 13$ nm upon O exposure and $\lambda \approx 17$ nm upon hydrogen exposure. Similar values were measured for a Pd-MOS device. These values are in good agreement with electron attenuation lengths of $17 \pm 3$ nm obtained from photoemission measurements on Pd/n-Si junctions using photon energies below 1.1 eV.[125]

By adding a thin oxide layer ($SiO_2$) between the metal and the semiconductor, Gergen et al.[116] showed that the modified MOS devices had an increased sensitivity to highly energetic gas species and showed a response to a variety of weakly interacting gas species not detectable on the standard MS Schottky diode sensors. The mechanisms for carrier generation and transport in these nanometer-thick devices were unclear at that point. The original explanation of chemicurrent (excitation followed by ballistic charge transport over the Schottky barrier[110,112-116]) appeared valid for the detection of species with large adsorption energies including atomic O and H, molecular NO and $NO_2$. However, the mechanism of carrier generation and detection on the same devices for weakly interacting species such as alkanes,

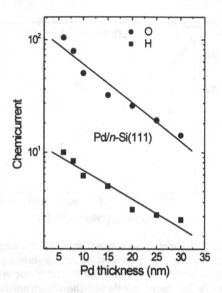

**FIGURE 7.** Thickness-dependent chemicurrent (log scale) for a Pd-MS diode upon exposure to atomic H and O. The attenuation constants of the linear fits are 13 nm for O→ Pd/Si and 17 nm for H → Pd/Si.

alkenes and Xe was less clear, since their average adsorption energies were below the average Schottky barrier.

In a recent work, Roldan *et al.*[123] proposed a new mechanism for current generation in the MOS diodes which includes the excitation of charge carriers in the metal and explains variations in the space-charge layer in the semiconductor without the need of gas particle penetration through the ultrathin gate metal film. Details on this new model and supporting experimental evidence are given below.

Figure 8 shows the phase-sensitive demodulated lock-in amplifier output from a Pd(6 nm)/SiO$_2$/$n$-Si(111) diode exposed to H, C$_2$H$_4$ and CO$_2$. Upon exposure to the reactive species (e.g., H), a decaying chemicurrent is observed in the Pd-MOS sensor (Fig. 8), similar to what was found for the standard Pd-MS device upon atomic oxygen exposure (Fig. 6). The current transients reproduce the chemical kinetics of the surface reaction, which may be modeled by spontaneous adsorption and abstraction mechanisms.[111] However, the chemicurrent signals from the much less reactive species (CO$_2$, C$_2$H$_4$ < 0.5 eV), not detectable using the Pd-MS diodes, are approximately constant over the exposure time. The direction of current relative to the incident beam can be inferred from the lock-in phase ($\varphi$ in Fig. 8). The highly energetic species such as H were found to have a negative phase, similar to the phase of the photocurrents measured upon diode exposure to visible monochromatic light. This reverse bias current is expected for a chemicurrent due to hot electron generation in the metal and transport over the Schottky barrier. However, for the low energetic species, this phase is positive, corresponding to a forward bias current not predicted by the original model.

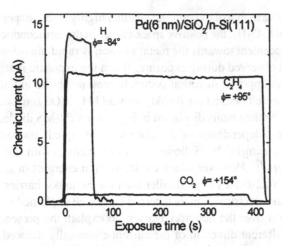

**FIGURE 8.** Lock-in detected chemicurrent signals from Pd(6 nm)/SiO$_2$(0.7 nm)/$n$-Si(111) diodes as a function of exposure time to H, C$_2$H$_4$ and CO$_2$. $T = 125$ K.

In order to get further insight into this new transport mechanism, time-averaged single-pulse currents were recorded during exposure of Pd-MOS diodes to C$_2$H$_4$ and O (Fig. 9). The signals were modulated at 4 Hz and the dotted line indicates the zero current base line. In both cases, the pressure in the beam chamber foreline (0.5 Torr) was used as a qualitative relative measure of the particle flux. For atomic oxygen, a continuous chemicurrent is observed, vanishing as soon as the beam shutter is closed. However, during C$_2$H$_4$ exposure, a displacement current similar to charging and discharging of a capacitor is measured. The direction of the current induced by the interaction of C$_2$H$_4$ (similar traces were observed for CO$_2$

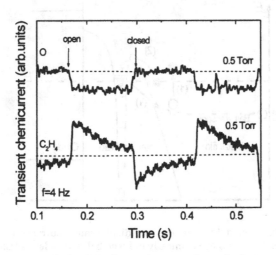

**FIGURE 9.** Time-averaged single-pulse currents measured during exposure of a Pd(6 nm)/SiO$_2$/$n$-Si(111) diode to O and C$_2$H$_4$. The signals were modulated at 4 Hz.

and Xe) is opposite to that of oxygen and all other highly reactive species. When the diode is exposed to $C_2H_4$, the positive space charge in the semiconductor is reduced by charge displacement towards the metal surface. A rapid, flux-dependent decay in the current is observed during exposure. When the exposure is stopped, charge movement in the opposite direction occurs to reestablish the initial equilibrium. Similar results were obtained for Au/Al$_2$O$_3$/n-Si(111) MOS sensors.[124]

In analogy to the results displayed in Fig. 7 for a Pd-MS diode, it was found that the thickness-dependency of the chemicurrent signals measured from low energetic species using Pd-MOS diodes is also in agreement with *ballistic transport* of charge carriers.[123] However, since the interaction energies of species such as $C_2H_4$ and $CO_2$ (0.2-0.5 eV) are smaller than the Schottky barrier height of the MOS devices, the electrical signals detected cannot be explained in terms of hot-electron transport over the Schottky barrier. To explain the presence of a signal at all and the different direction of the current chemically induced on the MOS diodes, Roldan *et al.*[123] proposed a dynamical model that takes into consideration the effect of the intermediate ultrathin oxide film. Based on capacitance-voltages measurements,[122] the presence of localized negative charges at the metal-oxide interface was inferred, and those charges were assumed to be present in acceptor-like interface trap states.

When electron-hole pairs are created by gas adsorption, the energy distribution of hot electrons and holes are approximately symmetric around the Fermi energy. For low interaction energies, the hot electrons have insufficient energy to overcome the Schottky barrier of the *n*-type diode and are scattered within the metal film and eventually thermalized. However, hot holes can travel ballistically across the metal film, reach the interface and recombine with electrons located at metal-oxide interface traps. This mechanism is illustrated in Fig. 10. The

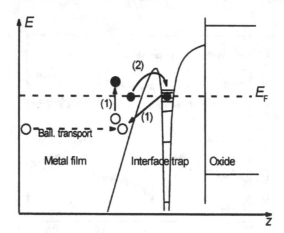

**FIGURE 10.** Dynamic model for the observed displacement currents. (1) Discharging of acceptor-type interface states by chemically induced ballistic holes. (2) Charging of interface states. [Reprinted figure with permission from B. Roldan Cuenya, Physical Review B, **70**, 115322 (2004). Copyright 2004 by the American Physical Society].

recombination occurs near the Fermi level by an Auger process (1) that creates an electron-hole pair that is rapidly thermalized. Subsequently, the unoccupied trap can be replenished by transferring metal electrons at the Fermi energy (pathway 2). The last process may happen either by tunneling or by thermoemission, depending on whether the potential barrier between the metal electron state and the trap state is small or large respectively. The occupation of the trap states changes when the beam shutter is opened and the diode exposed to the charge-generating molecules. The forward current disappears as soon as a detailed balance between the Auger process (1) and the recharging process (2) is reached. When the shutter is closed, process (1) is stopped but (2) will remain active, leading to the initial trap state occupation. Since the countercharge in the semiconductor side changes according to the variations of the trap charge, the on-and-off switching of process (1) leads to the observed flow of electrons in and out of the space layer of the semiconductor. In summary, the experimental observations obtained upon exposing MOS diodes to low energetic gas species can be explained by a model that combines chemically induced electron-hole pair generation, hot hole transport through nanometer-sized metal films, and discharging and recharging on localized states at the metal-oxide interface.

Energy conversion processes from catalytic reactions to hot electron currents have been recently investigated using Schottky nanodiodes by Park et al.[126,127]

## 3.2. Quasi-1D Nanostructured Oxides as a New Platform for Gas Sensing

In this section, we will concentrate on another platform for gas sensing, namely quasi-1D metal oxide semiconductors which are single crystals or polycrystalline nanostructures with an effective diameter on the order of 1-100 nm and lengths larger than a few micrometers. During the last few years, a great variety of metal oxide quasi-1D nanostructure prospects for gas sensing have been synthesized (Fig. 11) (see also recent reviews[55,56] and references therein). A growing number of groups have been testing their sensing performance as chemiresistors and chemi-field effect transistors (chemi-FET) versus the range of reducing and oxidizing gases.[49,51,52,128-133]

The modern amperometric gas sensors are based on granular or polycrystalline semiconducting oxide thin films where electron (hole) transport proceeds through the multiple percolation paths between interconnected grains[134,135] (Fig. 12). The resistance of such active elements is primarily determined by the connecting "necks" between the individual grains and, due to their nanometer size, is a sensitive function of the adsorption/desorption of the target gas. A deep understanding of the fundamental surface chemistry of oxides (see Refs.[136-139] and sources therein) as well as receptor and transduction functions of these kind of sensors was achieved in the past two decades.[140-142] This achievement has established a robust theoretical and technological background for implementation of modern nanotechnology in this field.

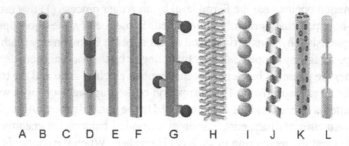

A  B  C  D  E  F    G      H     I  J  K   L

**FIGURE 11.** Demonstrated morphologies of quasi one-dimensional oxides prospective for sensing applications. A, simple homogeneous nanowires and faceted nanorods; B, metal-oxide, elemental semiconductor-oxide, oxide-oxide core-shell nanostructures; C, nanotubules; D, nanostructures with super lattices (compositional); E, homogeneous nanoribbons and nanobelts; F, nanotapes with compositionally different segments; G,H, hierarchical nanostructures; I, "conjugated" oxide nanoparticles or metal nanoparticles with oxide skin. J, nanosprings; K, nanoporous nanowires; L, nanostructures with segmented morphologies (encoded nanowires). Adapted from A. Komakov and M. Moskovits Annu. Rev. Mater. Res. **34**, 151, (2004) and reprinted, with permission, from the Annual Review of Materials Research, Volume 34 ©2004 by Annual Reviews www.annualreviews.

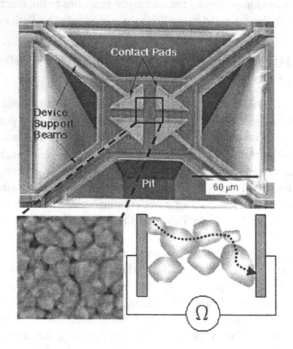

**FIGURE 12.** The state of the art micro-hot plate gas sensor based on thin film technology. The conductometric response of such nanostructured metal oxide thin films relies on percolation between the individual granules. (Reprinted from Sensors and Actuators B **77**, Semancik et al., p.579, (2001), with permission from Elsevier ).

### 3.2.1. Control Over the Surface Reactivity of Quasi-1D Metal Oxides

Following the ideas developed by Aigrain, Hauffe, Wolkenstein, Morrison and others on electronic theory of adsorption and catalysis (see[143,144] and references therein), the position of the Fermi level within the band gap and relative to the adsorbate levels is a crucial (though not necessarily sufficient) factor for the surface reactivity of the semiconductor (Figs. 2, 14). In particular, it controls the chemisorption probability of the adsorbate in ionic form and the availability of free carriers for the surface reactions. The position of the Fermi level in the band gap of the wired pristine nanostructure depends on (i) the doping level (ii) external electrostatic field (when configured as FET) (iii) temperature and photon flux. Therefore the reactivity and sensing performance of the nanostructures can be altered by tuning these parameters.

***3.2.1.1. The Effect of Doping.*** The doping effect on the sensing performance of metal oxide nanowires was directly demonstrated in Refs. 122 and 134. Provided that that doping in $In_2O_3$ is determined by the level of oxygen vacancies, authors were using two different oxygen concentrations (0.02% and 0.04%) in Ar during nanowire growth. Based on I-V and transconductance measurements, it was shown that a nanowire with a low density of oxygen vacancies (corresponding to a Fermi level located deeper in the band gap) behaved as an electron donor causing the NW's conductance to increase (Fig. 13 (b)) upon exposure to ammonia. With a higher oxygen vacancy density (the Fermi level approaching the bottom edge of the of conduction band) the conductance of the nanowires became lower under $NH_3$ exposure, which is a characteristic behaviour of the acceptor gas (Fig. 13(a)).

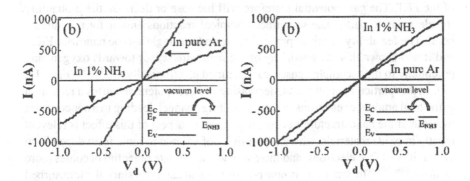

**FIGURE 13.** Alternating donor (right plot) vs acceptor (left plot) behavior of $NH_3$ adsorbate as a function of the doping level of an $In_2O_3$ nanowire. (Reused with permission from Daihua Zhang et al., Appl. Phys. Lett., **83**, 1845 (2003). Copyright 2003, American Institute of Physics).

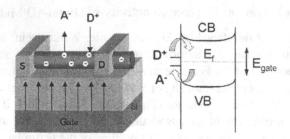

**FIGURE 14.** The effect of a global gate on surface adsorptivity of A-acceptor and B-donor molecules. The variation of the position of the Fermi level in the band gap will influence the charge status of the adsorbate and therefore the type and strength of the bond.

### 3.2.1.2. The Effect of the Electric Field. Global Gate Effect.

Due to the small size of the FET-configured nanostructure and facile penetration of the electrostatic field in to its interia, the gate potential (*global gate*) becomes an efficient tuning parameter for rational positioning of the Fermi level within the band gap of the nanowire. This then offers an electronic means of control over the adsorption, desorption and catalytic processes occuring at the 1D oxide nanostructure's surface (Fig. 14).

Namely, the gate potential in a FET configured nanowire can tune the relative position of the Fermi level within the band gap, and thus modulate the electron density $n$ in its conducting channel

$$n \cong \left( \frac{2\varepsilon\varepsilon_0}{eR^2 \ln(2h/R)} \right) (V - V_{Th}) \tag{8}$$

Here, $h$ and $\varepsilon$ are the thickness and dielectric constant of the gate oxide $SiO_2$ film, respectively, $R$ is the radius of the nanowire, and $V_{Th}$ is the threshold voltage of the FET. The gate potential therefore will increase or decrease the probability for electrons to participate in surface chemical reactions and/or form a surface bond. The feasibility of this approach was tested on single $n$-type nanowire FETs and it was shown that the *sensitivity* of the nanowire sensor towards oxygen, and potentially to more complex gases, can be tuned electronically[131,146,147] (Figs. 14, 15). In this particular application, decreasing the gate potential causes a reduction in the total amount of electrons in the conducting channel, leading to a concomitant increase of the nanostructure's sensitivity. Another aspect of this effect is relevent mostly in oxide materials, where adsorptivity of oxygen in ionic form depends on availability of free electrons, and increases when the gate potential becomes more positive.[146-148] The latter opens new possibilities in catalysis since the ionosorbed surface oxygen precursors $O_{\beta S}^{-\alpha}$ participate in many catalytic reactions (see as an example the reaction (5) in Section 2.2). Therfore, one can use the above general effect, not only for gas sensing, but also for an active control over the surface reactivity of the nanostructure. The gate effect on surface adsorptivity and reactivity was recently reported[146-148] when the CO oxidation rate and yield over

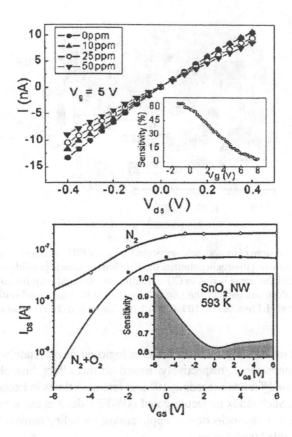

**FIGURE 15.** The dependence of the sensitivity of the ZnO (top)[131] and SnO$_2$ (bottom)[147] FET configured nanowires towards oxygen as a function of the gate potential [(Top) Reused with permission from Zhiyong Fan et al., Appl. Phys. Lett, 85, 5923 (2004). Copyright 2004, American Institute of Physics. (Bottom) Reprinted with permission from Y. Zhang et al., J. Phys. Chem. 109 (2005) 1923. Copyright (2005) American Chemical Society].

the SnO$_2$ nanowire surface were found to depend on the gate potential (Fig. 16). There is experimental evidence that, when only a few reaction channels for CO oxidation are available, the electronic gating is able to promote the channel that depends on oxygen ionosorption. These results demonstrate that in addition to the sensitivity, the *selectivity* of the sensor can be tuned electronically.

A very important result for potential applications of the global gate effect for selective gas sensing was demonstrated in Ref. 149. In this research, the individual semiconducting (Si, InP, GaN) nanowire, whose thin oxide "skin" was functionalized with redox active molecules (ferrocene, zinc tetrabenzoporphine, cobalt phtalocyanine), was configured as an FET. The negative/positive pulses of the global gate alternatively charge redox molecules positively or negatively, thus creating stable charge states on the nanowire's surface. This newly induced charge

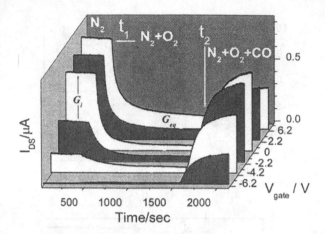

**FIGURE 16.** The degree of oxygen ionosorption $G_i$ by the FET configured nanostructures decreases with more negative gate potential until no ionsorption is possible at Vgate~ 6 eV. The uptake part of the curve is due to CO oxidation reaction (5) with ionosorbed oxygen which releases electrons back to the conduction band136. (Reprinted with permission from Y. Zhang et al., J. Phys. Chem. 109 (2005) 1923. Copyright (2005) American Chemical Society).

thereby maintains the NW-FETs in either a logic *on* or *off* state with high- or low-channel conductance, respectively. Based on this effect, bistable nanoscale switches with *on/off* ratios exceeding $10^4$ and retention times in excess of 20 min were obtained. Such redox molecule-gated NW-FET devices can serve as key elements in a range of nanoelectronics applications, including nonvolatile memory and programmable logic elements.

This line of research was further developed in Ref. 150. In this work, a 10 nm-diameter $In_2O_3$ nanowire was used as a sensitive conducting channel for an FET. The crucial step in the performance of the device was the careful selection of the redox active molecules (Fig. 17). Specifically, $Fe^{2+}$-terpyridin, which possesses multiple states for charge, was used.

Using precise values for the gate potential impulses, the charges were precisely placed (removed) at (from) up to eight discrete levels in redox active molecules self-assembled on single-crystal semiconducting nanowire field-effect transistors. Based on the sequential gate voltage pulses for writing operations, and current sensing for reading operations, multilevel molecular memory devices were achieved. As a result, up to three bits (eight levels) of data storage per individual cell having a retention time of 600 h was achieved. Despite the fact that these achievements are mostly related to nanoelectronics, this research demonstrates the great promise of the redox molecules as nanowire functionalization species for selective detection of specific target gas molecules.

For many years the strong ionosorption of the target molecules (which is required for their sensitive detection with conductometric sensors) was one of

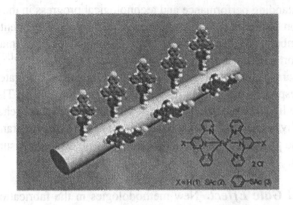

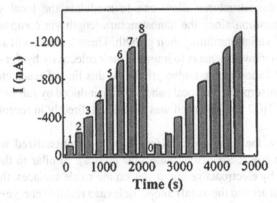

**FIGURE 17.** The demonstration of the 3 bits data storage (bottom panel) in the single $In_2O_3$ nanowire surface functionalized with Fe- terpyridin (top panel). [(Top) Reprinted with permission from Chao Li et al., J. Am. Chem. Soc. Comm. 126 (2004) 7750. Copyright (2004) American Chemical Society. (Bottom) Reused with permission from Chao Li et al., Appl. Phys. Lett., 84, 1949 (2004). Copyright 2004, American Institute of Physics].

the principal obstacles in manufacturing fast and reversible sensors operating at moderate (preferably at room) temperatures. It was recently demonstrated that the complete reversibility of single $SnO_2$ nanoribbon and ZnO nanowire field effect transistors chemical sensors toward adsorption/desorption cycles of $NO_2$ and $NH_3$ molecules can be achieved at room temperature. This can be done either by using UV photons for photo-desorption[49] or strong electrostatic fields for electro-desorption of adsorbates.[151] In the latter case it was shown that due to the small diameter of the nanowire sensor, the electric field applied over the back gate electrode is significant enough to refresh the sensor by an electrodesorption mechanism. In addition, different chemisorbed species were found to have different "refresh" threshold voltages and thus could be distinguished from each other based on this value and on the temporal response of the conductance. In spite of the proven outstanding performance and technological progress in the fabrication and sensitization of the FET configured chemical sensors, experimental research still faces a number of challenges. As an example, the recent AFM imaging of the electron transport in metal oxide nanowire FET sensors revealed[146,152] the significance of the parasitic effect of mobile charges at the surface of gate oxides on the temporal response, reproducibility and stability of the sensor. These effects were further studied in more details in Ref.[152] where the temporal behavior of the gate effect and hysteresis in the transconductance in ZnO nanowire transistors was explained by the presence of mobile surface charges on or near the surface of the ZnO nanowires.

### 3.2.1.3. Local Gate Effect.
New methodologies in the fabrication of metal oxide quasi-1D nanostructures allow one to modulate the local variations of the Fermi level position along the nanostructure length via compositional[154] or morphological[155] variations during their growth. These features will add new functionalities to the nanowire sensors to improve their selectivity by, for example, via site specific modulation of their adsorptivity. In this line of research, the dependence of surface adsorptivity on local band bending induced by sub surface charged impurities in the $TiO_2$ single crystal was visualized directly in recent UHV STM studies.[156,157]

Alternatively, the nanowire's surface can be functionalized with catalytic nanoparticles (or any redox active molecular groups). Similar to the local band bending induced by electroactive impurities on the oxide surfaces, the local band bending will occur around the metal nanoparticles as a result of charge exchange between the nanoparticles and nanowire oxide support. This nanoparticle–nanowire couple can be considered as a nanoscopic analog to the formation of the Schottky barrier.[158] Such a particle acts as a local gate that effectively influences the transport through the thin enough conducting channel (Fig. 18(a)). The adsorption of the target molecules on the surface of such a nanoparticle changes its work function or charge status and therefore modulates the degree of local depletion (accumulation) at the particle-oxide nanowire interface. The changes can be recorded via conduction change in the nanowire chemiresistor. In addition, catalyst nanoparticles can in turn initiate massive chemical processes that involve not only the catalyst particle itself but also the much larger area of the support around the particle via diffusion of activated species from the nanoparticle to the nanowire (Fig.18 (b)). This phenomenon is well known in catalysis and is called the *spillover effect*. For nanowires that are thin enough this is of crucial importance since the electron/hole transport through the conducting channel can be completely blocked. In fact, due to this enhanced action, it may be possible to monitor chemical processes over a *single* catalytic particle via simple conductometric measurements.

A recent comparative study by Kolmakov et al.[159] (see also Fig. 19) of the reactivity of pristine versus Pd-coated $SnO_2$ nanostructures demonstrated higher responsiveness and selectivity toward reducing gases for the Pd-functionalized

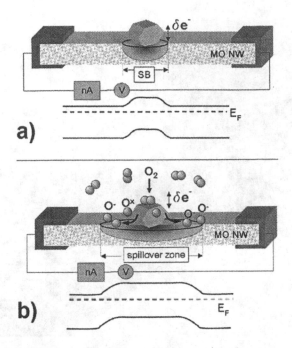

**FIGURE 18.** The different origins of the formation of local depleted regions (a) due to Schottky contact and (b) chemical action of catalyst particles. In the latter case the effect is more dramatic do to spillover (and back spillover) effects Ref 175.

nanoscaled oxides. The advantage of this approach is that the comprehensive knowledge achieved in catalysis studies can be exploited for improving the sensitivity and selectivity of quasi 1D metal oxides.

## 4. NEW SURFACE SCIENCE TRENDS FOR THE CHARACTERIZATION OF NANOSTRUCTURES

It is evident from the above discussion that further exploration of sensing and catalytic phenomena on nanostructured oxides will rely on experimental methods that are sensitive to *local* variations of the Fermi level, surface composition, and electronic structure along the length of 1D nano-objects. In addition, the strong dependence of the nanostructure's properties on diameter, facet orientation, stoichiometry, doping level, etc, precludes the interpretation and comparison of the experimental data obtained on different nanostructures. Therefore it is crucial for the measurements to be performed on *individual* well-characterized nanostructures. The latter implies that the spectroscopic techniques must be integrated into microscopic tools with lateral resolution not less than the diameter of the nanostructure.

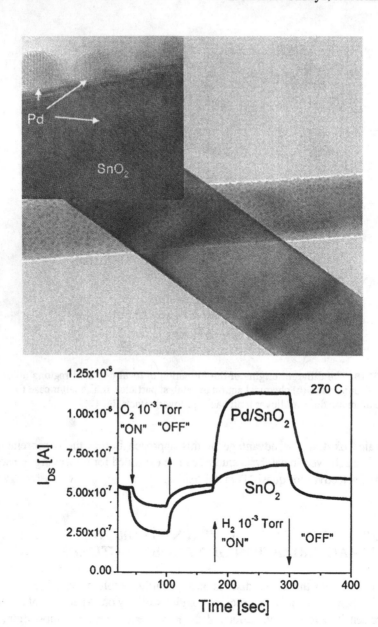

**FIGURE 19.** SnO$_2$ nanowire and nanobelt surface doped with Pd nanoparticles (top). The sensing performance of the SnO$_2$ nanobelt before (the curve close to the base line) and after (wider curve) Pd functionalization. (A. Kolmakov, *et al.* (unpublished)). Adapted with permission from A. Komakov Proc. SPIE 6370, 63700X-5 (2007). Copyright 2007 by SPIE.

The majority of the recent structural and compositional data on 1D nano-structures were obtained using scanning electron microscopy (SEM), transmission electron microscopy (TEM), high-resolution (HRTEM) and scanning transmission electron microscopy (STEM) (see Ref. 160 and sources therein). Being superior in spatial resolution, HRTEM and STEM-based chemical mapping methods are, however, much more sensitive to the chemical composition of the bulk atoms of the nanostructure rather than to the chemical state of the surface atoms, and routinely induce significant radiation damage in the nano object.[161] In the past, synchrotron radiation (SR) based photo electron microscopy (PEM) methods, such as X-ray photo electron emission microscopy (X-PEEM) and scanning photo electron microscopy (SPEM), demonstrated significant improvements in spatial resolution, overcoming the important thresholds of 10 nm[162] and 100 nm[163], respectively. The latter falls in the size domain of individual nanostructures. By applying PEEM or SPEM to individual nanostructures, the evolution of the fine details of the sur-faces valence/core level photoelectron spectra, along with NEXAFS spectra as a function of nanostructure diameter, faceting, morphology, type of adsorbate, tem-perature, time, etc., can be studied with 10–100 nm spatial resolution. In addition, these analytical tools are significantly less destructive. Coupling these experimen-tal tools with nanodevices such as a resistor or FET, the spectroscopic data can be combined with the entire range of electron (hole) transport measurements. For gas sensing applications and catalysis, these opportunities offer the comprehensive characterization of the interplay between surface and bulk electronic properties of individual nanostructures. In addition, the PES and NEXAFS spectra from the nanostructures can be collected as a function of the gas exposure of a target gas under different temperatures, gate potentials, bias voltages, etc. Ultimately, multi-component chemical reactions can be performed in situ over the surface of the wired nanostructures. The great potential of using spectromicroscopy to study 1D nanostructures has been recognized, and a few important reports, mostly on carbon nanotubes,[164-171] have been published. In spite of the significance of these measure-ments, there are apparent experimental challenges related to the implementation of SR-based spectromicroscopies to individual nanostructures and nanostructure de-vices. Namely, when the nanostructure (nanodevice) needs to be electrically wired and imaged/analyzed with PEM, the use of common planar $Si/SiO_2$ supports as a substrate material can cause severe charging artifacts due to low conductivity of the $SiO_2$ layer. These will manifest themselves as distorted and/or low-resolution images and shifted/broadened spectra[172,173]. Since the lateral resolutions of SPEM is sufficient enough to probe a spot comparable to the diameter of the nanostructure ($\sim$100 nm)[174], this advantage allows one to image and spectroscopically probe the operating nanostructure in its usual FET configuration without significant charg-ing of the surrounding $SiO_2$ gate oxide layer. However, the latter is not possible with PEEM, where the X-rays illuminate a large area on the sample. This exper-imental challenge can be overcome by using special supporting substrates such as high aspect ratio micro-walls[174]. These substrates can be made of dielectric materials with the top of the micro-walls being metallized. The charging of the

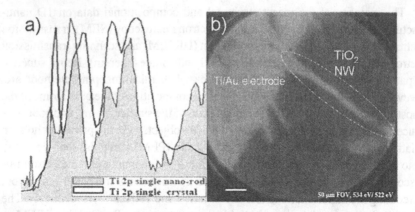

**FIGURE 20.** (a) A comparison of the NEXAFS spectra taken from the individual $TiO_2$ nanowire (shadowed) and from the rutile $TiO_2$ (solid curve) crystal. The energy shift in the spectra is presumably due to local charging of the nanostructure under soft X-ray radiation. (b) $50\mu$m FOV PEEM image of the suspended $TiO_2$ nanowire chemiresistor contacted by two metal electrodes. (From Ref. 174.)

depth of micro-walls is drastically hampered by their high aspect ratio, and yet the top of the electrically isolated walls provide a nearly planar conducting substrate. Fig. 20 demonstrates the feasibility of this approach by taking a PEEM image and a NEXAFS spectrum from an individual titania nanofiber bridged between two metallized micro-walls.

## 5. CONCLUDING REMARKS

This chapter provides some examples of new experimental trends in the fields of catalysis and chemical sensing. In particular, sensors based on chemically induced electronic phenomena on nanometer-scale metal films deposited on semiconductor junctions (MS and MOS devices), quasi-1D oxide chemiresistors and chemi-FET devices and chemically tunable magnetic nanostructures have been discussed.

We have shown that nanometer-thick MS and MOS devices can be used as highly selective gas sensors. Here, the detection is based on charge carriers (electron-hole pairs) created during nonadiabatic energy transfer to the metal from a surface reaction. Since the excited charge carriers travel ballistically through the metal film, an electrical signal can only be measured when the metal film thickness is less than the ballistic mean free path of the charge carriers generated. Ideal MS diodes are insensitive to reactions with energies lower than the Schottky barrier. However, if an oxide layer is present between the metal and the semiconductor, localized electron donor states become available, and charging and discharging of the device interface allows the detection of gas species with low interaction energies. The main limitation of these devices is that they require low temperature (125-200 K) operation in order to minimize thermal noise.

After less than a decade of development, chemiresistors and chemi-FET nanosensors made of quasi-1D oxides have demonstrated comparable or even superior performance over conventional bulk-sensors and are currently considered among the most promising sensing elements for the next generation of solid-state gas sensors. The research is now focused on improving the selectivity, reproducibility, and robustness of these devices by carefully crafting their active elements.

There are still plenty of unexplored lines of fundamental research, particularly in the size range where electron confinement dominates the electronic structure of nano-objects. Furthermore, there exists a strong necessity of further developing new spectroscopic and nondestructive microscopic techniques to probe surface properties of individual nanostructures, in particular, the ones that can be operated under realistic sensing conditions.

# ACKNOWLEDGMENT

B. Roldan Cuenya is grateful for the financial support of the National Science Foundation (Grant No. 0448491) and the American Chemical Society-Petroleum Research Foundation (Grant No. 42701-G5). A. Komakov also thanks the American Chemical Society Petroleum Research Foundation (Grant # 45842-G5) for suport of his research.

# REFERENCES

1. J.H. Sinfelt, Catalysis: An old but continuing theme in chemistry, in: *Proc. Am. Philos. Soc.* **143**, 388–399 (1999).
2. G.A. Somorjai and R.M. Rioux, High technology catalysts towards 100% selectivity: Fabrication, characterization and reaction studies, *Catal. Today* **100**, 201–215 (2005).
3. U. Heiz and U. Landman, Nanocatalysis, (Springer, Heidelberg, 2006).
4. M. Valden, X. Lai and D.W. Goodman, Onset of catalytic activity of gold clusters on titania with the appearance of nonmetallic properties, *Science* **281**, 1647–1650 (1998).
5. A. Naitabdi, L.K. Ono, and B. Roldan Cuenya, Local investigation of the electronic properties of size-selected Au nanoparticles by scanning tunneling spectroscopy, *Appl. Phys. Lett.* **89**, 043101 (2006).
6. D. Dalacu and L. Martinu, Optical properties of discontinuous gold films: Finite-size effects, *J. Opt. Soc. Am. B* **18**, 85–92 (2001).
7. M. Haruta, Size- and support-dependency in the catalysis of gold, *Catal. Today* **36**, 153–166 (1997).
8. H. Hakkinen, W. Abbet, A. Sanchez, U. Heiz and U. Landman, Structural, electronic, and impurity-doping effects in nanoscale chemistry: supported gold nanoclusters, *Angew. Chem. Int. Ed.* **42**, 1297–1300 (2003).
9. J.D. Aiken and R.G. Finke, A review of modern transition-metal nanoclusters: Their synthesis, characterization, and applications in catalysis, *J. Mol. Catal. A* **145**, 1–44 (1999).
10. C.T. Campbell, Ultrathin metal films and particles on oxide surfaces-Structural, electronic and chemisorptive properties, *Surf. Sci. Rep.* **27**, 1 (1997).

11. U. Heiz, S. Abbet, A. Sanchez, W.D. Schneider, H. Hakkinen and U. Landman, Chemical reactions on size-selected clusters on surfaces, in: *Proc. Nobel Symposium* **117** (E. Campbell, M. Larsson, eds.) (World Scientific, Singapore, 2001).

12. A. Kolmakov and D.W. Goodman, Size effect in catalysis by supported metal clusters, in: *Quantum Phenomena in Clusters and Nanostructures*, (A.W. Castleman Jr. and S.N. Khanna, eds.) (Springer, New York, 2003).

13. M. Haruta, Catalysis of gold nanoparticles deposited on metal oxides, *CATTECH* **6**, 102–115 (2002).

14. N. Lopez, T.V.W. Janssens, B.S. Clausen, et al., On the origin of the catalytic activity of gold nanoparticles for low-temperature CO oxidation, *J. Catal.* **223**, 232 (2004).

15. M. Mavrikakis, P. Stoltze and J. Norskov, Making gold less noble, *Catal. Lett.* **64**, 101–106 (2000).

16. R. Meyer, C. Lemire, S.K. Shaikhutdinov and H.J. Freund, Surface chemistry of catalysis by gold, *Gold Bulletin* **37**, 72 (2004).

17. S.K. Shaikhutdinov, R. Meyer, M. Naschitzki, M. Baumer and H.J. Freund, Size and support effects for CO adsorption on gold model catalysts, *Catal. Lett.* **86**, 211 (2003).

18. L.K. Ono, D. Sudfeld, B. Roldan Cuenya, In-situ gas-phase catalytic properties of TiC-supported size-selected gold nanoparticles synthesized by diblock copolymer encapsulation, *Surf. Sci.* **600**, 5041–5050 (2006).

19. J.D. Aiken, Y. Lin and R.G. Finke, A perspective on nanocluster catalysis: Polyoxoanion and n-$C_4H_9N^+$ stabilized Ir(O) nanoclusters "soluble heterogeneous catalysts", *J. Mol. Catal. A* **114**, 29–51 (1996).

20. O. Alexeev and B.C. Gates, Iridium clusters supported on $\gamma$-$Al_2O_3$: Structural characterization and catalysis of toluene hydrogenation, *J. Catal.* **176**, 310-320 (1998).

21. A.M. Argo and B.C. Gates, Support effects in alkene and hydrogenation catalyzed by well-defined supported rhodium and iridium clusters, *Abst. Papers Am. Chem. Soc.* **221**, U476–U476 (2001).

22. A.M. Argo, J.F. Odzak and B.C. Gates, Role of cluster size in catalysis: Spectroscopic investigation of $\gamma$-$Al_2O_3$-supported $Ir_4$ and $Ir_6$ during ethene hydrogenation, *J. Am. Chem. Soc.* **125**, 7107–7115 (2003).

23. A.T. Ashcroft, A.K. Cheetham, P.J.F. Harris, et al., Particle-size studies of supported metal-catalysts—a comparative-study by X-ray-diffraction, EXAFS and electron-microscopy, *Catal. Lett.* **24**, 47–57 (1994).

24. A. Berko and F. Solymosi, Effects of different gases on the morphology of Ir nanoparticles supported on the $TiO_2(110)$-(1 x 2) surface, *J. Phys. Chem. B* **104**, 10215–10221 (2000).

25. D.S. Cunha and G.M. Cruz, Hydrogenation of benzene and toluene over Ir particles supported on $\gamma$-$Al_2O_3$, *Appl. Catal. A* **236**, 55–66 (2002).

26. S.E. Deutsch, J.T. Miller, K. Tomishige, Y. Iwasawa, W.A. Weber and B.C. Gates, Supported Ir and Pt clusters: Reactivity with oxygen investigated by extended X-ray absorption fine structure spectroscopy, *J. Phys. Chem.* **100**, 13408–13415 (1996).

27. S.E. Deutsch, F.S. Xiao and B.C. Gates, Near absence of support effects in toluene hydrogenation catalyzed by MgO-supported iridium clusters, *J. Catal.* **170**, 161–167 (1997).

28. J.D. Grunwaldt, P. Kappen, L. Basini and B.S. Clausen, Iridium clusters for catalytic partial oxidation of methane—an in situ transmission and fluorescence XAFS study, *Catal. Lett.* **78**, 13–21 (2002).

29. S. Kawi and B.C. Gates, MgO-supported [Ir-6(CO)(15)]2-:Catalyst for CO hydrogenation, *J. Catal.* **149**, 317–325 (1994).

30. F.C.C. Moura, R.M. Lago, E.N. dos Santos and M.H. Araujo, Unique catalytic behavior of Ir-4 clusters for the selective hydrogenation of 1,5-cyclooctadiene, *Catal. Comm.* **3**, 541–545 (2002).

31. L. Stievano, S. Calogero, R. Psaro and F.E. Wagner, Advances in the application of Mössbauer spectroscopy with less-common isotopes for the characterization of bimetallic supported nanoparticles: [193]Ir Mössbauer spectroscopy, *Comm. Inorg. Chem.* **22**, 275–292 (2001).

32. M.S. Ureta-Zanartu, C. Yanez, G. Reyes, J.R. Gancedo and J.F. Marco, Electrodeposited Pt-Ir electrodes: Characterization and electrocatalytic activity for the reduction of the nitrate ion, *J. Solid. State Electrochem.* **2**, 191–197 (1998).

33. Z. Xu, F.S. Xiao, S.K. Purnell, et al., Size-dependent catalytic activity of supported metal-clusters, *Nature* **372**, 346–348 (1994).

34. A. Zhao and B.C. Gates, Toluene hydrogenation catalyzed by tetrairidium clusters supported on $\gamma$-$Al_2O_3$, *J. Catal.* **168**, 60–69 (1997).

35. O.B. Yang, S.I. Woo and Y.G. Kim, Comparison of platinum iridium bimetallic catalysts supported on gamma-alumina and hy-zeolite in n-hexane reforming reaction, *Appl. Catal. A* **115**, 229–241 (1994).

36. M. Frank and M. Baumer, From atoms to crystallites: Adsorption on oxide-supported metal particles, *Phys. Chem. Chem. Phys.* **2**, 3723–3737 (2000).

37. L.M.P. Gruijthuijsen, G.J. Howsmon, W.N. Delgass, D.C. Koningsberger, R.A. van Santen and J.W. Niemantsverdriet, Structure and reactivity of bimetallic FeIr/$SiO_2$ catalysts after reduction and during high-pressure CO hydrogenation, *J. Catal.* **170**, 331–345 (1997).

38. M.M. Bhasin, W.J. Bartley, P.C. Ellgen and T.P. Wilson, *J. Catal.* **54**, 120 (1978).

39. T. Fukushima, Y. Ishii, Y. Onda, M. Ichikawa, Promoting role of Fe in enhancing activity and selectivity of MeOH production from CO and $H_2$ catalyzed by $SiO_2$-supported Ir, *J. Chem. Soc. Chem. Comm.* **24**, 1752–1754 (1985).

40. J. R. Croy, S. Mostafa, Jing Liu, Yong-ho Sohn, B. Roldan Cuenya, Size-dependent study of MeOH decomposition over size-selected Pt nanoparticles synthesized via micelle encapsulation, *Catal. Lett.*, DOI:10.1007/S10562-007-9162-01 (in press 2007).

41. J. R. Croy, S. Mostafa, J. Liu, Yongho Sohn, H. Heinrich and B. Roldan Cuenya, Support dependence of MeOH decomposition over size-selected Pt nanoparticles, *Catal. Lett.* (accepted, 2007).

42. M. Valden and D.W. Goodman, Structure-activity correlations for Au nanoclusters supported on $TiO_2$, *Israel J. Chem.* **38**, 285–292 (1998).

43. M. Valden, S. Pak, X. Lai and D.W. Goodman, Structure sensitivity of CO oxidation over model Au/$TiO_2$ catalysts, *Catal. Lett.* **56**, 7–10 (1998).

44. A. Szabo, M.A. Henderson and J.T. Yates, Oxidation of CO by oxygen on a stepped platinum surface: Identification of the reaction site, *J. Chem. Phys.* **96**, 6191–6202 (1992).

45. C. Duriez, H. C.R. and C. Chapon, Molecular beam study of the chemisorption of CO on well-shaped palladium particles epitaxially oriented on MgO(100), *Surf. Sci.* **253**, 190 (1991).

46. M. Frank, S. Andersson, J. Libuda, et al., Particle size dependent CO dissociation on alumina-supported Rh: A model study, *Chem. Phys. Lett.* **279**, 92–99 (1997).

47. L.K. Ono and B. Roldan Cuenya, Effect of interparticle interaction on the low temperature oxidation of CO over size-selected Au nanocatalysts supported on ultrathin TiC films, *Catal. Lett.* **113**, 86–93 (2007).

48. E. Comini, A. Cristalli, G. Faglia and G. Sberveglieri, Light enhanced gas sensing properties of indium oxide and tin dioxide sensors, *Sens. Actuators B* **65**, 260–263 (2000).

49. M. Law, H. Kind, B. Messer, F. Kim and P.D. Yang, Photochemical sensing of $NO_2$ with $SnO_2$ nanoribbon nanosensors at room temperature, *Angew. Chem. Int. Ed.* **41**, 2405–2408 (2002).

50. Y. Cui, Q.Q. Wei, H.K. Park and C.M. Lieber, Nanowire nanosensors for highly sensitive and selective detection of biological and chemical species, *Science* **293**, 1289–1292 (2001).

51. E. Comini, G. Faglia, G. Sberveglieri, Z.W. Pan and Z.L. Wang, Stable and highly sensitive gas sensors based on semiconducting oxide nanobelts, *Appl. Phys. Lett.* **81**, 1869–1871 (2002).

52. A. Kolmakov, Y. Zhang, G. Cheng and M. Moskovits, Detection of CO and oxygen using tin oxide nanowire sensors, *Adv. Mater.* **15**, 997–1000 (2003).

53. D.H. Zhang, Z.Q. Liu, C. Li, et al., Detection of $NO_2$ down to ppb levels using individual and multiple $In_2O_3$ nanowire devices, *Nano Letters* **4**, 1919–1924 (2004).

54. C.M. Lieber, Nanoscale science and technology: Building a big future from small things, *MRS Bulletin* **28**, 486–491 (2003).

55. Y.N. Xia, P.D. Yang, Y.G. Sun, et al., One-dimensional nanostructures: Synthesis, characterization, and applications, *Adv. Mater.* **15**, 353–389 (2003).

56. Z.L. Wang, Functional oxide nanobelts: Materials, properties and potential applications in nanosystems and biotechnology, *Ann. Rev. Phys. Chem.* **55**, 159–196 (2004).

57. E. Rothenberg, M. Kazes, E. Shaviv and U. Banin, Electric field induced switching of the fluorescence of single semiconductor quantum rods, *Nano Letters* **5**, 1581–1586 (2005).

58. H.J. Freund, J. Libuda, M. Baumer, T. Risse and A. Carlsson, Cluster, facets, and edges: Site-dependent selective chemistry on model catalysts, *Chem. Record* **3**, 181–200 (2003).

59. L. Aballe, A. Barinov, A. Locatelli, S. Heun and M. Kiskinova, Tuning surface reactivity via electron quantum confinement, *Phys. Rev. Lett.* **93**, 196103 (2004).

60. N. Yamazoe, New approaches for improving semiconductor gas sensors, *Sens. Actuators B*, **5**, 7–19 (1991).

61. N.M. White and J.D. Turner, Thick-film sensors: Past, present and future, *Meas. Sci. Techn.* **8**, 1–20 (1997).

62. W. Göpel, Solid-state chemical sensors—Atomistic models and research trends, *Sens. Actuators* **16**, 167–193 (1989).

63. D. Kohl, Function and applications of gas sensors, *J. Phys.D* **34**, R125–R149 (2001).

64. P.T. Moseley, New trends and future-Prospects of thick-film and thin-film gas sensors, *Sens. Actuators B* **3**, 167–174 (1991).

65. A. Dieguez, A. Romano-Rodriguez, J.R. Morante, J. Kappler, N. Barsan and W. Gopel, Nanoparticle engineering for gas sensor optimization: Improved sol-gel fabricated nanocrystalline $SnO_2$ thick film gas sensor for $NO_2$ detection by calcination, catalytic metal introduction and grinding treatments, *Sens. Actuators B* **60**, 125–137 (1999).

66. N. Barsan, Conduction models in gas-sensing $SnO_2$ layers—Grain-size effects and ambient atmosphere influence, *Sens. Actuators B* **17**, 241–246 (1994).

67. G. Sberveglieri, Recent developments in semiconducting thin-film gas sensors, *Sens. Actuators B* **23**, 103–109 (1995).

68. A. Cabot, A. Dieguez, A. Romano-Rodriguez, J.R. Morante and N. Barsan, Influence of the catalytic introduction procedure on the nano-$SnO_2$ gas sensor performances—Where and how stay the catalytic atoms?, *Sens. Actuators B* **79**, 98–106 (2001).

69. N. Barsan and U. Weimar, Understanding the fundamental principles of metal oxide based gas sensors; the example of CO sensing with $SnO_2$ sensors in the presence of humidity, *J. Phys. Condens. Matter.* **15**, R813–R839 (2003).

70. K. Thompson, J.H. Booske, D.J. Larson and T.F. Kelly, Three-dimensional atom mapping of dopants in Si nanostructures, *Appl. Phys. Lett.* **87**, 052108 (2005).

71. R. Skomski, Nanomagnetics, *J. Phys. Condens. Matter.* **15**, 841–896 (2003).

72. S.D. Bader, Magnetism in low dimensionality, *Surf. Sci.* **500**, 172–188 (2002).

73. S.D. Bader, Colloquium: Opportunities in nanomagnetism, *Rev. Mod. Phys.* **78**, 1–15 (2006).

74. M.N. Baibich, J.M. Broto, A. Fert, et al., Giant magnetoresistance of Fe(001)/Cr(001), *Phys. Rev. Lett.* **61**, 2472–2475 (1988).

75. S.A. Wolf, D.D. Awschalom, R.A. Buhrman, et al., Spintronic: A spin-based electronics vision of the future, *Science* **294**, 1488–1495 (2001).

76. D.A. Thonson and J.S. Best, The future of magnetic data storage technology, *IBM J. Res. Develop.* **44**, 311 (2000).

77. B. Roldan Cuenya, A. Naitabdi,E. Schuster, R. Peters, M. Doi, and W. Keune, Epitaxial growth, magnetic properties and lattice dynamics of Fe nanoclusters on GaAs(001), (accepted in Phys. Rev. B 2007).

78. L. Fu, V.P. Dravid, K. Klug, X. Liu and C.A. Mirkin, Synthesis and patterning of magnetic nanostructures, *Europ. Cells and Mater.* **3**, 156 (2002).

79. T.F. Jaramillo, S.H. Baeck, B. Roldan Cuenya and E.W. McFarland, Catalytic activity of supported Au nanoparticles deposited from block copolymer micelles, *J. Am. Chem. Soc.* **125**, 7148–7149 (2003).

80. B. Roldan Cuenya, S.H. Baeck, T.F. Jaramillo and E.W. McFarland, Size and support dependent electronic and catalytic properties of $Au^0/Au^{3+}$ nanoparticles synthesized from block co-polymer micelles, *J. Am. Chem. Soc.* **125**, 12928–12934 (2003).

81. N. Cordente, C. Amiens, B. Chaudret, M. Respaud, F. Senocq and M.-J. Casanove, Chemisorption on nickel nanoparticles of various shapes: Influence on magnetism, *J. Appl. Phys.* **94**, 6358–6365 (2003).

82. H.G. Boyen, G. Kästle, K. Zurn, et al., A micellar route to ordered arrays of magnetic nanoparticles: From size-selected pure cobalt dots to cobalt-cobalt oxide core-shell systems, *Adv. Funct. Mater.* **13**, 359–364 (2003).

83. M. Giersig and M. Hilgendorff, The preparation of ordered colloidal magnetic particles by magnetophoretic deposition, *J. Phys. D* **32**, L111–L113 (1999).

84. J.C. Hulteen and R.P. Van Duyne, Nanosphere lithography: A materials general fabrication process for periodic particle array surfaces, *J. Vac. Sci. Technol. A* **13**, 1553–1558 (1995).

85. A. Naitabdi and B. Roldan Cuenya, Formation, thermal stability and surface composition of size-selected AuFe nanoparticles, *Appl. Phys. Lett.*, (accepted, 2007).

86. J. Chen, M. Drakaki and J.L. Erskine, Chemisorption-induced change in thin-film spin anisotropy: Oxygen adsorption on the p(1x1)Fe/Ag(100) system, *Phys. Rev. B* **45**, 3636–3643 (1992).

87. M.B. Knickelbein, Nickel cluster: The influence of adsorbates on magnetic moments, *J. Chem. Phys.* **116**, 9703–9311 (2002).

88. P.W. Selwood (Ed.), Chemisorption and magnetization (Academic Press, New York, 1975).

89. M. Primet, J.A. Dalmon and G.A. Martin, Adsorption of CO on well-defined $Ni/SiO_2$ catalysts in the 195–373 K range studied by infrared spectroscopy and magnetic methods, *J. Catal.* **46**, 25–36 (1977).

90. Q. Ge, S.J. Jenkins and D.A. King, Localization of adsorbate-induced demagnetization: CO chemisorbed on Ni(100), *Chem. Phys. Lett.* **327**, 125–130 (2000).

91. T. Hill, M. Mozaffari-Afshar, J. Schmidt, et al., Influence of CO adsorption on the magnetism of small CO nanoparticles deposited on $Al_2O_3$, *Chem. Phys. Lett.* **292**, 524–530 (1988).

92. S. Pick and H. Dreysse, Tight-binding study of ammonia and hydrogen adsorption on magnetic cobalt systems, *Surf. Sci.* **460**, 153–161 (2000).

93. M. Boudart, J.A. Dumesic and H. Topsoe, Surface, catalytic, and magnetic properties of small iron particles: The effect of chemisorption of hydrogen on magnetic anisotropy, in: *Proc. Natl. Acad. Sci. USA* **74**, 806–810 (1977).

94. I. Lundström, S. Shivaraman, C. Svensson and L. Lundkvist, A hydrogen-sensitive MOS field-effect transistor, *Appl. Phys. Lett.* **26**, 55 (1975).

95. K.I. Lundstrom, M.S. Shivaraman and C.M. Svensson, Hydrogen-sensitive Pd-gate MOS-transistor, *J. Appl. Phys.* **46**, 3876–3881 (1975).

96. M.C. Steele, J.W. Hile and B.A. Maciver, Hydrogen-sensitive palladium gate MOS capacitors, *J. Appl. Phys.* **47**, 2537–2538 (1976).

97. R. Morrison, Semiconductor gas sensors, *Sens. Actuators* **2**, 329–341 (1982).

98. D. Filippini, M. Rosch, R. Aragon and U. Weimar, Field-effect $NO_2$ sensors with group 1B metal gates, *Sens. Actuators B* **81**, 83–87 (2001).

99. S.J. Fonash, H. Huston and S. Ashok, Conducting MIS diode gas detectors—The Pd/SiOx/Si hydrogen sensor, *Sens. Actuators* **2**, 363–369 (1982).

100. H. Geistlinger, I. Eisele, B. Flietner and R. Winter, Dipole- and charge transfer contributions to the work function change of semiconducting thin films: Experiment and theory, *Sens. Actuators B* **34**, 499–505 (1996).

101. I. Lundström and L.G. Petersson, Chemical sensors with catalytic metal gates, *J. Vac. Sci. Technol. A* **14**, 1539–1545 (1996).

102. I. Lundström, Why bother about gas-sensitive field-effect devices?, *Sens. Actuators A* **56**, 75–82 (1996) and references therein.

103. M. Eriksson and L.G. Ekedahl, Hydrogen adsorption states at the Pd/SiO$_2$ interface and simulation of the response of a Pd metal-oxide-semiconductor hydrogen sensor, *J. Appl. Phys.* **83**, 3947–3951 (1998).

104. M. Johansson, I. Lundström and L.G. Ekedahl, Bridging the pressure gap for palladium metal-insulator-semiconductor hydrogen sensors in oxygen containing environments, *J. Appl. Phys.* **84**, 44–51 (1998).

105. D. Filippini, R. Aragon and U. Weimar, NO$_2$ sensitive Au gate metal-oxide-semiconductor capacitors, *J. Appl. Phys.* **90**, 1883–1886 (2001).

106. D. Filippini, L. Fraigi, R. Aragon and U. Weimar, Thick film Au-gate field-effect devices sensitive to NO$_2$, *Sens. Actuators B* **81**, 296–300 (2002).

107. D. Filippini and I. Lundström, Chemical images generated by large area homogeneous illumination of metal-insulator-semiconductor structures, *Appl. Phys. Lett.* **82**, 3791–3793 (2003).

108. W.-C. Liu, H.J. Pan, H.-I. Chen, K.-W. Lin and C.K. Wang, Comparative hydrogen-sensing study of Pd/GaAs and Pd/InP metal-oxide-semiconductor Schottky diodes, *Japn. J. Appl. Phys.* **40**, 6254–6259 (2001).

109. A. Spetz, M. Armgarth and I. Lundström, Hydrogen and ammonia response of metal-silicon dioxide-silicon structures with thin platinum gates, *J. Appl. Phys.* **64**, 1274–1283 (1988).

110. H. Nienhaus, S.J. Weyers, B. Gergen and E.W. McFarland, Thin Au/Ge Schottky diodes for detection of chemical reaction induced electron excitation., *Sens. Actuators B* **87**, 421–424 (2002).

111. H. Nienhaus, Electronic excitations by chemical reactions on metal surfaces, *Surf. Sci. Rep.* **45**, 3–78 (2002).

112. H. Nienhaus, H.S. Bergh, B. Gergen, A. Majumdar, W.H. Weinberg and E.W. McFarland, Direct detection of electron-hole pairs generated by chemical reactions on metal surfaces, *Surf. Sci.* **445**, 335–342 (2000).

113. H. Nienhaus, H.S. Bergh, B. Gergen, A. Majumdar, W.H. Weinberg and E.W. McFarland, Electron-hole pair creation at Ag and Cu surfaces by adsorption of atomic hydrogen and deuterium, *Phys. Rev. Lett.* **82**, 446–449 (1999).

114. H. Nienhaus, H.S. Bergh, B. Gergen, A. Majumdar, W.H. Weinberg and E.W. McFarland, Selective H atom sensors using ultrathin Ag/Si Schottky diodes, *Appl. Phys. Lett.* **74**, 4046–4048 (1999).

115. H. Nienhaus, H.S. Bergh, B. Gergen, A. Majumdar, W.H. Weinberg and E.W. McFarland, Ultrathin Cu films on Si(111): Schottky barrier formation and sensor applications, *J. Vac. Sci. Technol. A* **17**, 1683–1687 (1999).

116. B. Gergen, H. Nienhaus, W.H. Weinberg and E.W. McFarland, Chemically induced electronic excitations at metal surfaces, *Science* **294**, 2521–2523 (2001).

117. B. Gergen, S.J. Weyers, H. Nienhaus, W.H. Weinberg and E.W. McFarland, Observation of excited electrons from nonadiabatic molecular reactions of NO and O$_2$ on polycrystalline Ag, *Surf. Sci.* **488**, 123–132 (2001).

118. H.S. Bergh, B. Gergen, H. Nienhaus, A. Majumdar, W.H. Weinberg and E.W. McFarland, An ultrahigh vacuum system for the fabrication and characterization of ultrathin metal-semiconductor films and sensors, *Rev. Sci. Instr.* **70**, 2087–2094 (1999).

119. H. Nienhaus, B. Gergen, W.H. Weinberg and E.W. McFarland, Detection of chemically induced hot charge carriers with ultrathin metal film Schottky contacts, *Surf. Sci.* **514**, 172–181 (2002).

120. B. Gergen, H. Nienhaus, W.H. Weinberg and E.M. McFarland, Morphological investigation of ultrathin Ag and Ti films grown on hydrogen terminated Si(111), *J. Vac. Sci. Technol. B* **18**, 2401–2405 (2000).

121. B. Gergen, Observations of electronic excitations in gas-metal interactions, PhD thesis, (University of California Santa Barbara, 2001)

122. B. Roldan Cuenya and E.W. McFarland, Sensors based on chemicurrents, in: *Dekker Encyclopedia Nanosci. Nanotechnol.*, edited by James A. Schwarz, Cristian I. Contescu, Karol Putyera, p. 3527 (Marcel Dekker, New York, 2004).

123. B. Roldan Cuenya, H. Nienhaus and E.W. McFarland, Chemically induced charge carrier production and transport in Pd/SiO₂/n-Si(111) MOS Schottky diodes, *Phys. Rev. B* **70**, 115322 (2004).

124. X. Liu, B. Roldan Cuenya and E.W. McFarland, A MIS device structure for detection of chemically induced charge carriers, *Sens. Actuators B* **99**, 556–561 (2004).

125. C.R. Crowell, L.E. Howarth, W.G. Spitzer and E.E. Labate, Attenuation length measurements of hot electrons in metal films, *Phys. Rev.* **127**, 2006–2015 (1962).

126. J. Y. Park and G.A. Somorjai, The catalytic nanodiode: Detecting continuous electron flow at oxide-metal interfaces generated by a gas-phase exothermic reaction, *Chem. Phys. Chem.* **7**, 1409–1413 (2006).

127. J. Y. Park and G.A. Somorjai, Energy conversion from catalytic reaction to hot electron current with metal-semiconductor Schottky nanodiodes, *J. Vac. Sci. Technol. B* **24**, 1967–1971 (2006).

128. M.S. Arnold, P. Avouris, Z.W. Pan and Z.L. Wang, Field-effect transistors based on single semi-conducting oxide nanobelts, *J. Phys. Chem. B* **107**, 659–663 (2003).

129. C. Li, D.H. Zhang, X.L. Liu, et al., In₂O₃ nanowires as chemical sensors, *Appl. Phys. Lett.* **82**, 1613–1615 (2003).

130. Y.L. Wang, X.C. Jiang and Y.N. Xia, A solution-phase, precursor route to polycrystalline SnO₂ nanowires that can be used for gas sensing under ambient conditions, *J. Am. Chem. Soc.* **125**, 16176–16177 (2003).

131. Z.Y. Fan, D.W. Wang, P.C. Chang, W.Y. Tseng and J.G. Lu, ZnO nanowire field-effect transistor and oxygen sensing property, *Appl. Phys. Lett.* **85**, 5923–5925 (2004).

132. B.J. Murray, J.T. Newberg, E.C. Walter, Q. Li, J.C. Hemminger and R.M. Penner, Reversible resistance modulation in mesoscopic silver wires induced by exposure to amine vapor, *Anal. Chem.* **77**, 5205–5214 (2005).

133. D.J. Zhang, C. Li, X.L. Liu, S. Han, T. Tang and C.W. Zhou, Doping dependent NH₃ sensing of indium oxide nanowires, *Appl. Phys. Lett.* **83**, 1845–1847 (2003).

134. S. Semancik, R.E. Cavicchi, M.C. Wheeler, et al., Microhotplate platforms for chemical sensor research, *Sens. Actuators B* **77**, 579–591 (2001).

135. T. Rantala, V. Lantto and T. Rantala, Computational approaches to the chemical sensitivity of semiconducting tin dioxide, *Sens. Actuators B* **47**, 59–64 (1998).

136. V.E. Henrich and P.A. Cox, Surface science of metal oxides (Cambridge University Press, New York, 1996).

137. U. Diebold, The surface science of titanium dioxide, *Surf. Sci. Rep.* **48**, 53–229 (2003).

138. M. Batzill and U. Diebold, The surface and materials science of tin oxide, *Progr. Surf. Sci.* **79**, 47–154 (2005).

139. V. Lantto, T.T. Rantala and T.S. Rantala, Atomistic understanding of semiconductor gas sensors, *J. Europ. Ceram. Soc.* **21**, 1961–1965 (2001).

140. W. Göpel, J. Hesse and J. N. Zemel, Sensors: A comprehensive survey (VCH, Weinheim, 1995).

141. P.T. Moseley and B.C. Tofield., Solid-state gas sensors, (Adam Hilger, Bristol and Philadelphia, 1987).

142. G. Sberveglieri, Gas sensors: Principles, operation and developments, (Kluwer Academic, Boston, 1992).

143. T. Wolkenstein, Electronic processes on semiconductor surfaces during chemisorption (Springer, New York, 1991).

144. O.V. Krylov and V.F. Kiselev, Electronic phenomena in adsorption and catalysis on semiconductors and dielectrics, *Springer Series Surf. Sci.* **7** (Springer, Berlin, 1987).

145. C. Li, D.H. Zhang, B. Lei, S. Han, X.L. Liu and C.W. Zhou, Surface treatment and doping dependence of In₂O₃ nanowires as ammonia sensors, *J. Phys. Chem. B* **107**, 12451–12455 (2003).

146. A. Kolmakov and M. Moskovits, Chemical sensing and catalysis by one-dimensional metal-oxide nanostructures, *Ann. Rev. Mater. Res.* **34**, 151–180 (2004).

147. Y. Zhang, A. Kolmakov, Y. Lilach and M. Moskovits, Electronic control of chemistry and catalysis at the surface of an individual tin oxide nanowire, *J. Phys. Chem. B* **109**, 1923–1929 (2005).

148. Y. Zhang, A. Kolmakov, S. Chretien, H. Metiu and M. Moskovits, Control of catalytic reactions at the surface of a metal oxide nanowire by manipulating electron density inside it, *Nano Lett.* **4**, 403–407 (2004).

149. X.F. Duan, Y. Huang and C.M. Lieber, Nonvolatile memory and programmable logic from molecule-gated nanowires, *Nano Lett.* **2**, 487–490 (2002).

150. C. Li, W.D. Fan, B. Lei, et al., Multilevel memory based on molecular devices, *Appl. Phys. Lett.* **84**, 1949–1951 (2004).

151. Z.Y. Fan and J.G. Lu, Gate-refreshable nanowire chemical sensors, *Appl. Phys. Lett.* **86**, (2005).

152. S.V. Kalinin, J. Shin, S. Jesse, et al., Electronic transport imaging in a multiwire $SnO_2$ chemical field-effect transistor device, *J. Appl. Phys.* **98**, (2005).

153. J. Goldberger, D.J. Sirbuly, M. Law and P. Yang, ZnO nanowire transistors, *J. Phys. Chem. B* **109**, 9–14 (2005).

154. W.I. Park, G.C. Yi, M. Kim and S.J. Pennycook, Quantum confinement observed in ZnO/ZnMgO nanorod heterostructures, *Adv. Mater.* **15**, 526–529 (2003).

155. Y. Lilach, J.P. Zhang, M. Moskovits and A. Kolmakov, Encoding morphology in oxide nanostructures during their growth, *Nano Lett.* **5**, 2019–2022 (2005).

156. M. Batzill, K. Katsiev, D.J. Gaspar and U. Diebold, Variations of the local electronic surface properties of $TiO_2(110)$ induced by intrinsic and extrinsic defects, *Phys. Rev. B* **66**, 235401 (2002).

157. M. Batzill, E.L.D. Hebenstreit, W. Hebenstreit and U. Diebold, Influence of subsurface, charged impurities on the adsorption of chlorine at $TiO_2(110)$, *Chem. Phys. Lett.* **367**, 319–323 (2003).

158. V.P. Zhdanov, Nm-sized metal particles on a semiconductor surface, Schottky model, etc, *Surf. Sci.* **512**, L331–L334 (2002).

159. A. Kolmakov, D.O. Klenov, Y. Lilach, S. Stemmer and M. Moskovits, Enhanced gas sensing by individual $SnO_2$ nanowires and nanobelts functionalized with Pd catalyst particles, *Nano Lett.* **5**, 667–673 (2005).

160. Z.L. Wang, New developments in transmission electron microscopy for nanotechnology, *Adv. Mater.* **15**, 1497–1514 (2003).

161. A. Kolmakov, X.Chen and M. M. Moskovits, Functionalizing nanowires with catalytic nanoparticles for gas sensing applications, *J. Nanosci. Nanotech.* (in press) (2007).

162. B.H. Frazer, M. Girasole, L.M. Wiese, T. Franz and G. De Stasio, Spectromicroscope for the photoelectron imaging of nanostructures with X-Rays (SPHINX): Performance in biology, medicine and geology, *Ultramicroscopy* **99**, 87–94 (2004).

163. S. Gunther, B. Kaulich, L. Gregoratti and M. Kiskinova, *Prog. Surf. Sci.* **70**, 187–260 (2002).

164. J.W. Chiou, C.L. Yueh, J.C. Jan, et al., Electronic structure of the carbon nanotube tips studied by X-ray-absorption spectroscopy and scanning photoelectron microscopy, *Appl. Phys. Lett.* **81**, 4189–4191 (2002).

165. A. Goldoni, R. Larciprete, L. Gregoratti, et al., X-ray photoelectron microscopy of the C 1s core level of free-standing single-wall carbon nanotube bundles, *Appl. Phys. Lett.* **80**, 2165–2167 (2002).

166. S. Suzuki, Y. Watanabe, T. Ogino, et al., Extremely small diffusion constant of Cs in multiwalled carbon nanotubes, *J. Appl. Phys.* **92**, 7527–7531 (2002).

167. S. Suzuki, Y. Watanabe, T. Ogino, et al., Electronic structure of carbon nanotubes studied by photoelectron spectromicroscopy, *Phys. Rev. B* **66**, 035414 (2002).

168. J.W. Chiou, J.C. Jan, H.M. Tsai, et al., Electronic structure of GaN nanowire studied by X-ray-absorption spectroscopy and scanning photoelectron microscopy, *Appl. Phys. Lett.* **82**, 3949–3951 (2003).

169. I.H. Hong, J.W. Chiou, S.C. Wang, et al., Electronic structure of aligned carbon nanotubes studied by scanning photoelectron microscopy, *J. Phys. IV* **104**, 467–470 (2003).

170. S. Suzuki, Y. Watanabe, T. Ogino, et al., Observation of single-walled carbon nanotubes by photoemission microscopy, *Carbon* **42**, 559–563 (2004).

171. R. Larciprete, A. Goldoni, S. Lizzit and L. Petaccia, The Electronic properties of carbon nanotubes studied by high resolution photoemission spectroscopy, *Appl. Surf. Sci.* **248**, 8–13 (2005).

172. S. Gunther, A. Kolmakov, J. Kovac and M. Kiskinova, Artifact formation in scanning photoelectron emission microscopy, *Ultramicroscopy* **75**, 35–51 (1998).

173. B. Gilbert, R. Andres, P. Perfetti, G. Margaritondo, G. Rempfer and G. De Stasio, Charging phenomena in PEEM imaging and spectroscopy, *Ultramicroscopy* **83**, 129–139 (2000).

174. A. Kolmakov, U. Lanke, R. Karam, J. Shin, S. Jesse and S.V. Kalinin, Local origins of sensor activity in 1D oxide nanostructures: From spectromicroscopy to device, *Nanotechnology* 17 (16): 4014–4018 (2006)

175. A. Kolmakov, The effect of morphology and surface doping on sensitization of quasi-1D metal oxide nanowire gas sensor, *Proc. SPIE 6370*, 63700X-5(2007).

## QUESTION

The peak chemicurrent measured by a lock-in amplifier from the interaction of NO with a Ag(8 nm)/$n$-Si(111) Schottky diode is 35 pA. The sensor operates at 125 K and has a 0.7 cm$^2$ active surface area. Assuming a continuous gas flux of $2 \times 10^{12}$ NO molecules/s cm$^2$, and neglecting current attenuation due to ballistic electron transport across the Ag film, estimate how many incident gas molecules will give rise to a detectable electron.

# 7

# Nanostructured High-Anisotropy Materials for High-Density Magnetic Recording

## J. S. Chen,[1,2,3] C. J. Sun,[2,*] G. M. Chow[2]

[1]Data Storage Institute, Singapore 117608

[2]Department of Material Science and Engineering, National University of Singapore, Singapore 119260

[3]To whom correspondence should be addressed. J. S. Chen, Data Storage Institute, 5, Engineering Drive 1, Singapore 117608, E-mail: Chen_Jingsheng@dsi.a-star.edu.sg

[*]Materials Science and Technology Division, Oak Ridge National Laboratory, Oak Ridge, Tennessee 37831

## 1. INTRODUCTION

Magnetic recording is widely used in infocom technologies and consumer electronic products. The magnetic disk drive featuring a total storage capability of 5 MB at a recording density of 2 Kbit/in.$^2$ was invented at IBM in 1956 using longitudinal magnetic recording (LMR) mode, which aligned the magnetization of the recording bits horizontally parallel to the surface of the disk. The areal densities of hard disk drives increased over 100% per annum during the late 1990s as shown in Fig. 1.

The areal density for information storage on magnetic media, especially hard disk memory, has been increased more than 40 million-fold in modern disk drives. The 80-100 Gbit/in.$^2$ longitudinal media was commercialized in 2004. Compared to LMR, perpendicular magnetic recording (PMR) vertically aligns the magnetization of the recording bits perpendicular to the disk and has a lot of advantages over LMR, which allows higher recording densities. The advantages of PMR over LMR and the features of PMR have been previously reviewed.[1, 2, 3, 4, 5, 6, 7, 8, 9, 10] Recently, the hard disk drives of 170 Gb/in.$^2$ and 240 Gb/in.$^2$ have been demonstrated under the mode of PMR by Seagate in 2004[11] and Hitachi in 2005[12] as shown in Fig. 1.

345

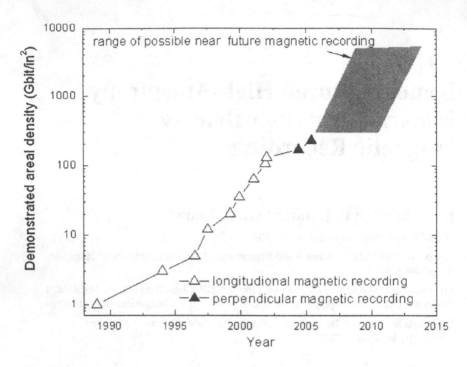

**FIGURE 1.** Demonstrated areal density trend (The data for longitudinal magnetic record-ing were reprinted with permission of Hitachi Global Storage technologies. From Page 19 of HDD Roadmap, http://www.hitachigst.com/hdd/hddpdf/tech/hdd_technology2003. pdf)

In this chapter, we present a review of research and development of ad-vanced materials for next-generation high-density magnetic recording based on thin film media including PMR and heat-assisted magnetic recording[4] (the shaded area marked in Fig. 1). The long-term proposed goals of fabrication of patterned media[13] and self-assembled magnetic structure[14] for advanced magnetic recording are beyond the scope of this current review.

The superparamagnetic effect (thermal stability) of the magnetic media ma-terials remains the main concern in realizing the high areal density.[3] A high areal density requires the reduction of both the average size and size distribution of the grains in each recording bit. Once the grain size is reduced to several nanome-ters, the thermal energy at room temperature becomes comparable to the magnetic anisotropy energy of the grain, resulting in thermal instability of the recording media. In order for the recording bits to be stable over a long period of time, the energy barrier $K_u V$ must significantly exceed the thermal energy $k_B T$, where $K_u$, $V$, $k_B T$ are anisotropy constant, grain volume, Boltzmann constant, and absolute temperature, respectively. The relaxation time $\tau$ is an exponential function of the grain volume, i.e.

$$\tau = 10^{-9} \exp\left(\frac{K_u V}{k_B T}\right)$$

For recording information that will be stable for several years (e.g., 10 years is used as a conventional benchmark), it requires $K_u V \geq 40 \sim 60 k_B T$. In high-density magnetic recording, signal-to-noise ratio (SNR) is proportional to the number of grains per bit. In order to maintain certain SNR and thus a certain number of grains in one recording bit, the grain size must decrease with increasing areal density, and materials with higher $K_u$ are required to maintain thermal stability. The magnetic properties of materials candidates for high-density magnetic recording have been reviewed (Table 1).[10] The higher the $K_u$, the smaller the minimal stable grain size ($D_p$) that can be realized.

Among these candidates, Co alloy–based media were the first to be investigated for perpendicular magnetic recording system. The advantages of these media included the known data on controlling grain size, the relationship between microstructure and magnetic properties, etc. Although the anisotropy of Co alloy increased by increasing Pt contents and decreasing Cr contents, stacking faults were induced as Pt contents exceeded certain a value due to the formation of face-centered-cubic (fcc) CoPt phase. The anisotropy of Co alloy–based media is yet not high enough to achieve thermal stability at very high density magnetic recording. Details of Co alloy–based perpendicular media can be found a recent review.[15]

The two highest anisotropy candidates are $Co_5Sm$ and FePt systems. They can provide thermally stable grains down to 2-3 nm, supporting recording densities of up to a Tbit/in.$^2$ regime in principle. FePt has a better chemical stability over $SmCo_5$.[16] As a result, $L1_0$ FePt media has been pursued as the most promising candidate for next-generation ultrahigh density magnetic recording. In this review, focus is placed on only nanostructured $L1_0$ FePt magnetic materials.

Nanostructured materials may exhibit unique properties due to the size effect and a large amount of interphase interface and surface.[17] The properties of nanostructured, multielement magnetic thin films may be tailored by the control of composition and texture. However, the structural and magnetic properties may not follow the traditional understanding of bulk materials; for example, the phase miscibility in nanoscale regime does not necessarily follow the prediction of conventional thermodynamics that does not consider the interfacial and surface effects. The particle size and interface effects on the structure and magnetic properties of nanostructured FePt alloy films will be addressed.

Practical applications of FePt media films face many technical challenges including desirable reduction of ordering temperature, control of the FePt (001) texture and decrease of media noise, as follows:

1. High temperature required for phase transformation from disordered fcc to ordered face-centered-tetragonal (fct) $L1_0$ phase: Phase transformation for bulk FePt alloy occurs above 1000°C, whereas that for FePt alloy thin films exceeds 500°C. Lower ordering temperature for $L1_0$ phase transformation is highly desirable for practical applications, especially for depositing magnetic media on glass substrate commonly used in media industries.

**TABLE 1.** Basic magnetic properties of magnetic materials for high-density magnetic recording. (Reprinted with permission from Ref.10. Copyright 2000 Institute of Electrical and Electronics Engineers.)

| Alloy system | Materials | $K_u$ ($10^7$ erg/cm$^3$) | $M_s$ (emu/cm$^3$) | $H_k$ (KOe) | $T_c$ (K) | $\delta_w$ (Å) | $\Upsilon$ (erg/cm$^3$) | $D_c$ (µm) | $D_p$ (nm) |
|---|---|---|---|---|---|---|---|---|---|
| Co alloys | CoCrPt | 0.20 | 298 | 13.7 | — | 222 | 5.7 | 0.89 | 10.4 |
| | Co | 0.45 | 1400 | 6.4 | 1404 | 148 | 8.5 | 0.06 | 8.0 |
| | Co$_3$Pt | 2.0 | 1100 | 36 | — | 70 | 18 | 0.21 | 4.8 |
| L1$_0$ phase | FePd | 1.8 | 1100 | 33 | 760 | 75 | 17 | 0.20 | 5.0 |
| | FePt | 6.6-10 | 1140 | 116 | 750 | 39 | 32 | 0.34 | 3.3-2.8 |
| | CoPt | 4.9 | 800 | 123 | 840 | 45 | 28 | 0.61 | 3.6 |
| | MnAl | 1.7 | 560 | 69 | 650 | 77 | 16 | 0.71 | 5.1 |
| Rare earth | Fe$_{14}$Nd$_2$B | 4.6 | 1270 | 73 | 585 | 46 | 27 | 0.23 | 3.7 |
| Transition metals | SmCo$_5$ | 11-20 | 910 | 240-400 | 1000 | 22-30 | 42-57 | 0.71-0.96 | 2.7-2.2 |

Magnetocrystalline anisotropy: $K_u$ (the first-order magetrocrystalline anisotropy constant $K_1$ was taken as $K_u$ for uniaxial systems.
Saturation magnetization: $M_s$
Curie temperature: $T_c$
Anisotropy field: $H_k = 2K_u/M_s$
Domain wall width: $\delta_w = \pi(A/K_u)^{1/2}$
Single particle domain size: $D_c = 1.4\gamma_w/M_s^2$
Exchange coupling constant: $A = 10^{-6}$ erg/cm
Minimal stable grain size: $D_P = (60k_B T/K_u)^{1/3}$

2. Control of FePt (001) texture: FePt films deposited by physical vapor deposition generally show the (111) preferred orientation since the (111) plane is closest packed with lowest surface energy. For FePt perpendicular media, it is desirable to achieve the (001) texture since the uniaxial magnetocrystalline anisotropy is aligned with the $c$-axis.

3. Reduction in media noise: The media noise mainly originates from the transition noise. The reduction of magnetic domain size to decrease media noise is therefore required.

In the following, the current status and progress of the above scientific and technological issues of $L1_0$ FePt film for ultrahigh density magnetic recording are discussed.

## 2. DEFINITION AND CHARACTERIZATION OF CHEMICAL ORDERING OF $L1_0$ FePt

### 2.1. Chemical Ordering of $L1_0$ FePt

In a random solid solution consisting of two elements, M and N, the atoms are randomly distributed in the lattice. For some solid solutions, the atoms M and N form random structure at high temperatures. When these solutions are cooled to below a critical temperature, M and N atoms arrange themselves in a certain order. The change in atomic arrangement upon ordering may alter the physical and chemical properties of the solid solution. The existence of ordering may be inferred from some of these changes. For example, FePt shows a chemical disordered fcc structure at high temperatures as shown in the phase diagram (Fig. 2).[18] At lower temperatures, the Fe and Pt atoms are ordered in an atomic multilayer structure, known as $L1_0$ chemical ordering. The temperature of fcc to $L1_0$ transformation is $\sim1300°C$. The magnetocrystalline anisotropies of fcc and $L1_0$ FePt are dramatically different. Although this is a first-order transition, the phase transformation can occur even the temperature is away from the equilibrium phase boundary and below the instability temperature for the disordered alloy.[19]

The $L1_0$ FePt phase belongs to the P4/mmm space group, with the configuration of alternating layers of Fe and Pt atoms when in a perfect chemical order (Fig. 3). It is a crystallographic derivative structure of the fcc structure. An important crystallographic feature of the $L1_0$ structure is its $c/a$ ratio. The lattice constant $a$ of $L1_0$ FePt is slightly larger than lattice constant $c$, where $c/a$ is equal to 0.968. While for fcc FePt, $c/a$ is equal to 1.

In real materials, $L1_0$ chemical structure usually departs from perfect order and this departure can be described by the long-range ordering parameter $S$ which is defined as follow:

$$S = \frac{R_A - C_A}{1 - C_A} \tag{1}$$

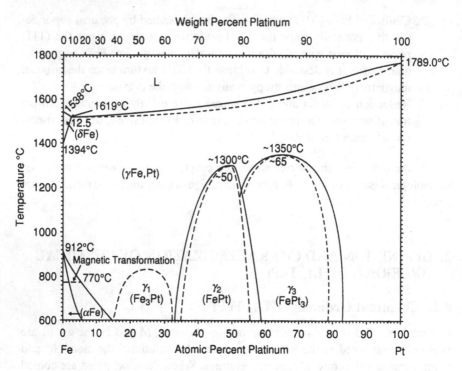

**FIGURE 2.** The FePt Phase diagram (Reprinted with permission of ASM International®. All rights reserved. www.asminternational.org)

where $R_A$ is the fraction of $A$ sites occupied by "right" atoms, i.e., $A$ atoms, and $C_A$ is the fraction of $A$ atoms in the alloy. For the $Fe_{50}Pt_{50}$ alloy, when the long-range order is perfect and Fe, Pt atoms occupy the "right" sites to form the alternating stack of Fe and Pt atoms ($R_{Fe}$ and $R_{Pt}$ are equal to 1), $S$ is equal

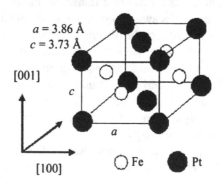

**FIGURE 3.** Illustration of chemically ordered FePt, Fe and Pt atoms are shown in open and solid circles, respectively.

to 1. When the atomic arrangement is completely random, $R_{Fe} = C_{Fe} = 50\%$ or $R_{Pt} = C_{Pt} = 50\%$ according to statistical theory, and thus $S$ equals 0.[20]

It is essential to qualitatively identify the existence and quantitatively characterize the extent of chemical order, for correlation of structure with properties of materials.

## 2.2. Characterization of L1$_0$ FePt Chemical Ordering

Among various experimental techniques for characterizing the chemical ordering of $L1_0$ FePt, X-ray diffraction (XRD) and selected area diffraction (SAD) using transmission electron microscopy (TEM) are the two common conventional methods. In this section, the characterization of $S$, and the investigation of crystallographic texture between substrate and epitaxial L1$_0$ FePt film using XRD[21] and TEM[22] are discussed.

### 2.2.1. Quantitative Evaluation of the Chemical Ordering of L1$_0$ FePt by XRD

In the L1$_0$-ordered FePt phase, the structure factor for fundamental and superlattice peaks are[23]:

$$F_f = 4(C_{Fe}f_{Fe} + C_{Pt}f_{Pt}) \quad \text{for fundamental diffraction peaks} \quad (2)$$

$$F_s = 2S(f_{Fe} - f_{Pt}) \quad \text{for superlattice peaks} \quad (3)$$

where, $f_{Fe}$ and $f_{Pt}$ are atomic form factors for Fe and Pt, respectively.

As the integrated intensity of XRD peaks are associated with the structure factor, integrated intensities for the fundamental and superlattice peaks are expressed as

$$I(hkl)_f = C_0 F_f^2 L_f P_f A_f e^{-2M} \quad \text{for fundamental diffraction peaks} \quad (4)$$

$$I(hkl)_s = C_0 F_s^2 L_s P_s A_s e^{-2M} \quad \text{for superlattice diffraction peaks} \quad (5)$$

Where $C_0$ is a constant, $A$ the area interacted with X-ray, $L$ the Lorentz factor, and $P$ the X-ray dependent polarization factor.

The Lorentz factors of polycrystalline films and expitaxial films may be expressed as follows:

$$L = 1/(\sin\theta \cdot \sin 2\theta) \quad \text{for polycrystalline films} \quad (6)$$

$$L = 1/(\cos\theta \cdot \sin\theta) \quad \text{for epitaxial films} \quad (7)$$

Therefore, using Eqs. 3-7, the values of $S$ can be calculated from the (001) superlattice and (002) fundamental diffraction peaks or the (110) and (220) peaks for epitaxial L1$_0$ FePt film and polycrystalline films using corresponding Lorentz factors. Therefore, we have

$$S = \left[ \left( \frac{I001}{I002} \right) \times \left( \frac{Ff}{Fs} \right) \frac{(L_f \times AC_f \times D_f)}{(L_s \times AC_s \times D_s)} \right]^{1/2} \quad (8)$$

where $AC$ and $D$ are the absorption correction factors and temperature factors, respectively.

For epitaxial films, the use of integrated intensity in powder scan alone is not sufficient for the actual integrated peak intensity. The actual value should be the integrated intensity of diffraction peaks multiplying the full-width at the half maximum of the rocking curve.[24]

Another formula for estimating the chemical-order parameter $S$ for epitaxial $L1_0$ FePt films was given in Eq. 9[25]:

$$S^2 = \frac{(I_{001}/I_{002})_{\text{meas}}}{(I_{001}/I_{002})_{\text{calc}}} \tag{9}$$

where $I_{001}$ and $I_{002}$ are the integrated intensities of (001) superlattice diffraction and (002) fundamental diffractions, $(I_{001}/I_{002})_{\text{meas}}$ and $(I_{001}/I_{002})_{\text{calc}}$ are the measured and calculated ratio of diffraction intensities. The $(I_{001}/I_{002})_{\text{calc}}$ is estimated to be 1.9-2.0 for film thickness ranging from 11 to 49 nm.

### 2.2.2. Characterization of Crystallographic Texture of $L1_0$ FePt by XRD

XRD is a useful characterization tool of the degree of $L1_0$ FePt chemical ordering. High-resolution XRD is a powerful tool to determine epitaxial relationship between single-crystal substrates and $L1_0$ FePt films. For example, $L1_0$ FePt (200) texture was expected to grow heteroepitaxially by the crystallographic relationship of Cu (002)[100]//fct FePt (200)[001].[21] This arose from a small lattice misfit of the Cu (002)/$L1_0$ FePt (200) planes (6.38%) in the out-of-plane direction, and that of the Cu (200) and $L1_0$FePt (002) planes (2.97%) in the in-plane direction.[26] The relative smaller misfit in the film plane of the latter case favored formation of $L1_0$FePt with the $c$ axis in the film plane.

$Fe_{50}Pt_{50}$ (30 nm) thin films were deposited by dc magnetron sputtering on Cu(001) single-crystal substrate at 400°C. The base pressure was $10^{-9} \sim 10^{-8}$ Torr, and Ar sputtering pressure was kept at 10 mTorr for FePt deposition. Magnetic properties of films were measured with a vibrating sample magnetometer. The structure of the films was investigated using high-resolution (HR) X-ray scattering.

Figure 4 shows a schematic illustration of the experimental X-ray scattering geometry. Since the fcc FePt (101) cannot be detected as the result of structure extinction, the $L1_0$ FePt (101) reflection was used to evaluate the existence of the $L1_0$ ordered fct structure. The epitaxial relationship between FePt (fct) and Cu was studied by observing the orientation of the off-specular reflections of FePt (101) and Cu (202) of the film with respect to that of the substrate on the azimuthal circles (Fig. 4).

The off-specular Cu (202) and fct FePt (101) reflection (45° and 45.9° away from the specular rod, respectively) was investigated by performing the azimuthal circle scans 1 and 2 (Fig. 4). The well-defined peaks in the azimuthal scans of Figs. 5a and 5b showed that the FePt (101) reflection followed the Cu (202)

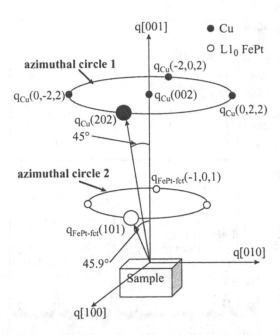

**FIGURE 4.** The epitaxial relationship between Cu (001) and $L1_0$ FePt (fct) was studied using the off-specular $q_{Cu}$ (2,0,2) and $q_{FePt\text{-}fct}$ (101) reflections of the film with respect to that of the substrate on the respective azimuthal circles. The reciprocal lattice points of $q_{Cu}$ (202) and $q_{FePt\text{-}fct}$ (101) are shown in larger circles. $q$ was the momentum transfer. (Reprinted with permission from C. J. Sun, et.al. Appl. Phys. Lett. 82, 1902 (2003), Copyright [2003] American Institute of Physics.)

reflection. This indicated that fct FePt was "epitaxially" grown on the Cu (001) substrate. Since there are two domains of fct FePt (001) perpendicular to each other due to the fourfold symmetry of the Cu (001) substrate, the sample was not truly epitaxial as a single-crystal thin film. The fourfold symmetric easy axis existed in the film. The epitaxial relationship can be revealed by the direction of the momentum transfer at the peak projected to the film plane relative to the in-plane orientation of the Cu substrate. From the relative directions of the $L1_0$ FePt and substrate crystalline axes, it was confirmed that the epitaxial relationship was $L1_0$ FePt (200)<001>// Cu (002) <100>.

## 2.2.3. Characterization of Crystallographic Texture of $L1_0$ FePt by TEM

Transmission electron microscopy is a useful tool for characterization of microstructures. The TEM electron diffraction patterns provide structural information of microstructures and correlation between substrate and epitaxial film.[27] Compared with XRD, the reduction in symmetry associated with $L1_0$ FePt phase results

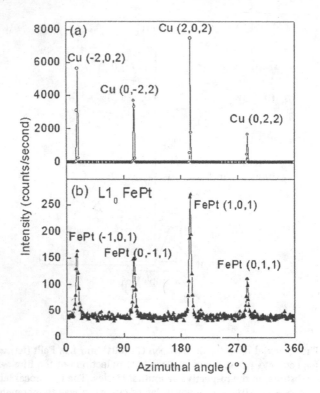

**FIGURE 5.** The intensity profiles of azimuthal scans of (a) the Cu (202), and (b) the $L1_0$ FePt (101). The application of the absorber during the Cu (202) measurement is responsible for the different backgrounds in the Fig. 5a and 5b. (Reprinted with permission from C. J. Sun, et.al. Appl. Phys. Lett. 82, 1902 (2003), Copyright [2003] American Institute of Physics.)

in extra diffraction spots (epitaxial films) or rings (polycrystalline films) in the electron diffraction patterns. These are normally used to form dark field images of the chemically ordered regions rather than quantitatively analyze the chemical order as in XRD. An example of TEM characterization of crystallographic texture between single crystal substrate and epitaxial $L1_0$ FePt films was reported.[22]

Fe$_{50}$Pt$_{50}$ (30 nm) thin films were deposited by dc magnetron sputtering on Cu (001) single crystal substrate at 400°C.[21] Figure 6 shows the SAD pattern from the planar view in Cu. The clear and strong $L1_0$ FePt (001) spot indicated that the zone axis for the SAD was the $a$-axis (long axis) of $L1_0$ FePt, and the $c$-axis (short axis) of $L1_0$ FePt was in the film plane. The FePt (002) and FePt (022) were overshadowed by Cu (200) and Cu (220) due to the strong Cu diffraction spots, respectively. The weak and spotty FePt (111) diffraction ring observed was due to some randomly oriented FePt crystallites.

Figure 7 shows the nano-beam diffraction (NBD) from cross-sectioned film in the Cu [100] direction. The clear and strong $L1_0$ FePt and (110) spot was observed,

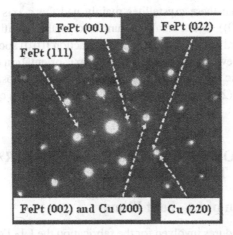

**FIGURE 6.** Electron diffraction pattern from planar view. (Reprinted with permission from C.J. Sun, et.al. J. Appl. Phys. 97, 10J103 (2005), Copyright [2005] American Institute of Physics.)

indicating that the zone axis for the NBD was the $c$-axis (short axis), and the $a$-axis (long axis) was along the surface normal of deposited film. The $L1_0$ FePt (220) and Cu (022) spots were superimposed. Further, the strong spot of FePt (100) and weak spot of FePt (010) in Fig. 7 showed that the zone axis of some crystallites was along the long axis, indicating the $c$-axis of these crystallites was either in the plane or along the surface normal of the film. Since the planar view of SAD (Fig. 6) indicated $a$-axis orientation, and magnetic observation of in-plane anisotropy,[22]

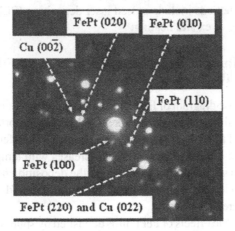

**FIGURE 7.** Electron diffraction pattern (NBD) from cross-sectioned view. (Reprinted with permission from C.J. Sun, et.al. J. Appl. Phys. 97, 10J103 (2005), Copyright [2005] American Institute of Physics.)

it was suggested that these crystallites mainly had the $c$-axis in the film plane (strong spot of (100)), and a small amount of crystallites with $c$-axis along the film's surface normal resulted in the weak diffraction (010) spot. The NBD from a cross-sectioned view further confirmed the $a$-axis orientation of $L1_0$ FePt films. This crystallographic texture correlation between Cu (001) substrate and $L1_0$ FePt films was $L1_0$ FePt (200) <001>//Cu (002) <100>, which is consistent with the XRD results.

## 3. PREPARATION OF $L1_0$ FePt FILMS AND PARAMETERS AFFECTING THE CHEMICAL ORDERING

### 3.1. Preparation of $L1_0$ FePt Films

There are two procedures involved for the fabrication the $L1_0$ FePt films. First, Fe and Pt atoms were condensed on the substrate to form the FePt alloy in thin film state. Then the chemically ordered $L1_0$ phase can be achieved by surface diffusion of deposited atoms via in situ heating, bulk diffusion in ex situ heating or other methods such as ion beam irradiation to promote interdiffusions of Fe and Pt.

Evaporation and sputtering are the most commonly used methods for condensing Fe and Pt atoms on the substrate. The surface diffusivity is orders of magnitude higher than the volume diffusivity. It is therefore expected that chemically ordered $L1_0$ FePt films would be fabricated with in situ heating at lower substrate temperature than that with ex situ heating. The $L1_0$ FePt films were fabricated by molecular beam epitaxy (MBE) (co-evaporation of Fe and Pt), co-sputtering of Fe and Pt targets, sputtering FePt alloy targets at elevated substrate temperature or ex situ annealing the sputtered FePt alloy film or Fe/Pt multilayer at high temperature. A detailed review of preparation of $L1_0$ FePt films by MBE and sputtering can be found.[28] Other methods include pulsed laser deposition (PLD),[29, 30, 31, 32] with either in situ heating or ex situ heating.

### 3.2. Effects of Temperature, Stoichiometry and Film Thickness on Chemical Ordering

According to the phase diagram of FePt, the temperature and stoichiometry are two important factors of the $S$. The film thickness is also very important since it affects atomic diffusion and thus the $S$ at a given temperature. The effects of temperature on $S$ of FePt were discussed.[28] Generally, $S$ increases with temperature. But $S$ at given temperature is significantly dependent on the deposition method, deposition rate and substrate, etc. Here the effects of stoichiometry and film thickness on chemical ordering are discussed.

The stoichiometric effects of FePt films deposited by sputtering on MgO(200) single-crystal substrate at 300°C were investigated.[33] The Fe concentration was varied from 19 to 68 at. %. The XRD (001) and (003) superlattice peaks were observed at a Pt-rich composition range, indicating formation of $L1_0$ ordered

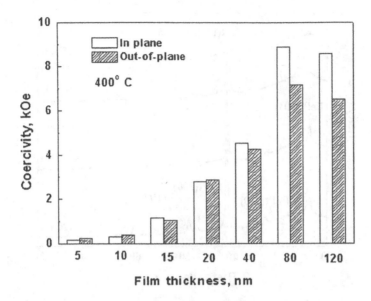

**FIGURE 8.** In-plane and out-of-plane $H_C$ for FePt thin films with different thicknesses deposited at 400°C.

structure. The maximum values of the $S$ and $K_u$ were 0.660.1 and 1.8 $\times 10^7$ erg/cm$^3$, respectively, for 38 at. % of Fe.

The effects of thickness of FePt film sputter-deposited at 400°C were also studied.[34] Figure 8 shows the thickness dependence of the in-plane and out-of-plane coercivities ($H_C$). For FePt films with thickness of 5 and 10 nm, both in-plane and out-of-plane $H_C$ were about two or three hundreds Oe. The low $H_C$ indicated dominance of disordered fcc FePt phase in thin FePt films on the glass substrate at 400°C. With thickness of 15 nm or greater, the $H_C$ increased to more than 1 kOe. With 80 nm thickness, FePt thin film showed maximum $H_C$ of 8 kOe in the in-plane direction; the $H_C$ decreased with further increasing film thickness.

Figure 9 shows the XRD scans of the series of FePt films with thicknesses varying from 15 to 80 nm. The 15 nm FePt film showed only the (111) peak and an amorphous hump at 20-25°, indicating a low degree of crystallization and dominance of disordered phase in the film, with a corresponding small coercivity. When the thickness exceeded 40 nm, the superlattice peaks such as (001) and (110) appeared. When the film thickness was between 80 and 120 nm, most of superlattice peaks appeared with stronger intensity, as shown for the 120 nm FePt film in Fig. 9.

The magnetic and microstructural properties of the FePt films deposited at 300°C with different thicknesses were investigated.[35, 36] The $H_C$ and $S$ showed strong dependence on the film thickness. The $S$ was nearly 0 for film thinner than 50 nm. The $S$ increased to 0.8 for a 300 nm thick film.

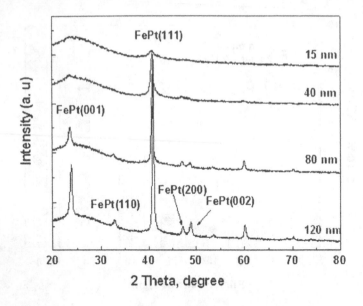

**FIGURE 9.** XRD scans for FePt thin films with different thicknesses.

## 3.3. Promotion of Chemical Ordering by Doping

Elements such as Cu and Ag are effective dopants to promote ordering of FePt. Addition of optimized concentration of 15% Cu into FePt alloy reduced the ordering temperature to 300°C.[37] The $H_C$ of $(Fe_{46.5}Pt_{53.5})_{85}Cu_{15}$ film exceeded 5 kOe, whereas undoped FePt film remained magnetically soft. The ternary alloy of FePtCu was formed[38] and the Fe site in the lattice was substituted by Cu. It was argued that the Gibbs free energy of FePtCu alloy was smaller than that of FePt, thus enhancing the driven force for the fcc-L1$_0$ transformation.[39] It was also reported that the ordering temperature of FePtCu film was reduced.[40] The $H_C$ was 8 kOe and the S was equal to 1 after annealing at 400°C. The $H_C$ and the S of undoped FePt films remained very low at the same temperature. The optimized composition was found to be $(FePt)_{96}Cu_4$. The Cu doping enhanced diffusion-driven phase transformation since it decreased the melting temperature of the alloy. The contradictory results however had also been reported.[41, 42] It was found that Cu decreased the $H_C$ and did not significantly improve the L1$_0$ ordering process. It was suggested that the decrease of ordering temperature by Cu doping was due to deviation from the equiatomic FePt composition which would alter the ordering kinetics.[43] A study using differential scanning calorimetry showed large dependence of ordering temperature on the ratio of Fe:Pt.[44]

The effects of Ag doping on the L1$_0$ ordering of thin FePt films have also been reported.[45, 46] The $H_C$ of the FePt-Ag increased with Ag doping. In-plane and out-of-plane $H_C$ as a function of the Ag content are shown in Fig. 10. When the Ag content was 30 vol. %, the in-plane $H_C$ was at a maximum. The increase

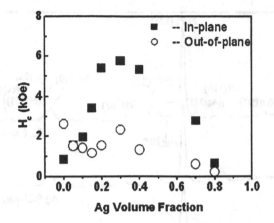

**FIGURE 10.** $H_C$ as a function of Ag volume fraction. (Reprinted with permission from Y. Z. Zhou, et al., J. Appl. Phys. **93**, 7577 (2003), Copyright [2003] American Institute of Physics.)

of the $H_C$ was attributed to both the increase of degree of ordering and the change in magnetization reversal from domain wall motion to rotation mode upon Ag addition. It was also reported that adding 15% Ag in FePt nanoparticles, the $H_C$ of the nanoparticles assembly exceeded 10 kOe after annealing at 500°C.[47] Without Ag addition, the FePt nanoparticles assembly was only 2 kOe. In this case both FePt and FePt-Ag nanoparticles were well separated. The increase of $H_C$ of FePt-Ag can therefore be attributed to increase of $L1_0$ ordering and magnetocrystalline anisotropy.

The promotion of $L1_0$ ordering with Ag doping may be explained as follow:[48, 49] Ag has a low solubility with both Fe and Pt. Furthermore, Ag has a low surface energy that promotes its segregation. When Ag is incorporated in FePt nanoparticles, it diffuses out of the particles upon heating, resulting in creation of vacancies in the FePt lattices. This in turn enhances the kinetics of $L1_0$ ordering and thus decreases the ordering temperature. However, it was found that when Ag overlayer was deposited on FePt films at 400°C, $L1_0$ ordering was also promoted.[50] Figure 11 shows the in-plane X-ray diffraction (XRD) scans of FePt thin films with different thicknesses of Ag overlayer, using an incident X-ray angle of 0.5°. The FePt film without Ag overlayer mostly showed the fundamental (111), (200) and (022) peaks of the FePt alloy and only a weak superlattice (110) peak of the $L1_0$ FePt phase in the XRD spectrum. Upon introduction of 1 nm Ag overlayer, the superlattice (110) peaks became stronger and other superlattice peaks such as (201) and the weak (221) and (003) peaks appeared. The hysteresis loops of FePt films with various Ag overlayer thickness are shown in Fig. 12.

The $H_C$ also increased from 1 kOe to 6.4 kOe with a deposited 4 nm Ag overlayer. The mechanisms of the $L1_0$ ordering promotion by the Ag overlayer should be quite different from that of the FePt-Ag composite films since there

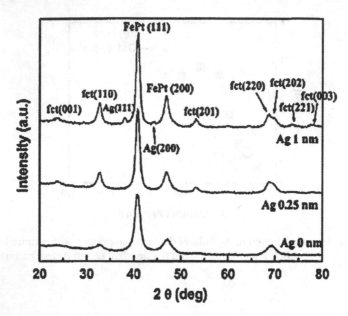

**FIGURE 11.** XRD FePt films with various Ag overlayer thickness. (Reprinted with permission from Z. L. Zhao, et al., Appl. Phys. Lett. 83, 2196 (2003), Copyright [2003] American Institute of Physics.)

was no vacancy created in the FePt lattice. Further investigations on the effects of co-deposited Ag and Ag overlayer on ordering of FePt are warranted.

Doping with Zr also enhanced the chemical ordering kinetics for annealed polycrystalline films prepared by sputtering.[51] However, in this case, the $H_C$ of the alloy only increased for films subject to short annealing time. It was proposed the larger covalent radius of Zr introduced lattice strain and defects in as-deposited films, enhancing diffusion of Fe and Pt and ordering kinetics. For longer annealing, Zr preferentially reacted with Pt to form an inter-metallic compound at the expense of the FePt $L1_0$ phase.

## 3.4. Strain- or Stress-Induced $L1_0$ Ordering

When the phase transformation from fcc to $L1_0$ phase occurs, the lattice constant $a$ increases and the lattice constant $c$ decreases. One can imagine that when a force is applied either parallel or perpendicular to the cube surface, the cube may deform to a tetragonal shape.

Based on this simple argument, various methods to promote the phase transformation from fcc to $L1_0$ phase were proposed.[52, 53, 54, 55, 56] The $L1_0$ ordered FePt film was deposited at 400°C using the high-pressure sputtering.[52] The compressive stress along the $c$-axis resulting from high gas pressure (100 Pa) was suggested to favor the formation of the $L1_0$ phase. The $L1_0$ FePt film was also prepared at 275°C

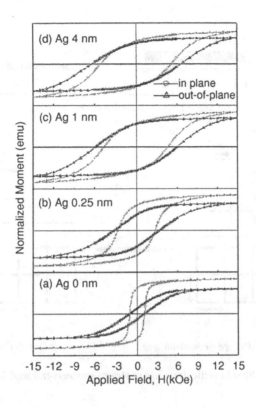

**FIGURE 12.** Hysteresis loops of FePt films with various Ag overlayer thickness. (Reprinted with permission from Z. L. Zhao, et al., Appl. Phys. Lett. **83**, 2196 (2003), Copyright [2003] American Institute of Physics.)

by the so-called dynamic stress.[56] In this method, FePt films were deposited on Cu underlayer/HF-cleaned Si (100) substrate, followed by post-deposition annealing. With increasing annealing temperature up to 250°C, the stress increased due to the difference in thermal expansion coefficients between Cu film and the Si substrate. The stress increased with further increasing temperatures due to formation of $Cu_3Si$. With the completion of reaction of Cu and Si, the stress tended to relax. It was demonstrated that the $L1_0$ ordering of FePt films was closely related to the change of the stress. Note that the ordering of FePt films associated with the lattice distortion was achieved by contraction along the [111], not the [001] direction.

When a thin film with certain lattice mismatch with the substrate is deposited, the lattice parameter of the film can either expand or shrink in the film direction depending on the negative or positive mismatch. For a negative lattice mismatch, i.e., the lattice constant of film is less than that of substrate; the lattice constant can be expanded in the film direction and shrunk in the film normal direction. The schematic drawings of this principle are shown in Fig. 13.

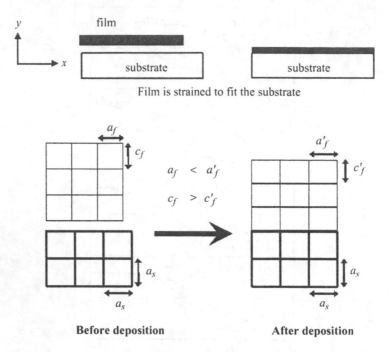

**FIGURE 13.** The schematic drawings of strain- or stress-induced $L1_0$ ordering.

There exists a critical film thickness for the lattice mismatch between the film and substrate. Beyond the critical thickness, the strain induced by the lattice mismatch would be relaxed by forming defects such as dislocations and the lattice constant of the top part of the film would revert to the un-strained value.

It has been suggested that the lattice mismatch at the interface between FePt film and underlayer may expand the $a$-axis and shrink the $c$-axis of the FePt film, favoring the ordering at low temperatures. The $L1_0$ FePt films were successfully deposited on Ag underlayer and CrX (X = Ru, Mo, W, Ti) underlayer at low temperatures.[57,22]

For further understanding, FePt films were deposited on 30 nm thick Pt, Cr, $Cr_{95}Mo_5$ and $Cr_{90}Mo_{10}$ intermediate layers/ MgO (100) substrate.[55] The intermediate layers were used to adjust the lattice mismatch. The relationships between the lattice mismatch and the chemical ordering of the FePt films, and their magnetic anisotropic constant were investigated.[55] The epitaxial relationships between the FePt films and/or intermediate layers and MgO substrates were investigated by off-specular phi scan, as shown in Fig. 14.[55] The off-specular MgO (202) and FePt (202) phi scans (45° and 44.1° away from the specular rod, respectively), MgO (202), Pt (202) and FePt (202) phi scans (45°, 45° and 44.1° away from the specular rod, respectively) and MgO (111), Cr (101) and FePt (111) phi scans (45°, 45° and 44.1° away from the specular rod, respectively) were investigated

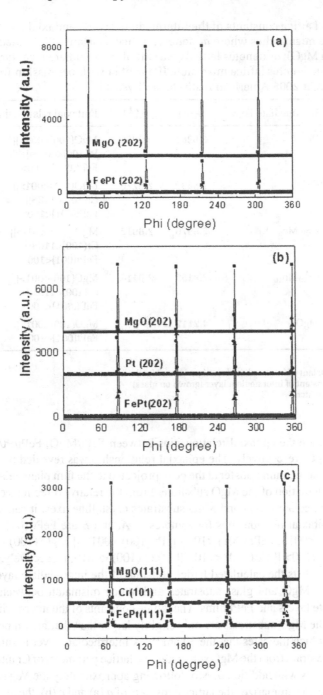

**FIGURE 14.** Off-specular phi scan of FePt films and MgO substrates (a) without intermediate layer; (b) with Pt intermediate layer (c) with Cr intermediate layer. (Reprinted with permission from Y. F. Ding, et al., J.Appl.Phys. 97, 10H303 (2005), Copyright [2005] American Institute of Physics.)

**TABLE 2.** Lattice constants of the intermediate layers, epitaxial relationships, and lattice mismatches where $a_1$ denotes lattice constant of intermediate layer (grown on MgO); $a_2$ denotes lattice constant of intermediate layer (grown on glass) and ε denotes lattice mismatch. (Reprinted with permission from Ref. 55. Copyright 2005 American Institute of Physics.)

| Sample | Intermediate layer | $a_1$(Å) | $a_2$(Å) | Epitaxial relationship | ε(%) |
|---|---|---|---|---|---|
| A | Pt | 3.9281 | 3.9231 | MgO(100)<001>\|\| Pt(100)<001>\|\| FePt(001)<100> | 2.23 |
| B | Cr | 2.8839 | 2.8776 | MgO(100)<001>\|\| Cr(100)<110>\|\| FePt(001)<100> | 5.88 |
| C | $Cr_{95}Mo_5$ | 2.8976 | 2.8912 | MgO (100)<001>\|\| Cr(100)<110>\|\| FePt(001)<100> | 6.33 |
| D | $Cr_{90}Mo_{10}$ | 2.9152 | 2.9114 | MgO(100)<001>\|\| Cr(100)<110>\|\| FePt(001)<100> | 6.89 |
| E | MgO | 4.2112 | — | MgO(100)<001>\|\| FePt(001)<100> | 8.86 |

*Notes:*
$a_1$: Lattice constant of intermediate layer (grown on MgO)
$a_2$: Lattice constant of intermediate layer (grown on glass)
ε : Lattice mismatch

to demonstrate the epitaxial relationships between FePt/MgO, FePt/Pt/MgO and FePt/Cr/MgO, respectively. The epitaxial relationship was revealed by the direction of the momentum transfer at the peak projected to the film plane relative to the in-plane orientation of the MgO substrate. From the relative directions of the FePt films, intermediate layers and MgO substrates crystalline axes, it can be derived that the epitaxial relationships for samples $E$, $A$, and $B$ are FePt (001) <100> \|\| MgO (100) <001>, FePt (001) <100> \|\| Pt (100) <001> \|\| MgO (100) <001> and FePt (001) <100> \|\| Cr (100) <110> \|\| MgO (100) <001>, respectively.

Table 2 lists the calculated lattice constant of the intermediate layers grown on both the MgO and glass substrates and lattice mismatch between different intermediate layers and FePt films. The lattice constants of the intermediate layers grown on the MgO substrates are only slightly larger than that grown on the glass substrate, which indicates that the 30 nm thick intermediate layer is sufficient for relaxing the strain from the MgO substrate. The lattice parameter of Cr intermediate layer increases with addition of Mo, following approximately the Vegard's law.

Figure 15 summarizes the lattice constants of $a$ (a) and $c$ (b), the tetragonality represented by $c/a$ (c), the long-rang order parameter $S$ (d), and the magnetocrystalline anisotropy energy, Ku (e), with respect to the lattice mismatch. The $S$ was characterized by the integrated intensity ratio between the FePt (001) and (002)

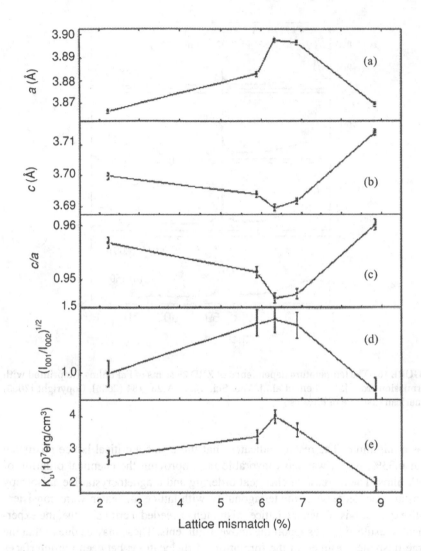

**FIGURE 15.** The lattice constants of $a$ (a) and $c$ (b), the tetragonality represented by $c/a$ (c), the long-rang order parameter $S$(d), and the magnetocrystalline anisotropy energy, $K_u$ (e) with respect to the lattice mismatch. (Reprinted with permission from Y. F. Ding, et al., J.Appl.Phys. **97**, 10H303 (2005), Copyright [2005] American Institute of Physics.)

peaks. The results showed that the $c$ value decreased with increasing lattice mismatch from about 2.23% to 6.33%. Upon further increase of lattice mismatch, $c$ increased in turn. On the other hand, the variation of a illustrated a reversal behavior to that of $c$ with the increased lattice mismatch. The value of $c/a$ was at its minimum ($c/a = 0.9466$) for lattice mismatch near 6.33%. The chemical ordering and Ku showed maximum values when lattice mismatch was 6.33% where $c/a$

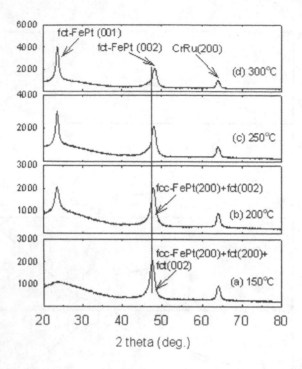

**FIGURE 16.** The temperature dependence of XRD 2θ scans of FePt films. (Reprinted with permission from J. S. Chen, et al., J. Vac. Sci. Tech. A **23**, 184 (2005), Copyright [2005] American Institute of Physics.)

was a minimum. The results indicated that there was a critical lattice mismatch near 6.33%, which was most favorable for improving the chemical ordering of FePt films. The increase of chemical ordering and magnetocrystalline anisotropy constant and decrease of the tetragonality with lattice mismatch were consistent with expectations. When the lattice mismatch exceeded a critical value, the experimental results did not support the above arguments. These may be due to that the large mismatch resulted in the formation of dislocations between the interfaces and thus the strain energy was released in the interface.

The effects of temperature on the texture and magnetic properties of FePt thin films with 4nm Pt buffer layer on optimized $Cr_{90}Ru_{10}$ (200)/glass have been reported.[58] The FePt films were deposited at various temperatures and CrRu and Pt layers at 350°C. Figure 16 shows the temperature dependence of XRD 2θ scans of FePt films. The (001) superlattice peak of the $L1_0$ phase was not observed at 150°C, indicating that the film was fcc. When the substrate temperature was greater than 200°C, the sharp (001) fct superlattice peak was clearly observed and the (001) peak shifted to a larger angle, and its intensity increased with increasing substrate temperature. The peak between 45° and 50° also shifted to a higher angle (smaller interplanar $d$ spacings, lattice constant $c$) with the increase of substrate

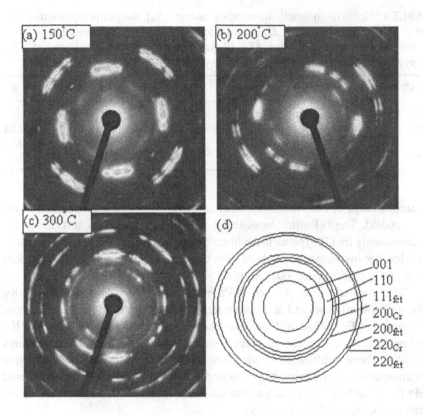

**FIGURE 17.** Selected area diffraction pattern of FePt films. (Reprinted with permission from J. S. Chen, et al., J. Vac. Sci. Tech. A 23, 184 (2005), Copyright [2005] American Institute of Physics.)

temperature. These results confirmed that the chemical ordering of the FePt films increased.[59]

Further investigation of the crystallographic structure of FePt films deposited at various temperatures were carried out using electron diffraction, as shown in Figs. 17(a)-(d). Figure 17(d) is a schematic diagram of the diffraction rings that exist in the above films. At a substrate temperature of 150°C, weak superlattice (001) and (110) diffraction rings are observed. The weak fundamental (111) diffraction ring is also found, which belongs to fct FePt phase according to the ratio of the squares of the diameters of the outer rings to that of the first ring (001). These indicate that there exists a few fct FePt grains in the fcc FePt matrix, although there is no superlattice peak in the corresponding XRD scan. With an increase of the substrate temperature (above 200°C), the weak (001) diffraction ring disappears as the c-axis orients more perpendicularly to the film plane.

In addition, the intensity of (110) diffraction ring increases with increasing substrate temperature, showing increased ordering. The out-of-plane and in-plane

**TABLE 3.** Uniaxial magnetic anisotropy energy ($K_u$), magnetic squareness ($M_{r\perp}/M_{S\perp}$), coercivity ($H_{C\perp}$, $H_{C//}$) and $S$ for the FePt films deposited at different substrate temperature ($T_s$). (Reprinted with permission from Ref. 58. Copyright 2005 American Institute of Physics.)

| $T_s(°C)$ | $K_u$ (erg/cc) | $M_{r\perp}/M_{S\perp}$ | $H_{C\perp}$ (Oe) | $H_{C//}$(Oe) | S |
|-----------|----------------|--------------------------|--------------------|----------------|------|
| 150       |                | 0.04                     | 164                | 436            |      |
| 200       | $7.6\times10^6$ | 0.91                    | 1317               | 842            | 0.53 |
| 250       | $1.0\times10^7$ | 0.98                    | 1435               | 1081           | 0.64 |
| 300       | $1.4\times10^7$ | 0.98                    | 1856               | 1599           | 0.76 |

hysteresis loops of FePt films deposited at different substrate temperatures have been studied. The FePt film deposited at 150°C is magnetically soft due to the predominantly fcc (200) phase in the film.[58] With increasing substrate temperature, the films became magnetically hard with the easy axis perpendicular to the film plane due to predominant (001) texture.

The magnetic properties of FePt films with a structure of Glass/CrRu (30nm)/Pt((4nm) deposited at different temperatures are listed in Table 3. It can be found that FePt films with uniaxial anisotropy constant higher than $1 \times 10^7$ erg/cm$^3$, and high magnetic squareness were obtained at substrate temperatures of 250°C. These results showed that based on the lattice mismatch induced phase transformation, the ordering began to occur even at 150°C, and highly ordered FePt films can be obtained at temperatures as low as 250°C, as indicated by the large magnetocrystalline anisotropy.

### 3.5. Other Approaches to Enhance L1$_0$ Ordering

Other approaches have also been investigated to decrease the ordering temperature and increase the $S$. The ordering process of FePt films was investigated using 130 keV He ion irradiation.[60] The $S$ increased up to 0.3 and 0.6, respectively, when the disorder ($S \sim 0$) and partially ordered ($S \sim 0.4$) FePt films were irradiated by He ion with fluences $4 \times 10^{16}$ ions/cm$^2$ at 280°C. The beam interaction led to atomic displacements and the heating promoted atomic re-arrangements and lattice relaxation. With increasing energy of He ion (2 MeV) and beam current (1.25-6 μA), highly ordered L1$_0$ FePt films were obtained.[61] The $H_C$ of the films exceeded 5700 Oe after the disordered FePt was irradiated at beam current 1.25 μA with ion fluencies of $2.4\times10^{16}$ ions/cm$^2$. The high beam current caused direct beam heating on the sample. In addition, the irradiation-induced heating provided efficient microscopic heating transfer and created excess point defects, significantly enhancing diffusion and thus promoting formation of L1$_0$-ordered FePt films. The L1$_0$-ordered FePt film can also be obtained by annealing at 350°C the FePt film deposited on AuCu underlayer.[62] The ordering of the AuCu underlayer at low temperature coherently induced the disorder-order transformation of FePt films.

# 4. INTRINSIC PROPERTIES OF THE L1$_0$ FEPT FILMS

## 4.1. Magnetocrystalline Anisotropy, Magnetization and Curie Temeperture of L1$_0$ FePt Films

The physical origin of the magnetocrystalline anisotropy is the coupling of the spin component of the magnetic moment to the electronic orbital shape and orientation (spin-orbital coupling), as well as the chemical bonding of the orbitals on a given atom with their local environment (crystalline electronic field). When a spin is coupled to the orientation of the orbital, the material shows a magnetic anisotropy that has the symmetry of the crystal field, specially, of the local atomic environment. As for the L1$_0$ FePt inermetallic alloy, the origin for the strong magnetocrystalline anisotropy is from the large spin-orbit coupling in Pt and 5d (Pt)-3d(Fe) hybridization.[63, 64] It is known that c/a in fcc FePt and $c/a$ in L1$_0$ FePt are 1, and 0.968, respectively. The magnetic anisotropy of L1$_0$ FePt will therefore be influenced by both chemical ordering and lattice distortion. Significant work has been done to calculate the magnetic anisotropy energy from first principles, employing local spin-density approximation. The magnetic anisotropy energy (MAE) was characterized by the total energy difference per Fe-Pt pair when the magnetization was orientated along [110] (or [100]) and [001] crystal axes (MAE = E[110] − E[001]).

The MAE calculated by linear muffin-tin orbitals method (LMTO) in atomic-sphere approximation (ASA) were 3.4meV,[65] 2.8 meV,[66] 3.5 meV,[63] and the augmented-spherical-wave (ASW) yielded 2.75 meV,[67] and full-potential LMTO yieds 3.9 meV[68] and 2.7 meV.[69] A value of 1 meV per FePt pair was equivalent to $5.68 \times 10^7$ ergs/cm$^3$. All these calculated results dramatically scattered from the experimental results of 1.3-1.4 meV for the bulk[70] and 1-1.2 meV for the films.[71, 72, 73]

The local spin-density approximation (LSDA) band theoretical method was used together with the full-potential relativistic linearized augmented plane-wave method (FP-LAPW) to calculate the MAE of L1$_0$ CoPt and FePt. The calculated MAE for L1$_0$ CoPt was in very good agreement with the experimental data. When accounting for the on-site Coulomb correlation effects by using LSDA method together with intrasite Coulomb repulsion U (LSDA+U) and choosing $U_{Pt}$ around 0.544 eV, the MAE showed good consistence with experimental data of FePt bulk. The effects of the chemical ordering and lattice distortion on the MAE by the first principle theory, using density functional electronic structure calculation and a mean field treatment of both compositional and magnetic "local moment" fluctuations, were studied.[64]

The MAE was enhanced by compositional order, however; the lattice distortion had little contribution to the large MAE of FePt comparing to the contribution of the compositional order. The dependence of MAE on the lattice distortion was also investigated.[74] The full potential local orbital method with consideration of relativistic effects was used and chemical order was kept complete, independent of $c/a$. With the decrease of $c/a$, the MAE also decreased, as shown in Fig. 18. In this

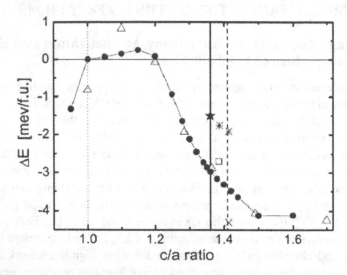

**FIGURE 18.** Calculated values of MCA energy $E$ of L1$_0$ FePt (closed symbols) as a function of $c/a$-ratio. (Reprinted with permission from J. Lyubina, et al., J. Phys: Condens. Matter 17, 4157 (2005), Copyright [2005] American Institute of Physics.)

figure, tP2 unit cell was used. In conventionally used tP4 pseudocell, $c' = c$, and $a' = \sqrt{2}a$ thus $c'/a' = 0.96$. Up to now, there is no experimental confirmation of the above theoretical calculation, due to the difficulty to experimentally separate the effects of the lattice distortion on MCA from that of chemical ordering.

Experimental investigations and measurements of the magnetocrystalline anisotropy of L1$_0$ FePt film were also conducted. The relationships between the magnetic anisotropy and chemical ordering of FePt films fabricated by MBE and magnetron sputtering on MgO (001) substrate or CrRu (200) underlayer were reported.[25, 58, 72, 75] The anisotropy, estimated by the area enclosed between the magnetization curves in applied field parallel and perpendicular to the film plane and a 45° method combining high field extrapolation, increased with chemical ordering. For FePt films with chemical ordering of 0.8 ~ 0.9 deposited on MgO single substrate the anisotropy was as high as $4 \times 10^7$erg/cc. It was reported that the anisotropy[71] and chemical ordering[35, 36] increased with FePt film thickness. The increase of anisotropy with film thickness was also attributed to the increase of chemical ordering. The anisotropy of the FePt films with different composition was investigated.[72] When the ratio of Fe:Pt was 52:48, both chemical ordering and magnetic anisotropy were the highest. However, as the ratio of Fe:Pt was 46:54, the chemical ordering was only slightly smaller than the highest value, but the magnetic anisotropy significantly decreased by 42%.

The above-measured anisotropy is referred to as the first-order anisotropy constant $K_1$ in the uniaxial anisotropy energy expression $E(\theta) = K_1 \sin^2 \theta + K_2 \sin^4 \theta + \cdots$, where $\theta$ denotes the angle of magnetization with respect to the easy axis. It is very difficult to accurately evaluate and separate the anisotropy constants ($K_1$ from $K_2$) of FePt films using conventional methods at low external field (such as toqure curve analysis and magnetization curve along a hard axis and easy axis) due to the extremely high anisotropy field of about 100 kOe.[76, 77, 78] An anomalous Hall effect method was developed to determine the anisotropy constant with high accuracy at low external field.[25, 79, 80, 81, 82] The magnetic anisotropy constant of the films can be determined by analyzing the normalized Hall voltage ($V_{HE}$) curves using the Sucksmith-Thompson method. From the equilibrium magnetization condition, the relation was deduced as

$$2K_1^{eff} + 4K_2\left(1 - m_{Z_z}^2\right) = \alpha H M_S \tag{10}$$

with

$$\alpha = \frac{m_Z \sin\theta_H - \sqrt{1 - m_{Z_z}^2}\cos\theta_H}{m_Z\sqrt{1 - m_{Z_z}^2}} \tag{11}$$

where $m_Z$ is the normalized magnetization ($M_S$) in film normal, $K_1^{eff}$ includes $K_1$ and demagnetization energy $2\pi M_S^2 \theta_H$ is the direction of the external field ($H$) and magnetization from the film normal. By plotting the $\alpha H M_S$ vs. $(1 - m_{Z_z}^2)$, one can determine both $K_1^{eff}$ and $K_2$. Using this method, the anisotropy constants of the L$1_0$ were evaluated and investigated.[25]

Figure 19 shows the dependence of chemical-ordering of $K_1$ and $K_2$ of 14 nm FePt films. With increasing chemical ordering, $K_1$ gradually increased. When

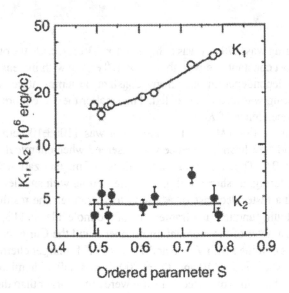

**FIGURE 19.** The first- and the second-order uniaxial anisotropy constant $K_1$ and $K_2$ at room temperature for 140 Å thick FePt films as a function of chemical-order parameters $S$. The solid lines are guides to the eye. (Reprinted figure with permission from S. Okamoto, et al., Phys. Rev. B 66, 024413 (2002). Copyright (2002) by the American Physical Society.)

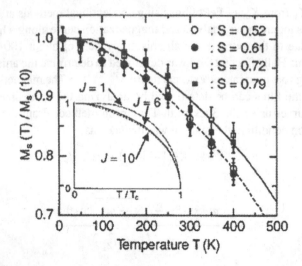

**FIGURE 20.** Temperature dependence of normalized magnetization $M_s(T)/M_s(10)$ for 140 Å thick FePt films with the chemical-order parameter $S = 0.52, 0.61, 0.72$ and $0.79$. The solid and broken lines are calculated $M_s(T)/M_s(0)$ curves using the Brillouin function with the molecular field approximation for the momentum quantum number $J = 6$ and the Curie temperature $T_c = 750$ K and for $J = 10$ and $T_c = 700$ K. The inset shows the calculated $M_s(T)/M_s(0)$ vs $T/T_c$ curves for $J = 1, 6,$ and $10$. (Reprinted figure with permission from S. Okamoto, et al., Phys. Rev. B **66**, 024413 (2002). Copyright [2002] by the American Physical Society.)

chemical ordering was 0.79, $K_1$ was as high as $4 \times 10^7$ erg/cc. On the other hand, $K_2$ remained almost constant and was about $5 \times 10^6$ erg/cc with increasing chemical ordering. The dependence of anisotropy constant on temperature with different chemical ordering was also investigated. The smaller the $S$ the more sensitive the temperature dependence of $K_1$.[25]

The magnetization ($M_S$) of L1$_0$ FePt film was 1100±100 emu/cc at room temperature and significant difference was observed when chemical ordering varied from 0.4 to 0.9. The temperature dependence of magnetization was sensitive to chemical ordering, as shown in Fig. 20. The sample with smaller chemical ordering showed a faster decrease of $M_S$ with temperature. The results were fitted using the Brillouin function in a framework of the molecular field approximation with fitting parameter of the momentum number $J$ and the Curie temperature $T_c$. The best fit was obtained with $J = 6$ and $T_c = 750$ K for larger chemical ordering 0.72 and 0.79, whereas $J = 10$ and $Tc = 700$ K for smaller chemical ordering of 0.52 and 0.61. The values of $J$ used in fitting were much larger than the usual value of Fe ($J = 1$).

The relationships between the temperature dependence of magnetic anisotropy and magnetization were investigated by a classical spin model pertinent to magnets with localized magnetic moments.[83, 84, 85] At low temperature

$$\frac{K(T)}{K(0)} \approx \left(\frac{M(T)}{M(0)}\right)^{\frac{l(l+1)}{2}}$$

and near Curie temperature

$$T_c, \frac{K(T)}{K(0)} \approx \left(\frac{M(T)}{M(0)}\right)^l$$

where $l$ is the order of the spherical harmonic describing the angular dependence of the local anisotropy, i.e., $l = 2$ and $l = 4$ for uniaxial and cubic system, respectively. The above formulas apply very well to the magnets where the magnetic moments are well localized, e.g., rare earth and oxide magnets. However, for itinerant ferromagnets such as FePt, it has been found that the above formula does not hold,[25, 72] and

$$\frac{K(T)}{K(0)} \approx \left(\frac{M(T)}{M(0)}\right)^n$$

where $n = 2$ instead of $n = 3$ over large range of temperature.

Figure 21 shows the relationship between

$$\frac{K(T)}{K(10)} \text{ and } \left(\frac{M(T)}{M(10)}\right)$$

for different degrees of chemical ordering. The slope of the least-square fitting solid line was 2.1. The slight deviation from the linear relation in high temperature region ($T \geq 300K$) was attributed to the presence of other contribution such as thermal lattice expansion and/or higher-order anisotropy.

The Curie temperature of bulk $Fe_{50}Pt_{50}$ alloy was reported to be 753K.[18] The Curie temperature of $Fe_{55}Pt_{45}$ films was estimated as 770K using Curie-Weiss law and extrapolating to $M = 0$.[72] The Curie temperature was reduced significantly with Ni doping to as low as 490 K for 30 at.% Ni. The effects of $Fe_{50}Pt_{50}$-$B_2O_3$ nanocompoiste with different FePt volume fraction on Curie temperature were studied.[86] Curie temperature as a function of FePt volume fraction is shown in Fig. 22 (a). Curie temperature for a pure FePt film was 733K. As FePt concentrations decreased from 100 % to 25 vol. %, the Curie temperature decreased almost linearly to 470 K.

Though this reduction correlated with the change in the $c/a$ ratio variation of FePt volume fraction as shown in Fig. 21(b), it was not related to the chemical ordering. It was suggested that finite-size and interface effects had little contribution to reduction in Curie temperature. The reduction in Curie temperature with decreasing volume fraction of FePt was attributed to the decrease of the interplane exchange parameter. The interplane exchange parameters is the Heisenberg exchange strength between the Fe atoms and Pt atoms in the fully chemically ordered FePt alloy (Fe and Pt atom layers alternatedly stacking). It was also reported that the

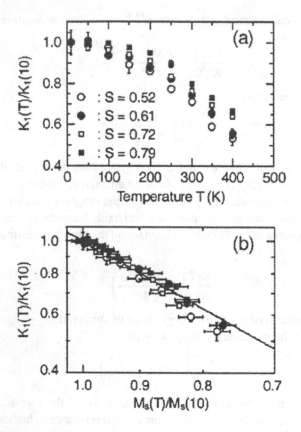

**FIGURE 21.** (a) Temperature dependences of $K_1(T)/K_1(10)$ for 140 Å thick FePt films with the chemical-order parameter $S = 0.52$, 0.61, 0.72, and 0.79. (b) log-log plot of $K_1(T)/K_1(10)$ vs $M_s(T)/M_s(10)$. The solid line in (b) is least-square fitting with a slope of 2.1. (Reprinted with permission from S. Okamoto, et.al., Phys. Rev. B 66, 024413 (2002). Copyright [2002] by the American Physical Society.)

Curie temperature decreased with particle size in Ni film and FePt/C nanocomposite films.[87, 88] However, the interplane exchange parameter effect cannot be excluded.

## 4.2. Effects of Size and Interface on Coercivity and Magnetization Reversal

In the previous section, the relations of magnetic anisotropy of L1$_0$ FePt films with long-range chemical ordering have been discussed. When the magnetic grain size is reduced to the nanometer regime, the magnetic anisotropy is eventually affected due to dimensional reduction along the $c$-axis. In addition, the effects of grain size reduction on and thermal fluctuation on coercivity cannot be neglected.[89, 90]

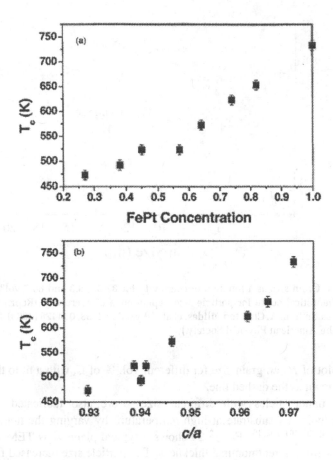

**FIGURE 22.** The Curie temperature $T_c$ as a function of (a) the FePt volume fraction and (b) the $c/a$ ratio. (Reprinted figure with permission from H. Zeng, et al., Phys. Rev. B 66, 184425 (2002). Copyright (2002) by the American Physical Society.)

In some cases the magnetic surface or interface anisotropy is strong enough to alter the magnetic behavior of the magnetic nano-grains.[91, 92] FePt nanoparticles in various matrixes such as carbon,[88, 93, 94, 95] $SiO_2$[96] and $Al_2O_3$[97, 98] were fabricated using multilayer precursors consisting of stacks of FePt/C ($SiO_2$ or $Al_2O_3$) bilayers, followed by heat treatment that disrupted the layer structure and resulted in formation of nanoparticles. The particle size was controlled by adjusting the multilayer structure and annealing time. It was found that coercivity as a function of the particle size (D) followed the relation below:

$$H_c(D) = 0.5 H_k \left[ 1 - \left( \frac{D_p}{D} \right)^{3/2} \right] \tag{12}$$

where $D_p$ is superparamagnetic grain size and $H_k$ the anisotropy field. Figure 23

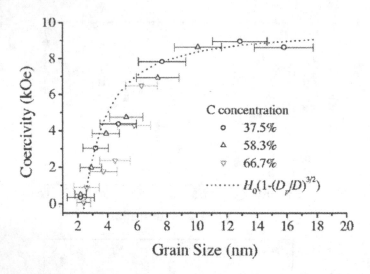

**FIGURE 23.** Grain size as a function of coercivity for 37.5, 58.3 and 66.7 vol% C along with the theoretical curve for particle size dependence of coercivity. (Reprinted figure with permission from A. Christodoulides, et al., Phys. Rev. B 68, 054428 (2003). Copyright (2003) by the American Physical Society.)

shows a plot of $H_c$ vs. grain size for different vol. % of $C$, with a fit to the above relation shown as the dashed line.

FePt nanoparticles with different sizes were also fabricated on MgO (100) single-crystal substrate at high temperature by varying the nominal film thickness.[99, 100, 101, 102, 103] Figure 24 shows a typical plane-view TEM image of FePt film with different nominal thickness. The particle size increased from several nanometers to several hundreds of nanometers with increasing nominal film thickness from 1 nm to 20 nm. At nominal thickness of 1 nm, the coercivity was 18 kOe. The coercivity increased with increasing thickness. As the nominal thickness was 5 nm, maximum coercivity of 70 kOe was achieved. Further increasing the thickness led to a decrease of coercivity. The relatively low coercivity of FePt nanoparticles with nominal thickness of 1 nm was attributed to the insufficient chemical ordering of the small particles. A similar trend in the dependence of chemical ordering on particle size was also reported.[88]

The decrease of coercivity with further increase of nominal thickness was attributed to the change in magnetization reversal. When the nominal thickness was below 8 nm, the initial magnetization curves as shown in Fig. 25 indicated the magnetization reversal followed the Stoner-Wohlfarth rotational mode, verified by angular dependence of remanent coercivity.[103] As the nominal thickness was in the range of 10-20 nm, both domain wall displacement and Stoner-Wohlfarth rotational mode occurred due to coexistence of single-domain particles and multidomain particles. When the nominal thickness exceeded 25 nm, only domain wall displacement dominated the magnetization reversal of FePt particles.

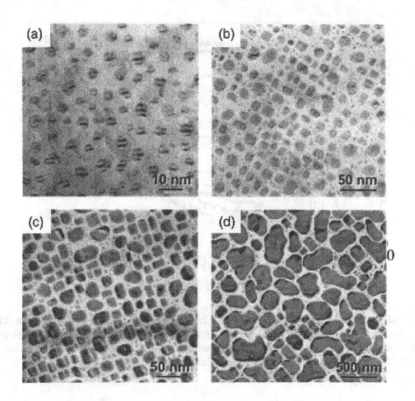

**FIGURE 24.** In-lane TEM bright field images for the FePt films with different thickness: $t_N = 1$ nm (a), 3 nm (b), 5 nm (c), and 20 nm (d). (Reprinted with permission from T. Shima, et al., Appl. Phys. Lett. 85, 2571 (2003), Copyright [2003] American Institute of Physics.)

The effects of interface on chemical ordering and magnetic properties of fully ordered FePt nanoparticles were investigated by covering the film with a thin $Al_2O_3$ layer.[99] When the ordered FePt nanoparticles were covered with $Al_2O_3$ at room temperature, a disordered layer around 2.5 nm was formed at the interface, which was attributed to interfacial strain. After post-deposition annealing at 700°C, the strain was released and disordered layer was not observed. The coercivity and magnetic anisotropy of FePt particles covered with $Al_2O_3$ layer at room temperature were much smaller than that after post-deposition annealing. However, the coercivity of FePt nanoparticles covered with $Al_2O_3$ after post-deposition annealing was still smaller than that without $Al_2O_3$ overlayer, which was ascribed to the residual strain at the FePt/ $Al_2O_3$ interface.

The effects of FePt nanoaprticles covered with Ag and Pt layer at room temperature on magnetic properties were also studied. Ag-coated sample exhibited higher coercivity than the Pt-coated one. The effective magnetic anisotropy ($K_u^{eff}(0)$) and the anisotropy field ($H_k(0)$) of Ag-coated sample at the temperature

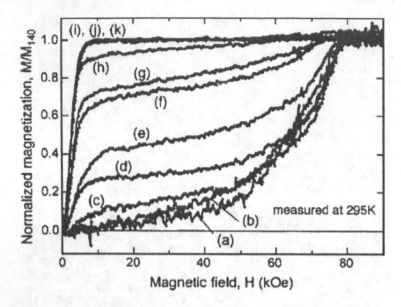

**FIGURE 25.** Initial magnetization curves for the FePt films with $t_n = 3$ nm (a), 5 nm (b), and 8 nm (c), and 10 nm (d), and 12 nm (e), and 18 nm (g), and 20 nm (h), 25 nm (i), 30 nm (j), and 40 nm (k). (Reprinted with permission from T. Shima, et al., Appl. Phys. Lett. 85, 2571 (2003), Copyright [2003] American Institute of Physics.)

of arbitrary zero was almost the same for fully ordered FePt single crystal. The Pt-coated sample exhibited smaller $H_k(0)$ and the $K_u^{eff}(0)$ was the same as that of the Ag-coated sample. It was considered that Pt atoms in contact with FePt were ferromagnetically polarized due to the proximity effect,[104, 105] which was the main reason for reduction of $H_k(0)$. It was also likely that $H_k(0)$ was somewhat underestimated due to the weak exchange coupling between FePt nanoparticles.

## 5. APPLICATION OF L1₀ FePt ALLOY THIN FILM FOR PERPENDICULAR MAGNETIC RECORDING

FePt films deposited by physical vapor depositions generally show the (111) preferred orientation since the (111) plane is closest packed with lowest surface energy. For FePt perpendicular media, it is desirable to achieve the (001) texture since the uniaxial magnetocrystalline anisotropy is aligned with the $c$-axis. Another important factor affecting the practical application of L1₀ FePt films for ultrahigh density magnetic media is the media noise. The media noise mainly originates from transition noise that is closely related to magnetic domain size. In this section, methods to control the FePt (001) texture and reduce the exchange coupling, grain size and thus media noise are reviewed.

## 5.1. Control of FePt (001) Texture

### 5.1.1. Epitaxial Growth

The epitaxial growth of FePt(001) textured films requires the substrates or under-layers to have similar atomic configuration with that of FePt (001) plane and small lattice mismatch. Substrates or underlayers normally used are MgO (100), SrTiO₃ (100), Cr (200), Ag (200) with eptitaxial relationship FePt (001) <100> || MgO (100) <001>, FePt (001) <100>||SrTiO₃ (100) <001>, FePt (001) <100> || Ag (100) <001>, and FePt (001) <100> || Cr (100) <110>. The lattice mismatch of FePt with MgO, SrTiO₃, Ag and Cr(200) are 8.5%, 2%, 7.1%, 5.8%, respectively. Although FePt (001) textured films had been produced using the MgO (100) and SrTiO₃(100) single substrates by molecular beam epitaxy (MBE), sputtering and laser ablation, they are not suitable for practical application because of the costly single substrates.[24, 106, 107, 108, 109] FePt (001) prefered films were also grown on Ag(100) underlayer/Si(100) substrate.[53, 57, 110]

Very thick Ag films made it very difficult for application of double-layered FePt perpendicular media. Since Ag has a low melting point and low surface energy, it is difficult to maintain small grain size at high temperatures. Furthermore the magnetic properties of FePt films on Ag (200) underlayer was not good. The MgO (100) and CrX (200) (X = Ru, Mo, W, Ti) underlayers are very promising for prac-tical applications. MgO thin films usually obtain the (200) texture when deposited at room temperature because of its NaCl structure. The fcc FePt (200) films were deposited on 10 nm MgO(200) underlayer and the $L1_0$ FePt (001) textured films were obtained by post-deposition annealing at high temperatures.[111, 112, 113, 114] The resultant texture after annealing depended on the film thickness. It was found that 5 nm FePt film on MgO underlayer exhibited FePt(001) texture after annealing. On the other hand, FePt films with the thickness of 20-40 nm showed predomi-nance of in-plane $c$-axes. FePt films with additive (such as Cu, Ag, Au) on MgO underlayer were also investigated.[45] It was found that Cu doping favored the FePt (001) orientation on MgO underlayer after post-deposition annealing at 650°C. As the thickness of the FePt-20 vol.% Cu film ranged from 5 nm to 20 nm, the FePt (001) texture was maintained, as shown in Fig. 26.

A good FePt (001) texture and perpendicular magnetic anisotropy were ob-tained by post-deposition annealing of FePt/MgO multilayers on glass substrate or SiO₂ seed layer.[115, 116, 117] The full-width-at-half maximum (FWHM) of the rocking curve of FePt (001) peak was 3.7°. When the surface roughness of sub-strate or seed layer increased from 0.74 nm to 2.35 nm, the FWHM increased to 4.2° correspondingly. Likewise, when doping a certain amount of Cu in FePt layer in optimized FePt/MgO multilater structure with subsequent annealing at high temperature, the FWHM of the rocking curve of FePt (001) peak was as narrow as 2.1°.[118] It was also reported that FePt (001) textured films were obtained by anneal-ing Pt(100)/Fe(100) bilayered films deposited on glass substrate. It was suggested that the initial growth layers that consisted of Fe oxides formed in the interface of substrate and Fe oxide layer affected the texture.[119] These were also confirmed by

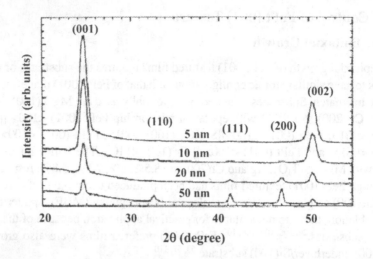

**FIGURE 26.** XRD scans of MgO (5 nm)/FePt+20% Cu films with varying total thickness annealed 650°C / 10 min. (Reprinted with permission from C. L. Platt, et al. , J. Appl. Phys. 92, (2002), Copyright [2002] American Institute of Physics.)

experiments that Fe-O underlayer was the main reason for FePt (001) texture.[120] FePt (001) textured films were also deposited on RuAl (001) underalyer.[121]

Cr underlayer can also be used to control the texture of FePT film. The development of Cr(200) texture is the key to obtain FePt (001) textured film. The driving force for texture development is the minimization of surface free energy by growing the lowest energy plane on the substrate surface. The lowest energy plane varies with the crystallographic structure. Cr is body-center cubic (bcc) and Cr films usually have (110) texture when deposited at room temperature. Although at elevated substrate temperature the Cr (200) texture was preferred, some of (110) oriented grains remained in the films.[122]

The Cr (200) texture was improved by application of substrate bias, providing the adatoms extra kinetic energy from the bombardment of the sputtering gas.[123] Using a Ta seed layer the Cr(200) texture can also be improved.[124, 125] Ta serves as an excellent wetting layer and provides enough lateral mobility for Cr adatoms on the surface to nucleate in the (200) direction. The FePt (001) texture was obtained on Cr(200)/Pt(200) underlayer induced by a NiTa seed layer and.[126] By exposing NiTa seed layer in oxygen gas, the FePt (001) texture was improved. The use of MgO as a seed layer to induce Cr(200) underlayer was also reported.[127, 128] The growth of FePt (001) textured films on Cr(200)/MgO(200) on glass substrate was studied.[52, 129] The lattice mismatch between Cr (200) and FePt (001) was ~5.8%.

In Sec. 3.4, it has been shown that when the lattice mismatch was 6.33%, FePt films had the largest magnetic anisotropy. The lattice parameter of Cr underlayer can be adjusted to achieve best chemical ordering. Ru, Mo, W and Ti were doped into Cr film to adjust the lattice constant. CrX (200) (X = Ru, Mo, W, Ti) textured

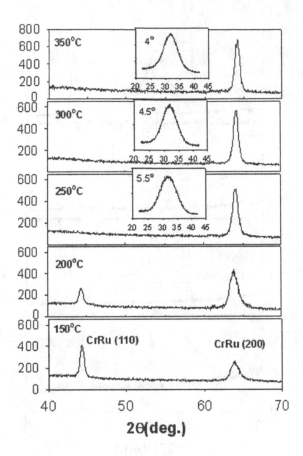

**FIGURE 27.** XRD θ-2θ scans of CrRu films with different temperatures. (Reprinted from J. Magn. Magn. Mater., 303, J. S. Chen, B. C. Lim, Y. F. Ding, G. M. Chow, Low temperature deposition of L1$_0$ FePt films for ultra-high density magnetic recording, 309, (2006), with permission from Elsevier).

film directly deposited on glass substrate were achieved.[54, 130] The substrate temperature and sputtering power were the key factors affecting the (200) texture of Cr$_{90}$Ru$_{10}$ film on glass.[131]

The thickness of CrRu films was 30 nm. The XRD θ-2θ scans of CrRu films, deposited with a fixed sputtering power of 200W at different temperatures, are shown in Fig. 27. The insets of the XRD spectra were the rocking curve of the Cr(200) peak. When the substrate temperature was 150°C, the Cr (110) peak predominated. With increasing temperature, the intensity of Cr(110) peak decreased and Cr(200) peak increased. When it was above 250°C, the Cr(110) peak disappeared and only Cr(200) peak was observed.

The FWHM of the rocking curve of Cr(200) peak increased with substrate temperature. The XRD scans of CrRu films, deposited at a constant substrate

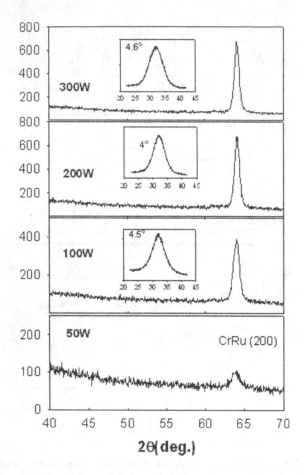

**FIGURE 28.** XRD θ-2θ scans of CrRu films with different sputtering powers. (Reprinted from J. Magn. Magn. Mater., 303, J. S. Chen, B. C. Lim, Y. F. Ding, G. M. Chow, Low temperature deposition of L1₀ FePt films for ultra-high density magnetic recording, 309, (2006), with permission from Elsevier).

temperature of 350°C with different sputtering powers, are shown in Fig. 28. The intensity of Cr(200) peaks increased and FWHM of the rocking curve of Cr(200) peak decreased with increasing sputtering power to 200W. Further increase of the sputtering power to 300W, the FWHM of the rocking curve increased. The evolution of Cr texture with temperature and sputtering power can be related to the increase of lateral adatom mobility. However, the deterioration of Cr (200) texture with sputtering power increased to 300 W could be attributed to the increase of energy of atom/ions in the film normal direction that may damage the crystallographic structure of the film. In addition to sputtering power and substrate temperature, base pressure or other impurities affected the texture of Cr (200).[58, 132, 133]

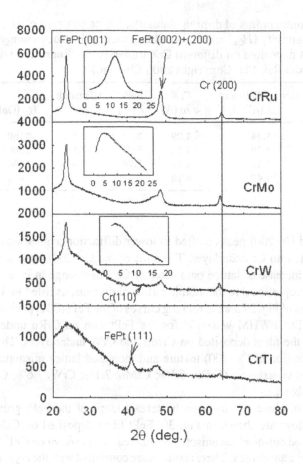

**FIGURE 29.** XRD θ-2θ scans of FePt films with the structure of FePt (20 nm)/Pt (4 nm) /Cr$_{90}$X$_{10}$ (30 nm)(X=Ru, Mo, W and Ti)/glass deposited at 350°C. (Reprinted from J. Magn. Magn. Mater., 303, J. S. Chen, B. C. Lim, Y. F. Ding, G. M. Chow, Low temperature deposition of L1$_0$ FePt films for ultra-high density magnetic recording, 309, (2006), with permission from Elsevier).

The effects of different CrX (X = Ru, Mo, W, Ti) underlayers on the structure and magnetic properties of FePt film were investigated.[134] Figure 29 shows the XRD scans of FePt films with the structure of FePt (20 nm)/Pt (4 nm) /Cr$_{90}$X$_{10}$(30 nm)(X = Ru, Mo, W and Ti)/glass deposited at 350°C. For the film grown on CrTi underlayer, only FePt (111) and (200) fundamental peaks were observed. When the CrW underlayer was used, the FePt (001) superlattice peak appeared and the intensity of FePt (111) fundamental peak decreased. For the FePt film on CrMo underlayer the intensity of FePt (001) and (111) peak further increased and decreased, respectively.

The FePt film on CrRu underlayer showed the strongest FePt (001) peak and the FePt (111) peak disappeared completely. The intensity of Cr (200) peaks

**TABLE 4.** Atomic radius of doping elements, lattice constant $c$, lattice mismatch, coercivity ($H_{C\perp}$) and uniaxial magnetic anisotropy energy ($K_u$), of the FePt films deposited on different CrX underlayers. (Reprinted with permission from Ref. 131. Copyright 2006 Elsevier.)

| Underlayer (CrX) | Atomic Radius of X (Å) | CrX $\sqrt{2}a$ (Å) | FePt $c$ (Å) | Mismatch (%) | $H_{c\perp}$ (Oe) | $K_u$ (erg/cc) |
|---|---|---|---|---|---|---|
| CrRu | 1.34 | 4.09 | 3.73 | 6.5 | 3160 | 1.80 |
| CrMo | 1.39 | 4.12 | 3.76 | 7.1 | 2960 | 1.3 |
| CrW | 1.41 | 4.14 | 3.78 | 7.6 | 2950 | 1.2 |
| CrTi | 1.47 | 4.16 | — | 7.9 | 1881 | — |

decreased and Cr (200) peaks shifted to lower diffraction angles with doping of Ru, Mo, W, Ti into Cr underlayer. The shift of the Cr (200) peak with dopants indicated the increase of lattice parameter of Cr. The change in lattice constants was linearly proportional to the atomic radius of dopants, as listed in Table 4.

The insets of Fig. 29 are the rocking curves of the FePt (001) peak on different underlayers. The FWHM was $\sim 5°$ for the FePt film on CrRu underlayer and increased for the films deposited on CrMo and CrW underlayers. These can be due to deterioration of Cr (200) texture and increased lattice mismatch between FePt films and underlayers (CrRu: 6.5%; CrMo: 7.1%; CrW: 7.6%; CrTi: 7.9%, as listed in Table 4.).

The out-of-plane and in-plane hysteresis loops of the FePt grown on different CrX alloys are shown in Fig. 30. FePt films deposited on CrX(X = Ru, Mo, W,) showed out-of-plane anisotropy, whereas films grown on CrTi underlayer showed in-plane anisotropy. There results were consistent with the crystallographic structure of the CrX alloy underlayers. The out-of-plane coercivity, magnetocrystalline anisotropy decreased and lattice constant $c$ increased with lattice mismatch (Table 2), consistent with FePt films on MgO single crystal with different intermediate layers.

The base pressure of deposition affected the texture of both CrRu underlayer and FePt magnetic layers.[132] The L1$_0$ FePt films exhibited (200) preferred orientation with longitudinal anisotropy at base pressure of $4 \times 10^{-6}$ Torr. Improving the base pressure (below $9 \times 10^{-7}$ Torr), the L1$_0$ FePt films showed the (001) texture with perpendicular anisotropy. The XRD spectra of FePt films deposited at various base pressures are shown in Fig. 31.

A thick CrX (200) underlayer was required to optimize the FePt(001) texture.[135] In double-layered perpendicular recording media, if the soft magnetic layer is below the underlayer, the recording resolution will be drastically deteriorated due to increase in head field-gradients. If a soft magnetic layer is above the underlayer, the lattice match with underlayer such as FeSi or FeCo with (100) orientation is needed, limiting the choice of the soft magnetic layer with the best magnetic properties.

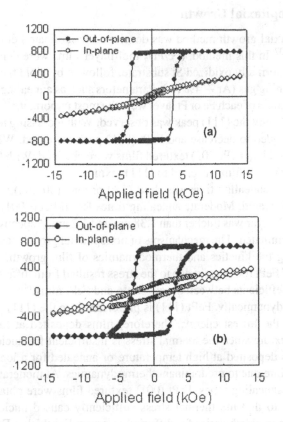

**FIGURE 30.** The out-of-plane and in-plane hysteresis loops of the FePt grown on different CrX alloys (a) $Cr_{90}Ru_{10}$, (b) $Cr_{90}Mo_{10}$, (c) $Cr_{90}W_{10}$, (d) $Cr_{90}Ti_{10}$, (Reprinted from J. Magn. Magn. Mater., 303, J. S. Chen, B. C. Lim, Y. F. Ding, G. M. Chow, Low temperature deposition of L10 FePt films for ultra-high density magnetic recording, 309, (2006), with permission from Elsevier).

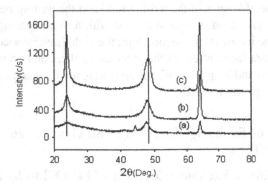

**FIGURE 31.** XRD spectra of Glass/CrRu (80 nm)/Pt(4 nm)/$Fe_{52}Pt_{48}$(20 nm) films deposited at base pressure of (a) $4 \times 10^{-6}$ Torr (b) $9 \times 10^{-7}$ Torr (c) $5 \times 10^{-9}$ Torr. (substrate temperature 400°C). (Reprinted with permission from J. S. Chen, et al., Appl. Phys. Lett. 81, 1848 (2002), Copyright [2002] American Institute of Physics.)

### 5.1.2. Nonepitaxial Growth

The nonepitaxial growth method was developed to fabricate FePt (001) textured films.[96, 136, 137] In this method, $(Fe/Pt)_n$ multilayer films were deposited on glass substrate or thermally oxidized Si substrate, followed by rapid thermal annealing (RTA) in forming gas (Ar + 4% $H_2$). Parameters such as annealing time, temperature and thickness of each Fe or Pt layer were the most important.[138] At temperature below 350°C, only the (111) peak was observed. With increasing temperature, the (111) peak tended to decrease and FePt (001) peak increased. When 550°C was exceeded, good $L1_0$ FePt(001) textured films were obtained. With further increase to above 750°C, the films reverted to (111) texture.

When the annealing time was short (2 s) or long (30 min),[139] a strong FePt (111) peak appeared. Moderate annealing times led to FePt (001) texture. When each Fe and Pt layer was thicker than 1.38 nm and 1.2 nm, respectively, FePt (111) texture predominated. The mechanisms of nonepitaxial growth can be understood by combining the kinetics and thermodynamics of film growth, with the initial nucleation of FePt (001) grain due to the stress resulted from difference in thermal expansion coefficients between the substrate and FePt films.[140]

Thermodynamically, FePt (111) is preferred since the (111) plane is closely packed with the lowest energy. Therefore, films deposited at low temperature have (111) texture since the thermal stress is insufficient for nucleation of (001) grains. Films deposited at high temperature or annealed for a long time favored the (111) texture due to predominant thermodynamics. At moderate temperatures or moderate annealing times, FePt (001) textured films were obtained. This was possibly due to that the thermal stress sufficiently caused nucleation of (001) grains with favorable kinetics for diffusion to occur. For thicker Fe and Pt layers, the (111) preferred orientation could be due to insufficient diffusion length at that temperature and annealing time.

The advantage of nonepitaxial growth is the use of a thinner layer between soft underlayer and magnetic recording layer in the double-layer perpendicular media, where the efficiency of the writing field and the writing field gradient can be dramatically enhanced compared to that with a thick spacing layer. On the other hand, in the industrial production line, the yield (several second per disk) is the most important factor affecting the practical application of a technology. The mutilayer process and long time thermal annealing are the main disadvantages of the nonepitaxial growth method.

## 5.2. Control of Exchange Coupling and Grain Size of FePt Films

### 5.2.1. Reduction of Exchange Coupling of FePt Films by Diffusion of Nanomagnetic Overlayers

Breaking the exchange coupling decreased the media noise, with an optimum state of a maximum SNR value.[141] The diffusion coefficient in grain boundaries can be

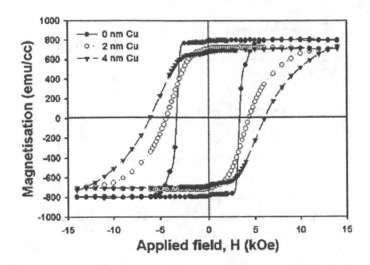

**FIGURE 32.** Out-of-plane hysteresis loops of FePt films with Cu overlayer. (Reprinted from J. Magn.Magn. Mater., 284, J. S. Chen, J. P. Wang, Structural and Magnetic Properties of FePt Film with Cu Top Layer Diffusion, 123, (2004), with permission from Elsevier.)

orders of magnitude higher than in the bulk of grain, as grain boundaries have lower activation energy in diffusion.[142] It is therefore possible to have significant diffusion of nonmagnetic materials from adjacent layers into the grain boundaries of the magnetic film to magnetically decouple the grains, whereas only a small amount diffuses into the bulk of the magnetic grains. In such a case, the magnetic anisotropy of magnetic grains does not deteriorate with diffusion. The diffusion of overlayer or underlayer to magnetic grain boundaries have been reported experimentally and theoretically to be a promising approach to increase isolation of magnetic grains in Co based media.[143, 144, 145]

The effects of Cu overlayer diffusion on magnetic isolation of FePt grains were also reported.[146] Compositional depth profiles using X-ray photoelectron spectroscopy (XPS) showed that the depth of Cu diffusion for the sample with 2 nm and 4 nm Cu overlayer was about 6 nm, 16 nm, respectively. The out-of-plane hysteresis loops of FePt films with Cu overlayer are shown in Fig. 32. With introduction of the 2 nm Cu overlayer, $H_C$ increased from 3160 Oe to 4300 Oe and $M_s$ decreased from 815 emu/cc to 711 emu/cc. For the sample with 4 nm Cu overlayer, the $H_C$ further increased up to 6030 Oe and $M_s$ slightly decreased to 700 emu/cc. The $H_c$ was increased by 91% without Cu overlayer and with 4 nm Cu overlayer, whereas the $M_S$ was only decreased by 14%, suggesting the weak bulk diffusion.

The slope of the M-H loop at coercivity, $(\frac{dM}{dH})_{Hc}$, decreased significantly after deposition of the 2 nm Cu overlayer, and then decreased slowly with further increase of the Cu overlayer. Generally, the decrease of slope of the M-H loop and increase of coercivity are mainly due to two reasons: deterioration of texture of

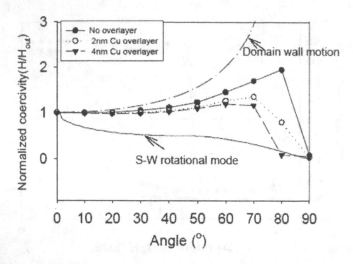

**FIGURE 33.** The angular variation of the coercivity. (Reprinted from J. Magn. Magn. Mater., 284, J. S. Chen, J. P. Wang, Structural and Magnetic Properties of FePt Film with Cu Top Layer Diffusion, 423, (2004), with permission from Elsevier.)

FePt layer and a decrease of exchange coupling between grains.[5, 147] However, the rocking curve measurement indicated that the degree of alignment of FePt layer had no obvious change after a 4 nm overlayer was deposited.

The decreased slope and increased coercivity were therefore primarily due to a decrease of exchange coupling between FePt grains. The angular variation of coercivity, as shown in Fig. 33, shows that without Cu overlayer the magnetization reversal process was dominated by domain wall motion. With the increase of Cu overlayer thickness, the magnetization reversal approached the rotational mode due to improvement of magnetic isolation of FePt grains. The thicker the Cu overlayer, the higher the concentration of Cu and thus a better driving force of Cu diffusion that allowed for better magnetic isolation. Similar experimental results were obtained when Ag overlayers were deposited.[148] CrMn and Zn overlayer were also used to decouple the FePt grains by grain boundary diffusion.[114] The characteristic of reduction of exchange coupling such as increase of coercivity, decrease of $M_s$ and $(\frac{dM}{dH})_{Hc}$ and large deviation from domain motion mode were also observed. The above results demonstrated that overlayer diffusion was effective to reduce the exchange decoupling.

## 5.2.2. Control of Grain Size and Exchange Coupling by Nonmagnetic Additives

In order to control grain size as well as to achieve magnetic isolation among grains, various nonmagnetic additives were investigated in the FePt system. Materials such as C,[149, 150] AlO$_x$,[97, 151] AlN,[152] SiN$_x$,[153] BN,[154] Zr,[155] ZrO$_x$,[156] Cr,[157] Ag,[158]

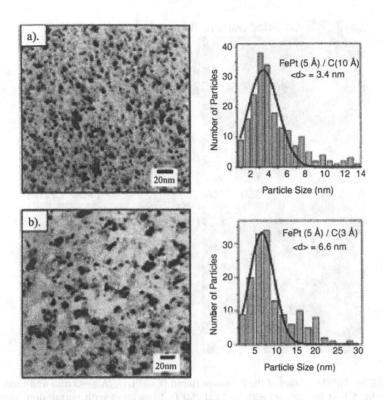

**FIGURE 34.** Microstructure of FePt/C films with (a) carbon thickness 10 Å and (b) carbon thickness 3 Å, both annealed at 700°C for 10 min. (Reprinted with permission from J. A. Christodoulides, et al., J. Appl. Phys. 87, 6938 (2000), Copyright [2000] American Institute of Physics.)

W[159], Ti,[159] HfO$_2$[160] were doped into FePt films by post-deposition annealing of co-sputtered or multilayered (FePt +additives) films. It was found that C, AlO$_x$, ZrO$_x$, SiN$_x$, HfO$_2$ and Ag were effective to suppress the growth of FePt grains during subsequent annealing. Other additives such as B,[161, 162] and MnO only reduced magnetic coupling and showed little effect on control of grain size. The typical TEM images of FePt films doped with C and AlO$_x$ are shown in Fig. 34 and 35.

In films with high $C$ contents (10 Å), no significant growth of FePt grains was observed after annealing. The average particle size for films annealed at 700°C was 3.4 nm. When the thickness of $C$ was 3 Å, the average particle size significant increased to 6.6 nm. For AlO$_x$ additive, the 2 nm FePt grains in as-deposited FePt-AlO$_x$ film were well separated. With increasing annealing temperature, co-alescence of FePt grains occurred, resulting in increase of particle size from 3 nm (at 500°C) to 11 nm (at 750°C). The exchange coupling was examined by the δM curve. For noninteracting single-domain particles, $M_d = 1-2M_r$ is expected

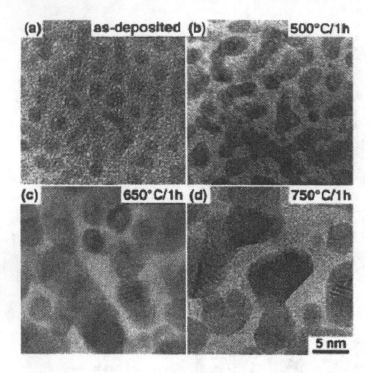

**FIGURE 35.** HREM images of the (a) as-sputtered $(Fe_{55}Pt_{45})_{63}Al_{37}$–O film and those annealed for 1 h at (b) 500, (c) 650, and (d) 750°C. (Reprinted with permission from M. Watanabe, et al., Appl. Phys. Lett. **76**, 3971 (2000), Copyright [2000] American Institute of Physics.)

and the deviation ($\delta M$) from that indicates the presence of interaction, where $M_d$ and $M_r$ are the reduced dc remanence and isothermal remanence, respectively. Positive $\delta M$ ($\delta M = M_d - 1 + 2M_r$) indicates the presence of exchange interactions while negative $\delta M$ means dipolar interactions. $\delta M$ plots with different $C$ contents in films are shown in Fig. 36. For the FePt-C films with lower $C$ content, the exchange coupling was present. Increase of $C$ content the interaction between FePt grains approached the dipolar interaction.

Although the grain size and exchange coupling were reduced in the above-mentioned investigation, the texture of FePt films was random and not useful for magnetic recording application. Granular FePt-MgO films with (001) texture were obtained by post-deposition annealing of $(FePt/MgO)_n$ multilayers.[163] Good (001) textured FePt-B$_2$O$_3$ or FePt-C nanocomposite films with well-separated grains (less than 10 nm in diameter) were fabricated by post-deposition annealing of FePt/B$_2$O$_3$ and FePt/C multilayers.[136, 164] FePt/SiO$_2$ multilayers showed a small grain size of 10 nm but poor orientation.[96] The grain size and exchange coupling strongly depended on concentration of nonmagnetic additives and annealing time and temperature.

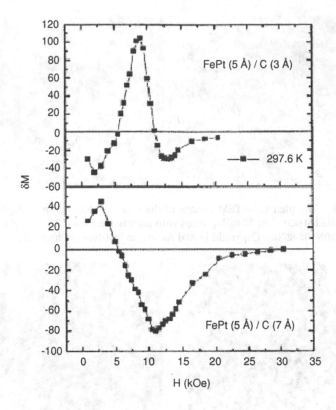

**FIGURE 36.** δM plots in FePt/C films with different C thickness. (Reprinted with permission from J. A. Christodoulides, et al., J. Appl. Phys. **87**, 6938 (2000), Copyright [2000] American Institute of Physics.)

Granular (001) textured FePt-C films with grain size of 4-5 nm were deposited on MgO (200) single-crystal substrate at elevated temperature by co-sputtering FePt and $C$ targets.[165] It was reported that the MgO doping in FePt film by in situ elevated temperature deposition on CrX alloy or MgO underlayer only reduced the domain size of FePt film without formation of granular microstructures.[166, 167]

Figure 37 shows the plane-view TEM images of $(FePt)_{90}$-$MgO_{10}$ films, as-deposited and post-deposition annealed at 600°C for 30 s. Insets are the electron diffraction patterns of corresponding films. The grains of both films were indistinguishable and the FePt films were not granular after the 10% MgO addition. The absence of FePt (001) diffraction spots or rings and the presence of (110) and (200) diffraction spots or rings indicated that the FePt films were (001) textured.

Figure 38 shows the images of magnetic force microscopy (MFM) of ac-demagnetized $(FePt)_{100-x}$-$MgO_x$ films. The maze-like domain was observed in pure FePt film and the domain size decreased with increasing concentration of MgO. After annealing at 500°C for 30 s, the domain size of $(FePt)_{90}$-$MgO_{10}$ films

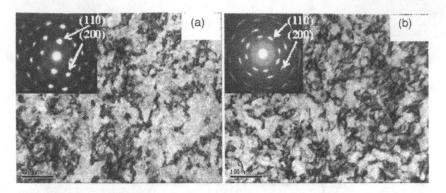

**FIGURE 37.** The plan view TEM images of the $(FePt)_{90}$-$MgO_{10}$ film as-deposited and post-annealed at 600°C for 30 s.(Reprinted with permission from J. S. Chen, et al., J. Appl. Phys. 97, 10N108 (2005), Copyright [2005] American Institute of Physics.)

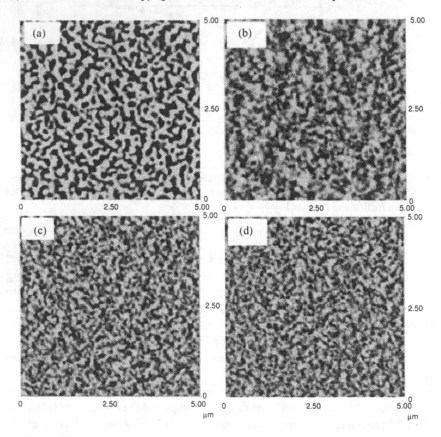

**FIGURE 38.** MFM images of (a) as-deposited FePt film at 350°C, (b) as-deposited $(FePt)_{95}(MgO)_{5}$ film at 350°C, (c) as-deposited $(FePt)_{90}(MgO)_{10}$ film at 350°C, and (d) post-annealed $(FePt)_{90}(MgO)_{10}$ film at 500°C for 30s. (Reprinted with permission from J. S. Chen, et al., J. Appl. Phys. 97, 10N108 (2005), Copyright [2005] American Institute of Physics.)

was reduced. Granular FePt-SiO$_2$ and FePt-Al$_2$O$_3$ films were prepared by in situ deposition on MgO underlayer. The substrate temperatures for deposition of FePt-SiO$_2$ and FePt-Al$_2$O$_3$ were higher than that of FePt-MgO and pure FePt film for the same chemical ordering.[168]

### 5.2.3. Control of Grain Size by Modification of Underlayer and Surface

Underlayer easily forming small grain size were used to control the grain size of magnetic layers.[169] RuAl and NiAl (B2 structure) were reported to have strong atomic bonding and low atomic mobility, yielding a smaller grain size.[170, 171] In addition, the lattice misfit between RuAl, NiAl and FePt (001) is below 10%. FePt films with $L1_0$(001) texture, mean grain size of 6.6 nm was successfully induced using a RuAl underlayer.[121] NiAl was used to be an intermediate layer to reduce the grain size and maintain the (001) texture of FePt on CrRu (200) underlayer.[172] Sputtered NiAl intermediate layer on CrRu underlayer grew easily in small isolated and uniform sized grains (average 12 nm) as shown in Fig. 39 (a). The images of atomic force microscopy (AFM) of FePt films without and with a 4 nm of NiAl intermediate layer are shown in Fig. 39 (b) and (c), respectively. The sample with 4 nm of NiAl intermediate layer shows a clearly separated FePt grains that were approximately 18 nm in size. For the sample without NiAl intermediate layer, the FePt grains ($\sim$ 35 nm) were twice as large and not clearly separated, as compared to that with a 4 nm of NiAl intermediate layer.

Surface roughness, affecting the wetting factor of critical nucleation energy, influences nucleation rate. The effects of Ar-ion etching on the topography of Pt seed layers as well as the overlying FePt films were reported.[173] As can be seen in Figs. 40 (a), (d) and (g), the Ar-ion etching initially resulted in a significant roughening of the Pt seed layer, compared to the relatively smooth and featureless as-deposited film. After the seed layer was etched for 60 s (about 1 nm remaining thickness), the feature size became significantly smaller and the surface smoother. An apparent grain size of FePt film was estimated from AFM results by setting a threshold at half the difference between the average grain height and the average trench depth. For the FePt film deposited onto the as-grown Pt seed layer described above, small granular features were detected in the AFM images. The apparent grain diameter ranged from 20 nm to 35 nm, with a mean diameter of 29.6 and a standard deviation of 5.3 nm. After 20 s Ar-ion etching, the apparent grain remained approximately constant (29.3 nm) with a slightly narrower size distribution ($\pm$4.4 nm) (Fig. 39f). Only after 60 s etching, the mean grain size distribution shifted to significantly smaller grain sizes and a narrower grain size distribution of 15.0 $\pm$ 2.2 nm.

### 5.2.4. Control of Exchange Coupling by Domain Wall Pinning

Defects such as crystal imperfections and inhomogeneities can serve as pinning sites to break the exchange coupling and reduce the domain size in magnetic thin

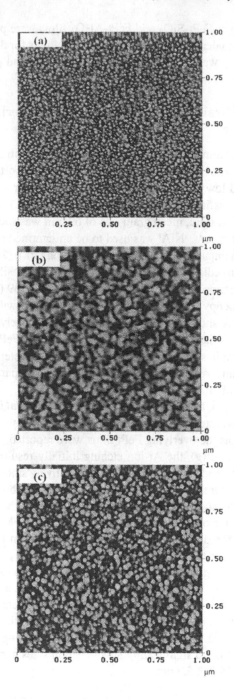

**FIGURE 39.** AFM images of sputtered NiAl intermediate layer on CrRu underlayer (a), the FePt films without (b) and with 4 nm of NiAl intermediate layer (c). (Reprinted with permission from J. S. Chen, et al., J. Appl. Phys. 93, 8167 (2003), Copyright [2003] American Institute of Physics)

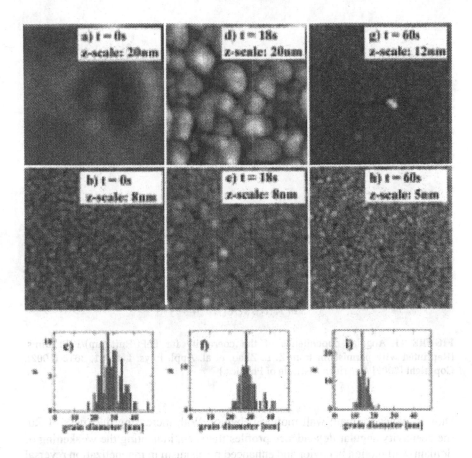

**FIGURE 40.** AFM topography scans (1 μm × 1 μm) of Pt seed layers etched with Ar-ions for 0 s (a), 18 s (d), and 60 s (g), topography scans of the FePt films grown on these Pt seed layers, respectively (b, e, h), and the grain size distribution obtained from the scans of the FePt (c, f, i). (Reprinted with permission from J.-U Thiele, et.al. IEEE Trans Magn. 37, 1271 (2001). © [2001] Institute of Electrical and Electronics Engineers.)

film. The pining effect is most pronounced when the size of pinning sites is comparable to the domain wall width.[174] The narrow wall width (δ) of FePt is estimated as 4 nm using the equation of $\delta \sim \pi(A/K_u)^{1/2}$, where $A$ is exchange stiffness and $K_u$ is magnetic anisotropy. Hence small structural defects, as pinning sites, may significantly inhibit the domain size and domain wall movements. A thin Ru layer (0-1 nm) inserted between FePt films grown on CrRu (200) underlayer was used to introduce pinning sites.[148, 175] The FePt (001) texture remained when the inserted Ru layer was less than 0.5 nm. In order to understand the magnetization reversal mechanism after inserting the Ru layer with different thickness, the angular dependence of the coercivity for FePt/Ru(t nm)/FePt films was investigated, as shown in Fig. 41. The coercivity angular dependence for the film without Ru pinning layer

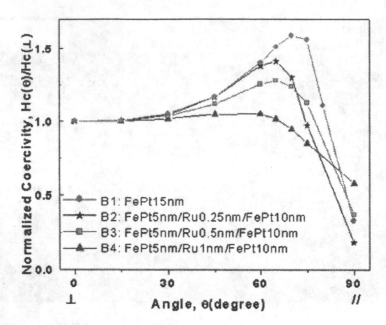

**FIGURE 41.** Angular dependence of the coercivity for FePt/Ru(t nm)/FePt films. (Reprinted with permission from Z. L. Zhao, et al. Appl. Phys. Lett. 81, 3612 (2002), Copyright [2002] American Institute of Physics.)

showed typical domain-wall motion behavior. With increasing thickness of Ru, the coercivity angular dependence profiles flattened, indicating the weakening of domain-wall motion behavior and enhanced mechanism in magnetization reversal process by a nucleation related rotation mode.

The relative concentration of defects was determined by comparing the high field behavior of the magnetization curve using the law of approach to saturation (LATS).[176] To investigate the defects for thin films, initial magnetization curves were measured by VSM. Figure 42 shows the normalized magnetization *vs.* inverse field for obtaining the magnetic hardness coefficient of the samples with various Ru thickness.[148] Parameter $a$ in Fig. 42 is the hard magnetic coefficient in the equation, $M = M_s(1-a/H-b/H^2) + kH$. Parameters $b$ and $k$ were neglected when the applied field was selected (in the range of approach to saturation field) from 9 to 15 k Oe.

It was suggested that a larger hard coefficient was associated with larger defects. As expected, the film with 1 nm Ru inserted had the largest hardness coefficient of the three samples, indicating potentially large defects. The smaller defects in film introduced by the thinner Ru layer can form pinning sites to enhance the coercivity; however, the larger defects in the film introduced by the thicker Ru layer can serve as nucleation sites to decrease coercivity.

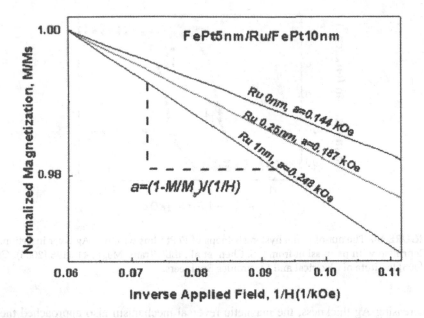

**FIGURE 42.** Normalized magnetization vs. inverse field for films with different thickness of Ru layer. The approach to saturation field ranged from 9 kOe to 15 kOe. (Reprinted with permission from Z. L. Zhao, et al. Appl. Phys. Lett. 81, 3612 (2002), Copyright [2002] American Institute of Physics.)

Since Ru is hexagonal close-packed (hcp), it deteriorates the FePt (001) epitaxial growth if the Ru layer becomes too thick. An inserted Ag layer between the FePt layers was used to introduce defects.[177] Ag (100) plane has similar atomic configuration with that of FePt (001) plane and the lattice mismatch between the two planes is 7.1%. It is expected when a thin Ag layer is inserted into FePt layers, Ag itself and structural defects of FePt layers caused by inserted Ag layer may pin the domain wall and decrease the domain size and media noise, while maintaining the magnetic properties of FePt layer and the FePt (001) texture. FePt films with the structure of substrate/$Cr_{90}Ru_{10}$(30 nm)/Pt (4 nm)/FePt(6 nm)/Ag($t$ nm)/FePt(6 nm) were fabricated, where the total FePt thickness was maintained at 12 nm and $t = 0, 0.5, 1, 2$.

The FePt (001) texture was retained regardless of Ag layer thickness. The out-of-plane hysteresis loops of FePt films with one Ag layer insertion are shown in Fig. 43. The coercivity increased linearly from 1.93 kOe to 3.2 kOe with increasing Ag thickness from 0 to 2 nm. The slope of the hysteresis loop at coercivity $(\frac{dM}{dH})_{Hc}$ decreased monotonically, indicating the exchange coupling was reduced. The angular variation of coercivity, as indicated that without Ag insertion, the magnetic reversal mechanism was close to the domain wall motion mode and with

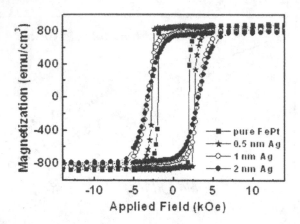

**FIGURE 43.** The out-of-plane hysteresis loops of FePt films with one Ag layer insertion. (Reprinted with permission from J. S. Chen, et al., IEEE Trans. Magn. 41 3196 (2005). © [2005] Institute of Electrical and Electronics Engineers. )

increasing Ag thickness, the magnetic reversal mechanism also approached the rotation mode.

The magnetic force microscopy (MFM) images of ac-demagnetized FePt films with different thickness of inserted Ag layer are shown in Fig. 44. The domain size decreased with increasing Ag pinning layer thickness, indicating domain wall pinning by Ag. Corresponding to the decrease of the slope of M-H loops with the increase of the Ag layer thickness, the domain size also decreased, further indicating the decrease of lateral exchange coupling.

A fcc FePt layer was used as a nucleation layer to reduce the domain size.[178] A fcc FePt layer with thickness up to 3 nm was deposited 50°C, followed by a $L1_0$ FePt layer deposited at 375°C. The FePt (001) texture remained and $L1_0$ FePt layer was homoepitaxially grown on the fcc FePt layer. Similarly, the slope of hysteresis loop at coercivity decreased and domain size reduced with increasing thickness of fcc FePt nucleation layer, as shown in Fig. 45 and Fig. 46, respectively.

## 5.3. Recording Performance of $L1_0$ FePt Perpendicular Media

Recording tests have been performed on several kinds of FePt perpendicular media, which may be categorized into continuous, pinning typed and granular FePt perpendicular media.

### 5.3.1. Continuous FePt Perpendicular Recording Media

The thickness effect of continuous FePt double layered perpendicular media on recording performance were studied.[179, 180] The MFM images of recorded patterns written by single-pole writing head on contact tester are shown in Fig. 47. The

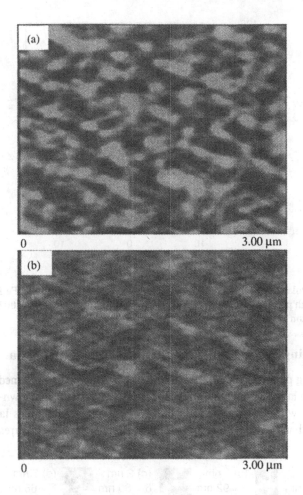

**FIGURE 44.** MFM images of 10 nm FePt films with one-layer inserted Ag thickness of (a) 0 nm; (b) 2 nm.

7.5 nm thick medium exhibited a fully saturated recorded state, judging from the high contrast of recorded patterns. A high linear density recorded pattern of 500 kFRPI was also observed. The somewhat irregular transition lines are thought to arise from strong lateral exchange-coupling as shown by the AC-demagnetized domain patterns. However, the degree of such irregularity does not seem to depend on the linear recording density. In the case of the 3 nm thick medium, a large number of reversed domains are observed at the density of 30 kFRPI, because the perpendicular squareness (SQ) is less than unity. Since the demagnetizing field is smaller at higher linear densities, the 200 kFRPI recorded pattern is clearly observed. The 5 nm thick medium shows intermediate features between the 7.5 nm thick and 3 nm thick media.

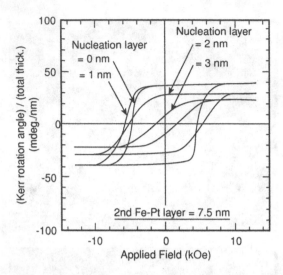

**FIGURE 45.** Polar-Kerr loops whose rotation angles were divided by total Fe-Pt thickness. (Reprinted with permission from T. Suzuki et al., J. Appl. Phys. 91, 8079 (2002), Copyright [2002] American Institute of Physics.)

### 5.3.2. Pinning Typed FePt Perpendicular Recording Media

The recording performance of FePt double layered perpendicular media with fcc FePt pinning layer were reported by Toshio Suzuki *et al.*, as shown in Fig. 48. The medium with the 1 nm nucleation layer exhibited 1.13 times larger signal output level than that of a medium without the nucleation layer, corresponding to

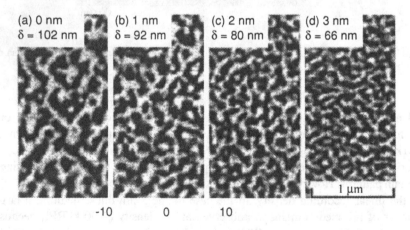

**FIGURE 46.** Domain patterns and sizes ($\delta$) for double-layered media with various thick nucleation layers. Second Fe-Pt layer thick. −7.5 nm. (Reprinted with permission from T. Suzuki et al., J. Appl. Phys. 91, 8079 (2002), Copyright [2002] American Institute of Physics.)

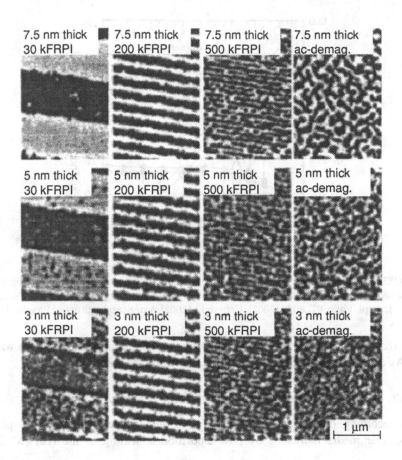

**FIGURE 47.** MFM images of recorded patterns for double-layered media. (Reprinted from J. Magn. Magn. Mater., 235, T. Suzuki, T. Kiya, N. Honda, K. Ouchi, Fe-Pt perpendicular double-layered media with high recording resolution, 312, (2001), with permission from Elsevier.)

an increase in the total thickness of the ordered FePt layer. Recording resolution, $D_{50}$, was around 250 kfrpi for both media, whose value should be limited by a shield gap length of the GMR reproducing head. On the other hand, medium noise of the medium with the nucleation layer was lower than that of the medium without the nucleation layer. Consequently, the signal-to-noise ratio (SNR) of the former medium was 5.7 dBpp/rms for the output at 500 kfrpi (frequency band width: 0.5–50 MHz), 4 dB higher than that of the latter.

The FePt single layered perpendicular media with inserted Ag pinning layers were also tested a Guzik spin-stand (1701B) using a 30 Gb/in$^2$ commercial ring head.[177] The media noise and SNR of the FePt media with different thickness of Ag are shown in Fig. 49 and Fig. 50, respectively. With increasing Ag thickness up to 2 nm, media noise was effectively reduced and SNR was remarkably enhanced.

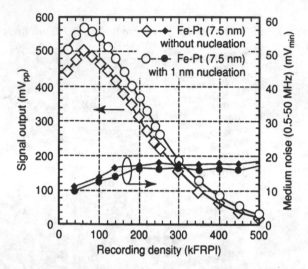

**FIGURE 48.** Density responses for double-layered media without and with 1 nm nucleation layer. Second Fe-Pt layer thick. −7.5 nm. (Reprinted with permission from T. Suzuki et al., J. Appl. Phys. 91, 8079 (2002), Copyright [2002] American Institute of Physics.)

The reduction in media noise was mainly due to the pinning of the domain wall by Ag and structural defects in FePt layer caused by Ag insertion.

### 5.3.3. Granular Typed FePt Perpendicular Recording Media

The recording performance of FePt-MgO granular-type double-layered perpendicular media fabricated by post-deposition annealing of the (FePt/MgO)

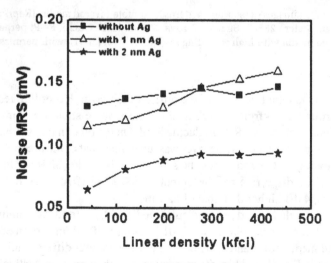

**FIGURE 49.** Noise as a function of linear density for FePt samples with different one-layer Ag thickness. (Reprinted with permission from J. S. Chen, et al., IEEE Trans. Magn. 41 3196 (2005). © (2005) Institute of Electrical and Electronics Engineers. )

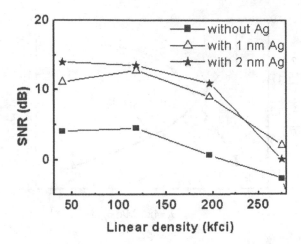

**FIGURE 50.** Signal-to-noise (SNR) as a function of linear density for FePt samples with different one-layer Ag thickness. (Reprinted with permission from J. S. Chen, et al., IEEE Trans. Magn. 41 3196 (2005). © (2005) Institute of Electrical and Electronics Engineers.)

multilayer were reported.[181, 182, 183] The media were tested by spin-stand with a commercialized recording head for 50 Gbits/in$^2$ longitudinal recording. Figure 51 shows the $SN_mR$ of the disks with different multilayer structures, where $SN_m R$ is defined as the zero-to-peak signal at 66 kfci divided by the integrated rms noise over 150 MHz bandwidths after writing at 790 kfci. It was observed that the initial

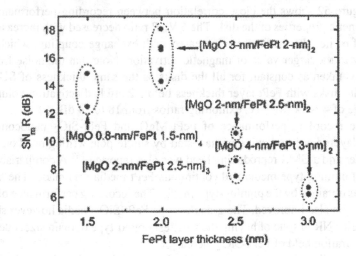

**FIGURE 51.** Summary of SNmR ratio for various disks with different initial multilayer structures. For a same multilayer structure, the scattered data points are corresponding to different annealing conditions. (Reprinted with permission from Ref. 181, T.Suzuki, et al., IEEE Trans. Magn. 41, 555 (2005). © (2005) Institute of Electrical and Electronics Engineers.)

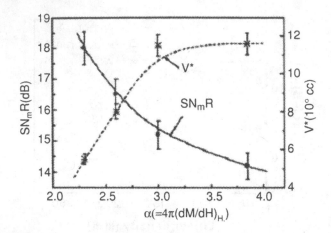

**FIGURE 52.** $SN_mR$ ratio and magnetic activation volume verse out-of-plane hysteresis loop slope for granular-type FePt-MgO disks with SUL. (Reprinted with permission from Ref. 181, T.Suzuki, et al., , IEEE Trans. Magn. 41, 555 (2005). © (2005) Institute of Electrical and Electronics Engineers.)

FePt/MgO multilayer structures strongly affected the final $SN_m R$. The highest value of $SN_m R$ was obtained for the 2 nm FePt layer thickness sample. As the FePt grain size and activation volume decreased pronouncedly with decreasing FePt thickness, a lower transition noise was therefore expected with decreasing FePt thickness.

Figure 52 shows the close correlation between recording performance and the magnetic properties of the disk. The $SN_m R$ ratio decreased with increasing the value of $\alpha$, i.e., the media noise increased with exchange coupling, which leads possibly to the larger value of magnetic activation. Note that the noise from the SUL was taken as constant for all the disks as the same thickness of SUL was used. The disks with FePt layer thickness below 2 nm had activation volumes in the range of $4 \sim 7 \times 10^{-19}$ cc, exhibiting ratios from 18 to 14 dB.

The recording performance of FePt-MgO and FePt-SiO$_2$ nanocomposite double-layered perpendicular media tested by single-pole writing head on a contact slider and a GMR reproducing head was also reported.[168] A comparison with fcc FePt pinning-type media and continuous FePt media was made. The highest SNR was obtained by the pinning-type media. The recording performance of FePt-MgO media also improved. The granular type FePt-SiO$_2$ media however showed the lowest SNR in spite of having the smallest dotted-type domain size, due to the large saturation field of FePt-SiO$_2$ media.

The media noise can be decreased by introducing pinning sites and reducing magnetic domain size. Yet the challenge to obtain uniform pinning sites using this approach remains. In addition, the domain size is proportional to pinning sites caused by crystalline defects. The crystalline defects in FePt crystals will, however, decrease the anisotropy and deteriorate magnetic properties. When the areal density is 1 Tbits/in.$^2$ and above, the domain wall pinning-based method results in thermal instability of recording bits due to deterioration of magnetocrystalline anisotropy caused by crystalline defects. More work for optimizing the granular type media for ultrahigh density magnetic recording media is therefore warranted.

## 6. SUMMARY AND OUTLOOK

In this chapter, we have reviewed the current status of the study of $L1_0$ FePt alloy films for the application for ultrahigh density magnetic recording. For practical realization of $L1_0$ FePt as recording media, issues such as decreasing preparation temperature, easy axis control, and reduction in media noise must be solved. We reviewed the parameters such as stoichiometry and film thickness that affect chemical ordering of FePt as well as the methods to reduce the ordering temperature. The Ag doping and strain-inducing methods are the most promising to lower the ordering temperature. However, the films using Ag doping and strain-inducing method usually showed lower coercivity due to significant amount of defects in the film. Two methods, epitaxial growth and nonepitaxial growth, were introduced to control the easy axis and each method had its advantages and disadvantages. Different methods to reduce the noise were also reviewed. For the recording media beyond 1 Tbits/in.$^2$, the traditional way to reduce the noise such as small grain size and suitable exchange coupling was the most promising. More efforts are still required to further lower the ordering temperature, reduce the grain size and thus reducing media noise. To date significant amount of work on FePt magnetic media for high areal density recording applications has been mainly focused on the engineering approach to improve and optimize the properties. Many scientific issues, such as the distribution of dopants in the crystal lattice, types of defects, mechanisms of dopant-enhanced ordering, remain unclear. The rational design of these advanced engineering materials would be optimized by better understanding of some of these critical scientific issues.

## REFERENCES

1. R. W. Wood, J. Miles and T. Olson, Recording technologies for terabit per square inch systems, *IEEE Trans. Magn.* **38**, 1711(2002).
2. R. Wood, Y. Sonobe, Z. Jin and B. Wilson, Perpendicular recording: the promise and the problems, *J. Magn. Magn. Mater.* **235**, 1 (2001).
3. R. Wood, The feasibility of magnetic recording at 1 Terabit per square inch, *IEEE Trans. Magn.* **36**, 36 (2000).
4. A. Moser, K. Takano, D. T. Margulies, M. Albrecht, Y. Sonobe, Y. Ikeda, S. H. Sun and Eric E. Fullerton, Magnetic recording: advancing into the future, *J. Phys. D* **35**, R157 (2002).
5. N. Honda, K. Ouchi and S. Iwasaki, Design consideration of ultrahigh-density perpendicular magnetic recording media, *IEEE Trans. Magn.* **38**, 1615 (2002).

6. D. Litvinov, M. H. Kryder and S. Khizroev, Recording physics of perpendicular media: hard layers, *J. Magn. Magn. Mater.* **241**, 453 (2002).

7. S. Khizroev, M. H. Kryder and D. D. Litvinov, Next generation perpendicular systems, *IEEE Trans. Magn.* **37**, 1922 (2001).

8. D. Litvinov, M. H. Kryder and S. Khizroev, Recording physics of perpendicular media: soft underlayer, *J. Magn. Magn. Mater.* **232**, 84 (2001).

9. M. H. Kryder and R. W. Gustafson, High-density perpendicular recording: advances, issues, and extensibility, *J. Magn. Magn. Mater.* **287**, 449 (2005).

10. D. Weller, A. Moser, L. Folks, M. E. Best, W. Lee, M. F. Toney, M. Schwickert, J. U. Thiele and M. F. Doerner, High $K_u$ materials approach to 100 Gbits/in.$^2$, *IEEE Trans. Magn.* **36**, 10 (2000).

11. www.seagate.com

12. www.hgst.com

13. C. A. Ross, Patterned magnetic recording media, *Annu. Rev. Mater. Res.* **31**, 203 (2001).

14. B. D. Terris and T. Thomson, Nanofabricated and self-assembled magnetic structures as data storage media, *J. Phys. D* **38**, R 199 (2005).

15. J. H. Judy, Past, present, and future of perpendicular magnetic recording, *J. Magn. Magn. Mater.* **287**, 16 (2005).

16. K. Ouchi, Review on recent developments of perpendicular recording media, *IEICE Trans. Electron.* **E84C**, 1121 (2001).

17. G. M. Chow, W. C. Goh, Y. K. Hwu, T. S. Cho, J. H. Je, H. H. Lee, H. C. Kang, D. Y. Noh, C. K. Lin and W. D. Chang, Structure determination of nanostructured Ni-Co films by anomalous x-ray scattering, *Appl. Phys. Lett.* **75**, 2503 (1999).

18. T. B. Massalski, H. Okamoto, P. R. Subramanian and L. Kacprzak, *Binary Alloy Phase Diagrams*, ASM International, Materials Park, OH (1990).

19. D. E. Laughlin, K. Srinivasan, M. Tanase and L. Wang, Crystallographic aspects of L1$_0$ magnetic materials, *Scripta. Mater.* **53**, 383 (2005).

20. B. D. Cullity and S. R. Stock, *Elements of X-ray Diffraction*, 3$^{rd}$ Edition, Prentice-Hall International, London (2001).

21. C. J. Sun, G. M. Chow and J. P. Wang, Epitaxial L1$_0$ FePt magnetic thin films sputtered on Cu (001), *Appl. Phys. Lett.* **82**, 1902 (2003).

22. C.J. Sun, B. H. Liu, J. P. Wang and G. M. Chow, Sputtered FePt films with uniform nanoscale grain size on Cu (001) single crystal, *J. Appl. Phys.* **97**, 10J103 (2005).

23. A. Cebollada, R. F. C. Ferrow and M. F. Toney, *Magnetic Nanostructure*, H. S. Nalwa ed., American Scientific Publisher, Stevenson Ranch, California (2002), p. 98

24. R. F. C. Farrow, D. Weller, R. F. Marks, M. F. Toney, A. Cebollada and G. R. Harp, Control of the axis of chemical ordering and magnetic anisotropy in epitaxial FePt films, *J. Appl. Phys.* **79**, 5967 (1996).

25. S. Okamoto, N. Kikuchi, O. Kitakami, T. Miyazaki, Y. Shimada and K. Fukamichi, Chemical-order-dependent magnetic anisotropy and exchange stiffness constant of FePt (001) epitaxial films, *Phys. Rev. B* **66**, 024413 (2002).

26. JCPDS 02-1167 for fct FePt, 03-1005 for Cu, and 29-0717 for fcc FePt were taken as references.

27. J. W. Edington, in *Practical Electron Microscopy in Materials Science*, Van Nostrand Reinhold, New York (1976).

28. A. Cebollada, R. F. C. Ferrow and M. F. Toney, *Magnetic Nanostructure*, H. S. Nalwa, ed., American Scientific Publisher, Stevenson Ranch, California (2002), p. 106.

29. M. Weisheit, L. Schultz and S. Fahler, Textured growth of highly coercive L1$_0$-ordered FePt thin films on single crystalline and amorphous substrates, *J. Appl. Phys.* **95**, 7489 (2004).

30. M. Weisheit, L. Schultz and S. Fahler, On the influence of composition on laser-deposited Fe-Pt films, *J. Magn. Magn. Mater.* **290-291**, 570 (2005).

31. K. H. Kang, T. Yang and T. Suzuki, Structural and magnetic properties of FePt-Ag composite film with perpendicular magnetic anisotropy, *IEEE Trans. Magn.* **38**, 2039 (2002)

32. T. Yang, K. H.Kang, G. H.Yu and T. Suzuki, Structural and magnetic properties of (001)-oriented FePt/Ag composite film, *J. Phys. D- Appl. Phys.* **35**, 2897 (2002).

33. T. Seki, T. Shima, K. Takanashi, Y. Takahashi, E. Matsubara and K. Hono, $L1_0$ ordering of off-stoichiometric FePt(001) thin films at reduced temperature, *Appl. Phys. Lett.* **82**, 2461 (2003).

34. Z. L. Zhao, Ph.D Thesis, "Magnetic Properties of $L1_0$ FePt Thin Films with Additional Ultrathin Nonmagnetic Layers Data Storage Institute and National University of Singapore (2005).

35. Y. K. Takahashi, M. Ohnuma and K. Hono, Ordering process of sputtered FePt films, *J. Appl. Phys.* **93**, 7580 (2003).

36. K. Takahashi and K. Hono, Ordering process and size effect of FePt magnetic thin films, *J. Magn. Soc. Jpn* **29**, 72 (2005).

37. T. Maeda, T. Kai, A.Kikitsu, T. Nagase and J. I. Akiyama, Reduction of ordering temperature of an FePt-ordered alloy by addition of Cu, *Appl. Phys. Lett.* **80**, 2147 (2002).

38. T. Kai, T. Maeda, A. Kikitsu, J. Akiyama, T. Nagase and T. Kishi, Magnetic and electronic structures of FePtCu ternary ordered alloy, *J. Appl. Phys.* **95**, 609 (2004).

39. T. Maeda, A. Kikitsu, T. Kai, , T. Nagase, H. Aikawa and J. Akiyama, Effect of added Cu on disorder-order transformation of $L1_0$-FePt, *IEEE Trans. Magn.* **38**, 2796 (2002).

40. Y. K.Takahashi, M. Ohnuma and K. Hono, Effect of Cu on the structure and magnetic properties of FePt sputtered film, *J. Magn. Magn. Mater.* **246**, 259 (2002).

41. K. W. Wierman, C. L. Platt and J. K. Howard, Impact of stoichiometry on $L1_0$ ordering in FePt and FePtCu thin films, *J. Magn. Magn. Mater.* **278**, 214 (2004).

42. X. C. Sun, S. S. Kang, J. W. Harrell, D. E. Nikeles, Z. R. Dai, J. Li and Z. L. Wang, Synthesis, chemical ordering, and magnetic properties of FePtCu nanoparticle films, *J. Appl. Phys.* **93**, 7337 (2003).

43. K. W. Wierman, C. L. Platt, J. K. Howard and F. E. Spada, Evolution of stress with $L1_0$ ordering in FePt and FeCuPt thin films, *J. Appl. Phys.* **93**, 7160 (2003).

44. K. Barmak, D. C. Berrya, B. J. Kima, K. W. Wierman, E. B. Svedberg and J. K. Howard, Engineering Conference International, Copper Mountain, CO, August 15-20, 2004.

45. C. L. Platt, K. W. Wierman, E. B. Svedberg, R. van de Veerdonk, J. K. Howard, A. G. Roy and D. E. Laughlin, $L1_0$ ordering and microstructure of FePt thin films with Cu, Ag, and Au additive, *J. Appl. Phys.* **92**, 6104 (2002).

46. Y. Z. Zhou, J. S. Chen, G. M. Chow and J. P. Wang, Structure and magnetic properties of in-plane oriented FePt-Ag nanocomposites, *J. Appl. Phys.* **93**, 7577 (2003).

47. S. S. Kang, D. E. Nike and J. W. Harrell, Synthesis, chemical ordering, and magnetic properties of self-assembled FePt-Ag nanoparticles, *J. Appl. Phys.* **93**, 7178 (2003).

48. C. Chen, O. Kitakami, S. Okamoto and Y. Shimada, Ordering and orientation of $CoPt/SiO_2$ granular films with additive Ag, *Appl. Phys. Lett.* **76**, 3218 (2000).

49. O. Kitakami, Y. Shimada, Y. Oikawa, H. Daimon and K. Fukamichi, Low-temperature ordering of $L1_0$-CoPt thin films promoted by Sn, Pb, Sb, and Bi additives, *Appl. Phys. Lett.* **78**, 1104 (2001).

50. Z. L. Zhao, J. Ding, K. Inaba, J. S. Chen and J. P. Wang, Promotion of $L1_0$-ordered phase transformation by the Ag top layer on FePt thin films, *Appl. Phys. Lett.* **83**, 2196 (2003).

51. S. R. Lee, S. Yang, Y. K. Kim and J. G. Na, Rapid ordering of Zr-doped FePt alloy films, *Appl. Phys. Lett.* **78**, 4001(2001).

52. T. Suzuki, K. Harada, N. Honda and K. Ouchi, Preparation of ordered Fe-Pt thin films for perpendicular magnetic recording media, *J. Magn. Magn. Mater.* **193**, 85 (1999).

53. Y.-N. Hsu, S. Jeong, D. Laughlin and D. N. Lambeth, Effects of Ag underlayers on the microstructure and magnetic properties of epitaxial FePt thin films, *J. Appl. Phys.* **89**, 7068 (2001).

54. Y. F. Xu, J. S. Chen and J. P. Wang, In situ ordering of FePt thin films with face-centered-tetragonal (001) texture on $Cr_{100-x}Ru_x$ underlayer at low substrate temperature, *Appl. Phys. Lett.* **80**, 3325 (2002).

55. Y. F. Ding, J. S. Chen, E. J. Liu, C. J. Sun and G. M. Chow, Effect of lattice mismatch on chemical ordering of epitaxial $L1_0$ FePt films, *J. Appl. Phys.* **97**, 10H303 (2005).

56. C. H. Lai, C. H. Yang, C. C. Chiang and T. K. Tseng, Dynamic stress-induced low-temperature ordering of FePt, *Appl. Phys. Lett.* **85**, 4430 (2004).

57. Y.-N. Hsu, S. Jeong, D. N. Lambeth and D. Laughlin, In situ ordering of FePt thin films by using Ag/Si and Ag/Mn$_3$Si/Ag/Si templates, *IEEE Trans. Magn.* **36**, 2945 (2000).

58. J. S. Chen, B. C. Lim and T. J. Zhou, Effect of ultrahigh vacuum on ordering temperature, crystallographic and magnetic properties of L1$_0$ FePt (001) film on CrRu underlayer, *J. Vac. Sci. Tech. A* **23**, 184 (2005).

59. J. S. Chen, Y. F. Xu and J. P. Wang, Effect of Pt buffer layer on structural and magnetic properties of FePt thin films, *J. Appl. Phys.* **93**, 1661 (2003).

60. D. Ravelosona, C. Chappert, V. Mathet and H. Bermas, Chemical order induced by ion irradiation in FePt (001) films, *Appl. Phys. Lett.* **76**, 236 (2000).

61. C. H. Lai, C. H. Yang and C. C. Chiang, Ion-irradiation-induced direct ordering of L1$_0$ FePt phase, *Appl. Phys. Lett.* **83**, 4550 (2003).

62. Y. Zhu and J. W. Cai, Low-temperature ordering of FePt thin films by a thin AuCu underlayer, *Appl. Phys. Lett.* **87**, 32504 (2005).

63. G. H. O. Daalderop, P. J. Kelly and M. F. H. Schurmanns, Magnetocrystalline anisotropy and orbital moments in transition-metal compounds, *Phys. Rev. B*, **44**, 12054 (1991).

64. S. Ostanin, S. S. A. Razee, J. B. Staunton, B. Ginatempo and E. Bruno, Magnetocrystalline anisotropy and compositional order in Fe0.5Pt0.5: calculations from an ab initio electronic model, *J. Appl. Phys.* **93**, 453 (2003).

65. I. V. Solovyev, P. H. Dederichs and I. Mertig, Origin of orbital magnetization and magnetocrystalline anisotropy in TX-ordered alloys (where $T$ = Fe, Co and $X$ = Pd, Pt), *Phys. Rev. B* **52**, 13419 (1995).

66. A. Sakuma, First principle calculation of the magnetocrystalline anisotropy energy of FePt and CoPt ordered alloys, *J. Phys. Soc. Jpn.* **63**, 3053 (1994).

67. P. M. Oppeneer, Magneto-optical spectroscopy in the valence-band energy regime: relationship to the magnetocrystalline anisotropy, *J. Magn. Magn. Mater.* **188**, 275 (1998).

68. I. Galanakis, M. Alouani and H. Dreysee, Perpendicular magnetic anisotropy of binary alloys: a total-energy calculation, *Phys. Rev. B* **62**, 6475 (2000).

69. P. Ravindran, A. Kjekshus, H. Fjellvåg, P. James, L. Nordström, B. Johansson and O. Eriksson, Large magnetocrystalline anisotropy in bilayer transition metal phases from first-principles full-potential calculations, *Phys. Rev. B* **63**, 144409 (2001).

70. O. A. Ivanov, L. V. Solina, V. A. Demshira and L. M. Magat, *Phys. Met. Metallogr.* **35**, 92 (1973).

71. J.-U. Thiele, L. Folks, M. F. Toney and D. K. Weller, Perpendicular magnetic anisotropy and magnetic domain structure in sputtered epitaxial FePt (001) L1$_0$ films, *J. Appl. Phys.* **84**, 5686 (1998).

72. J. U. Thiele, K. R. Coffey, M. F. Toney, J. A. Hedstrom and A. J. Kellock, Temperature-dependent magnetic properties of highly chemically ordered Fe$_{55-x}$Ni$_x$Pt$_{45}$L1$_0$ films, *J. Appl. Phys.* **91**, 6595 (2002).

73. A. B. Shick and O. N. Mryasov, Coulomb correlations and magnetic anisotropy in ordered L1$_0$ CoPt and FePt alloys, *Phys. Rev. B* **67**, 172407 (2003).

74. J. Lyubina, I. Opahle, K. H. Müller, O. Gutfleisch, M. Richter, M. Wolf and L. Schultz, Magnetocrystalline anisotropy in L1$_0$ FePt and exchange coupling in FePt/Fe$_3$Pt nanocomposites, *J. Phys: Condens. Matter* **17**, 4157 (2005).

75. T. Shima, T. Moriguchi, S. Mitani and K. Takanashi, Low-temperature fabrication of L1$_0$ ordered FePt alloy by alternate monatomic layer deposition, *Appl. Phys. Lett.* **80**, 288 (2002).

76. H. Nishio, H. Taguchi, S. Hashimoto, K. Yajima, A. Fukuno and H. Yamamoto, A comparison of magnetic anisotropy constants and anisotropy fields of permanent magnets determined by various measuring methods, *J. Phys. D: Appl. Phys.* **29**, 2240 (1996).

77. Y. Uesaka, Y. Nakatani, N. Hayashi, H. Fukushima and N. Inaba, Accuracy of 45° torque method for obtaining anisotropy constant of 2D random films, *IEEE Trans. Magn.* **35**, 2673, (1999).

78. M. Takahashi, T. Shimatsu, M. Suekane, M. Miyamura, K.Yamaguchi and H.Yamasaki, Magnetization reversal mechanism evaluated by rotational hysteresis loss analysis for the thin-film media, *IEEE Trans. Magn.* **28**, 3285 (1992).

79. Y. Endo, O. Kitakami, S. Okamoto and Y. Shimada, Determination of first and second magnetic anisotropy constants of magnetic recording media, *Appl. Phys. Lett.* **77**, 1689 (2000).

80. S. Okamoto, K. Nishiyama, O. Kitakami and Y. Shimada, Enhancement of magnetic surface anisotropy of Pd/Co/Pd trilayers by the addition of Sm, *J. Appl. Phys.* **90**, 4085 (2001).

81. W. Sucksmith and J. E.Thompson, *Proc. R. Soc. London*, A **225**, 362 (1954).

82. S. De haan, C. Lodder T. J. A. Popma, *J. Magn. Soc. Jpn* **15 (S2)**, 349 (1991).

83. N. Akulov, *Z. Phys.* **100**, 197 (1936).

84. C. Zener, Classical theory of the temperature dependence of magnetic anisotropy energy, *Phys. Rev.* **96**, 1335 (1954).

85. H. B. Callen and E. Callen, The present status of the temperature dependence of magnetocrystalline anisotropy, and the $l(l+1)/2$ power law, *J. Phys. Chem. Solids* **27**, 1271 (1966).

86. H. Zeng, R. Sabirianov, O. Mryasov, M. L. Yan, K. Cho and D. J. Sellmyer, Curie temperature of FePt : $B_2O_3$ nanocomposite films, *Phys. Rev. B* **66**, 184425 (2002).

87. R. Burgholz and U. Gradmann, Structure and magnetism of oligatomic Ni(111)-films on Re(0001), *J. Magn. Magn. Mater.* **45**, 389 (1984).

88. J. A. Christodoulides, M. J. Bonder, Y. Huang, Y. Zhang, S. Stoyanov and G. C. Hadjupanayis, Intrinsic and hysteresis properties of FePt nanoparticles, *Phys. Rev. B* **68**, 054428 (2003).

89. M. P. Sharrock, Time-dependent magnetic phenomena and particle-size effects in recording media, *IEEE Trans. Magn.* **26**, 193 (1990).

90. D. Weller and A. Moser, Thermal effect limits in ultrahigh-density magnetic recording, *IEEE Trans. Magn.* **35**, 4423 (1999).

91. C. Chen, O. Kitakami and Y. Shimada, Particle size effects and surface anisotropy in Fe-based granular films, *J. Appl. Phys.* **84**, 2184 (1998).

92. M. Jamet, M. Négrier, V. Dupluis, J. T. Combes, P. Mélion, A. Pérez, W. Wernsdorfer, B. Barbara and B. Baguenard, Interface magnetic anisotropy in cobalt clusters embedded in a platinum matrix, *J. Magn. Magn. Mater.* **237**, 293 (2001).

93. S. Stavroyiannis, I. Panagiotopoulos, D. Niarchos, J. A. Christodoulides, Y. Zhang and G. C. Hadjupanayis, Investigation of CoPt/M (*M* = Ag, C) films for high-density recording media, *J. Magn. Magn. Mater.* **193**, 181 (1999).

94. J. A. Christodoulides, P. Farber, M. Daniil, H. Okumura, G. C. Hadjupanayis, V. Skumryev and D. Weller, Magnetic, structural and microstructural properties of FePt/M (*M* = C, BN) granular films, *IEEE Trans. Magn.* **37**, 1292 (2001).

95. J. A. Christodoulides, Y. Huang, Y. Zhang, G. C. Hadjupanayis, I. Panagiotopoulos and D. Niarchos, CoPt and FePt thin films for high-density recording media, *J. Appl. Phys.* **87**, 6938 (2000).

96. C. P. Luo and D. J. Sellmyer, Structural and magnetic properties of FePt: $SiO_2$ granular thin films, *Appl. Phys. Lett.* **75**, 3162 (1999).

97. M. Watanabe, T. Masumoto, D. H. Ping and K. Hono, Microstructure and magnetic properties of FePt-Al-O granular thin films, *Appl. Phys. Lett.* **76**, 3971 (2000).

98. M. Matsumoto, A. Morisako and N. Katayama, Magnetic properties of FePt and FePt-$Al_2O_3$ granular films by post-annealing, *J. Appl. Phys.* **93**, 7169 (2003).

99. Y. K. Takahashi and K. Hono, Interfacial disorder in the $L1_0$ FePt particles capped with amorphous $Al_2O_3$, *Appl. Phys. Lett.* **84**, 383 (2004).

100. T. Miyazaki, S. Okamoto, O. Kitakami and Y.Shimada, Fabrication of two-dimensional assembly of $L1_0$FePt nanoparticles, *J. Appl. Phys.* **93**, 7759 (2003).

101. T. Shima, K. Takanashi, Y. K. Takahashi and K. Hono, Preparation and magnetic properties of highly coercive FePt films, *Appl. Phys. Lett.* **81**, 1050 (2002).

102. T. Shima, K. Takanashi, Y. K. Takahashi and K. Hono, Coercivity exceeding 100 kOe in epitaxially grown FePt sputtered films, *Appl. Phys. Lett.* **85**, 2571 (2004).

103. S. Okamoto, O. Kitakami, N. Kikuchi, T. Miyazaki and Y. Shimada, Size dependences of magnetic properties and switching behavior in FePt $L1_0$ nanoparticles, *Phys. Rev. B* **67**, 094422 (2003).

104. J. L. Pérez-Díaz and M. C. Muñoz, Induced spin polarization on Fe/nonmagnetic metal interfaces, *J. Appl. Phys.* **75**, 6470 (1994).

105. W. J. Ante, M. M. Schwickert, T. Lin, W. L. O'Brien and G. R. Harp, Induced ferromagnetism and anisotropy of Pt layers in Fe/Pt(001) multilayers, *Phys. Rev. B,* **60**, 12933 (1999).

106. B. M. Lairson and B. M. Clemens, Enhanced Magnetooptic kerr rotation in epitaxial PtFe(001) and PtCo(001) thin films, *Appl. Phys. Lett.* **63**, 1438 (1993).

107. M. R. Visokay and R. Sinclair, Direct formation of ordered CoPt and FePt compound thin films by sputtering, *Appl. Phys. Lett.* **66**, 1692 (1995).

108. T. Yang, E. Ahmad and Y. Suzuki, FePt-Ag nanocomposite film with perpendicular magnetic anisotropy, *J. Appl. Phys.* **91**, 6860 (2000).

109. Y. F. Ding, J. S. Chen and E. Liu, Epitaxial $L1_0$ FePt films on $SrTiO_3$ (1 0 0) by sputtering, *J. Crystal Growth.* **276**, 111 (2005).

110. Y.-N. Hsu, S. Jeong, D. E. Laughlin and D. N. Lambeth, The effects of Ag underlayer and Pt intermediate layers on the microstructure and magnetic properties of epitaxial FePt thin films, *J. Magn. Magn. Mater.* **260**, 282 (2003).

111. S. Jeong, Y.-N. Hsu, D. E. Laughlin and M. E. McHenry, Magnetic properties of nanostructured CoPt and FePt thin films, *IEEE Trans. Magn.* **36**, 2336 (2000).

112. S. Jeong, Y.-N. Hsu, D. E. Laughlin and M. E. McHenry, Atomic ordering and coercivity mechanism in FePt and CoPt polycrystalline thin films, *IEEE Trans. Magn.* **37**, 1299 (2001).

113. S. Jeong, M. E. McHenry and D. E. Laughlin, Growth and characterization of $L1_0$ FePt and CoPt <001> textured polycrystalline thin films, *IEEE Trans. Magn.* **37**, 1309 (2001).

114. S. Jeong, T. Ohkubo, A. G. Roy, D. E. Laughlin and M. E. McHenry, In situ ordered polycrystalline FePt $L1_0$ (001) nanostructured films and the effect of CrMn and Zn top layer diffusion, *J. Appl. Phys.* **91** 6863 (2002).

115. K. Kang, Z. G. Zhang, C. Papusoi and T. Suzuki, (001) oriented FePt-Ag composite nanogranular films on amorphous substrate, *Appl. Phys. Lett.* **82**, 3284 (2003).

116. K. Kang, Z. G. Zhang, C. Papusoi and T. Suzuki, Composite nanogranular films of FePt-MgO with (001) orientation onto glass substrates, *Appl. Phys. Lett.* **84**, 404 (2004).

117. Z. G. Zhang, K. Kang and T. Suzuki, FePt (001) texture development on an Fe-Ta-C magnetic soft underlayer with $SiO_2$/MgO as an intermediate layer, *Appl. Phys. Lett.* **83**, 1785 (2003).

118. C. L. Platt and K. W. Wierman, Use of film thickness and Cu additive to improve (001) texture in MgO/FePtCu(C) bilayers, *J. Magn. Magn. Mater.* **295**, 241 (2005).

119. S. Nakagawa and T. Kamiki, Highly (001) oriented FePt ordered alloy thin films fabricated from Pt(100)/Fe(100) structure on glass disks without seed layers, *J. Magn. Magn. Mater.* **287**, 204 (2005).

120. A. Yano, T. Koda and S. Matsunuma, FePt fct phase ordered alloy thin film prepared by 30-s annealing with Fe-O under-layer, *IEEE Trans. Magn.* **41**, 3211 (2005).

121. W. K. Shen, J. H. Judy and J. P. Wang, In situ epitaxial growth of ordered FePt(001) films with ultra small and uniform grain size using a RuAl underlayer, *J. Appl. Phys.* **97**, 10H301 (2005).

122. S. L. Duan, J. O. Artman, B. Wong and D. E. Laughlin, The dependence of the microstructure and magnetic properties of Conicr/Cr thin films on the substrate temperature, *IEEE Trans. Magn.* **26**, 1587 (1990).

123. Y. C. Feng, D. E. Laughlin and D. N. Lambeth, Formation of crystallographic texture in Rf sputter-deposited Cr thin films, *J. Appl. Phys.* **76**, 7311 (1994).

124. H. Kataoka, T. Kanbe, H. Kashiwase, E. Fjita, Y. Yahisa and K. Furasawa, Magnetic and recording characteristics of Cr, Ta, W and Zr precoated glass disks, *IEEE Trans. Magn.* **31**, 2734 (1995).

125. M. Mirzamaani, X. P. Bian, M. F. Doerner, J. Li and M. Parker, Recording performance of thin film media with various crystallographic preferred orientations on glass substrates, *IEEE Trans. Magn.* **34**, 1588 (1998).

126. T. Maeda, Fabrication of highly (001) oriented $L1_0$ FePt thin film using NiTa seed layer, *IEEE Trans. Magn.* **41**, 3331 (2005).

127. L. L. Lee, B. K. Cheong, D. E. Laughlin and D. N. Lambeth, MgO seed layers for CoCrPt/Cr longitudinal magnetic recording, *Appl. Phys. Lett.* **67**, 3638 (1995).

128. L. L. Lee, D. E. Laughlin and D. N. Lambeth, Seed layer induced (002) crystallographic texture in NiAl underlayers, *J. Appl. Phys.* **79**, 4902 (1996).

129. T. Suzuki, N. Honda and K. Ouchi, Preparation and magnetization properties of sputter-deposited Fe-Pt thin films with perpendicular anisotropy. *J. Magn. Soc. Japan*, **21-s2**, 177 (1997).

130. Y. F. Ding, J. S. Chen and E. Liu, Structural and magnetic properties of FePt films grown on the $Cr_{1-x}M_{ox}$ underlayers, *Appl. Phys. A*, **81**, 1485 (2005).

131. J. S. Chen, B. C. Lim, Y. F. Ding and G. M. Chow, Low-temperature deposition of $L1_0$ FePt films for ultra-high density magnetic recording, *J. Magn. Magn. Mater.* **303**, 309 (2006).

132. J. S. Chen, B. C. Lim and J. P. Wang, Controlling the crystallographic orientation and the axis of magnetic anisotropy in $L1_0$ FePt films, *Appl. Phys. Lett.* **81**, 1848 (2002).

133. B. C. Lim, J. S. Chen and J. P. Wang, Crystallographic orientation control in $L1_0$ FePt films on CrRu underlayer, *Surf. Coat. Technol.* **198**, 296 (2005).

134. Y. F. Ding, Ph.D Thesis, "Development of $L1_0$ FePt (001) Films for Perpendicular Magnetic Recording," Data Storage Institute and Nanyang Technological University (2005).

135. Y. F. Ding, J. S. Chen, E. Liu and J. P. Wang, Dependence of microstructure and magnetic properties of FePt films on $Cr_{90}Ru_{10}$ underlayers, *J. Magn. Magn. Mater.* **285**, 443 (2005).

136. C. P. Luo, S. H. Liou, L. Gao, Y. Liu and D. J. Sellmyer, Nanostructured FePt: $B_2O_3$ thin films with perpendicular magnetic anisotropy, *Appl. Phys. Lett.* **77**, 2225 (2000).

137. H. Zeng, M. L. Yan, N. Powers and D. J. Sellmyer, Orientation-controlled nonepitaxial $L1_0$ CoPt and FePt films, *Appl. Phys. Lett.* **80**, 2350 (2002).

138. M. L. Yan, N. Powers and D. J. Sellmyer, Highly oriented nonepitaxially grown $L1_0$ FePt films, *J. Appl. Phys.* **93**, 8292 (2003).

139. M. L. Yan, H. Zeng, N. Powers and D. J. Sellmyer, $L1_0(001)$-oriented FePt: $B_2O_3$ composite films for perpendicular recording, *J. Appl. Phys.* **91**, 8471 (2002).

140. P. Rasmussen, X. Rui and J. E. Shield, Texture formation in FePt thin films via thermal stress management, *Appl. Phys. Lett.* **86**, 191915 (2005).

141. S. J. Greaves, H. Muraoka, Y. Sugita and Y. Nakamura, Intergranular exchange pinning effects in perpendicular recording media, *IEEE Trans. Magn.* **35**, 3772 (1999).

142. I. Kaur and W. Gust, Fundamentals of grain and interphase boundary diffusion, 2nd Edition, Ziegler Press, Stuttgart (1989), Chapter 1-2.

143. J. Zou, B. Bian, D. E. Laughlin and D. N. Lambeth, Improved grain isolation of $Co_{80}Pt_{20}$ films via grain boundary diffusion of Mn, *IEEE Trans. Magn.* **37**, 1471 (2001).

144. J. P. Wang, L. P. Tan, M. L. Yan and T. C. Chong, Co alloy longitudinal thin film media with ultrahigh coercivity, *J. Appl. Phys.* **87**, 6352 (2000).

145. D. Jin, J. P. Wang and H. Gong, Theoretical study of Cr diffusion in Co-Cr alloy thin film recording media, *J. Vac. Sci. Technol. A* **20**, 7 (2002).

146. J. S. Chen and J. P. Wang, Structural and magnetic properties of FePt Film with Cu top layer diffusion, *J. Magn. Magn. Mater.* **284**, 423 (2004).

147. L. H. Lewis, T. R. Thurston, V. Panchanathan, U. Wildgruber and D. O. Welch, Spatial texture distribution in thermomechanically deformed 2-14-1-based magnets, *J. Appl. Phys.* **82**, 3430 (1997).

148. Z. L. Zhao, J. P. Wang, J. S. Chen and J. Ding, Control of magnetization reversal process with pinning layer in FePt thin films, *Appl. Phys. Lett.* **81**, 3612 (2002).

149. Y. Huang, H. Okumura, G. C. Hadjipanayis and D. Weller, Perpendicularly oriented FePt nanoparticles sputtered on heated substrates, *J. Magn. Magn. Mater.* **242**, 317 (2002).

150. J. A. Christodoulides, Y. Huang, Y. Zhang, G. C. Hadjipanayis, I. Panagiotopoulos and D. Niarchos, CoPt and FePt thin films for high-density recording media, *J. Appl. Phys.* **87**, 6938 (2000).

151. D. H. Ping, M. Ohnuma, K. Hono, M. Watanabe, T. Iwase and T. Masumoto, Microstructures of FePt-Al-O and FePt-Ag nanogranular thin films and their magnetic properties, *J. Appl. Phys.* **90**, 4708 (2001).

152. S. C. Chen, P. C. Kuo, C. T. Lie and J. T. Hua, Microstructure and coercivity of granular FePt-AlN thin films, *J. Magn. Magn. Mater.* **236**, 151 (2001)

153. C. M. Kuo and P. C. Kuo, Magnetic properties and microstructure of FePt-$Si_3N_4$ nanocomposite thin films, *J. Appl. Phys.* **87**, 419 (2000).

154. M. Daniil, P. A. Farber, H. Okumura, G. C. Hadjipanayis and D. Weller, FePt/BN granular films for high-density recording media, *J. Magn. Magn. Mater.* **246**, 297 (2002).

155. S. R. Lee, S. H. Yang, Y. K. Kim and J. G. Na, Microstructural evolution and phase transformation characteristics of Zr-doped FePt films, *J. Appl. Phys.* **91**, 6857 (2002).

156. K. R. Koffey, M. A. Parker and J. K. Howard, High anisotropy $L1_0$ thin films for longitudinal recording, *IEEE Trans. Magn.* **31**, 2737 (1995).

157. P. C. Kuo, Y. D. Yao, C. M. Kuo and H. C. Wu, Microstructure and magnetic properties of the $(FePt)_{100-x}Cr_{-x}$ thin films, *J. Appl. Phys.* **87**, 6146 (2000).

158. S. C. Chen, P. C. Kuo, A. C. Sun, C. T. Lie and W. C. Hsu, Granular FePt-Ag thin films with uniform FePt particle size for high-density magnetic recording, *Mater. Sci. Eng. B* **88**, 91 (2002).

159. C. M. Kuo, P. C. Kuo, W. C. Hsu, C. T. Li and A. C. Sun, Effects of W and Ti on the grain size and coercivity of Fe50Pt50 thin films, *J. Magn. Magn. Mater.* **209**, 100 (2000).

160. C. L. Platt, K. W. Wierman, J. K. Howard, A. G. Roy and D. E. Laughlin, A comparison of FePt thin films with $HfO_2$ or MnO additive, *J. Magn. Magn. Mater.* **260**, 487 (2003).

161. N. Li, B. M. Lairson and O. H. Kwon, Magnetic characterization of intermetallic compound FePt and FePtX (X = B, Ni) thin films, *J. Magn. Magn. Mater.* **205**, 1 (1999).

162. N. Li and B. M. Lairson, Magnetic recording on FePt and FePtB intermetallic compound media, *IEEE Trans. Magn.* **35**, 1077 (1999).

163. K. Kang, T. Suzuki, Z. G. Zhang and C. Papusoi, Structural and magnetic studies of nanocomposite FePt-MgO films for perpendicular magnetic recording applications, *J. Appl. Phys.* **95**, 7273 (2004).

164. M. L. Yan, R. F. Sabirianov, Y. F. Xu, X. Z. Li and D. J. Sellmyer, $L1_0$ ordered FePt: C composite films with (001)texture, *IEEE Trans. Magn.* **40**, 2470 (2004).

165. H. S. Ko, A. Perumal and S. C. Shin, Fine control of $L1_0$ ordering and grain growth kinetics by C doping in FePt films, *Appl. Phys. Lett.* **82**, 2311 (2003).

166. T. Suzuki and K. Ouchi, Sputter-deposited (Fe-Pt)-MgO composite films for perpendicular recording media, *IEEE Trans. Magn.* **37**, 1283 (2001).

167. J. S. Chen, T. J. Zhou, B. C. Lim, Y. F. Ding and B. Liu, Microstructure and magnetic properties of rapidly annealed FePt (001) and FePt-MgO (001) films, *J. Appl. Phys.* **97**, 10N108 (2005).

168. T. Suzuki, Nanostructured $L1_0$ Fe-Pt based thin films for perpendicular magnetic recording, *Materials Transactions*, **44**, 1535 (2003).

169. S. Yoshimura, D. D. Djaaprawira, T. K. Kong, Y. Masuda, H. Shoji and M.Takahashi, Grain size reduction by utilizing a very thin CrW seed layer and dry-etching process in CoCrTaNiPt longitudinal media, *J. Appl. Phys.* **87**, 6860 (2000).

170. L. L. Lee, D. E. Laughlin and D. N. Lambeth, NiAl underlayers for cocrta magnetic thin films, *IEEE Trans. Magn.* **30**, 3951 (1994).

171. K. W. Liu and F. Mucklich, Synthesis and thermal stability of nano-RuAl by mechanical alloying, *Mater. Sci. Eng., A* **329**, 112 (2002).

172. J. S. Chen, B. C. Lim and J. P. Wang, Effect of NiAl intermediate layer on structural and magnetic properties of $L1_0$ FePt films with perpendicular anisotropy, *J. Appl. Phys.* **93**, 8167 (2003).

173. J.-U Thiele, M. E. Best, M. F. Toney and D. Weller, Grain size control in FePt thin films by Ar-ion etched Pt seed layers, *IEEE Trans. Magn.* **37**, 1271 (2001).

174. R. Skomski and J. M. D. Coey, *Permanent Magnetism*, Institute of Physics Publishing Ltd, Philadelphia (1999), p. 185.

175. Z. L. Zhao, J. Ding, J. S. Chen and J. P. Wang, Coercivity enhancement by Ru nonmagnetic pinning layer in Fe-Pt thin films, *J. Appl. Phys.* **93**, 7753 (2003).
176. W. F. Brown, The effect of dislocations on magnetization near saturation, *Phys. Rev.* **60**, 139 (1941).
177. J. S. Chen, Y. Z. Zhou, B. C. Lim, T. J. Zhou, J. Zhang and G. M. Chow, Improvement of recording performance in FePt perpendicular media by Ag pinning layer, *IEEE Trans. Magn.* **41** 3196 (2005).
178. T. Suzuki and K. Ouchi, Ordered Fe-Pt(001) thin films by two temperature step depositions for recording media, *J. Appl. Phys.* **91**, 8079 (2002).
179. T. Suzuki, T. Kiya, N. Honda and K. Ouchi, High-density recording on ultrathin Fe-Pt perpendicular composite media, *IEEE Trans. Magn.* **36**, 2417 (2000).
180. T. Suzuki, T. Kiya, N. Honda and K. Ouchi, Fe-Pt perpendicular double-layered media with high-recording resolution, *J. Magn. Magn. Mater.* **235**, 312 (2001).
181. T. Suzuki, Z. G. Zhang, A. K. Singh, J. H. Yin, A. Perumal and H. Osawa, High-density perpendicular magnetic recording media of granular-type (FePt/MgO)/soft underlayer, *IEEE Trans. Magn.* **41**, 555 (2005).
182. Z. G. Zhang, A. K. Singh, J. H. Yin, A. Perumal and T. Suzuki, Double-layered perpendicular magnetic recording media of granular-type FePt-MgO films, *J. Magn. Magn. Mater.* **287**, 224 (2005).
183. J. H. Yin, A. K. Singh and T. Suzuki, Recording performance of granular-type FePt-MgO perpendicular media, *IEEE Trans. Magn.* **41**, 3208. (2005).

# 8

# High-Resolution Transmission Electron Microscopy for Nanocharacterization

## Helge Heinrich*

*Advanced Materials Processing and Analysis Center (AMPAC), and Department of Physics, University of Central Florida, 4000 Central Florida Blvd., Eng 1, #381, Orlando, FL 32816, E-mail: hheinric@mail.ucf.edu

## 1. INTRODUCTION

The study of nanomaterials is not only limited to the characterization of their properties as an ensemble of nanoparticles, but also often extends to the study of individual nanoparticles. Variations in size, shape and internal structure of nanoparticles may influence the macroscopic properties of these materials. Therefore, research in nanotechnology is frequently aimed at developing materials with uniform size and shape. In some cases periodic arrangements of uniform particles are developed. These requirements pose significant technological challenges for the preparation of devices incorporating nanostructured materials. Testing of the desired uniformity or periodicity of nanomaterials cannot be done by optical inspection as the resolution of optical methods is not sufficient for the characterization of nanomaterials. While some structural properties can be inferred from the macroscopic properties of the whole device or the ensemble of nanoparticles, scattering methods (using X-rays or neutrons) measure structural properties by averaging over the irradiated volume.

Direct imaging is, however, essential to analyze deviations from the average structure and helps to identify problems in the production process. High-resolution techniques used for the characterization of nanostructures include surface sensitive methods such as scanning tunneling microscopy (STM), atomic force microscopy (AFM) (Neddermeyer and Hanbüchen, 2003) and scanning electron microscopy, as well as methods to study the interior of materials such as tomographic atomic probe field ion microscopy (APFIM) (Al-Kassab *et al.*, 2003) and transmission electron microscopy (TEM).

For TEM, sample thicknesses significantly lower than 1 μm are required. For high-resolution micrographs, thicknesses even below 10 nm are necessary. Therefore, nanomaterials have ideal sizes for TEM studies. However, sample preparation for TEM is very important to obtain high-quality data. Heterogeneous etching of different phases, surface contamination, radiation damage, and structural and chemical changes during the preparation process are often challenges requiring extended testing and adjustments for sample preparation.

Advanced sample preparation techniques have been developed alongside the progress in transmission electron microscope techniques. Cutting and thinning of bulk materials results in TEM samples from arbitrary locations within the material, while cross-sectional cutting and embedding methods yield samples from surface region. The focused ion beam (FIB) technique allows for the selected preparation of thin TEM foils from specific areas of a material. This technique is especially important for the study of electronic devices where specific nanoscaled device structures are targeted in TEM analysis.

A variety of techniques can be applied in a transmission electron microscope (Fig. 1) to obtain structural and chemical information. Three different operation modes of a transmission electron microscope have to be distinguished:

1. In the normal imaging mode an area of the sample is irradiated with electrons and a magnified image is formed by the electron optical system below the sample. The direct imaging of the crystalline structure

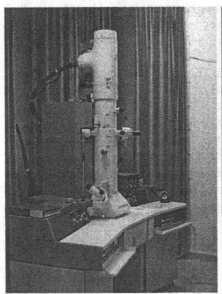

**FIGURE 1.** A 200 kV JEOL (left) and a 300 kV FEI (right) transmission electron microscope. (Courtesy Roland Wessicken, ETH Zürich, Switzerland).

in projection is called high-resolution transmission electron microscopy (HRTEM), while for conventional defect imaging with the bright-field (BF) or the dark-field (DF) technique the resolution is limited to a few nanometers.

2. In the scanning mode of a transmission electron microscope a small electron probe (formed by the condenser system) is scanned across a part of the sample, and the intensity of electrons scattered to different angles is measured as a function of the position of the electron probe. With advanced microscopes equipped with a field-emission electron source, atomic-column resolution is obtained both in the normal imaging mode and in scanning transmission electron microscopy (STEM).

3. In the diffraction mode of a transmission electron microscope a cut through the reciprocal space of the sample is recorded. The analysis of diffraction patterns yields information on lattice parameters, on crystal symmetries and on the arrangement of atoms in the unit cell of a crystal.

Data on the local composition of a sample can be obtained from energy-dispersive X-ray spectroscopy (EDS) and from electron energy loss spectroscopy (EELS). Both analytical methods are typically employed in the STEM mode to measure the local composition at selected locations of a sample or at an individual nanoparticle deposited on a thin amorphous carbon film. Line scans and area scans revealing the distribution of elements are obtained by a stepwise motion of the electron probe to the next position after each acquisition. Alternatively, with an electron energy-loss imaging filter, micrographs for different electron energy losses can be acquired. Elemental distribution maps are obtained by acquiring two micrographs for background extrapolation at energy losses smaller than the energy loss characteristic for an element and one micrograph at the absorption edge for the selected element.

A combination of different methods in TEM is typically applied in the characterization of nanostructures. The analytical methods like EDS and EELS provide chemical information. Electron diffraction is used to measure lattice parameters, and the imaging methods help to identify stacking faults, dislocations, twinning, grain boundaries, and interphase boundaries. The size distribution and shape of nanoparticles is studied with imaging methods. Quantum dots for electronic applications are often characterized in cross-sectional TEM. Thin TEM samples are cut perpendicular to the original surface. Therefore, lattice mismatch, growth conditions, orientation relations and diffusion gradients can be directly analyzed.

## 2. SAMPLE PREPARATION

For an ideal TEM sample the micrographs and spectra taken are representative for the whole material. In the study of nanoparticles with TEM it is especially difficult to obtain statistically relevant results. Some particles may be in the wrong orientation for high-resolution lattice imaging, other particles may be too thick or too thin for good imaging contrast. For the study of nanomaterials special care should be taken that a consistent representative selection of particles is studied.

An ideal TEM sample from bulk materials or from surfaces has uniform thickness. Good TEM samples should be free of contamination and stable in the electron beam, and they should be self-supporting, conducting and nonmagnetic (Goodhew, 1985).

Electrons accelerated in the transmission electron microscope interact with both the electrons and the nuclei in the sample. Absorption and scattering are strong: The intensity of 100 kV electrons is reduced to $1/e$ of its original intensity just after passing about 100 nm of an aluminum sample (atomic number $Z = 13$). The interaction of electrons with matter is approximately proportional to $\lambda^2 Z \approx Z/U$. Only very thin samples significantly less than 1 μm in thickness can be studied in TEM.

For bulk samples, four final TEM preparation techniques are commonly used: tripod polishing, electropolishing, ion milling, and focused ion beam cutting (Roberts *et al.*, 2001). Before the final preparation steps, other initial preparation steps are usually necessary. From a bulk sample a disk with 3 mm diameter is cut using spark erosion (electro discharge machining), a wire saw or a diamond wheel saw. By mechanical polishing these disks are typically thinned down to about 100 μm in thickness. A diamond powder or aluminum oxide suspension on a short nap polishing cloth is the final step of the mechanical thinning to provide a smooth surface for a further reduction in sample thickness.

If sample surfaces and interfaces in layered structures are studied, cross-sections are of special interest. In cross-sectional samples the electron beam is transmitted parallel to the interfaces through the TEM foil to yield projections of the layered structure. The samples are cut in smaller rods with 2 mm width and 1 mm height. Two of these rods are inserted in a 3 mm metal tube filled with epoxy (Fig. 2). The surfaces of interest of the sample face each other in the center of the tube. After curing the epoxy, disks of 0.6 mm in thickness are cut from the embedded material and polished to about 100 μm in thickness.

Dimple grinding is used to reduce the sample thickness in the center to about 10 to 20 μm (Fig. 3). This is important for the following preparation steps: removal of materials with ion milling is relatively slow, and for electrolytic thinning it is

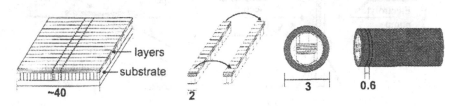

**FIGURE 2.** Step for cross-sectional sample preparation, the numbers indicate dimensions in millimeters. (Reprinted with permission from M. Terheggen, 2003.)

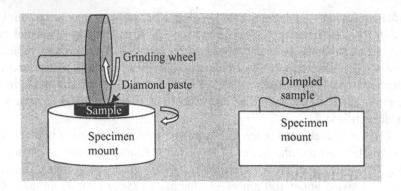

**FIGURE 3.** Schematic of the dimple grinding process.

advantageous to form a dimple in the center of the sample to ensure that the final hole is also centered. A rotating wheel is pushed with a few 10 mN on the sample, which is glued on a rotating specimen mount.

The last step in many sample preparation procedures is plasma cleaning (Roberts *et al.*, 2003). Specimen contamination is a major problem in TEM, especially if it occurs during the experiment around the irradiated area of the sample. This contamination layer of organic hydrocarbons should be reduced or removed in a low-energy plasma (1-20 eV) with a mixture of 10% $O_2$ in the Ar gas used. Plasma cleaning before a TEM session can reduce the contamination rates in the transmission electron microscope by up to one order of magnitude.

## 2.1. Electropolishing

Thinning of a metallic sample can be done with an electropolisher. The sample is clamped in a plastic holder with Pt contacts for the positive electrode (Fig. 4(a)).

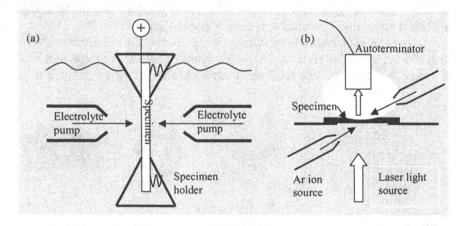

**FIGURE 4.** (a) Schematic of an electropolishing apparatus. (b) Schematic of an argon ion thinning system.

The sample holder has two holes (about 2 mm in diameter on each side) for direct contact of the sample with the electrolyte. The holder is submerged in the electrolyte bath where an electric current removes material from the sample. To optimize the material removal, an electrolyte flow is pointed to the two sample surfaces. This allows a rapid removal of the metal ions from the surfaces and warrants relatively stable etching conditions. With a light source focussed on the sample and an LED detector on the other side of the sample, the system is shut off when light is transmitted through the holder and perforation of the sample is detected.

Electrolytic thinning can yield large transparent areas, but the process optimization for new materials often requires extended tests of the different parameters involved: The electrolyte composition (viscosity, acids and solvents) and temperature have to be adjusted. The applied voltage and the flow are critical to obtain flat and large thin areas in the wedge-shaped regions around the hole in the center of the sample.

## 2.2. Ion-beam Milling

For many nonmetallic materials ion-beam milling is a standard technique for TEM sample preparation. Argon ions with a few keV in energy are used to sputter material from the surfaces. The best milling parameters are obtained for small incident angles (Fig. 4(b)). This reduces preferential etching in a multiphase material. The thickness of the surface layer suffering irradiation damage by the ion milling process is reduced and large thin areas are obtained for small angles (2°-15°) of the incident Ar ions. During milling the sample is rotated to obtain homogeneous sputtering rates. For cross-sectional samples the sample should be preferentially oriented with the interface perpendicular to the ion beam. This reduces preferential etching, which would remove one phase at the interface faster than the other. With this sector etching, a relatively uniform thickness across the interfaces can be maintained. Low topographic variations are essential for high-resolution studies of the layers and their interfaces. The thinning process is terminated when laser light is transmitted through the sample. However, the sensitivity of the detector has to be carefully adjusted for samples with different optical absorption properties. Amorphization of the surfaces can be avoided if low-energy ion milling below 1.5 kV at low incidence angles is employed as the last preparation step (Barna et al., 1998).

## 2.3. The Focused Ion-beam Technique

With the advancement of micro- and nanotechnologies in recent years, new methods were needed to study specifically selected regions of a device. The focused ion-beam (FIB) technique is operating in a similar way as a scanning electron microscope, but instead of electrons, gallium ions are focused on the sample surface. A liquid metal source provides Ga atoms. The gallium reservoir is in contact with a sharp tungsten needle and wets its tip. A high extractor voltage (around 10 kV) induces a high electric field of more than $10^{10}$ V/m at the tip. A sharp cone of Ga atoms forms at the tip and Ga atoms are ionized and emitted. Beam deflectors are used to scan the Ga beam across the sample. The probe current can be varied from a few tens of pA to several nA by adjusting the electrostatic lenses and the aperture sizes. A secondary electron detector is typically employed for imaging. The Ga ions not only remove atoms from the sample surface, but additionally, secondary electrons are emitted which are used for imaging. Alternatively, in a dual-beam FIB a scanning electron microscope and a focused ion-beam system are combined and the surface as the ion-beam milling progresses can be monitored with the electron beam, thus avoiding removal of material during imaging.

The Ga ions of a few tens of keV in energy hit the sample and sputter material from the surface. Material is removed at specific locations using beam diameters as small as 5 nm for the smallest probe current. While high probe currents remove large quantities of material and dig trenches of several μm in depth into the sample, the·small probe currents are used in the final preparation steps for TEM sample preparation. To protect the sample area of interest a Pt layer is deposited in the FIB system (Fig. 5(a)). This is accomplished by a gas injection needle, which

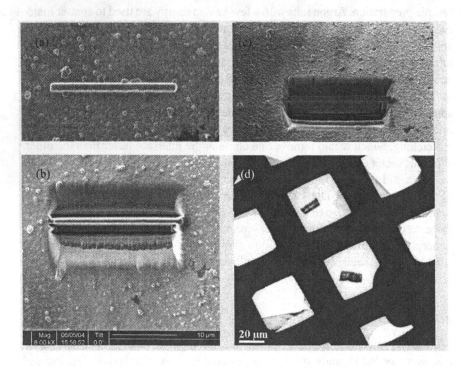

**FIGURE 5.** Different steps of an ex situ lift-out process. (a) Pt deposition, (b) cutting of the sides to form a thin section, (c) side view of the section with the bottom cut, (d) low-magnification TEM image of sample sections deposited on a carbon film on a Cu grid.

is positioned close to this sample region. An organometallic precursor gas from the injection system decomposes in the Ga ion beam and Pt is deposited. After removing the gas injection needle, the cutting of a sample for TEM starts. Around the stripe with the protective Pt layer the sample is cut to a depth of a few μm (Fig. 5(b)). Two different approaches are used to obtain a sample thin enough for TEM.

In the ex situ lift-out method the Ga beam size is successively reduced and the sample is thinned down to below 1000 nm. After tilting the sample to 45° the cut section is viewed from the side and the bottom of the section is cut (Fig. 5(c)). Further thinning at both sides of the section is done with ion currents below 100 pA. When a thickness between 50 to 100 nm is reached the sides of the section are cut off. The specimen is removed from the FIB system and placed under an optical microscope. A thin quartz needle attached to a micromanipulator is approached to the cut region. The thin section usually sticks to the tip of the quartz needle and can be removed from the specimen. In the next step, the quartz needle is approached to a fresh Cu grid with an amorphous carbon film. When the quartz needle is touching the carbon film the thin sample section sticks typically easily to the carbon film. The thin section on the Cu grid is then ready for TEM (Fig. 5(d)).

The in situ lift-out method is the second technique to obtain TEM samples with a FIB system (Kempshall et al., 2004). After the first side cuts, the sample is tilted and the bottom is cut. A micromanipulator is used to approach a Mo needle to the section (Fig. 6(a)). With the Pt source in the FIB the section is attached to the needle (Fig. 6(b)). After cutting of the sides, the section is now only attached to the needle (Fig. 6(c)). The needle is then retracted and a thin Cu half-disk is introduced. The section on the needle is approached to the half-disk and attached using the Pt gas injection system (Fig. 6(d)). The needle is then cut off from the section and retracted. Now the sample can be further thinned down as described above. If the section is thin enough the half-disk can be removed from the FIB system and placed in the transmission electron microscope. If the section is too thick for high-quality TEM the half-disk with the attached sample can even be placed back in the FIB system for further thinning.

The FIB technique provides cross-sectional samples for TEM. The great advantage of this method is the targeted preparation, which allows for a selective analysis of specific regions from devices or from samples with critical interfaces and coatings (Liu et al., 2006). Low ion currents in the last thinning steps reduce ion implantation and the amorphous layers on the two sample surfaces can be limited to thicknesses below 5 nm. The FIB method is an important tool used for quality control of the processing steps in the electronics industry.

## 2.4. Tripod Polishing

Tripod polishing is a mechanical polishing technique. Specimens are thinned to a wedge shape until an electron transparent area is obtained. This method was originally introduced for Si-based devices and layered systems. It can be applied

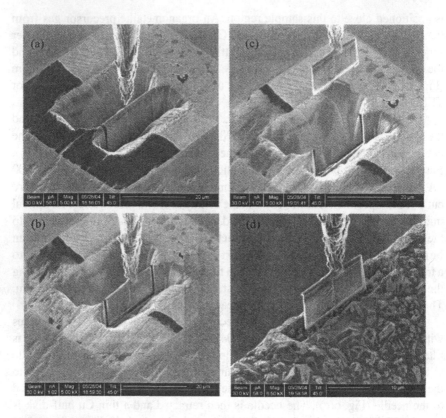

**FIGURE 6.** Steps for the in situ lift-out process for the study of interfaces on bond-coats and thermal barrier coatings for Ni-based superalloys (Liu *et al.*, 2006): After the bottom is cut a Mo needle is attached to the section. The section is attached and the edges are cut. The section is removed from the specimen and attached to the edge of a Cu half-grid for further thinning and for TEM. (Images courtesy of J. Liu, University of Central Florida.)

to study bulk materials as well as cross-sections from layered surfaces (Andersen and Klepeis, 1997). Ceramics and semiconductors are frequently prepared with tripod polishing yielding large electron transparent areas. The sample is glued on a glass rod that is mounted on the tripod polisher as one of the three legs. Using finer and finer lapping films a smooth surface is prepared. The sample is finally flipped and glued from the other side on the glass rod. The same procedure with increasingly finer lapping materials is used until color changes at the edge of the sample indicate thin sample areas.

## 2.5. Powders and Suspensions

For powders of nanoparticles or nanoparticles in a solution the preparation of TEM samples is especially easy. A droplet of such a solution is dried on a Cu grid, which is coated with a thin amorphous carbon film. Two kinds of carbon film are used,

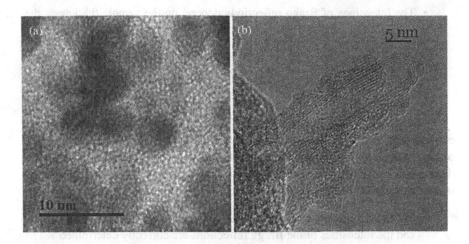

**FIGURE 7.** (a) Au nanoparticles on a thin continuous carbon film. (b) A carbon nanotube decorated with Pd nanoparticles on a holey carbon film. The nanotube in this image extends over a hole in the carbon film and the contrast is therefore not influenced by the carbon film. (Sample courtesy of D. Bera and S. Seal, AMPAC, University of Central Florida.)

continuous carbon films (Fig. 7(a)) and holey carbon films. If the area of interest is on top of a hole in the latter films, contrast contributions from the carbon film can be avoided. Figure 7(b) shows an example of a carbon nanotube extending into a hole of the carbon film (Bera *et al.*, 2006).

## 3. PRINCIPLES OF IMAGE FORMATION

The contrast in transmission electron micrograph is influenced by the scattering in the sample and by the properties of the microscope and its lenses. Elastic electron scattering can be described by two approaches: kinematical scattering and dynamical scattering. Inelastic scattering events in the sample lead to energy losses for the incident electrons and are used to gain information on the local composition of samples. Before going into details of the scattering processes responsible for image formation, the properties of kinematical and dynamical elastic scattering theories and of inelastic scattering are summarized.

The *kinematical scattering* theory is useful if the intensity of the incident radiation is constant and much larger than the intensity of all diffracted beams together.

- Only single scattering from the undiffracted beam to any diffracted beam has to be considered.
- The locations of the diffraction spots in a diffraction pattern are correct, but their intensities are incorrect if the sample thickness reaches or exceeds the extinction length.

- The intensities of Bragg reflections depend on the sample thickness, the Fourier components of the crystal potential and on the deviation parameters, i.e., the distances of the Ewald sphere from the exact reciprocal lattice points.
- The symmetry of Laue zones in the diffraction pattern can be used for sample orientation.

The strong electron-matter interaction is responsible for multiple scattering effects even for very thin samples of light atoms. This limits the applicability of the kinematical scattering theory, and the elastic electron scattering is highly dynamical. In **dynamical scattering** theory the intensity of the undiffracted beam is not constant (extinction) and multiple scattering is considered.

- There is an exchange of intensity (electrons) between all Bragg reflections and the intensities of the Bragg reflections are correctly determined.
- Thickness contour lines occurring in bright-field and dark-field images of wedge-shapes samples are correctly described.
- Bending contour lines in the bright-field mode are used to orient thin crystalline samples.

*Inelastic scattering* processes (absorption) reduce the number (intensity) of electrons, which maintain their original energy.

- Energy losses of electrons in the sample are used for a local compositional analysis by electron energy loss spectroscopy (EELS).
- Irradiation by high-energy electrons gives rise to characteristic X-rays emitted from the atoms in the sample. This radiation can also be used for a local chemical analysis using energy-dispersive X-ray spectroscopy (EDS).
- Inelastically scattered electrons suffer diffraction at lattice planes. This leads to Kikuchi lines for thick crystalline samples. Kikuchi lines in the diffraction patterns are used to orient a sample by tilting.

### 3.1. The Transmission Electron Microscope

Transmission electron microscopes are operated at acceleration voltages $U$ ranging from 60 kV to 3 MV. For conventional transmission electron microscopy typical voltages are 100 kV, 200 kV and 300 kV. The electron wavelength is determined by the acceleration voltage, with a relativistic correction necessary for acceleration voltages higher than a few 10 kV:

$$\lambda = \frac{h}{\sqrt{2m_0 e U \left(1 + \frac{eU}{2m_0 c^2}\right)}} \tag{1}$$

For a 200 kV electron microscope, the electron wavelength is 0.02508 Å, while electrons accelerated in a 300 kV field have a wavelength of 0.01969 Å. The electron wavelength is therefore about 100 times smaller than typical distances of atoms in solids.

### 3.1.1. Electron Lenses

Electrostatic lenses modify the electron wavelength by local electric fields. Similar to light optics the rule of Willebrord van Roijen Snell on the refractive index $n$ is applicable (Snell's law). Electrostatic lenses with an additional potential $V_A$ change the wavelength according to:

$$\sqrt{(U + V_A)/U} = \lambda/\lambda_{Lens} = n$$

However, focussing with electrostatic lenses requires high fields. Therefore, electrostatic lenses are not used in electron microscopy, except for the acceleration of the electrons and for the extractor unit of a field emission source.

Electrons in a homogeneous magnetic field $\underline{B}$ move on helical trajectories with fixed radius $r$. Magnetic electron lenses, however, generate inhomogeneous magnetic fields. Magnetic lenses consist of coils, magnetic circuits (yokes) and polepieces. The polepieces have a rotationally symmetric hole in which electrons are deflected (Fig. 8(a)). The field distribution along the (optical) axis of the lens is (unfortunately) closely related to the field distribution in the whole bore, i.e., we cannot produce perfect aberration-free lenses for charged particles. The components $B_z$ and $B_r$ of the magnetic field provide the angular ($v_\phi$) and radial ($v_r$) components of the electron velocity vector. Rotationally symmetrical magnetic and electric lenses both are convex lenses for charged particles. For magnetic lenses the image is rotated with respect to the object as shown in Fig. 8(b)).

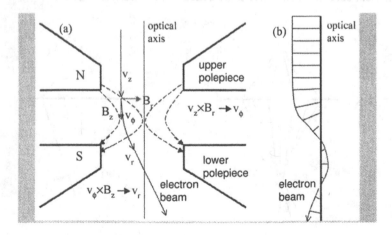

**FIGURE 8.** (a) The electron path in the inhomogeneous field of a magnetic lens with rotational symmetry. (b) Rotation of the image with respect to the object in a magnetic lens.

### 3.1.2. Lens Errors

In light optics, spherical and chromatic aberrations can be corrected by combining convex and dispersing lenses. That is not possible in electron optics using lenses with rotational symmetry. There are no rotationally symmetrical dispersing lenses possible for charged particles as shown by Scherzer (1936). Furthermore, the field distribution in a lens can not be freely adjusted, so electron lenses suffer from aberrations. But there are new developments in electron optics allowing for a correction of the spherical aberration. Rose (1990), proposed a correction system breaking rotational symmetry, which was developed (Haider *et al.*, 1995) and recently installed (Haider *et al.*, 1998) on a growing number of transmission electron microscopes.

**Spherical aberration:** This is the most important lens error in electron optics. Rays parallel to the optical axis in front of the lens are deflected such that they cut the optical axis in the back focal plane (in the back focal point) of the lens. In the case of a lens with spherical aberration this focal length depends on the distance of the parallel ray from the optical axis when it enters the lens. In the standard case of positive spherical aberration the focal length decreases when the distance of the entering ray from the optical axis increases. The image of a point is not anymore a point in the ideal back focal plane, i.e., in the Gaussian image plane (Fig. 9(a)). A point is imaged as a disk with a radius depending on the aperture angle $\alpha$ of the lens and on the parameter of spherical aberration: $\Delta r_s = C_s \alpha^3$. The spherical aberration reduces the "point resolution" of a lens. For a perfect lens without spherical aberration, the point resolution would be limited only by diffraction effects from the finite lens aperture $\Delta r_o = 0.6\lambda/\alpha$. Here, $\Delta r_o$ is the minimum distance of two objects leading to separable images similar to light microscopy. In transmission electron microscopy $\Delta r_o$ and the contribution of spherical aberration $\Delta r_s$ have to be added: $\Delta r = 0.6\lambda/\alpha + C_s \alpha^3$.

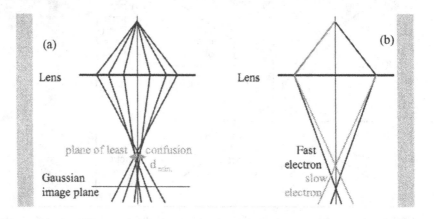

**FIGURE 9.** (a) Imaging using a lens with spherical aberration leads to a spread of image information. (b) The effect of spherical aberration for different wavelengths.

The point resolution in TEM is mainly determined by the diffraction error and the spherical aberration. A compromise is necessary to optimize the point resolution, for example, by inserting an aperture in a diaphragm (Carter and Williams, 1996), which limits the angular distribution of electrons. The optimum angle $\alpha_{opt} = \sqrt[4]{0.2\lambda/C_s}$ yields an optimum resolution $\Delta r_{opt} \approx \sqrt[4]{C_s\lambda^3/4}$.

With $C_s$ being typically in the order of mm, the point resolution is in the Ångstrom range. For a microscope operated at 300 kV (2 pm wave length) having an objective lens with $C_s = 1.0$ mm, the point resolution is about 2 Å. The point resolution (the minimum distance of two objects in the object plane, which results in distinguishable the images of neighboring objects in the image plane) is limited by the wavelength of the radiation used. Additionally, the resolution is limited by the finite size of lenses (diffraction error), lens aberrations, incoherent radiation, the intensity of the radiation used and the scattering contrast of the objects.

**Chromatic aberration:** Electron lenses have different focal lengths for different particle energies and wavelengths (Fig. 9(c)). In light optics the lens error can be compensated by selecting appropriate lens materials. In electron optics, that is not possible, and we have to use highly monochromatic electrons. The energy distribution in front of the sample is (depending on the electron source and on the stability of the high voltage) in the range of 0.2-5 eV for electron energies above 100,000 eV. Additionally, fluctuations in the lens currents cause variations of the focal length.

**Distortion:** This is a typical lens error occuring when imaging off-axis objects. In electron microscopy, a combination of pincushion and spiral distortion (Fig. 10) appears especially for low magnifications.

**Coma:** In spherical lenses, different parts of the lens surface exhibit different degrees of magnification. This gives rise to an aberration known as coma. Each concentric zone of a lens forms a ring-shaped image called a comatic circle. This causes

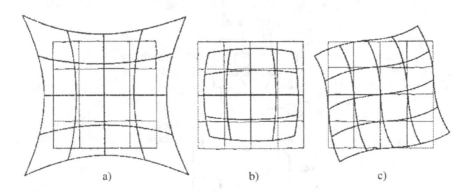

**FIGURE 10.** Different types of image distortion. (From Transmission Electron Microscopy, 1989, Transmission Electron Microscopy, Reimer, L., with kind permission of Springer Science and Business Media).

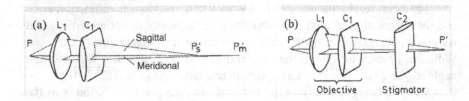

**FIGURE 11.** (a) Lens with astigmatism: different focal lengths for different electron paths through the lens. (b) Stigmator for correction of astigmatism. (From Transmission Electron Microscopy, 1989, Transmission Electron Microscopy, Reimer, L., with kind permission of Springer Science and Business Media).

blurring in the image plane (surface) of off-axis object points. An off-axis object point is not a sharp image point, but it appears as a characteristic cometlike flare.

   **Astigmatism:** Pole-piece apertures are relatively small in TEM (< 1 mm). Therefore, it's impossible to produce lenses with perfect rotational symmetry and the field distribution in the lens becomes accordingly asymmetric. Additionally to the effect of spherical aberration the focal length depends on the angle around the optical axis, where an off-axis electron beam enters the lens (Fig. 11). Astigmatism can be corrected by small asymmetric and tuneable octupole lenses (stigmators) that introduce a compensating field to balance the inhomogeneities causing the astigmatism. Coma and astigmatism can and should be corrected by the operator before high-resolution micrographs are acquired.

### 3.1.3. The Electron Source

A thermionic electron source in a transmission electron microscope consists of a heated cathode (filament) emitting electrons (Fig. 12). A Wehnelt cylinder (at a

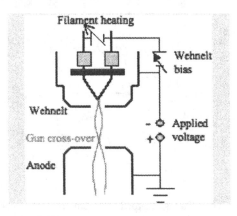

**FIGURE 12.** Schematics of a conventional thermionic electron source.

TABLE 1. Characteristics of the different electron sources. (Data from Fultz and Howe, 2002.)

| Cathode | Tungsten | LaB$_6$ | Thermal FEG | Cold FEG |
|---|---|---|---|---|
| Temperature (K) | 2800 | 1700 | 1400-1600 | 300 |
| Brightness (kA/(cm$^2$sr)) | 100 | 1000 | 100,000 | 100,000 |
| Energy width (eV) | 1-2 | 0.5-2 | 0.3-1 | 0.2-0.4 |
| Vacuum required (Pa) | $10^{-3}$ | $10^{-4}$ | $10^{-7}$ | $10^{-8}$ |
| Lifetime | Weeks | Months | 2-3 years | Months |

more negative potential than the cathode) bundles electrons in a "cross-over", and the anode finally accelerates the electrons. The anode is on ground potential while the cathode is at a physically negative voltage.

Three types of cathodes (Table 1) can be distinguished:

**Tungsten filaments:** These filaments, with a crossover size of about 50 μm, are operated at 2800 K to emit electrons. The high operating temperature and the large diameter of the tip gives rise to a broad electron energy distribution. These thermionic cathodes are well suited for conventional transmission electron microscopy at lower magnifications.

**LaB$_6$ cathodes:** These cathodes have a low work function for the emission of electrons and can be used around 1700 K. More important than the current density $J$ is the brightness $\beta$ of the source, i.e., the current density per unit solid angle $\alpha$ of the source, which is given by $\beta = J/\alpha^2$. To get a high electron current density on a small area of the sample, electrons emitted from the source at higher angles $\alpha$ are defocused by lens errors and have to be removed by apertures.

**Field-emission source (FEG):** A field-emission gun has a very fine tungsten tip with a tip radius $r$ below 100 nm. When applying a potential of typically $U = 4.5$ kV to the tip, the electrical field $E = U/r$ considerably reduces the local work function at the tip surface. FEGs can be, depending on the type, operated at 300 K or 1600 K. The brightness and the current density of FEGs are much higher than for thermal sources.

## 3.1.4. Components of a Transmission Electron Microscope

The gun (electron source) consists of a cathode, a Wehnelt cylinder and an anode (Fig. 12). The condenser lens system (the illumination system) with at least two lenses reduces the area illuminated by the electron beam (Fig. 13), it essentially demagnifies the electron probe. A condenser aperture selects electrons near the optical axis for illumination of the sample. This aperture of the condenser diaphragm is in the front-focal plane conjugate to the diffraction plane of the specimen.

In normal operation mode of a TEM the upper part of the objective lens generates a parallel beam entering the sample (telefocal system). The sample is centered within the objective lens. In scanning mode a small convergent electron

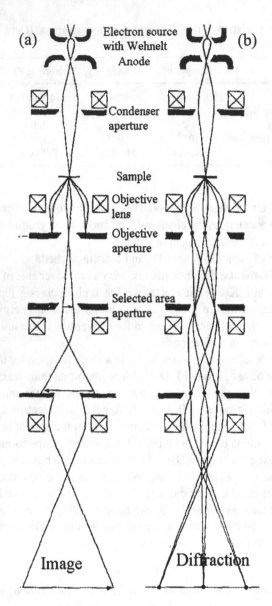

(a)    Electron source    (b)
       with Wehnelt
       Anode

Condenser
aperture

Sample

Objective
lens

Objective
aperture

Selected area
aperture

Image                    Diffraction

**FIGURE 13.** (a) Imaging and (b) diffraction mode of the transmission electron microscope dependent on the excitation of the intermediate lens (Reprinted from Electron Microscopy of Materials, von Heimendahl, Electron Microscopy of Materials, (1980), with permission from Elsevier).

probe is moved over the sample. Deflection coils are used to move the probe across the sample.

The lower part of the objective lens is used to generate a magnified image. Additionally, a diffraction pattern is formed in the back focal plane of the lens

(Fig. 13). Each Bragg reflection is either a spot (parallel illumination) or a disk for a finite beam convergence angle on the sample. An objective aperture in this back focal plane (it's also called the diffraction plane, as a diffraction patterns is formed there) can be used to select one or more Bragg reflections and to form magnified images of the sample only with contributions from Bragg reflections transmitted through this aperture.

The imaging system consists of several lenses depending on the type of microscope. They are called diffraction lens, intermediate lens, and projector lens. These lenses are used to obtain additional magnification of the sample image or of the diffraction pattern. In the plane of the first image (formed by the objective lens) of the sample a selected area aperture can be inserted, which allows only electrons from a limited area of the sample to pass. By different excitations of the intermediate lens either the imaging plane or the diffraction plane of the sample forms a magnified image on the viewing screen of the microscope.

For the camera system a fluorescent material like ZnS is used on a viewing screen. This screen can be tilted up to take micrographs on photographic films, on a CCD camera, on imaging plates, to use a TV-rate camera for imaging. Alternatively, electrons can also be brought on other detectors, like the bright-field (BF) and dark-field (DF) detectors or the high-angle annular dark-field (HAADF) detector in the scanning transmission mode (STEM), or the electron can be transferred to an electron energy-loss spectrometer.

In a CCD camera of a TEM, light is generated by the electrons impinging on a scintillator material like Ce-doped yttrium-aluminum garnet (YAG). The photons are transferred by a fiberoptics system on the pixels of the CCD array. A CCD camera consists of an array of MOS devices, usually $1024 \times 1024$ pixels or $2048 \times 2048$ pixels. Generated by incoming light, charges are accumulating in a pixel as the pixels are electrically insulated from each other by a potential well. The charges can be read out by lowering the potential wells sequentially. This way charges are transferred from one pixel to the next one until they reach an output node. During read-out, exposure of the CCD array is switched off by beam deflection.

Additional detectors can be an EDX-detector for measuring characteristic X-rays, a secondary electron detector and an electron energy-loss spectrometer.

The sample itself is mounted on a sample holder, which is held by a goniometer in the gap between the upper and the lower polepiece of the objective lens. In materials science double-tilt holders allow for sample tilts in different directions.

As electrons strongly interact with matter, we need a "good" vacuum system. Oil diffusion pumps are effective from 0.1 Pa to $10^{-9}$ Pa, but they may contaminate the electron microscope with hydrocarbon oil. UHV conditions can also be reached with turbomolecular pumps. For the critical volume of the electron source ion getter pumps (IGP) are used. Good information on the principles of pumps and holders can be found in Chapter 8 of the book by Williams and Carter (1996).

### 3.1.5. Electron Matter Interactions

The following interactions can be distinguished:

(a) The electrons are unscattered (zero-loss).

(b) Electrons are elastically scattered at the atoms of the crystal lattice planes (zero-loss).

(c) Electrons excite phonons. The energy changes $\Delta U < 0.1$ eV are too small to be analyzed in a transmission electron microscope (zero-loss).

(d) Electrons excite plasmons ($\Delta U \approx 20$ eV), i.e., the "gas" of free electrons in a metal vibrates with respect to the lattice of the atomic nuclei.

(e) Formation of bound electron-hole pairs (excitons) with energies in the range of 10 eV.

(f) Excitation of electrons in the material leading to transitions from bound electron states (bound to individual atoms) to free-energy states above the Fermi level including the emission of secondary electrons. X-rays or Auger electrons are generated, when the bound state is filled again with an electron. The primary (incoming) electrons suffer energy losses, which are specific for the material (10 eV to 80 keV).

### 3.1.6. X-ray Analysis and Electron Energy Loss Spectroscopy

Both energy dispersive X-ray analysis (EDX) and electron energy loss spectroscopy (EELS) measure chemical compositions in small sample volumes (Fig. 14). Using a field emission electron source the analyzed area can be smaller

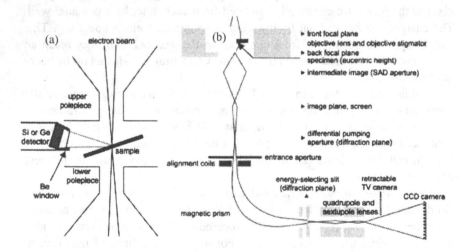

**FIGURE 14.** (a) Sample geometry for analysis with an EDX detector. (b) Principle of a post-column electron energy loss spectrometer with the option of energy-filtered imaging. (Reprinted with permission from R. Erni, 2003.)

than 1 $nm^2$. A chemical analysis of regions a few tens of nanometers in diameter is possible in transmission electron microscopes with W- or $LaB_6$-cathodes.

For the analysis of characteristic X-rays emitted from the irradiated area, a cooled semiconducting detector like Si(Li) or Ge is used (Fig. 14(a)). Electron-hole pairs are generated when X-rays hit the detector area. The number of charges produced depends on the energy of the incoming X-ray. In an X-ray spectrum, the number of X-rays is plotted as a function of their X-ray energies. Absorption has to be considered especially for X-rays of low energy. X-rays are absorbed in the sample itself, but also in the window material of the detector.

Light elements up to oxygen cannot be quantitatively evaluated by EDX, but EDX is well suited to the analysis of heavy elements. The sample is typically tilted about 20° towards the detector to minimize absorption of low-energy X-rays in the sample (Fig. 14(a)). Unfortunately, the X-rays and secondary electrons generated in the volume illuminated by the electron probe can themselves excite atoms and cause the emission of X-rays. Sample regions not illuminated by the electron probe or even parts from the microscope can therefore contribute to the EDX spectrum. Therefore, special care has to be taken in quantitative EDX analysis to account for a possible generation of X-rays in neighboring regions.

The energy losses of electrons in the sample can be analyzed by electron energy-loss spectroscopy (EELS). In a homogeneous magnetic field, electrons are forced on a circular trajectory with a radius according to their velocity. This allows the separation of different electron energies by their location in an energy-selective slit (Fig. 14(b)) after passing through this field or the acquisition of a complete energy-loss spectrum on a CCD (Fig. 15).

EELS is a method well suited for light elements (except for H and He) and for elements of medium atomic weight; therefore, it is complementary to EDX. If an incoming electron excites a bound electron in the sample to an energy level above the Fermi level, the transferred energy is subtracted from the original energy of the primary electron. The electron energy-loss spectrum consists of a "zero-loss" peak (Fig. 15(a)), a "plasmon peak," and a signal background (from multiple excitations and Bremsstrahlung) exponentially decreasing with increasing energy loss. On top of this background, ionization edges (Fig. 15(b)) with increased intensity are found at energies characteristic for the elements in the analyzed sample volume. In the imaging mode of an EELS instrument an energy-selecting slit can be inserted into the electron path in the energy-selective plane. A magnified sample image or a diffraction pattern is then formed with electrons of a specific interval in electron energy. This technique is called energy-filtered transmission electron microscopy (EFTEM).

## 3.2. The Ewald Construction and the Reciprocal Space

Scattering of radiation by a material is used to extract average structural data. The wavelength of the radiation remains unchanged if we assume elastic scattering only.

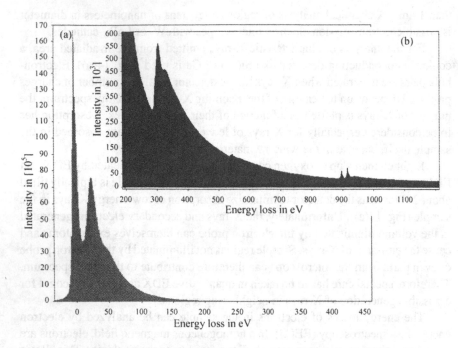

**FIGURE 15.** Electron energy loss spectrum of a $CeO_2$ nanoparticle. (a) Zero-loss and plasmon peak. (b) Absorption edges: carbon K-edge at 284 eV energy loss, oxygen K-edge at 532 eV, cerium M-edge at 887 eV.

Furthermore, it's a very convenient approximation, if we can apply *Fraunhofer diffraction*, i.e., a coherent flat wavefront interacts with the object. This corresponds to a radiation source far away from the object (at infinity). Correspondingly, in Fraunhofer diffraction, the detector for measuring the scattered radiation is also in the far-field of the sample. So, we only have to deal with scattering angles and not with distances from the radiation source or with distances to the detector. In the case of near-field diffraction, we do not consider flat wavefronts. This mathematically more complicated (but also more realistic) case is called *Fresnel diffraction*.

Diffraction is an effect of *coherent interference* of waves. The total interference pattern of these waves is built by the sum of all their amplitudes. If the waves are incoherent, the total intensity is simply the sum of all their intensities.

In transmission electron microscopy wave vectors $\underline{k}$ are usually used whose length is defined by $|\underline{k}| = 1/\lambda$ (and not $2\pi/\lambda$). This corresponds to the definition $h\underline{k}$ for the momentum. Therefore, the wave functions are written as: $\psi = A \exp(2\pi i \underline{k} \cdot \underline{r} - i\omega t)$. With this definition it is easy to calculate lengths in reciprocal space. If we calculate a coherent interference patterns of waves scattered at two pointlike objects at positions $\underline{r}_0 = (0,0,0)$ and $\underline{r} = (x, y, z)$ using the Fraunhofer approximation, the difference $\Delta l$ in the path lengths of these two waves for diffraction from the incident

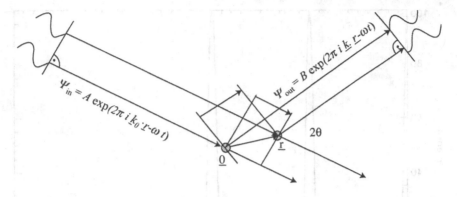

**FIGURE 16.** Schematics of the phase shift introduced by scattering at two objects.

wave $\underline{k}_0$ to the diffracted wave $\underline{k}$ is

$$\Delta l = \frac{k \cdot r}{|k|} - \frac{k_0 \cdot r}{|k_0|}$$

and the phase shift $\Delta \phi$ between the outgoing waves (Fig. 16) is

$$\Delta \phi = \frac{2\pi \Delta l}{\lambda} = 2\pi(\underline{k} - \underline{k}_0) \cdot \underline{r}. \tag{2}$$

Assuming, that the two pointlike objects have the same scattering behavior, the sum of the two scattered waves is $\Psi_{out} = B \exp(-i\omega t)\left[1 + exp\left\{2\pi i(\underline{k} - \underline{k}_0) \cdot \underline{r}\right\}\right]$. The scattered intensity $I = \Psi\Psi^*$ depends on the relative location of the scattering objects, on their scattering strength $f = B/A$, and on the scattering vector $\underline{K} = \underline{k} - \underline{k}_0$.

The next step is to calculate scattering from a periodic arrangement of atoms, a crystal lattice. The easiest crystal structure is simple cubic (lattice parameter $a$) with only one atom at position 0,0,0 in each unit cell. Here, we have to sum over all unit cell positions given by the vectors $\underline{r}_j = [am_x, am_y, am_z]$, where $m_x, m_y, m_z$ are natural numbers. For a crystal with $M_x M_y M_z$ unit cells and an atomic scattering factor $f_j$ we get

$$\psi_{out}(\underline{K}) = f_j \sum_{m_x=0}^{M_x-1} \exp(2\pi i K_x a m_x) \sum_{m_y=0}^{M_y-1} \exp(2\pi i K_y a m_y) \sum_{m_z=0}^{M_z-1} \exp(2\pi i K_z a m_z)$$

These sums are geometrical rows, and the total scattered intensity $I = \Psi\Psi^*$ from the finite crystal is then:

$$I(\underline{K}) = f(\underline{K})f^*(\underline{K})\frac{\sin^2\left(\pi M_x K_x a\right)}{\sin^2\left(\pi K_x a\right)}\frac{\sin^2\left(\pi M_y K_y a\right)}{\sin^2\left(\pi K_y a\right)}\frac{\sin^2\left(\pi M_z K_z a\right)}{\sin^2\left(\pi K_z a\right)} \tag{3}$$

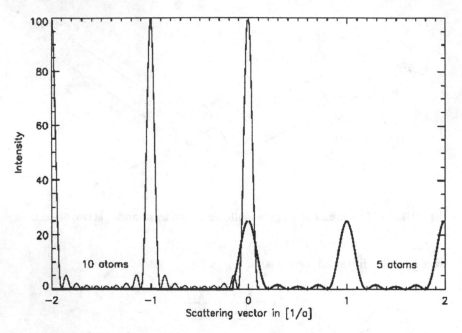

**FIGURE 17.** Comparison of the scattering intensity distribution for scattering at 5 and at 10 atoms in a row.

### 3.2.1. Bragg Reflections

The scattering intensity is strongly modulated as shown in Fig. 17. The more atoms are scattering in a crystal, the higher is the modulation frequency. Strong maxima in the intensity evolve for $Ka$ being natural numbers. Their intensities are proportional to the number of scattering atoms squared. With increasing number of atoms in a row, the width of these maxima decreases. The first side minima are at distances $1/(Ma)$ from the respective maxima.

If we would increase the number of atoms in a row to infinity, the maxima would become delta functions with infinite intensity. These maxima are called **Bragg reflections**. Ideally, they are delta functions in reciprocal space (in $K$ space the axes have dimensions of $m^{-1}$). For the simple cubic lattice, the reciprocal lattice is periodic with equal distances of the reciprocal lattice points, the possible Bragg reflections, in the three reciprocal crystallographic directions.

We only get a Bragg reflection, if there is a *reciprocal lattice vector $\underline{G}$* connecting two reciprocal lattice points as the difference vector between the incoming and the outgoing wave vectors: $\underline{G} = \underline{K}$. For scattering, a zero-point in the reciprocal lattice is defined as the end-point of the incoming wave vector $\underline{k}_0$.

A reciprocal lattice vector $\underline{G}$ represents a vector perpendicular to its corresponding lattice plane in real space, so the scalar product between any vector in this real-space plane and the reciprocal lattice vector is zero. *Example*: Take two vectors in real space such as $\underline{r} = a[1,1,1]$ and $\underline{s} = a[1,1,0]$. A vector perpendicular to these two vectors is $[1,-1,0]$ or any multiples of that. For the cubic lattice, the reciprocal lattice has orthogonal bases of length $1/a$. A vector in the reciprocal lattice with the indices $\underline{G} = [1,-1,0]$ has a length of $|\underline{G}| = \sqrt{1^2 + (-1)^2 + 0^2}/a$.

The length of any reciprocal lattice vector of a cubic lattice with indices $h,k,l$ is $|\underline{G}| = \sqrt{h^2 + k^2 + l^2}/a$. These indices $h,k,l$ of a reciprocal lattice vector are called *Miller indices*. Miller indices and reciprocal lattice vectors have a correspondence in real space, the *lattice plane distances*, which for the cubic lattice is $d_{h,k,l} = a/\sqrt{h^2 + k^2 + l^2}$.

Let's use the example again: We now look for lattice planes with indices $1,-1,0$. The distance of the planes is $a/\sqrt{2}$. So, all neighboring $(1,-1,0)$ planes have this distance from each other. There is a phase shift of $2\pi$ when comparing waves scattered at atoms in one plane compared to waves scattered at atoms in the neighboring plane. For our example, we may take one atom at position $\underline{r}_1 = [0, 0, 0]a$ in a $(1,-1,0)$ plane and another atom at position $\underline{r}_2 = [1,0,0]a$ in the neighboring $(1,-1,0)$ plane. Using Eq. 2, the phase shift for scattering at these two planes is $2\pi$.

What is the meaning of a reciprocal vector like $[2,-2,0]$? The interplanar spacing in real space is now cut in half: $a/\sqrt{8}$. If we take one plane through $[0,0,0]$ the neighboring plane is shifted by $[a/2,-a/2,0]$. But for the simple cubic lattice, there are no atoms in this neighboring $(2,-2,0)$ plane. There are only scattering contribution from every other lattice plane with Miller indices $2,-2,0$. For the next nearest lattice plane, the phase shift is obviously $4\pi$, so we have positive interference of scattering contributions from atoms in next nearest $(2,-2,0)$ planes.

The Ewald construction (Fig. 18) is a graphical description for the reciprocal space and depicts the directions of incoming (wave vector $\underline{k}_0$) and outgoing ($\underline{k}$) radiation with respect to the reciprocal lattice. If the surface of the Ewald sphere

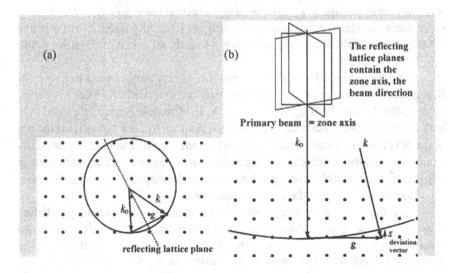

**FIGURE 18.** Schematics of the Ewald construction for x-rays (a) and for electrons (b). The deviation parameter s is negative.

with a radius $|\underline{k}| = g|\underline{k}_0|$ intersects a reciprocal lattice point the *Bragg condition* for elastic scattering holds:

$$\underline{G} = \underline{K} = \underline{k} - \underline{k}_0 \qquad (4)$$

or expressed differently:

$$2d_{hkl} \sin(\Theta) = \lambda \qquad (5)$$

These equations are valid for elastic scattering. For monochromatic radiation a Bragg reflection can only occur, if the surface of the Ewald sphere hits a reciprocal lattice point. The total scattering angle $2\Theta$ can be directly determined from Ewald's construction, it is the angle between the wave vectors of the diffracted and the undiffracted beam. This angle is twice the Bragg angle $\Theta$, which is given by $^1\!/_2 |\underline{G}| = g|\underline{k}| \sin(\Theta)$. In TEM the wavelength of the electrons (Eq. (1) is about 100 times smaller than the distance of the atoms. So, the radius of the Ewald sphere is approximately 100 times larger than the distances between the reciprocal lattice points. The curvature of the Ewald sphere and the scattering angles are therefore small (Fig. 18(b)).

Especially simple diffraction patterns are found, if the incoming electron beam is oriented along a highly symmetrical crystallographic direction. Such a direction is often called a *zone axis*, *i.e.*, a direction that is common to many low-indexed lattice planes. For a beam along a zone axis $\underline{k}_0 = [m,n,o]$ many lattice planes with indices $(h,k,l)$ fulfil $\underline{k}_0 \cdot \underline{G} = mh + nk + ol = 0$. These reciprocal lattice points are in the zero-order Laue zone.

As the Ewald sphere has a small curvature, the reciprocal lattice points in the zero-order Laue zone are not exactly intersected by the Ewald sphere. The *deviation vector* $\underline{s}$ or deviation parameter (the length of $\underline{s}$ parallel to $\underline{k}_0$) describes a vector from the reciprocal lattice point to the surface of the Ewald sphere. This vector is usually taken parallel to $\underline{k}_0$. So, the Bragg equation reads now $\underline{G} = \underline{k} - \underline{k}_0 + \underline{s}$. In transmission electron microscopy we still get scattering intensity for a Bragg reflection, even if the Ewald sphere does not exactly hit a reciprocal lattice point.

Why is that so?

According to Eq. (3) for a limited number of scattering objects in a row, the Bragg reflections are not delta functions, they are extended: the more scattering objects, the sharper the Bragg reflections. A TEM sample is usually extended in both lateral directions, but it has to be thin (a few nm up to a few 100 nm) in electron beam direction to be transparent for the electrons. A TEM sample is essentially a thin disk, and the reciprocal lattice points corresponding to the crystal lattice in this thin disk can therefore be represented as needle-shaped as shown in Fig. 19.

How do we calculate the scattered intensity?

In the theory of kinematical scattering the intensity of the incoming radiation is assumed to be large in comparison to the intensities of all diffracted beams together. So, only a small fraction of the electrons is scattered, and the incoming beam is not significantly influenced by diffraction effects. The fraction of electrons scattered away from the incoming beam is small. Therefore, scattering is only

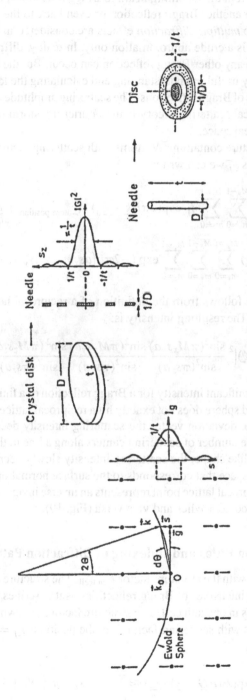

**FIGURE 19.** Schematics of a needle-shaped scattering strength around reciprocal lattice points caused by the thin TEM foil (From Transmission Electron Microscopy, Reimer, 1989, with kind permission of Springer Science and Business Media).

assumed to take place from the incoming beam to Bragg reflections, but not from one Bragg reflection to another Bragg reflection or even back to the direction of the incoming beam. No *multiple diffraction* effects are considered in kinematical scattering theory. This is a crude approximation only. In reality, diffraction from any Bragg reflection to any other Bragg reflection can occur. But the kinematical scattering theory is very useful in understanding and calculating the locations and approximate intensities of Bragg reflections. The scattering amplitude of the Bragg reflections in kinematical scattering theory is the Fourier transform of the object scattering strength in real space.

For a crystal structure containing $N$ atoms with scattering factors $f_j$ in each unit cell at the positions $\underline{r}_j$ we can write:

$$\psi_{\text{out}}(\underline{K}) = \sum_{n=1}^{N} \sum_{m_x=0}^{M_x-1} \sum_{m_y=0}^{M_y-1} \sum_{m_z=0}^{M_z-1} f_j \, \exp(2\pi i(\underline{k}-\underline{k}_0) \cdot (\underline{r}_{\text{atom position}} + \underline{r}_{\text{unitcell}}))$$

$$= F_{\text{unitcell}}(\underline{K}) \sum_{m_x=0}^{M_x-1} \sum_{m_y=0}^{M_y-1} \sum_{m_z=0}^{M_z-1} \exp\left(-2\pi i \left(as_x m_x + as_y m_y + as_z m_z\right)\right)$$

This last equation follows from the definition of a reciprocal lattice vector: $\underline{G} \cdot \underline{r}_{\text{unitcell}}$ is an integer. The resulting intensity is:

$$I(\underline{K}) = \left| F_{\text{unitcell}}(\underline{K}) \right|^2 \frac{\sin^2(\pi M_x s_x a)}{\sin^2(\pi s_x a)} \frac{\sin^2(\pi M_y s_y a)}{\sin^2(\pi s_y a)} \frac{\sin^2(\pi M_z s_z a)}{\sin^2(\pi s_z a)} \tag{6}$$

There is even a significant intensity for a Bragg reflection of a finite crystal, if the surface of the Ewald sphere does not exactly hit a reciprocal lattice point. With increasing length of the deviation vector the scattering intensity decreases. This decrease depends on the number of scattering centers along a line in the crystal. If the crystal has a plate-like shape, the scattering intensity slowly decreases in the direction in reciprocal space that corresponds to the surface normal of the sample in real space. Each reciprocal lattice point represents an inverse image of the crystal in real space. Plates become needles and vice versa (Fig. 19).

### 3.2.2. The Extinction Rules and Indexing of Diffraction Patterns

The next step will deal with the structure factor $F_{\text{unitcell}}$. The structure factor introduces variations of the intensities of Bragg reflections as it describes the interference effects of all atoms in one unit cell. The structure factor as shown in Eq. (6) is a function of the atoms with scattering factor $f_n$ at the positions $\underline{r}_j = (x_n, y_n, z_n)$ $a$ in the unit cell:

$$F_{\text{unitcell}}(\underline{K}) = \sum_{n=1}^{N} f_n \exp(2\pi i(\underline{G} - \underline{s}) \cdot \underline{r}_n) \approx \sum_{n=1}^{N} f_n \exp(2\pi i(hx_n + ky_n + lz_n))$$

$$\tag{7}$$

**TABLE 2.** Left: Extinction rules for the bcc lattice. Right: Extinction rules for CsCl (B2 structure).

| $h$ | $k$ | $l$ | $e^{\pi i(h+k+l)}$ | $F_{bcc}$ | $I_{bcc}$ | $h$ | $k$ | $l$ | $e^{\pi i(h+k+l)}$ | $F^{\alpha}_{CsCl}$ | $I^{\alpha}_{CsCl}$ |
|---|---|---|---|---|---|---|---|---|---|---|---|
| 0 | 0 | 0 | 1 | $2f$ | $4f^2$ | 0 | 0 | 0 | 1 | 72 | 5184 |
| 1 | 0 | 0 | $-1$ | 0 | 0 | 1 | 0 | 0 | $-1$ | 38 | 1444 |
| 2 | 0 | 0 | 1 | $2f$ | $4f^2$ | 2 | 0 | 0 | 1 | 72 | 5184 |
| 3 | 0 | 0 | $-1$ | 0 | 0 | 3 | 0 | 0 | $-1$ | 38 | 1444 |
| 1 | 1 | 0 | 1 | $2f$ | $4f^2$ | 1 | 1 | 0 | 1 | 72 | 5184 |
| 2 | 1 | 0 | $-1$ | 0 | 0 | 2 | 1 | 0 | $-1$ | 38 | 1444 |
| 1 | 1 | 1 | $-1$ | 0 | 0 | 1 | 1 | 1 | $-1$ | 38 | 1444 |
| 2 | 1 | 1 | 1 | $2f$ | $4f^2$ | 2 | 1 | 1 | 1 | 72 | 5184 |

If all atoms in a unit cell are the same, there may be some "forbidden reflections", i.e., the structure factors of these reflections is zero. The extinction rules determine, which reflections are "allowed" and which are "forbidden". The easiest example is the body-centered cubic (bcc) lattice. There are two atoms, at position 0,0,0 and at position $a/2, a/2, a/2$ in the unit cell. The structure factor $F$ for the bcc structure is:

$$F_{\text{unitcell}}(\underline{G}) = f\left[\exp(\pi i 0) + \exp(\pi i\,(h + k + l))\right] = f\left[1 + \exp(\pi i\,(h + k + l))\right]$$

In Table 2 the structure factors and the relative intensities for different Bragg reflections are listed. We can easily find the rule for "allowed" reflections: $h + k + l$ has to be even for Bragg reflections from the bcc lattice. Reflections with an odd sum $h + k + l$ are not present, they are "forbidden".

With this, we are able to construct and index electron diffraction patterns from crystals with bcc structure. The radius of the Ewald sphere is large in comparison to the distances of the reciprocal lattice points. Near the 000 reflection (the undiffracted beam) we find only Bragg reflections, if $\underline{k}_0 \cdot \underline{G} = 0$ is fulfilled. The distance of these reflections from the 000 position is inversely proportional to the distance of the corresponding lattice planes. The angle between two Bragg reflections is given by the scalar product of their normalized indices. Furthermore, we have to consider the extinction rules. The full indexing of a diffraction patterns can be done by applying linear algebra for addition of vectors. Fig. 20 shows some examples for electron diffraction patterns of the bcc structure.

### 3.2.3. Superstructure Reflections

In long-range ordered structures of a multielement material, extra reflections are present. They appear for integers $h,k,l$, where a forbidden reflection for a single-element material with equivalent basic structure is expected. One example for a long-range ordered structure is CsCl. The CsCl structure is also called the B2 structure (see, e.g., Villars and Calvert, 1991). It is derived from the bcc structure by placing one kind of atom in the center and the other type at the corner of the unit cell as shown in Fig. 21.

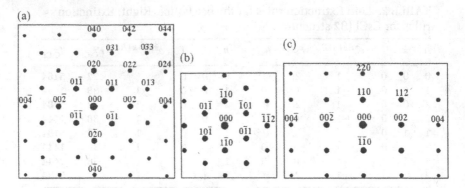

**FIGURE 20.** Calculated and indexed electron diffraction pattern of the bcc structure for electron beam directions (a) [100], (b) [111], and (c) [1-10]. The JEMS software from Jouneau and Stadelmann (1998) was used for the simulations.

For simplicity it is often assumed that the scattering amplitude of atoms is approximately proportional to the atomic number. Then, the extinction rules for CsCl are:

$$F_{CsCl}(\underline{G}) = f_{Cs} + f_{Cl} \exp(\pi i (h + k + l)) \propto 55 + 17 \exp(\pi i (h + k + l)).$$

For the B2 structure all integer reflections of the cubic lattice appear, for the *fundamental reflections* the structure factor is the sum of the atomic scattering factors, for *superstructure reflections* the structure factor is the difference of the atomic scattering factors (Table 2). As the atomic scattering factors are

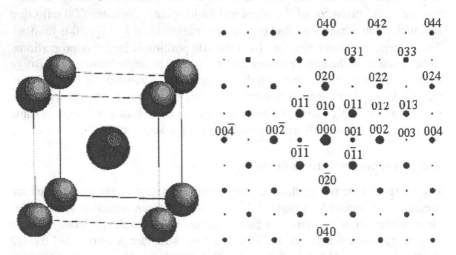

**FIGURE 21.** Left: Structural image of the B2 unit cell. Right: Electron diffraction pattern of the B2 structure for a [100] zone axis. The JEMS software from Jouneau and Stadelmann (1998) was used for the simulations.

approximately proportional to the atomic number of an element, the relative intensities of superstructure and fundamental reflections depend on the difference between the atomic numbers of the two elements in this material.

The structures of many materials can be found in the *EMS On Line* website (Jouneau and Stadelmann, 1998). This website offers a variety of tools to generate and display structures, to calculate electron diffraction pattern and to determine structure factors. The extended commercial Java-based version of this software offers full simulation capabilities. The book *International Tables for X-ray Crystallography*, Vol. 1 (Henry and Lonsdale, 1969) contains a list of all 230 three-dimensional space groups with their symmetries. The "Strukturbericht" (structure report) name of the bcc structure is A2. To learn more about the crystal structures of the phases in "your" material, you can also look up the crystal lattice structures website at http://cst-www.nrl.navy.mil/lattice/index.html, and you can get the Wyckoff positions and the symmetry operations of crystal structures at http://www.cryst.ehu.es/ (Aroyo *et al.*, 2006).

### 3.2.4. Laue Zones

Laue zones represent planes with reciprocal lattice points in the reciprocal lattice. These planes are perpendicular to the wave vector $\underline{k}_0$ of the incident beam. For all reciprocal lattice points in a Laue zone the scalar product $\underline{k}_0 \cdot \underline{G}$ is the same and determines the index of the zone. When irradiating a crystal along a highly symmetrical direction, the Bragg reflections in each Laue zone form concentric circles in the electron diffraction patterns with the 000 beam as the center (Fig. 22). In every Laue zone the reciprocal lattice points closest to the Ewald sphere (smallest deviation parameter) appear as especially bright Bragg reflections.

Laue zones are very useful to orient a nanocrystal when working in the diffraction mode. If a sample is not oriented with a highly symmetrical direction parallel to the incoming electron beam, the centers of the Laue circles are not anymore at the 000 position (Fig. 23). Still, bright Bragg reflections (including the 000 reflection) are visible on segments of circles. To tilt a sample towards a highly symmetrical orientation, the circles have to be centered about the 000 position. This method of sample tilting is especially important for nanoparticles when individual crystallites have to be oriented along a certain crystallographic direction for high-resolution lattice imaging.

### 3.2.5. Bright-field and Dark-field Imaging

The objective aperture is located in the back focal plane of the objective lens. When one of the apertures in this diaphragm is centered about a selected Bragg reflection in the diffraction pattern only this beam can contribute to the formation of an image. All other beams are blocked by this diaphragm. When switching back from the diffraction mode, where the diffraction pattern with the aperture in the back-focal plane of the objective lens is magnified, to the imaging mode, this

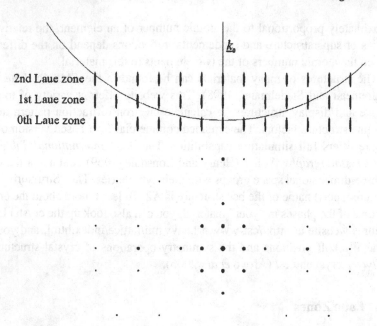

**FIGURE 22.** (Top) Ewald construction for symmetrical crystal orientation and (bottom) rings of bright Bragg reflections. The JEMS software from Jouneau and Stadelmann (1998) was used for the simulations.

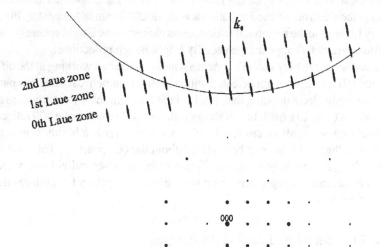

**FIGURE 23.** (Top) Ewald construction for asymmetric crystal orientation and (bottom) rings of bright Bragg reflections. The JEMS software from Jouneau and Stadelmann (1998) was used for the simulations.

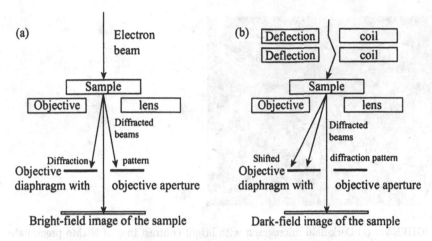

**FIGURE 24.** (a) The bright-field and (b) dark-field imaging techniques.

aperture is still blocking all other beams. Then, the magnified image of the sample contains only information propagated through this aperture.

If the objective aperture is centered about the undiffracted beam, we have *bright-field* conditions (Fig. 24(a)). Only the undiffracted beam or, for larger apertures, also a few of the low-indexed Bragg reflections are then contributing to the first image in the imaging plane of the objective lens. This way, the contrast in the micrograph is enhanced when compared to an image without an aperture. Using this technique, crystal defects can be identified, as they locally modify the *diffraction contrast*, i.e., how many electrons are scattered to other Bragg reflections and are lost to the undiffracted beam.

In *dark-field* imaging (Fig. 24(b)) the incident electron beam is tilted with respect to the optical axis of the objective lens. Therefore, the electron diffraction pattern in the back-focal plane is shifted. The electron beam is tilted in a way that one Bragg reflection is on the optical axis. The objective aperture is then centered about this Bragg reflection. All other reflections are excluded by the diaphragm and do not contribute to the image. In imaging mode only those regions appear bright which contribute to the Bragg reflection in the objective aperture (Fig. 25). All other regions appear dark.

### 3.2.6. Ring Patterns (Debye-Scherrer Patterns)

Nanoparticles deposited from a solution on a carbon film have random orientation. For electron diffraction of a nanocrystalline material, each crystallite will generate its individual diffraction pattern. If the image of several nanoparticles is contained in the selected area aperture the diffraction pattern will be comprised of the sum of the diffraction patterns from all crystallites imaged within the back-projection of the selected area aperture. Each crystallite contributes to a few Bragg reflections in the whole diffraction pattern. The distance of a Bragg reflection from the 000

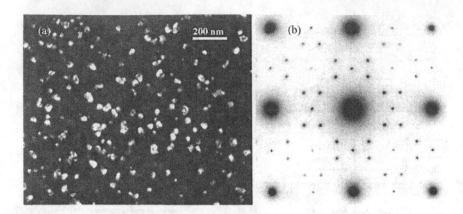

**FIGURE 25.** (a) Dark-field micrograph with bright contrast in a of $Ni_4Mo$ precipitates coherently embedded in the Ni-rich fcc matrix of a Ni-Al-Mo alloy. One superstructure reflection unique for $Ni_4Mo$ precipitates was used for imaging. (b) Corresponding electron diffraction pattern along the [001] direction of the Ni-rich matrix.

reflection is proportional to $2\Theta$, but the location of this reflection in the pattern depends on the actual crystallite orientation. So, for a polycrystalline material, the Bragg reflections appear only on fixed rings with radius $2\Theta$ about the incident beam in the diffraction pattern (Fig. 26). The more crystallites are contributing to the

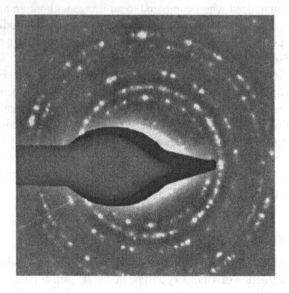

**FIGURE 26.** Debye-Scherrer pattern of $CeO_2$ nanoparticles. (Sample courtesy of S. Deshpande and S. Seal, University of Central Florida.) The bright undiffracted beam is blocked by a mechanical beam stop inserted in the electron path. Ring patterns from only a few nanoparticles.

diffraction pattern, the more Bragg reflections are found on each *Debye-Scherrer ring*. Finally, if the crystallite size is very small and the selected area aperture relatively large, the individual Bragg reflections cannot anymore be distinguished on a Debye-Scherrer ring. From the radii of the Debye-Scherrer rings and from their relative intensities structures and lattice parameters of nanomaterials can be determined with an accuracy of typically 0.5%.

Additionally, the extinction rules for the Debye-Scherrer rings help to identify the crystal structure. But it is equally desirable to obtain and evaluate electron diffraction patterns from individual nanoparticles.

### 3.2.7. Convergent-beam Electron Diffraction

With convergent-beam electron diffraction (CBED) the electron probe can be focussed on individual particles. Instead of diffraction spots we get diffraction disks. Within each diffraction disk, intensity modulations occur which are caused by changes in deviation parameters for the slightly different incident beam directions displayed in each disk. Additionally, excitations of higher-order Laue zone (HOLZ) reflections occur as sharp dark lines in the disks of the zeroeth Laue zone. The books of Tanaka and Terauchi (1985) and Tanaka et al. (2002) provide a valuable source for data analysis with CBED.

The evaluation of HOLZ line positions in the 000 disk allows lattice parameters to be determined with accuracies in the range of $\Delta a/a = 10^{-4}$ or better. This method is frequently applied to determine lattice parameters and crystal symmetries in electron crystallography and for strain analysis. CBED is applied to measure the lattice mismatch of $\gamma'$ precipitates in the Ni-rich matrix of Ni-based superalloys. These precipitates are typically several 100 nm in diameter. However, for nanoparticles below 20 nm the fine structure in the diffraction disks completely disappears (Carter and Williams, 1996). This is caused by a widening of the HOLZ lines with reduced sample thickness in combination with thickness variations of the nanoparticles within the illuminated area. Therefore, the determination of lattice parameters from individual small nanoparticles is only possible by measuring the distances and angles between the diffraction disks in a CBED pattern. This limits the accuracy of these measurements to about 0.5%.

### 3.2.8. Kikuchi Patterns

Kikuchi lines are observed in diffraction patterns of individual single crystalline particles and bulk samples with thicknesses above about 50 nm. The incident electrons are inelastically scattered exciting, e.g., plasmons and excitons. The incident electrons lose energy in the range of (10–40) eV. Therefore, the electron wavelength remains almost unchanged. These inelastic scattering processes also change the directions of the incident electrons, but these changes are small. Electrons are preferentially scattered in forward direction nearly parallel to their original (incident

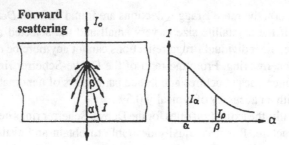

**FIGURE 27.** Schematic intensity distribution of inelastic scattering (Reprinted from Electron Microscopy of Materials, von Heimendahl, Electron Microscopy of Materials, (1980), with permission from Elsevier).

beam) direction (Fig. 27). With higher the inelastic scattering the electron intensity in this direction decreases.

The inelastically scattered electrons also can be diffracted elastically at the lattice planes in a second scattering event. But, as the thick sample itself acts as a source of electrons (the inelastically scattered electrons are emitted from the sample with a preferential propagation direction parallel to the original beam direction) these electrons can fall onto a set of parallel lattice planes from any direction in the crystal. For a specific set of parallel lattice planes, the Bragg condition is fulfilled, when the angle between these planes and the direction of the electrons is the Bragg angle.

The directions for which the Bragg condition is fulfilled for a special set of lattice planes are lying on Kossel cones. These cones are found on both sides of the parallel lattice planes (Fig. 28(a)). If the electron direction is lying on a Kossel cone, Bragg diffraction occurs and these electrons are scattered from their direction after inelastic interaction towards a new direction after Bragg diffraction. So, an exchange of (electron) intensity is occurring, if the Bragg condition is fulfilled. Intensity is partly exchanged between electrons entering the lattice planes from the two opposing sides (Fig. 28(b)).

If no Bragg condition is fulfilled for any set of parallel lattice planes, this intensity exchange does not occur. This gives rise to a diffuse background intensity in the diffraction pattern which decreases with increasing inelastic scattering angle.

If, however, the Bragg condition is fulfilled, intensity exchange between the Kossel cones occurs. More electrons are scattered towards higher scattering angles by this second scattering event and therefore, intensity is missing for smaller total scattering angles. As the intersection line between a flat Kossel cone and the diffraction plane (the plane where the electron diffraction pattern is obtained) is almost a straight line, the Kikuchi pattern consists of sets of two mutually parallel Kikuchi lines. The Kikuchi line with a smaller angle to the incident beam direction is less intense than the diffuse background, while the line at higher scattering angles appears brighter than the background (Fig. 29(b)).

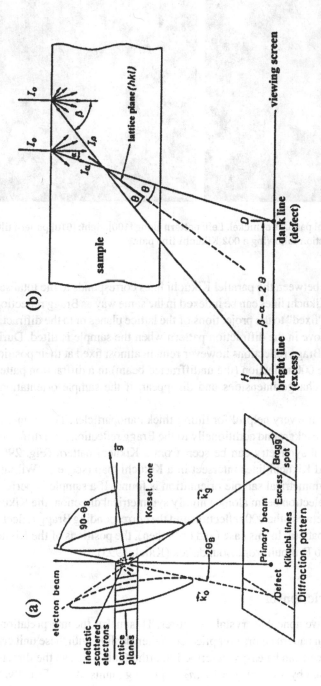

**FIGURE 28.** (a) The Bragg condition is fulfilled for electrons propagating along directions on Kossel cones (From Transmission Electron Microscopy, Reimer, 1989, with kind permission of Springer Science and Business Media). (b) Defect and excess lines in the inelastic background of the diffraction pattern of a single crystal (Reprinted from Electron Microscopy of Materials, von Heimendahl, Electron Microscopy of Materials, (1980), with permission from Elsevier).

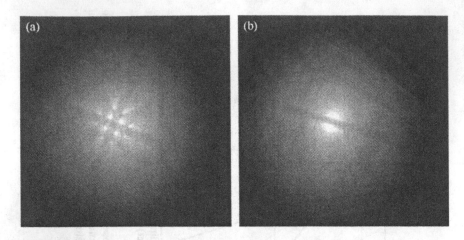

**FIGURE 29.** Kikuchi patterns of nickel. Left: pattern along [100], right: [610] pattern tilted from the [100] direction following a 002 Kikuchi line pair.

The distance between the parallel Kikuchi lines corresponds to the total scattering angle $2\Theta$. Kikuchi lines can be indexed in the same way as Bragg reflections. Kikuchi lines are "fixed" to the projections of the lattice planes onto the diffraction plane, and they move in the diffraction pattern when the sample is tilted. During sample tilting the Bragg reflections however remain almost fixed at their positions with respect to the 000 reflection (the undiffracted beam) in a diffraction pattern. Bragg reflections change intensities and disappear, if the sample orientation is changed.

Kikuchi lines are very helpful for tilting thick nanoparticles. These lines appear in the diffuse background additionally to the Bragg reflections in a diffraction pattern. The crystal symmetry can be seen from a Kikuchi pattern (Fig. 29(a)). When low-indexed Kikuchi lines intersect in a Kikuchi map (see, e.g., Williams, 1996), a highly symmetrical sample orientation is found. If a sample is perfectly oriented with the electron beam along a highly symmetrical direction, the Kikuchi lines are exactly between the 000 reflection and the corresponding Bragg reflection in the diffraction pattern. In this case (and only then), the positions of the Kikuchi lines correspond to Brillouin zone boundaries (Kittel, 1995).

### 3.2.9. The Metric Tensor

Many materials have noncubic crystal structures. This makes the interpretation of electron diffraction patterns more complicated. When dealing with these unit cells, the axes of a unit cell can be easily described in orthogonal units, so the direction of the a axis is given by a vector $\underline{a} = [a_1, a_2, a_3]$ in, e.g., units of nm. Equally, the two other axes $b$ and $c$ are given. In reciprocal space the directions and lengths of the axes are given by the vectors $\underline{a}^*$, $\underline{b}^*$ and $\underline{c}^*$.

The equations

$$\underline{a}^* = \frac{\underline{b} \times \underline{c}}{\underline{a} \cdot (\underline{b} \times \underline{c})}, \quad \underline{b}^* = \frac{\underline{c} \times \underline{a}}{\underline{a} \cdot (\underline{b} \times \underline{c})}, \quad \text{and} \quad \underline{c}^* = \frac{\underline{a} \times \underline{b}}{\underline{a} \cdot (\underline{b} \times \underline{c})}$$

are used to find the axes in reciprocal space. (*Remark*: If the length of the wavevector is defined by $2\pi/\lambda$, each reciprocal vector has to be multiplied by $2\pi$.)

The metric tensor $\underline{\underline{M}}$ can be used to calculate the length $|[x,y,z]|$ of any vector $\underline{R} = [x,y,z]$ given in units of the axial vectors $\underline{a}, \underline{b}$ and $\underline{c}$ of the unit cell:

$$|\underline{R}| = \sqrt{(x,y,z) \cdot \begin{pmatrix} a^2 & ab \cos\gamma & ac \cos\beta \\ ab \cos\gamma & b^2 & bc \cos\alpha \\ ac \cos\beta & bc \cos\alpha & c^2 \end{pmatrix} \cdot \begin{pmatrix} x \\ y \\ z \end{pmatrix}}$$

$$= \sqrt{(x,y,z) \cdot \underline{\underline{M}} \cdot \begin{pmatrix} x \\ y \\ z \end{pmatrix}}$$

The angle $\tau$ between two vectors $\underline{R}$ and $\underline{S}$ is determined by

$$\cos \tau = \frac{\underline{R} \cdot \underline{\underline{M}} \cdot \underline{S}^T}{|\underline{R}| |\underline{S}|}.$$

The length of vectors $\underline{Q} = [h,k,l]$ in reciprocal space given in units of the reciprocal unit vectors $\underline{a}^*, \underline{b}^*$ and $\underline{c}^*$ can be determined by $|\underline{Q}| = \sqrt{\underline{Q} \cdot \underline{\underline{M}}^{-1} \cdot \underline{Q}^T}$.

The angle $\Omega$ between vectors $\underline{Q}$ and $\underline{P}$ in reciprocal space is found from

$$\cos \Omega = \frac{\underline{Q} \cdot \underline{\underline{M}}^{-1} \cdot \underline{P}^T}{|\underline{Q}| |\underline{P}|}.$$

The inverse matrix of the metric tensor is:

$$\underline{\underline{M}}^{-1} = \frac{\begin{pmatrix} b^2c^2 \sin^2\alpha & abc^2 (\cos\alpha \cos\beta - \cos\gamma) & ab^2c (\cos\alpha \cos\gamma - \cos\beta) \\ abc^2 (\cos\alpha \cos\beta - \cos\gamma) & a^2c^2 \sin^2\beta & a^2bc (\cos\gamma \cos\beta - \cos\alpha) \\ ab^2c (\cos\alpha \cos\gamma - \cos\beta) & a^2bc (\cos\gamma \cos\beta - \cos\alpha) & b^2a^2 \sin^2\gamma \end{pmatrix}}{a^2b^2c^2 (1 - \cos^2\alpha - \cos^2\beta - \cos^2\gamma + 2\cos\alpha \cos\beta \cos\gamma)}$$

If we want to find a vector $\underline{R}$ perpendicular to a plane with indices $(h,k,l)$ in real space, we can use the relation $\underline{R} = \underline{\underline{M}} \cdot (h,k,l)$, see, e.g., De Graef, 2003. A similar relation, $(h,k,l) = \underline{\underline{M}}^{-1} \cdot \underline{R}$, is used to find indices for planes perpendicular to a certain direction. To determine the spacing of a lattice plane in real space, it is easiest to use the inverse matrix to calculate the length of the corresponding reciprocal vector. Using the above equations the lattice spacing for an orthorhombic unit cell is therefore:

$$d_{hkl} = 1 \left/ \sqrt{(h/a)^2 + (k/b)^2 + (l/c)^2} \right.$$

For the hexagonal lattice ($a = b, \gamma = 120°$) one gets

$$d_{hkl} = 1 \left/ \sqrt{\frac{4\left(h^2 + hk + k^2\right)}{3a^2} + \left(\frac{l}{c}\right)^2} \right.$$

Frequently, four indices for hexagonal lattices are used making the indexing of real-space and reciprocal-space vectors more symmetrical. The fourth index $i$ makes indexing of diffraction patterns very convenient. This fourth index is given by $h + k + i = 0$. But there is a rotation by 30° about the $c$-axis when transforming three-index axes to four-index axes:

$$d_{hkl} = 1 \left/ \sqrt{\frac{2\left(h^2 + k^2 + i^2\right)}{3a^2} + \left(\frac{l}{c}\right)^2} \right. .$$

## 3.3. Scattering Theory

### 3.3.1. The Column Approximation

The amplitudes and phases of Bragg reflections can be calculated as a function of the position of the beam entering the sample using the column approximation. For contrast simulations of a thin TEM sample, the illuminated area is divided in columns with constant contrast. The column diameter is well defined by the Huygens principle (see, e.g., Fultz and Howe, 2002). The amplitude and phase of the electron wave at any point in the exit plane of a sample of thickness $t$ is determined by scattering of electrons for a sample column above that point. Other contributions from higher-angle scattering are neglected in the column approximation, as electron scattering appears predominantly in forward direction. Additionally, contributions from higher angles add only incoherently to the image contrast (transversal incoherence). Within a column in the sample, the path differences of rays are limited to a maximum of $\lambda/2$ (Fig. 30). This conical volume is called the first Fresnel zone. Therefore, the column radius is $\sqrt{t\lambda}$.

The image intensity at a point in the image of a sample is determined by the volume of the corresponding column. The neighborhood of a column can be neglected. Within a column linear scattering theory can be used as a first

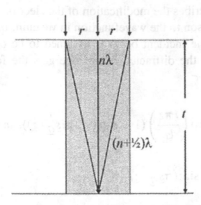

**FIGURE 30.** Schematics of the column approximation used for contrast simulations.

approximation to determine the wave functions of Bragg reflections as a function of the sample thickness $t$ (or, more accurately, as a function of the depth $z$ within the sample parallel to the beam direction). The following overview on scattering theory can be directly applied to contrast simulations for HRTEM and STEM.

### 3.3.2. Kinematical Scattering

The kinematical scattering theory is especially simple as it considers an incident beam with constant intensity. All diffracted beams get exclusively scattering contributions from this incident beam, and there is no exchange of electrons between different diffracted beams considered. The calculation can be limited to the wave-functions of the incident beam $\psi_0$ and one diffracted beam $\psi_G$ as a representative of any diffracted beam, and for a beam in the zeroeth Laue zone one gets the following differential equations:

$$\frac{d\psi_0}{dz} = \frac{i\pi}{\xi_0}\psi_0 \quad \text{and} \quad \frac{d\psi_G}{dz} = \frac{i\pi}{\xi_0}\psi_G + \frac{i\pi}{\xi_G}\psi_0 \exp(-2\pi i s_G \cdot z) \tag{8}$$

The extinction length $\xi$ of a Bragg reflection is related to the structure factor by

$$\xi_G = \frac{\pi V_{\text{Unitcell}}}{\lambda F(G)}$$

where $V_{\text{Unitcell}}$ is the volume of a crystal unit cell. The wave function of the un-diffracted beam is given by: $\psi_0(z) = \psi_0(z = 0)\exp(i\pi z/\xi_0)$. Here, the extinction length can be expressed by $\xi_0 = \lambda/2$, but it is more accurate to use

$$\xi_0 = \frac{\pi V_{\text{Unitcell}}}{\lambda F(0)}$$

This relation describes the modification of the electron wavefunction within the sample in comparison to the wavefunction in vacuum. In kinematical scattering, the intensity of the incident beam is assumed to be constant, and absorption is neglected. For the diffracted beam, one gets the following solution of Eq. (8):

$$\psi_{\underline{G}}(\underline{z}) = \frac{\psi_{\underline{0}}(0)}{2s_{\underline{G}}\xi_{\underline{G}}} \exp\left(\frac{i\pi z}{\xi_{\underline{0}}}\right) (1 - \exp(-2\pi i \underline{s}_{\underline{G}} \cdot \underline{z})), \text{ and for the intensity}$$

$$I_{\underline{G}}(\underline{z}) = \frac{|\psi_{\underline{0}}(0)|^2}{(s_{\underline{G}}\xi_{\underline{G}})^2} \sin^2(\pi \underline{s}_{\underline{G}} \cdot \underline{z}) \tag{9}$$

This result is equivalent to Eq. (6), the intensity of the diffracted beam is a squared sine function of deviation parameter $s_G$ and sample thickness $z$. The intensity of the diffracted beam shows periodic modulations with sample thickness and in dark-field imaging thickness contour lines are visible. The sample thickness changes by one extinction length when going from one thickness contour line to the neighboring line.

For $s_G = 0$ the intensity of the diffracted beam increases proportional to the square of the sample thickness. This last point shows that the kinematical scattering theory is a poor approximation. The total intensity of all beams can not be higher than the intensity of the incident beam. Multiple scattering as outlined below has to be used to quantitatively describe electron scattering.

### 3.3.3. Dynamical Scattering and Bloch Waves

The wave functions described in the dynamical scattering theory can also be represented by linear combinations of Bloch waves. Every diffracted and undiffracted wave can be described by superimposed Bloch waves $\psi_G$. The total standing wavefield in the crystal has the periodicity of the crystal lattice:

$$\Psi\left(\underline{k}, \underline{r}\right) = \sum_{\underline{G}} \psi_{\underline{G}} \exp\left(2\pi i \left(\underline{k}_0 + \underline{G} + \underline{s}_{\underline{G}}\right) \cdot \underline{r}\right) \tag{10}$$

The scattering of electrons in a periodic crystal potential $\phi(\underline{r}) = \phi(\underline{r} + \underline{a}) = eV(\underline{r})$ is described using the Schrödinger equation $\Delta\Psi = -2me\hbar^{-2}(V(\underline{r}) + U)\Psi$. For the periodic potential in the crystal an additional local acceleration voltage in the sample is used:

$$V\left(\underline{r}\right) = \sum_{\underline{G}} V_{\underline{G}} \exp\left(2\pi i \underline{G} \cdot \underline{r}\right) \tag{11}$$

The coefficients $V_G$ are the Fourier coefficients of the crystal potential. Therefore, the Schrödinger equation is:

$$\Delta \Psi = \sum_{\underline{G}} \left( \Delta \psi_{\underline{G}} + 4\pi i \left( \underline{k}_0 + \underline{G} + \underline{s}_{\underline{G}} \right) \cdot \underline{\nabla} \psi_{\underline{G}} + 4\pi^2 \left( \underline{k}_0 + \underline{G} + \underline{s}_{\underline{G}} \right)^2 \psi_{\underline{G}} \right)$$

$$\exp \left( 2\pi i \left( \underline{k}_0 + \underline{G} + \underline{s}_{\underline{G}} \right) \cdot \underline{r} \right)$$

$$= -2me\hbar^{-2} \left( \sum_{\underline{G'}} V_{\underline{G'}} \exp \left( 2\pi i \underline{G'} \cdot \underline{r} \right) + U \right)$$

$$\sum_{\underline{G''}} \psi_{\underline{G''}} \exp \left( 2\pi i \left( \underline{k}_0 + \underline{G''} + \underline{s}_{\underline{G''}} \right) \cdot \underline{r} \right)$$

Bloch waves only change little along the beam direction, so the term $\Delta \psi_{\underline{G}}$ can be neglected. With the wavevector given by $|\underline{k}_0 + \underline{G} + \underline{s}_{\underline{G}}|^2 = |\underline{k}_0|^2 = 2meU\hbar^{-2}$ and using the condition that the equation has to be fulfilled for any $\underline{G} = \underline{G'} + \underline{G''}$, so that equal coefficients in the exponential functions can be compared, we find:

$$\frac{d\psi_{\underline{G}}}{dz} = \sum_{\underline{G'}} \frac{2\pi i me}{\hbar^2} \lambda V_{\underline{G'}} \psi_{\underline{G}-\underline{G'}} \exp \left( 2\pi i \left( \underline{s}_{\underline{G}-\underline{G'}} - \underline{s}_{\underline{G'}} \right) \cdot \underline{z} \right)$$

Multiple diffraction effects are therefore responsible for an exchange of intensities between all Bragg reflections. Electrons can also be scattered back from the direction of a Bragg reflection to the undiffracted 000 beam. With the extinction length defined as

$$\xi_{\underline{G}} = \frac{\pi V_{\text{Unitcell}}}{\lambda F \left( \underline{G} \right)} = \frac{h^2}{2me\lambda V_{\underline{G}}},$$

this system of coupled linear equations of $n$ different beams can also be represented by:

$$\frac{d\psi_{\underline{G}}}{dz} = \sum_{H=\underline{G}_1}^{\underline{G}_n} \frac{i\pi}{\xi_{\underline{G}-\underline{H}}} \exp \left( 2\pi i \left( \underline{s}_{\underline{H}} - \underline{s}_{\underline{G}} \right) \cdot \underline{z} \right) \psi_{\underline{H}} \qquad (12)$$

Equation (12) can be rewritten using the substitution $\psi_{\underline{G}} = \Phi_{\underline{G}} \exp \left( -2\pi i \underline{s}_{\underline{G}} \cdot \underline{z} \right)$:

$$\frac{d\Phi_{\underline{G}}}{dz} = 2\pi i \underline{s}_{\underline{G}} \Phi_{\underline{G}} + \sum_{H=\underline{G}_1}^{\underline{G}_n} \frac{i\pi}{\xi_{\underline{G}-\underline{H}}} \Phi_{\underline{H}} \qquad (13)$$

For this set of n linear differential equations a vector notation can be introduced, where each wavefunction $\Phi_G$ is one component of a vector $\underline{\Phi}$. Using a $n \times n$ matrix $\underline{\underline{W}}$ with components $W_{\underline{G},\underline{H}} = \delta_{\underline{G},\underline{H}} 2\pi s_{\underline{G}} + \frac{\pi}{\xi_{\underline{G}-\underline{H}}}$, Eq. (13) is replaced by

$$\frac{d\underline{\Phi}}{dz} = i \underline{\underline{W}} \cdot \underline{\Phi} \qquad (14)$$

For a defect-free sample of thickness $z$, Eq. (14) can be solved by determining the eigenvalues $\kappa_l$ and eigenvectors of the matrix $\underline{\underline{W}}$ (Heinrich and Kostorz, 2000). The eigenvectors can be written as elements of a matrix $\underline{\underline{T}}$. Another matrix $\underline{\underline{F}}$ contains only diagonal elements of the form $\exp(i\kappa_l z)$. The eigenvalues $\kappa_l$ of $\underline{\underline{W}}$ are inversely proportional to the wavelengths of the different Bloch waves propagating through the crystal. The matrix $\underline{\underline{T}}$ contains the contributions of the Bloch waves to each diffracted beam $\Phi_G$:

$$\underline{\Phi}(z) = \underline{\underline{T}} \cdot \underline{\underline{F}} \cdot \underline{\underline{T}}^{-1} \cdot \underline{\Phi}(z = 0). \tag{15}$$

### 3.3.4. The Two-beam Case

The large system of equations can be solved using the above Bloch-Wave approach of Eq. (15), but for the case of only two intense beams (the undiffracted and one diffracted beam) the wavefunctions can still be determined easily. For the two-beam imaging condition, one diffracted beam and the undiffracted beam are on or close to the corresponding Kikuchi lines. The deviation parameter for the diffracted beam is small, while deviation parameters for all other diffracted beams are large. When imaging materials using the two-beam condition, defects like dislocations and stacking faults show a strong contrast. The two-beam case is described by the Howie-Whelan equations which are easily found from Eq. (12):

$$\frac{d\psi_0}{dz} = \frac{i\pi}{\xi_0}\psi_0 + \frac{i\pi}{\xi_G}\psi_G \exp\left(2\pi i\underline{s_G} \cdot \underline{z}\right) \quad \text{and}$$

$$\frac{d\psi_G}{dz} = \frac{i\pi}{\xi_0}\psi_G + \frac{i\pi}{\xi_G}\psi_0 \exp\left(-2\pi i\underline{s_G} \cdot \underline{z}\right) \tag{16}$$

As for the diffracted beam in the kinematical case, one gets now a "Pendellösung" (pendulum solution) for both the undiffracted and the diffracted beam. Both beams show oscillations of their intensities as a function of sample thickness. For $s_G = 0$ the repeat distance of these oscillations corresponds to the extinction length $\xi_G$, but for a finite deviation parameter the repeat distance is given by an effective extinction distance $\xi_{G,\text{eff.}} = 1/\sqrt{\xi_G^{-2} + s_G^2}$. The solutions of Eqs. (16) are

$$\psi_0(z) = \psi_0(0)\exp\left(\frac{i\pi z}{\xi_0}\right)\exp\left(\pi i\underline{s_G} \cdot \underline{z}\right)\left(\cos\left(\frac{\pi z}{\xi_{G,\text{eff.}}}\right)\right.$$

$$\left. - i\underline{s_G}\sin\left(\frac{\pi z}{\xi_{G,\text{eff.}}}\right)\right) \quad \text{and}$$

$$\psi_G(z) = \frac{i\xi_{G,\text{eff.}}\psi_0(0)}{\xi_G}\exp\left(\frac{i\pi z}{\xi_0}\right)\exp\left(-\pi i\underline{s_G} \cdot \underline{z}\right)\sin\left(\frac{\pi z}{\xi_{G,\text{eff.}}}\right)$$

These periodic variations of the intensities of reflections in the two-beam case give rise to thickness contour lines (lines of equal sample thickness) when using the objective aperture to select one Bragg reflection for imaging. With increasing deviation parameter the effective extinction length $\xi_{G,\text{eff.}}$ decreases and thickness contour lines become more dense in a micrograph. If absorption of electrons is considered in contrast calculations, in addition to extinction lengths, absorption lengths can be introduced to replace $i/\xi_G$ by $i/\xi_{G,e}-1/\xi_{G,a}$.

As shown above the wavefunctions in the dynamical scattering theory can be represented by linear combinations of Bloch waves. Every diffracted and undiffracted wave can be described by superimposed Bloch waves. The total standing wavefield in the crystal has the periodicity of the crystal lattice and is described by Eq. 10. Intensity variations as a function of sample thickness can be explained by interference of the Bloch waves. If two Bloch waves of different wavelength show interference, modulated signals (a beat) for both the sum as well as for the difference between the Bloch waves occur. One Bloch wave has its maxima at the atomic positions and therefore, the electrons contributing to this Bloch wave "feel" a lower local crystal potential than electrons of a second Bloch wave with maxima between the atomic nuclei (Fig. 31(b)). This difference in crystal potential causes a difference in the electron wavelength. These differences in electron wavelength lead to the beat and to the intensity modulations as a function of sample thickness in the two-beam case.

For the two-beam case the Ewald construction has to be redrawn to account for the electrons with different wavelengths in the sample. The lengths of the wavevectors are adjusted as we have two different lengths according to the two

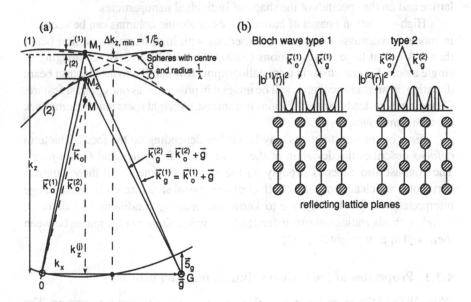

**FIGURE 31.** (a) Ewald construction with two dispersion surfaces for the two-beam case. (b) The two Bloch waves have the periodicity of the crystal lattice with maxima at different positions in the lattice. (From Transmission Electron Microscopy, Reimer, 1989, with kind permission of Springer Science and Business Media.)

Bloch states. The starting points of the wavevectors are located now on two **dispersion surfaces**. The dispersion surfaces represent the solutions of the Schrödinger equation for different sample orientations (deviation parameters) in a graphical way (Fig. 31(a)). The gap between the two dispersion surfaces is related to the inverse of the extinction length $\xi_G$. If more than two beams are considered in a contrast simulation, we have to use more Bloch functions. The number of Bloch waves corresponds to the number of diffracted beams considered. These Bloch waves do not correspond to beams in certain Bragg reflections, they are standing waves in the crystal. Only their superposition (as given by the matrix $\underline{T}$ in Eq. (15)) with phase shifts depending on the actual Bragg reflection and on the deviation parameters defines the different Bragg reflections.

# 4. IMAGING OF NANOSTRUCTURED MATERIAL

## 4.1. High-resolution Transmission Electron Microscopy (HRTEM)

While bright- and dark-field micrographs reveal the arrangement and sizes of nanoparticles, imaging without objective aperture can also be used, but a significant loss of contrast often results for nanocrystals composed of light atom. These micrographs help to gain data on the average size and size distribution of particles. High-resolution micrographs, however, provide direct information on the crystal lattice and on the specifics of the shape of individual nanoparticles.

High-resolution images of lattice planes or atomic columns can be acquired in modern transmission electron microscopes with high lateral resolution better than 2 Å and at large magnifications (>200,000). HRTEM is not the imaging of single atoms. Several atoms in a crystallographic row parallel to the electron beam direction, in an atomic column, will be imaged in projection as one contrast feature. Changing focus leads to modifications in contrast, as bright spots may become dark spots in a micrograph.

Finer or coarser structures may be visible depending on the focus. The term *defocus* describes the deviation of the actual focus from the ideal *Gauss focus*. The contrast also depends strongly on the sample thickness. All those contrast variations, which are not a direct effect of the crystal structure itself, make image interpretation difficult. We have to know the imaging conditions like defocus, wavelength, aberrations, etc, to understand if atomic columns or the spaces between them will appear bright (Fig. 32).

## 4.1.1. Properties of the high-resolution imaging process

With HRTEM the phase contrast of thin objects is mainly used for imaging. The phase contrast of the electron waves stems from the local variations of the crystal potential. A thin HRTEM foil or a small nanoparticle with a few nm in thickness is

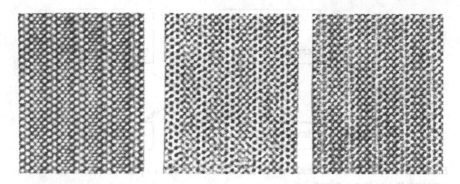

FIGURE 32. Contrast inversion for different defocus values for a $Bi_2Sr_2CaCu_8$ high-temperature superconductor imaged along [110] with a Philips CM30 operated at 300 kV. Each micrograph of 7 nm width was taken from the same sample region.

essentially a **phase object**, while the reduction of the electron intensity (**amplitude contrast**) by absorption is less important. For HRTEM samples more than 10 nm in thickness, the contrast is not anymore interpretable and the image resolution is strongly reduced by the resolution limit given in the column approximation. The phase contrast from the sample is transformed into an amplitude modulation by the spherical aberration of the objective lens and by defocusing.

The contrast in a HRTEM image depends on many microscope parameters like acceleration voltage (electron wavelength), spherical and chromatic aberration, astigmatism, coma, defocus stability of voltage, energy distribution of electron source, stability of lens currents causing variations in focus, brightness of the electron source, selection of objective and condenser apertures, beam convergence, sample drift, magnification, acquisition time, sensitivity or type of camera used, and external influence like microscopic stray fields, mechanical vibrations, and noise. The sample thickness and orientation, lattice spacings, atomic species and their interaction with electrons, are important for the formation of a HRTEM image. But high-quality micrographs are often prevented by limited electrical conductivity or by magnetization of the sample, contamination of the sample surfaces, radiation damage, and inelastic interactions.

We can gain structural information from HRTEM about:

1. atomic arrangements (example Ti-Al), about superlattices;
2. atomic-scale defects (vacancy clusters, several atoms on "wrong" atomic sites);
3. the sample thickness;
4. lattice distortions, dislocations, stacking faults;
5. grain boundaries, coherent and incoherent interfaces, amorphous layers;
6. nanoparticles and small precipitates, their shape, structure and size.

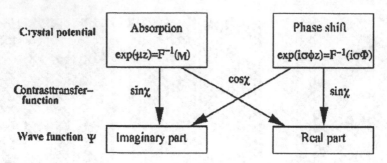

**FIGURE 33.** Information transfer for an amplitude object and a phase object into the wave function in the image plane.

But we learn nothing about:

(a) very light atoms near heavy atoms;
(b) lattice parameters cannot be measured; accuracy worse than 1%;
(c) single atomic defects, vacancies;
(d) small individual lattice distortions below 0.1 Å,
(e) overlapping lattices of an interface in projection;
(f) everything else.

### 4.1.2. Image Formation in HRTEM

Electrons "feel" the local potential $\phi(x,y,z)$ in the crystal (Fig. 33(b)). This modifies the wavelength from $\lambda$ to $\lambda_{\mathrm{cryst.}} \approx \sqrt{2me(U + \phi)}$. For simulations of high-resolution TEM images, the local potential $\phi(x,y,z)$ is assumed constant as $\phi(x,y)$ for a short distance along the $z$ direction. From the refractive index $n = \lambda/\lambda_{\mathrm{cryst.}}$ a phase shift $\varepsilon(x, y) = \sigma\phi(x, y)z$ after traveling a distance $z$ in the crystal is obtained with $\sigma = 2\pi me\lambda\, h^{-2}$ as interaction constant of electrons. In reality there is also a small contribution $\mu$ from absorption and the incident electron waves $\psi(x,y)$ experience a modification $q(x, y) = \exp(i\sigma\phi(x, y)z - \mu(x, y)z)$ in amplitude and phase.

The objective lens produces a Fourier transform $\mathbf{F}$ of the modified wave function $q$ at the exit plane of the sample: $Q(u, v) = \mathbf{F}(q(x, y)) \approx \delta(u, v) + i\sigma\Phi(u, v) - M(u, v)$ with the Fourier transforms of the projected potentials $\Phi(u, v) = \mathbf{F}(\phi(x, y)z)$ and $M(u, v) = \mathbf{F}(\mu(x, y)z)$.

For parallel illumination, this Fourier image appears in the back-focal plane of the objective lens. But the objective lens modifies this Fourier image by the contrast transfer function (CTF) $\chi(u,v)$ which introduces additional phase shifts as a function of distance $\sqrt{u^2 + v^2}$ of diffracted beams from the undiffracted beam in reciprocal space:

$$\chi(u, v) = \frac{\pi}{2\lambda} \left(C_s\lambda^4(u^2 + v^2)^2 + 2\Delta f(u^2 + v^2)\lambda^2\right) \tag{17}$$

with the spherical aberration $C_s$ and the defocus $\Delta f$. The contrast transfer function describes an additional phase shift of electron waves introduced by lens aberrations and defocussing. This phase shift increases with increasing off-axis components $u,v$ of the electron wave vector. The contrast transfer function determines how the Fourier coefficients of details of a wave function, which is altered by the sample are transferred to the image. The CTF takes into account, that the resolution of a microscope mainly depends on the wavelength, the spherical aberration, the size of the objective lens and the defocus.

The objective aperture can be inserted to limit the area in the diffraction pattern contributing to the image. Only information contained in this aperture will affect the image. So we use $A(u, v) = 1$ for information transfer inside the aperture and $A(u, v) = 0$ outside. The wave function in the image plane is then given by the back-Fourier transform

$$\psi(x, y) = \mathbf{F}^{-1}[Q(u, v)\exp(i\chi(u, v))A(u, v)]$$

The image intensity $I(x, y) = \psi(x, y)\psi^*(x, y)$ then results in:

$$I(x, y) \approx 1 - 2\sigma\mathbf{F}^{-1}[\Phi(u, v)\sin(\chi(u, v))A(u, v)] \\ - 2\mathbf{F}^{-1}[M(u, v)\cos(\chi(u, v))A(u, v)] \quad (18)$$

For the *Gauss focus* (defined by $\Delta f = 0$) the CTF for small $(u,v)$ only slightly differs from zero. Therefore, the Gauss focus is not suited for pure phase objects, where the sine-term is relevant (Fig. 33). For negative defocus values (underfocus) $\chi(u,v)$ first becomes negative, and at $\chi(u, v) = -\pi/2$ the information transfer is maximized.

In *Scherzer defocus* $\Delta f_S = -\sqrt{C_S\lambda}$ the value $\chi(u, v) = -\pi/2$ is reached for the first time (Fig. 34(a), this focus frequently yields optimized contrast conditions and determines the *point resolution*. As already derived in Section 3.1.2,

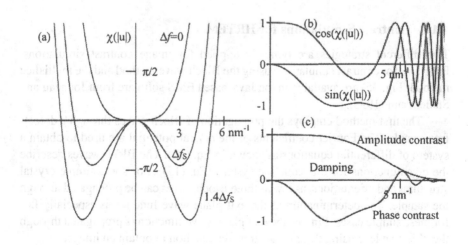

**FIGURE 34.** (a) The contrast transfer function for different values of defocus; (b) sine and cosine parts of the contrast transfer without damping; (c) damping of the contrast transfer.

information is transferred to the image for a projected distance of objects larger than the point resolution. The same result is obtained when using the Scherzer focus for imaging and determining the first zero-point in the contrast transfer function (Fig. 34(b)). The reciprocal value of this first zero point in the CTF is called point resolution:

$$\Delta r_{opt} \approx \sqrt[4]{C_s \lambda^3 / 4}. \tag{19}$$

The contrast transfer of phase objects reaches large negative values for projected distance of objects larger than the point resolution. The resolution limit (information limit) $\Delta r_{Inf.}$ for imaging of lattice planes is better than the point resolution. The information limit is determined by additional damping terms for the CTF caused by chromatic aberration, voltage and lens current variations, beam coherence and beam convergence. According to Fig. 34(c) the CTF is lowered for high spatial frequencies, i.e., small distances. Objects at smaller projected distances can not be distinguished. For field-emission sources the damping effect is small, as a small tip area produces highly monochromatic electrons. For typical 300 kV electron microscopes with field emission sources the point resolution is in the range of 0.2 nm, while information is contained in the micrographs for distances below 0.14 nm. The damping envelope according to Carter and Williams (1996) is given by

$$E(u) = \exp\left[\frac{-\pi^2 \lambda^2 C_C^2 \sqrt{4(\Delta I/I)^2 + (\Delta E/eU)^2 + (\Delta U/U)^2} u^4}{2}\right].$$

Fluctuations of the lens currents $\Delta I$ and of the acceleration voltage $\Delta U$ as well as the energy spread of the electrons from the source $\Delta E$ reduce the contrast transfer at high spatial frequencies.

### 4.1.3. Contrast Simulations for HRTEM

Two different strategies are typically applied for image contrast simulations. Programs for contrast simulations using the Bloch wave method and the multislice method (Fig. 35) are available in the Java-based EMS software from Jouneau and Stadelmann (1998).

The first method employs the propagation of Bloch waves through a defect-free sample. The Fourier coefficients of the crystal potential are used to obtain a system of differential equations as shown in Eq. (12). The Bloch waves describe the eigenfunctions for this equation system (Eq. (15)). For some simple crystal structures and orientations, as few as three Bloch waves can be propagated through the sample. The determination of the exit plane wave function is especially fast for these simple cases. Finally, the exit plane wave function is propagated through the objective lens using the contrast transfer function to obtain an image.

The second method is the multislice technique (Fig. 35), which is used when the contrast of defects in a crystal structure or of nonperiodic objects is simulated.

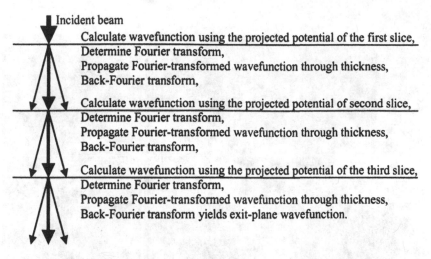

Incident beam
Calculate wavefunction using the projected potential of the first slice,
Determine Fourier transform,
Propagate Fourier-transformed wavefunction through thickness,
Back-Fourier transform,

Calculate wavefunction using the projected potential of second slice,
Determine Fourier transform,
Propagate Fourier-transformed wavefunction through thickness,
Back-Fourier transform,

Calculate wavefunction using the projected potential of the third slice,
Determine Fourier transform,
Propagate Fourier-transformed wavefunction through thickness,
Back-Fourier transform yields exit-plane wavefunction.

**FIGURE 35.** Schematic of the multislice method with three slices.

The model structure is sliced into thin layers. For each layer a projected potential is determined using all atoms in the corresponding layer. The wavefunction at the entrance plane of each slice is modified by the projected potential resulting in a new wavefunction after this slice. This realspace wavefunction is transformed in Fourier space and multiplied there with a propagator function. This accounts for additional phase shifts for electrons inclined to the incident beam direction while they move through each slice. The result is back-Fourier transformed and yields a new real-space wavefunction to be transmitted through the next slice. The idea of using a projected potential for each slice instead of the real 3D potential requires thin slices to be chosen. The quality of the contrast simulations increases with decreasing thickness of the slices.

### 4.1.4. Imaging of Nanostructures with HRTEM

With high-resolution transmission electron micrographs particle sizes and shapes of nanoparticles can be determined. Additionally, defects like stacking faults and surface layers are typically studied. Besides contrast inversion as shown in Fig. 32, a major problem is image delocalization. This effect, visible in Fig. 36, depends strongly on defocus and is especially pronounced for highly coherent electrons from a field emission source.

The delocalization of information is proportional to the gradient of $\chi$(Eq. 17) with respect to a reciprocal vector $\underline{g} = [u, v]$: $D = |\nabla\chi(g)| = |C_s\lambda^3 g^3 + 2\Delta f\lambda^3 g|$ (Rosenauer, 2003). The energy spread of a thermionic electron source dampens the contrast transfer function at space frequencies above the point resolution. For field emission sources, however, the information limit is at much higher values for the space frequency, as the dampening effect on the contrast transfer function is smaller. However, as delocalization is proportional to

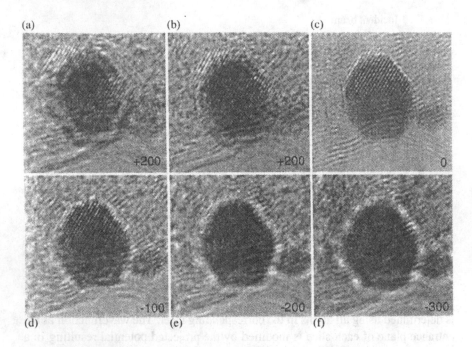

**FIGURE 36.** The effect of defocus on the measurement of the size of a gold particle. The defocus values are given in nm with a dark fringe surrounding the particle for overfocus (positive defocus) and a bright fringe for underfocus. (From Zandbergen and Træholt, Handbook of Microscopy: Application, 708, (1997) reused with permission of VCH, Weinheim, Germany).

the gradient of $\chi$, micrographs of nanoparticles acquired in a transmission electron microscope with field emission source are especially affected. The effects of defocus and delocalization have to be taken into account for the determination of particle sizes, as interfaces appear to move in the images if the focus is changed (Fig. 36).

The structures of carbon nanotubes were first identified by Iijima (1991) using HRTEM. These materials offer new prospects for electronic, catalytic and structural applications. Therefore, TEM plays a major role in the study of nanotubes (see, e.g., Zhou, 2001) and their relations to other materials attached to be nanotubes (Bera *et al.*, 2004). TEM is used to measure the number of concentric graphitic walls, termination of layers, defects in the walls, and in the caps of the nanotubes (Fig. 37). These studies will help to improve production processes for defect-free carbon nanotubes, which are highly desirable for many novel applications. The TEM studies also reveal other carbonaceous materials (Bera *et al.*, 2004) which are simultaneously produced in arc-discharge processes (Fig. 38).

Single-walled carbon nanotubes can have either metallic or semiconduction electronic properties depending on diameter and helicity. High-resolution TEM does not provide appropriate information on the helicity as the hexagons in the wall can not be imaged directly. However, electron diffraction and corresponding

**FIGURE 37.** Caps of multiwall carbon nanotubes. Left: Surface contamination with amorphous carbon on a nanotube. Right: A Pd nanoparticle on top of a carbon nanotube. Even though the Pd crystallite appears in the region of the empty core of the nanotube, sample tilt reveals that the Pd nanoparticles are attached to the surface. (Samples courtesy of D.Bera and S. Seal, AMPAC, University of Central Florida.)

simulations have provided the necessary evidence to identify the way how the nanotubes are built (Cowley *et al.*, 1997; Qin, 2001).

Radiation damage can be significant for some nanomaterials. Especially for materials with low atomic number and low melting temperatures, electron irradiation can cause significant structural changes. Figure 39 shows an example

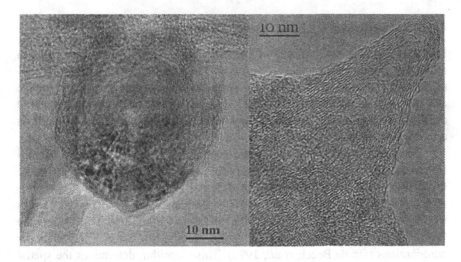

**FIGURE 38.** Nano-onions and nanohorns produced as by-products in an arc-discharge process. (Samples courtesy of D. Bera and S. Seal, AMPAC, University of Central Florida.)

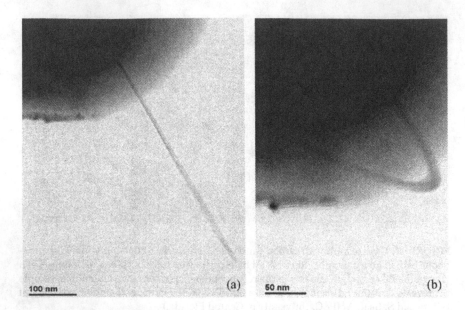

**FIGURE 39.** (a) The tip of a nanotube (first micrograph acquired) appears blurred as it vibrates in the electron beam. (b) The bending of the same carbon nanotube is induced by a few seconds of electron irradiation at 300 kV. (Sample courtesy of Lee Chow, Physics Department, University of Central Florida.)

of a carbon nanotube where bending is caused by only a few seconds of electron irradiation.

### 4.1.5. Quantitative High-resolution TEM

In quantitative high-resolution TEM the amplitude and phase of the wave function at the exit plane of the sample is measured. Three different methods can be employed to retrieve this information: focal series reconstruction, off-axis holography, and tilt-series reconstruction. The exit plane wave function, however, cannot be directly correlated to the structure in the sample. Simulation and iteration methods are necessary to retrieve this structure.

Focal series reconstruction eliminates the effects of defocus and delocalization and allows the retrieval of amplitude and phase information up to the information limit of the microscope. A series of 20-30 micrographs of the same region is acquired with focal increments in the range of 3 nm. The focal series reconstruction uses two algorithms: The parabolic method (PAM) considers the linear contributions to the image formation process and corrects for nonlinear contributions (Op de Beeck *et al.*, 1996). This algorithm determines the spatial frequencies of the exit plane wave affected by the transmission cross coefficient for scattering from $\psi_0$ to $\psi_k$. In the second step of the reconstruction this exit

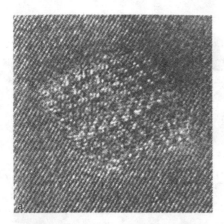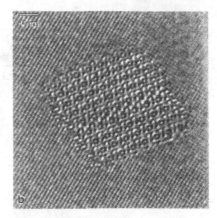

FIGURE 40. (a) HRTEM image of a Mg-Si precipitate in an Al matrix acquired at −70 nm defocus in a 300 kV microscope. (b) Phase image of the exit plane wave function obtained from focal series reconstruction (From Zandbergen and Træholt, Handbook of Microscopy: Application, 708, (1997) reused with permission of VCH, Weinheim, Germany).

plane wave function is used as input for the MAL algorithm (maximum likelihood method) which accounts for nonlinear image contributions of the transmission cross coefficient (Coene *et al.*, 1996).

The phase image in Fig. 40(b) retrieved from a focal series of 20 micrographs shows a sharp interface between the precipitate and the Al-matrix in a Al-Mg-Si alloy (Zandbergen *et al.*, 1997). Also, a defect in the stacking sequence is visible in the center of the precipitate. The contrast in an individual HRTEM micrograph (Fig. 40(a)), however, does not reveal the exact location of the interface.

For many electronic devices it is desirable to measure the composition of each atomic column individually. In multilayered systems interdiffusion may influence the composition across the interfaces. A method applied by Schwander *et al.* (1993), QUANTITEM (quantitative analysis of the information from transmission electron micrographs), allows for the determination of thickness (Rosenauer, 2003) and/or composition of individual atomic columns in the projected image of a crystalline material.

The QUANTITEM method is essentially a black-box approach that does not consider the details of the image formation. The micrograph is split in different image unit cells. All image unit cells of a micrograph can be constructed from a set of eigenimages that represent typical image features. Depending on the number of Bloch waves responsible for the formation of the image a limited set of eigenimages can be used. Each image unit cell can now be represented by a linear combination of these eigenimages (Fig. 41). The amplitudes of each eigenimage forming an image unit cell can be represented by a reference vector. The concentration for each image unit cell can be determined, if the imaging conditions are chosen in a way that the reference vector changes linearly with changes in composition. The

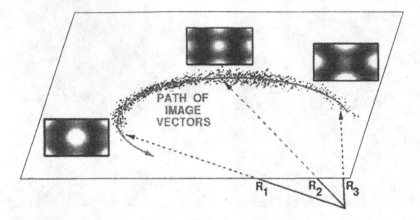

**FIGURE 41.** Image unit cells in quantitative lattice imaging are represented by a linear combination of eigenimages. The components of the reference vector are given by the amplitudes of each eigenimage. Different eigenimages represent different compositions (Reprinted figure with permission from Schwander et al., 1993, Phys. Rev. Lett. 71, 4150-4153. Copyright (1993) by the American Physical Society).

data evaluation in Fig. 42(b) shows the result of a QUANTITEM analysis for a layered Si-SiGe-Si nanostructure (Fig. 42(a)).

Strain analysis for thin films and for quantum dots is important for an understanding of the electronic properties of these devices. High-resolution TEM yields direct images of the two-dimensional atomic positions averaged along the beam direction as shown in the left inset of Fig. 43, and the same methods to identify the lattice sites are used to obtain a reference grid for the image unit cells

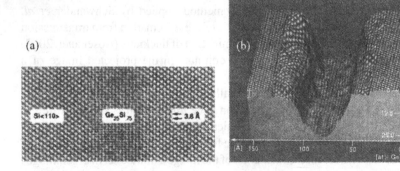

**FIGURE 42.** (a) Experimental HRTEM micrograph along [110] of a Si/SiGe/Si layered structure. (b) Interpolation of the sample thickness (inset) across the SiGe layer is used in combination with the QUANTITEM method to determine the Ge concentration in each atomic column (Reprinted figure with permission from Schwander et al., 1993, Phys. Rev. Lett. 71, 4150-4153. Copyright (1993) by the American Physical Society).

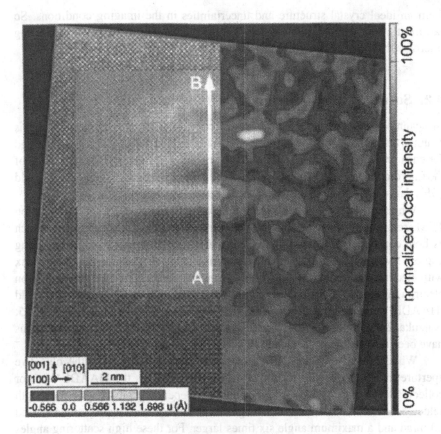

**FIGURE 43.** HRTEM image of an (In,Ga) As layer (bottom) and a Ga(Sb, As) layer on top. Left inset: map of the displacement field on the right side, right inset: normalized local intensity of the 002 reflections representative of composition (Reprinted from Journal of Alloys and Compounds, 382, Neumann et al., Quantitative TEM analysis of quantum structures, 2-9, 2004, with permission from Elsevier).

(Hÿtch *et al.*, 1998). The small deviations of the real lattice positions from the positions of the reference grid can be analyzed to get displacement vectors for each image unit cell (Rosenauer, 2003). Additionally, in special cases with homogeneous sample thickness, the amplitude of the intensity modulations yields chemical information (normalized local intensity) as shown on the right of Fig. 43.

The quantitative analysis of HRTEM images is based on contrast simulations. The TEM contrast from a "supercell" with a model structure including the defect structure is simulated and compared with the real micrograph or with the amplitude and phase of the exit plane wave function. Fitting algorithms are used to iteratively improve the structural models (Möbus and Rühle, 1994; Hofmann and Ernst, 1994). Those contrast simulations and the fitting procedures consider the effects of noise (electron beam shot noise, surface contamination and radiation damage), deviations from an ideal crystal structure and uncertainties in the imaging conditions. So far, most of these quantitative structural refinements have been performed using micrographs from special grain boundaries.

## 4.2. Scanning Transmission Electron Microscopy

In scanning transmission electron microscopy (STEM) the electron probe is scanned across the sample using deflection coils and the detector signal is recorded for each probe position. The STEM detectors are used in the diffraction mode of the microscope to collect specific parts of the diffraction pattern on the bright-field (BF), dark-field (DF) and the HAADF detector (Fig. 44(a)).

While conventional bright-field and dark-field TEM as well as high-resolution TEM are strongly affected by multiple scattering, dynamical scattering effects such as bent contours can be minimized using electrons scattered to high scattering angles. Two methods allow for a quantitative study of precipitates in a matrix with electrons scattered to high scattering angles: conical dark-field transmission electron microscopy (Hattenhauer *et al.*, 1993) and high-angle annular dark-field (HAADF) scanning transmission electron microscopy (Hillyard and Silcox, 1995; Ishizuka, 2002; Shiojiri, 2004). Methods for HAADF-STEM contrast simulations have been developed by Kirkland (1998) and Erni *et al.* (2003b).

While the BF and the DF detectors can be used similar to the objective aperture in direct imaging to utilize diffraction contrast, the HAADF detector collects electrons scattered to higher angles. Depending on the camera length selected electrons are detected in a range with a minimum collection angle around 50 mrad and a maximum angle six times larger. For these high scattering angles the scattered signal is in a first approximation proportional to the atomic number squared ($\alpha = 2$). The intensity in an atomic column is given by

$$I = I_{bg} + I_c + \beta\left(n A Z_A^\alpha + n_B Z_B^\alpha\right). \tag{20}$$

While $I_{bg}$ and $I_c$ describe the detector signal without electron beam, and the contribution from a contamination layer, respectively, $n_A$ and $n_B$ are the number of atoms of the two species A and B in each column, and $Z_A$ and $Z_B$ are their corresponding atomic numbers, while $\beta$ is a contrast factor. The contrast factors $\beta$, $I_{bg}$ and $I_c$ depend on the contrast brightness setting of the detector, on the electron probe intensity, and on the collection angles. For ideal Rutherford scattering the exponent $\alpha$ in Eq. (20) is two, but in reality high-angle scattering does not only occur at the nuclei of the atoms but also at the electrostatic potential induced by the bound electrons of each atom. Therefore, the actual elastic scattering intensity is lower than expected for ideal Rutherford scattering (Fig. 44(b)) and the exponent $\alpha$ is smaller than two. This atomic number contrast can be used to quantitatively determine local compositions for binary alloys with significant differences in the atomic number $Z$ of the elements.

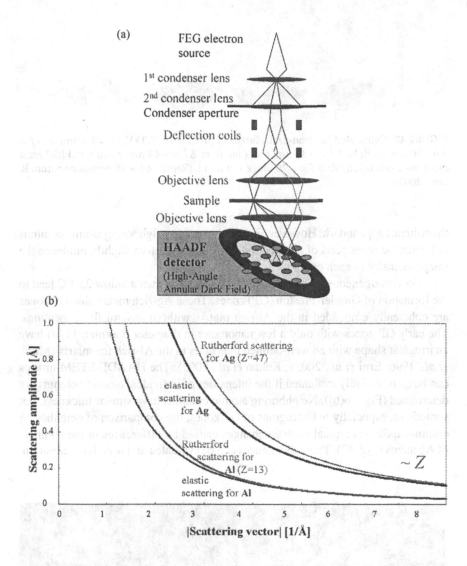

**FIGURE 44.** (a) Schematic of the imaging process in scanning transmission electron microscopy with an HAADF detector. (b) Schematic of the angular dependence of the electron scattering factors for Al and for Ag. (Reprinted with permission from R. Erni, 2003.)

## 4.2.1. Quantitative High-resolution HAADF-STEM

The diameter of the electron probe is determined by the contrast transfer function (Eq. (17)) and by the beam semi-convergence $\varepsilon$ which depends on the condenser aperture diameter and on the spot size. If the electron probe (Fig. 45) is small enough individual atomic columns can be distinguished. In high-resolution HAADF-STEM images of atomic columns appear bright while the regions between

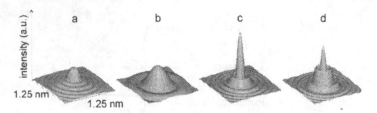

**FIGURE 45.** Calculated electron-probe intensity profiles for 300 kV, $C_s = 1.2$ mm. a: $\Delta f = 0$ nm, $\varepsilon = 9$ mrad; b: $\Delta f = -48$ nm, $\varepsilon = 3$ mrad; c: $\Delta f = -48$ nm, $\varepsilon = 9$ mrad (Scherzer incoherent condition); d: $\Delta f = -70$ nm, $\varepsilon = 9$ mrad. (Reprinted with permission from R. Erni, 2003.)

the columns appear dark. However, Fig. 45 shows that neighboring atomic columns influence the outer parts of the electron probe and therefore slightly influence the image intensity in each column.

For homogeneous Al-rich Al-Ag alloys heat treatments below 250°C lead to the formation of Guinier-Preston (GP) zones. These Ag-rich metastable GP zones are coherently embedded in the Al-rich matrix without any misfit dislocations. The early GP zones with only a few nanometer in diameter (Guinier, 1949) have an irregular shape with no well-defined interfaces to the Al-rich fcc matrix (Malik *et al.*, 1996; Erni *et al.*, 2003a; Konno *et al.*, 2005). The HAADF-STEM images can be quantitatively evaluated if the intensities of individual atomic columns are determined (Fig. 46(b)). Neighboring atomic columns have similar thickness, but sometimes, especially in the regions of GP zones, the comparison of neighboring columns indicates typical contrast changes caused by differences in the numbers of Ag atoms (Fig. 47). These differences can be evaluated and used for calibration.

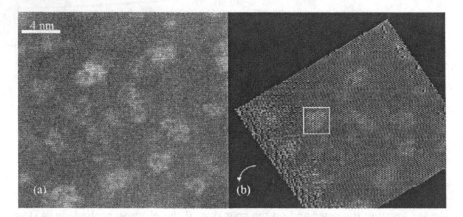

**FIGURE 46.** (a) High-resolution HAADF-STEM images of GP zones in Al-Ag. (b) Rotated micrograph with atomic column intensities extracted.

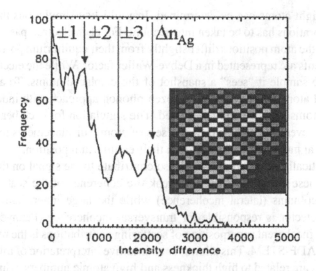

**FIGURE 47.** The inset shows the data from the GP zones marked in Fig. 46(b). The histogram of intensity differences between neighboring atomic columns reveals maxima indicating a difference of one, two or three Ag atoms when comparing these columns.

Image simulations are necessary to determine the exponent $\alpha$ in Eq. (20). These simulations are especially time consuming as the multislice method has to be applied for each electron probe on the sample separately. For each pixel in a STEM image, a full multislice simulation has to be performed (Fig. 48(a)). The multislice simulation can however be limited to regions where the convergent electron probe has significant intensity. With these two calibrations for HAADF-STEM the micrographs can be directly evaluated and the number of heavy atoms in a

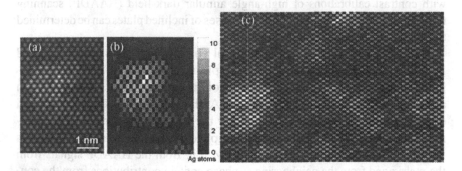

**FIGURE 48.** (a) HAADF-STEM contrast simulation using a model crystal from Malik *et al.*, 1996. (Reprinted from Ultramicroscopy 94, Erni et al., Quant. char. of chem. inhom. using Z-cont. STEM, 125-133, (2003), with permission from Elsevier). (b) Number of Ag atoms in each atomic column for the model crystal and (c) from the experimental HAADF image area evaluated.

matrix of light atoms can be determined. In multislice simulations the effect of thermal vibrations has to be taken into account. For each electron passing through the sample the atom positions differ slightly from their equilibrium position. These displacements are represented in a Debye-Waller factor. While the electron travels through the sample it "sees" a snapshot of the displaced atoms. To account for the thermal atomic displacements a frozen phonon approach with independently vibrating atoms (Einstein model) is used. The simulation for each beam position is repeated several times with different sets of atomic displacements representing the statistical motion of the atoms from their equilibrium position.

Inelastically scattered electrons also contribute to the signal on the HAADF detector. These electrons efficiently break the coherence of the scattered waves along the columns (lateral incoherence) while the large inner diameter of the HAADF detector is responsible for transversal incoherence (Pennycook *et al.*, 1997). The incoherent superposition of scattering contributions is the main advantage of HAADF-STEM. This allows for an intuitive interpretation of micrographs: bright areas are related to high thickness and high atomic numbers, while electron scattering is weaker in dark regions. Potentials for elastic Bragg scattering and for inelastic thermal diffuse scattering have to be used (Yamazaki *et al.*, 2004) in contrast simulations as described in the book of Kirkland (1998).

### 4.2.2. Quantitative Low-resolution HAADF-STEM

For later stages of the decomposition in Al-Ag alloys plates of the γ-phase form at the expense of the metastable GP zones (Howe *et al.*, 1987). These γ-plates with a composition of Ag-40 at.% Al have a hexagonal structure. The γ-plates appear bright in HAADF micrographs as they contain 60 at.% Ag, while the matrix is depleted of silver. The limited tilt angles in a transmission electron microscope often prevent to view all platelike precipitates edge-on, thus the measurement of the plate thickness by direct high-resolution imaging is impossible. However, with contrast calibrations of high-angle annular dark-field (HAADF) scanning transmission electron micrographs thicknesses of inclined plates can be determined as shown in Heinrich *et al.* (2006).

The calibration of the HAADF signal employs two micrographs acquired from different orientations of the same sample area (Fig. 49). In one of the micrographs a plate (continuous through the whole plate thickness) is shown edge-on as in Fig. 49(a) next to the arrow. The HAADF micrograph of this plate imaged edge-on yields the HAADF intensities for the γ-phase and for the Al-rich matrix independently, if the $Ag_2Al$ plates completely penetrate the sample thickness from the top to the bottom of the thin TEM foil. Both the HAADF signals from the matrix and from the neighboring γ-phase contain contributions from the contamination layer ($I_c$), and they are significantly influenced by the contrast and brightness settings of the HAADF detector. Therefore, the HAADF-STEM calibrations and experiments have to be conducted at fixed contrast-brightness settings yielding a constant offset $I_{bg}$ of the HAADF intensity. The HAADF intensities $I_{Al}$

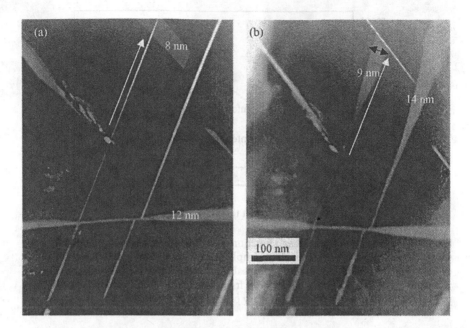

**FIGURE 49.** (a) HAADF micrograph of γ-plates parallel to the beam direction [211] (parallel to the white arrow). The thicknesses of the different γ-plates as determined from the contrasts in this micrograph are indicated in nm. (b): HAADF micrograph of the same plates viewed along [112].

for the Al-matrix and $I_\gamma$ for the γ-phase depend on the projected thickness of the sample $(t_{Al}, t_\gamma)$ and the materials contrast $(C_{Al}, C_\gamma)$ measured in counts per nanometer of the projected materials thickness (Fig. 49): $I_{Al} = I_{bg} + I_c + t_{Al}C_{Al}$ and $I_\gamma = I_{bg} + I_c + t_\gamma C_\gamma$ (Heinrich *et al.*, 2006).

For the calibration of the sample thickness $t$ a HAADF micrograph is taken with the electron beam inclined to the γ-plate. In a wedge-shaped sample area, the projection of the γ-plate forms a wedge of bright contrast in the HAADF micrograph (Fig. 49(b)). For known parameters of the sample geometry, i.e., plate normal, overall surface normal of the crystal and the two electron beam directions for the two HAADF micrographs, the sample thickness can be determined from the wedge angle or the width of the contrast as indicated in Fig. 49(b) by the double arrow.

The contamination layers on the top and the bottom surfaces of the TEM sample are assumed to be independent of the sample position. Scanning the small electron probe over the sample however causes an increase of the contamination layer by diffusion of mobile H, C, and O atoms on the surfaces. This effect is obvious if the electron beam is stopped for several seconds on one position. Then, the HAADF intensity can increase locally by sometimes more than 10% of the signal from the Al matrix. Therefore, the electron beam has to be deflected when

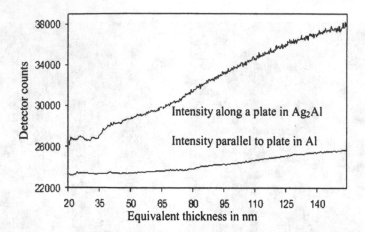

**FIGURE 50.** HAADF intensities as a function of sample thickness along the plate in Fig. 49(a). For this geometry the equivalent sample thickness represents 1.77 times the width of the wedge indicated in Fig. 49(b). The real sample thickness measured at the inclined $\gamma$-plate locally differs slightly (up to 10%) from this equivalent thickness as determined from the lines drawn to mark the boundary of contrast in Fig. 49(a).

the sample is not scanned. HAADF micrographs are recorded with scanning times in the range of 10 seconds to prevent any significant additional contamination by exposing each measuring position to the electron beam only for durations well below 1 ms. Orientation effects can also change the intensities measured with the high-angle annular detector. To minimize this effect, the smallest available camera length (80 mm) for a Tecnai F30 microscope operating at 300 kV was used.

In the case of Fig. 49(b) with the $\gamma$-plate having a $[-111]$ plate normal, and the imaging direction along the $[112]$ direction, the projected electron path through a $[-111]$ plate is 1.87 times the plate thickness. The sample thickness is calibrated measuring the width of the plate contrast. Along the white arrow in Fig. 49(b) the thickness increases almost linearly as the width of the plate contrast grows. From Fig. 49(b) with the plate viewed edge-on along the $[112]$ direction, the corresponding HAADF intensities within the same $\gamma$-plate and outside the plate are determined along the arrow in Fig. 49(a). The local sample thickness and the HAADF intensities for the edge-on micrograph are correlated in line profiles taken along the plate and outside the plate as shown in Fig. 50. The slope of the intensities as a function of sample thickness yields the contrasts (in units of counts per nanometer) for the $\gamma$-phase and for the Al-matrix for the specific settings of the microscope.

With this calibration, the thicknesses of all inclined plates in Fig. 49 are directly determined by measuring the increase in contrast at the edge of each inclined plate. The plate thicknesses determined by this method are indicated in the micrographs. Using the HAADF contrast calibration the thickness of the $\gamma$-plate marked by the arrow as measured from Fig. 49(b) is found to be $(9\pm1)$ nm in good agreement with the direct measurement.

Dynamical scattering effects are found not to be significant for these low-resolution HAADF-STEM micrographs. The direct measurement the sample thickness with HAADF-STEM may also be applied to other nanoparticles on a carbon film or to samples prepared by the focused ion-beam technique. This method provides three-dimensional information of the shape of these particles and on sample thicknesses.

While the HAADF signal is typically very noisy (only a few of all electrons are scattered to high angles), conventional HRTEM images usually have a good signal-to-noise ratio. The acquisition times for STEM images are typically several seconds to several tens of seconds. This is a possible source for sample drift during acquisition, which is more significant than for conventional HRTEM. Also, scan distortions influence the micrographs and lattice distortions from local strain fields cannot be evaluated in the STEM mode. The linear dependence of the HAADF signal on thickness and scattering strength of the atoms however yields directly interpretable data, whereas conventional HRTEM shows ambiguities in the determination of atomic column positions (defocus dependence) and in the determination of thicknesses (similar contrast for different thicknesses).

## 4.3. Electron Holography

Electron holography is a method to determine amplitude and phase of the electron wave after it has passed the sample. Electron holography was suggested by Gabor (1948) as a technique to avoid the effects of lens aberrations limiting the resolution. However, holographic methods were first introduced in light optics when coherent laser sources became available. In TEM a highly coherent field-emission electron source is available, and a "biprism," a charged Au-coated wire, is used for the acquisition of electron holograms (Fig. 51).

### 4.3.1. High-resolution Holography

The distance between interference fringes in a hologram can be adjusted by changing the potential of the biprism. The distances of the fringes have to be significantly smaller than the distances of the objects in the images. For high-resolution holography interference fringe distances correspond to about 0.5 Å distances in the sample, i.e., three times smaller than the details that are to be resolved. This provides enough distance between the autocorrelation and the sidebands in the Fourier image of a hologram (Fig. 52). The sideband of a hologram is selected and a back-Fourier transform yields amplitude and phase of the electron wave at the exit plane of the sample (see, e.g., Lehmann et al., 1999).

While the wavefunction for the reference wave remains constant, the sample modifies the wavefunction of the object wave to $A(x,y)\exp(i\Phi(x,y))$. In the overlap region of the hologram both waves interfere with a carrier frequency $q_c$ (determined by the distance of the fringes) and a contrast $\mu$ of the interference fringes. The

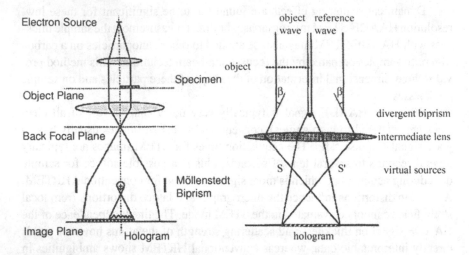

**FIGURE 51.** (a) Setup for high-resolution holography (Reprinted from Materials Characterization 42, Lehmann et al., Electr. hologr. at atomic dimension – present state, 249-263, (1999), with permission from Elsevier), (b) set-up for off-axis holography with negatively charged biprism and a Lorentz lens to measure electrical and magnetic fields (Reprinted from Ultramicroscopy 67, Beeli et al., Measure of reman. magn. of nanowires by TEM hologr., 143-151, (1997), with permission from Elsevier). A hologram is formed in the overlapping region of the object and the reference wave.

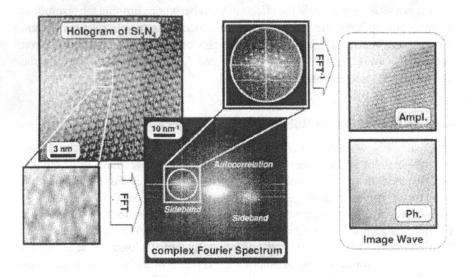

**FIGURE 52.** Procedure for the reconstruction of the exit plane wave function. A sideband from the Fourier transform of the hologram is selected to obtain amplitude and phase of the electron wavefunction (Reprinted from Materials Characterization 42, Lehmann et al., Electr. hologr. at atomic dimension – present state, 249–263, (1999), with permission from Elsevier).

intensity of a hologram is therefore given by

$$I_H(x, y) = 1 + A^2(x, y) + 2\mu A(x, y) \cos\left(2\pi \underline{q}_c \cdot \underline{r} + \Phi(x, y)\right) \qquad (21)$$

The Fourier transform of this intensity has three parts: the autocorrelation $F(1 + A^2(x, y))$ and the two sidebands with $\mu F(A(x, y) \exp(i\Phi(x, y)) \otimes d(q - q_c)$, where the symbol $\otimes$ denotes a convolution. After correction for coherent wave aberrations in the Fourier space by using the contrast transfer function $\exp(-i\chi(q))$, one of the sidebands (Fig. 52) in the Fourier pattern is selected and centered. The back-Fourier transform directly yields amplitude and phase of the electron wave. Voelkl *et al.* (1997), have introduced the software package HOLOWORKS for automatic sideband reconstruction.

Image distortions can cause a bending and stretching of the fringes in a hologram. Especially for low magnifications or if an energy filter is used for imaging, these distortions have to be considered. The finite size of the Au-coated quartz wire acting as biprism causes intensity modulations of the fringes in a hologram (Lichte *et al.*, 1996). These effects can be measured by acquiring an empty reference hologram from a hole in the sample. Procedures to correct holograms of real samples for these effects by using a reference hologram are described in, e.g., Rosenauer (2003).

High-resolution holography directly yields amplitude and phase of the electron wave in the exit plane of the sample. Image information can be extracted down to the information limit of the microscope if the interference fringe distances are small enough. The fringe contrast is another important parameter which is limited by the stability of the electron lenses and the electron source, and by the energy distribution for the electrons. For optimum resolution and small fringe distances of 0.5 Å the modulation amplitude of the fringes is still below 10% of the average intensity (Lehmann *et al.*, 1999). Additionally, the number of pixels on a CCD camera limits the field of view. Electron holography however suffers less from image delocalization than conventional high-resolution TEM. Compared with focal series reconstruction, holographic methods require less computational work in the data evaluation. The requirements on exposure time, sample drift and radiation damage are also less demanding for holograms with typical acquisition times of about 4 s in comparison to focal series reconstruction and STEM.

### 4.3.2. Holography with a Biprism and a Lorentz Lens

In Lorentz microscopy the objective lens is switched off and only a weaker Lorentz lens is used for imaging. The sample region of the microscope is free of larger magnetic fields induced by the electron lenses. The resolution of the Lorentz lens is typically limited to about 2 nm, but this is sufficient to map the phase shifts of electrons caused by electrical and magnetic fields of the sample and by the mean inner potential of the material.

Figure 53(a) shows a conventional TEM image of a high-coercivity magnetic force microscope (MFM) tip. The CoPt layer deposited on the tip is not anymore

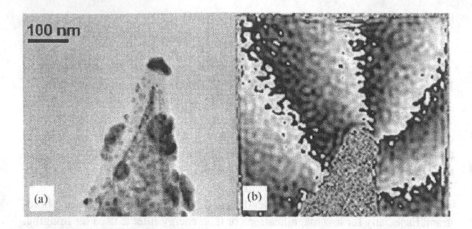

**FIGURE 53.** (a) Conventional transmission electron micrograph of the tip of a magnetic force microscope. (b) Distribution of the magnetic field around the tip. (Reprinted from J. of Magnetism and Magnetic Materials 272–276, Signoretti et al., Electron hologr. quant. Measure on MFM probes, pp. 2167-2168, (2004), with permission from Elsevier).

continuous and several CoPt crystallites are visible. The magnetic field around the tip displaces the electron beam and causes an additional phase shift with respect to the reference beam as shown in Fig. 53(b). The actual stray field from the tip depends on the arrangement of the CoPt nanoparticles on the surface. Different MFM tips reveal different field distributions. The phase image in Fig. 53(b) cannot be directly correlated with the field distribution around the tip. The micrograph is a two-dimensional map of the phase shift, but the stray field around the tip is three-dimensional. However, simulations of the phase shift induced by three-dimensional field distribution models have shown a reasonable match with the experimental data (Signoretti *et al.*, 2004). This holographic method to measure magnetic field distributions has also been applied on different magnetic nanoparticles (see, e.g., Signoretti *et al.*, 2003).

The mean inner potential changes the wavelength of the electrons as they pass through the sample. If Bragg reflections are only weekly excited, any additional phase shift by dynamic diffraction can be neglected and the phase shift of the electron beam increases linearly with increasing mean inner potential and increasing sample thickness. In contrast to TEM samples from bulk materials, where the local sample thickness cannot be directly measured in TEM, nanorods with rotational symmetry are ideal samples with known local thickness. Therefore, the mean inner potential can be directly determined from the phase shift of the electron wave in a hologram (Müller *et al.*, 2005) as shown in Fig. 54.

The phase shift of the electron wave is proportional to the electrostatic potential and the sample thickness. The correlation between phase shift and the damping of the electron wave by inelastic scattering results in a spiral form of the complex wavefunction as a function of sample thickness (Fig. 55(a)).

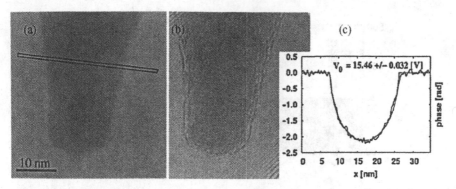

**FIGURE 54.** (a) Phase image of the hologram (b) from a ZnO nanorod. (c) Line scan of the phase and fitted curve for a mean inner potential of 15.46 eV (Reprinted with permission from E. Müller et al., Applied Physics Letters, 86, 154108 (2005). Copyright [2005], American Institute of Physics).

Holography also yields information on changes in the electrostatic potential within a sample. With increasing sample thickness, the holographic fringe contrast decreases due to incoherent scattering while the phase shift caused by the electrostatic potential increases with thickness. Rau *et al.* (1999) found an optimum thickness range of 200-400 nm for measurements of the electrostatic potential.

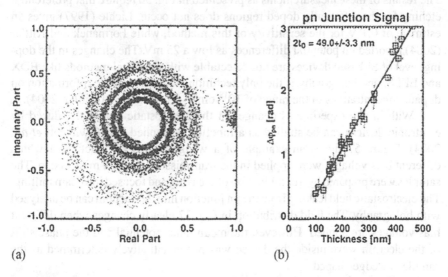

**FIGURE 55.** (a) Argand plot (plot of the distribution of amplitude and phase for a whole hologram) of the complex wave from a hologram of a wedge-shaped sample with an abrupt *pn*-junction. The damping of the spiral is caused by inelastic scattering. (b) Relative change of the phase across the *pn*-junction for different sample thicknesses. The extrapolation to zero phase shift yields the thickness of the dead layers at both surfaces (Reprinted with permission from Rau et al., 1999, Phys. Rev. Lett. 82, 2614-2617. Copyright (1999) by the American Physical Society).

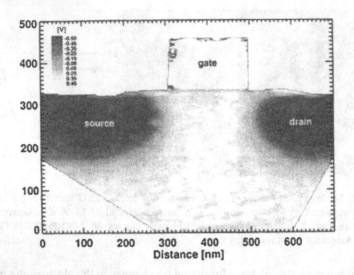

**FIGURE 56.** Map of the potential in the depletion region of a *p*-MOS transistor. The electrostatic potential changes by 0.9 ± 0.12 V across the *pn* junctions. (Reprinted with permission from Rau et al., 1999, Phys. Rev. Lett. 82, 2614-2617. Copyright (1999) by the American Physical Society).

The results of these measurements as presented in Fig. 56 require that preferential etching for the differently doped regions does not occur. Lichte (1997) gives an estimate of 0.1 V for the sensitivity of this method, while Formánek and Kittler (2004) reported on potential differences as low a 25 mV. The changes in the doping level of Si-based devices are not detectable with analytical methods like EDX and EELS, and holography is the only technique that can provide information on dopant concentrations in the range of $10^{18}$ cm$^{-3}$ (Formánek and Kittler, 2004).

With in situ experiments changes in the electrostatic potential induced by electronic devices can be studied as a function of applied bias (Twitchett *et al.*, 2004). Figure 57 shows an example of a wedge-shaped *pn* junction at which different bias voltages were applied in the transmission electron microscope. The samples were prepared by a combination of cleaving and focused ion beam milling. The electrostatic field leaking from the pn junction into the vacuum can be analyzed with holography. The field distribution in Fig. 57 clearly changes when different bias voltages are applied. However, the mean inner potential and the phase shift of the electron wave inside the device was not quantitatively determined as the sample is wedge-shaped.

## 5. ANALYTICAL ELECTRON MICROSCOPY

Qualitative and quantitative information on the local chemical composition of a material is obtained by analytical electron microscopy (e.g., Kohler-Redlich and

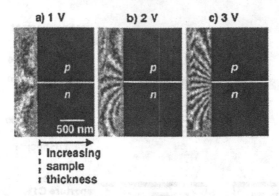

**FIGURE 57.** Phase image of the region outside a *pn* junction for different bias voltages (From Twitchett et al., Journal of Microscopy, 214, Off-axis electron holography of electrostatic potentials in unbiased and reverse biased focused ion beam milled semiconductor devices, (2004), reused with permission from Blackwell Publishing).

Mayer, 2003). For EDX analysis the goniometer with the sample has to be tilted about 15° towards the EDX detector as shown in Fig. 14(a). Otherwise, the X-rays generated in the irradiated volume are reabsorbed in the sample before they can reach the detector.

For EELS analysis it is best to avoid tilting the sample to maintain a short electron path through the material. However, for some sample geometries, e.g., for cases where concentration variations across interfaces are studied, sample tilt is necessary to warrant that the electron beam passes the material parallel to the interface. This provides highest resolution for line scans across an interface. However, excitation of X-rays also occurs outside the volume, which is irradiated with electrons. Secondary electrons and X-rays are emitted from the irradiated area and they hit other regions of the sample and possibly nearby parts of the electron microscope. The secondary electrons and X-rays can excite X-rays in other parts of the sample. Therefore, the lateral resolution in EDX analysis is not the same as the electron beam diameter on the sample. In STEM the electron probe size and intensity as well as the special geometry of the sample within the microscope determine the lateral resolution. The EDX signal always contains contributions from neighboring areas not directly hit by the incident electron beam. If nanoparticles on an amorphous carbon film deposited on a Cu grid are studied, element-specific X-rays from copper are detected even though the particles may contain no Cu.

While electron probe sizes as small as 0.2 nm can be obtained in the STEM mode of most transmission electron microscopes with field emission source, the low intensity of the electron beam leads to low count rates especially for EDX analysis and for EELS spectroscopy at high energy losses. Surface contamination during acquisition of spectra using a small STEM probe may cause additional

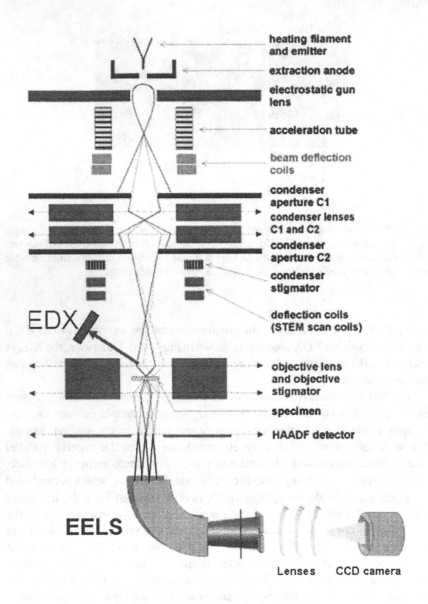

heating filament and emitter

extraction anode

electrostatic gun lens

acceleration tube

beam deflection coils

condenser aperture C1

condenser lenses C1 and C2

condenser aperture C2

condenser stigmator

deflection coils (STEM scan coils)

EDX

objective lens and objective stigmator

specimen

HAADF detector

EELS

Lenses　　CCD camera

**FIGURE 58.** Schematics of a field emission transmission electron microscope with STEM unit, HAADF detector, EDX detector and post-column electron energy loss spectrometer. All three signals (HAADF, EDX, EELS) can be acquired simultaneously. (Reused with permission from M. Terheggen, 2003.)

problems for the analysis. When using the HAADF detector in the STEM mode, Z contrast micrographs can be acquired and EDX as well as EELS spectra can be acquired simultaneously at specific points of the sample (Fig. 58). Analytical techniques in TEM are:

   (a) Point measurements for one small illuminated area of the sample.

   (b) *Line scans*: The small electron probe is sequentially positioned on points along a line. For each electron probe position an EDX spectrum or an EELS spectrum is acquired.

   (c) *Maps*: The small electron probe is sequentially positioned on points on a two-dimen-sional grid. Line scans and maps require the use of the scanning mode for the electron beam.

   (d) Energy-filtered transmission electron microscopy (EFTEM) is used to obtain maps of the distribution of elements in the sample (see, e.g., Kothleitner and Hofer, 2003). In the imaging mode of the microscope an electron beam illuminates a larger area of the sample and, for the electron energy filter element, specific energy windows are selected. Only electrons within such an energy window are transmitted on a CCD camera to form an image.

## 5.1. Electron Energy-loss Spectroscopy

### 5.1.1. The Components of an Electron Energy-loss Spectrum

A typical electron energy-loss spectrum is shown in Fig. 17. Each channel in this energy-loss spectrum has a width of 1 eV. The "zero-loss" peak contains the electrons that have passed the sample without energy loss or with energy changes too small to be detected from phonon excitations (a few 10 meV). The next peak at and energy loss of about 20 eV in Fig. 17 is caused by excitation of plasmons. The plasmon energies between 10 and 40 eV are characteristic for the materials studied and depend on the density $\rho$ of the conduction electrons which are collectively excited: $E_{\text{Plasmon}} = \hbar e \sqrt{4\pi\rho/m_e}$.

For thicker samples an incident electron may excite several plasmons and multiple peaks can occur (Fultz and Howe, 2002). Plasmon excitation can be used to measure the sample thickness $z$ from the ratio of the plasmon intensity $I_P$ and the zero-loss intensity $I_0$: $z = \bar{\lambda} \ln(I_P/I_0)$ (Malis *et al.*, 1988). However, the mean-free inelastic path $\bar{\lambda}$ has to be measured or calculated by an independent method (Egerton, 1996).

For higher energy losses the signal exponentially decreases with increasing energy loss. These energy losses are caused by multiple plasmon excitations, but predominantly by acceleration and deceleration of the incident electrons in the crystal potential of the material. On top of this continuous bremsstrahlung spectrum element specific absorption edges are found. While plasmon excitations do not contain much useful information on the atoms in the sample, these absorption edges occur at element specific energy losses and can be used for a quantitative analysis of the composition. The incident electrons can excite an electron bound to a specific core level of an atom in the sample. An electron in a bound state of an atom can only take over energy from the incident electron if the energy transfer is large enough to reach a free energy level above the Fermi energy of the material. Therefore, element specific absorption edges are observed in EELS. For the example in Fig. 15(b) core losses with absorption edges at 284 eV, 528 eV and 884 eV are found. They are specific for the C-K edge, the O-K edge and the Ce-M edge, respectively. The Ce-M edge contains two high-intensity (white) lines, which are caused by different energy levels within the M shell of the Ce atom.

The energy-loss near-edge structure (ELNES) contains information on the density and energy of unoccupied energy levels above the Fermi level of a material. The EELS data can be used to identify the oxidation state of metal atoms in a material (Disko, 1992). The extended energy-loss fine structure (EXELFS) yields data on the distances and types of atoms surrounding the central atom from which an excited electron is emitted. The excited electron from that atom obtained some additional kinetic energy from the incident electron and can be reflected at neighboring atoms. This leads to interference effects of the excited electron wave at the central atom and the scattering probability increases or decreases depending on the atomic distances and as a function of the wavelength and phase shift of the excited electron. Similar to EXAFS (extended X-ray absorption fine structure) an EXELFS spectrum can evaluated for characteristic distances of neighboring atoms (see, e.g., Fultz and Howe, 2002).

### 5.1.2. Quantitative Analysis of an Electron Energy-loss Spectrum

The integrated intensity of an absorption edge can be quantitatively evaluated. In a first step the bremsstrahlung background is extrapolated from low energy losses (pre-edge) to the post-edge region, which is influenced by the absorption edge (Fig. 59). After the background signal is subtracted, the integrated intensity of the edge for a certain energy window is determined.

For quantitative analysis the ratio of atomic concentrations $c_A$ and $c_B$ is determined using the ionization cross-sections $\sigma(\Delta, \beta)$ of the corresponding absorption edges:

$$\frac{c_A}{c_B} = \frac{I_A(\Delta)\sigma_B(\Delta, \beta)}{I_B(\Delta)\sigma_C(\Delta, \beta)} \tag{22}$$

The cross-sections in Eq. (22) depend on the energy of the incident electrons, the energy window used in the data evaluation and on the collection angle of electrons given by the camera length and the entrance aperture of the energy filter (see, e.g., Wang, 2001).

A series of EELS data can be acquired when the electron probe is moved across the sample while spectra are acquired for each scan position. Line scans and area scans (maps) can be acquired. However, for maps the acquisition time becomes critical as high-quality EELS data have to be obtained for each pixel in a map. For a $50 \times 50$ pixel map the acquisition time can be well above one hour depending on the energy range selected. The sequential acquisition of electron energy-loss spectra yields a three-dimensional data cube with the scan positions in the $x$- and $y$-axis and the energy-loss scale in the $z$ direction (Fig. 60). The energy

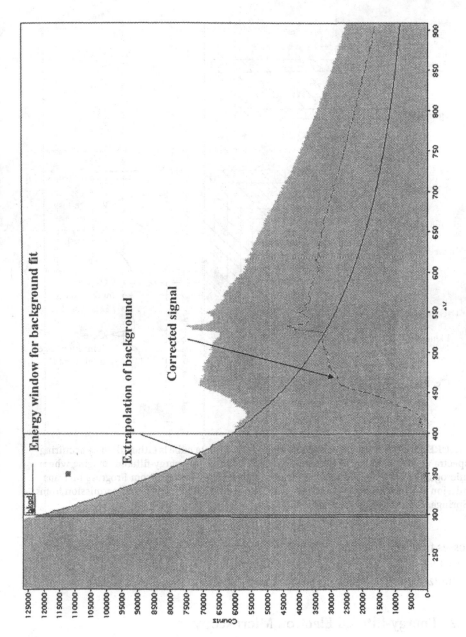

**FIGURE 59.** Background extrapolation and corrected signal from the Ti absorption edge.

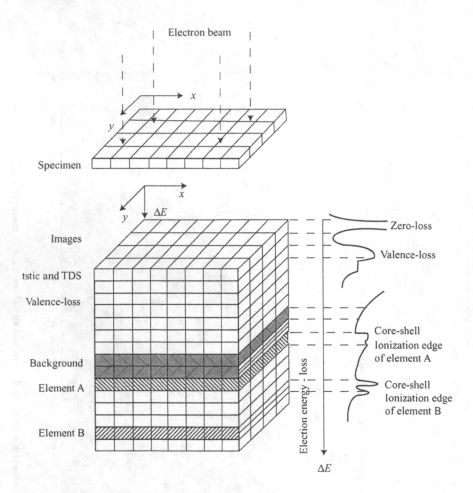

**FIGURE 60.** Schematics of energy-filtered TEM. The data cube is either filled by acquiring spectra at each position of the sample using STEM or by energy-filtered imaging where micrographs are acquired for a series of specific energy losses (From Progress in Transmission Electron Microscopy I, 2001; Z.L. Wang, pp. 113-159, with kind permission from Springer Science and Business Media).

losses from core excitation are highly localized and chemical as well as electronic structure information can be obtained from individual atomic columns using an optimized electron probe (Pennycook *et al.*, 1995).

## 5.2. Energy-filtered Electron Microscopy

In energy-filtered TEM (EFTEM) images are acquired with an energy-selective slit inserted in the in-column filter (from LEO) or the post-column filter (from GATAN) of a modern transmission electron microscope. This slit is located in the

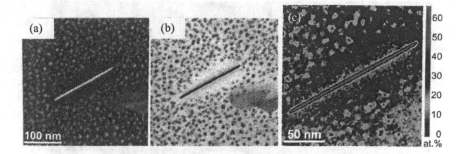

**FIGURE 61.** (a) Jump-ratio EFTEM image of the Ag distribution in an Al-3 at.% Ag alloy. (b) Jump-ratio image of the Al distribution. (c) Ag concentration as evaluated from the two jump-ratio micrographs. For contrast calibration the Al- 60 at.% Ag plate was used. (Reprinted with permission from R. Erni, 2003.)

energy-selective plane of the microscope. An energy filtered image of the sample with a selected energy loss and width of the energy window is obtained with the CCD camera. A series of these micrographs for different energy losses can be used to fill a data cube as shown in Fig. 60.

For a direct estimate of the concentrations of the different elements in a material two different methods are employed. Both aim at removing the influence of thickness variations on the compositional EFTEM maps. The first method is the two-window method which measures the jump ratio, i.e., the ratio between the post-edge signal and the pre-edge signal for each pixel in the micrographs (Figs. 61(a,b)). The second method is the three-window method to acquire elemental distribution maps. Two maps are acquired at energy losses below the absorption edge and one above. The two pre-edge micrographs are used for background extrapolation (Fig. 62).

In the three-window method the pre-edge signals are used for background subtraction. However, the EFTEM maps cannot be directly interpreted in a quantitative

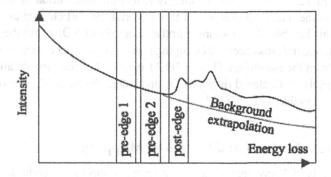

**FIGURE 62.** Schematics of the three-window method applied to each pixel of the three EFTEM images.

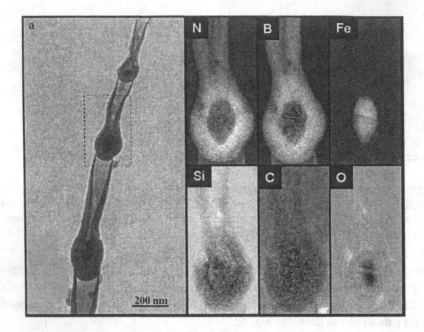

**FIGURE 63.** (a) Bright-field TEM image of BN nanobamboo structures. The EFTEM maps for the different elements are shown on the right-hand side. (From Fan *et al.*, 2005. Reprinted with permission from Blackwell Publishing.)

way. The model for background subtraction is only based on one (jump-ratio) or two data points in the pre-edge region obtained by integration over an energy window of 10 eV or more. Therefore, the model used for background extrapolation has to the treated with care. Jump-ratio images and EFTEM maps reveal where the different elements are located in a sample. Only with appropriate calibration as for Fig. 61(c), the local compositions can be obtained.

The EFTEM maps in Fig. 63 clearly reveal the distribution of nitrogen and of boron in the multiwalled layers of this material. Fe particles act as nucleation sites for this bamboolike structure. Another example of $VO_x$ nanotubes is shown in Fig. 64. Special cross-sectional sample preparation was necessary to obtain thin sections from the nanotubes. Only EFTEM reveals the structural details of these nanomaterials. Both closed rings of $VO_x$ and rolled sheets of $VO_x$ nanotubuli are directly observable with EFTEM.

## 5.3. X-ray Analysis and Chemical Mapping

The methods of X-ray analysis are frequently employed to obtain data on local compositions. The spectra contain element specific peaks with a FWHM of about 150 eV for 10 kV photons. For lower X-ray energies, the peak widths are smaller,

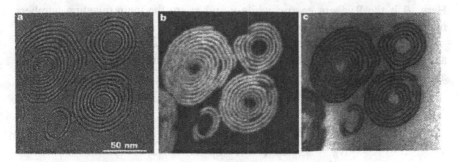

**FIGURE 64.** (a) Bright-field TEM image of $VO_x$ tubuli in cross-sectional view. (b) EFTEM maps from $VO_x$ at the $V$ edge (b) and at the C edge (From Krumeich et al., Zeitschrift fuer Anorganische Allgemeine Chemie, 626, 2211, (2000), reused with permission from Wiley Interscience).

but still peak overlap poses a severe problem for many materials (Fig. 65). The Moseley law gives approximate values for the relationship between the energy $E$ of the characteristic X-rays and the atomic number Z:

$$E = R(Z - 1)^2(1/m^2 - 1/n^2).$$

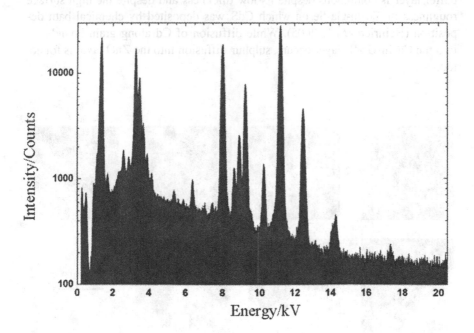

**FIGURE 65.** Energy dispersive X–ray spectrum of a Cu-In-Ga-Se thin-film solar cell. Contributions from neighboring phases lead to a variety of other elements detected in the sample.

Here, $R$ is the Rydberg energy, while $n$ and $m$ are the main quantum numbers involved in the transitions of the excited electrons.

For quantitative EDX analysis reference standards with similar composition are necessary. The internal standards used in the acquisition software yield only approximate data which should be considered carefully. For thin samples the ratio of the counts from the peaks in an EDX spectrum is given by:

$$\frac{N_A}{N_B} = \frac{1}{k_{AB}} \frac{c_A}{c_B} \left( 1 - \frac{\text{cosec}(\zeta)}{2\rho} (\mu_A - \mu_B) z \right) \tag{23}$$

In contrast to Eq. (22) for EELS evaluation the relative heights of the peaks in an EDX spectrum described by Eq. (23) depend on the sample thickness $z$. If spectra are quantitatively evaluated the thickness and the take-off angle $\zeta$ for the X-ray detector-sample orientation has to be known. Also, the Cliff-Lorimer factor $k_{AB}$, sample density and absorption coefficients have to be known (Cliff and Lorimer, 1972), especially for low-energy photons.

As for EELS, EDX point-measurements of spectra as well as linescans and maps employ the STEM mode. The distribution of the elements in a Cu-In-Ga-Se solar cell with CdS buffer layer, ZnO as transparent conducting oxide and a Ni and Al front contact is shown in Fig. 66. The 50 nm thin CdS buffer layer is continuous despite its low thickness and despite the high surface roughness of Cu-In-Ga-Se on which CdS was deposited by chemical bath deposition (Heinrich *et al.*, 2005). While diffusion of Cd along grain boundaries into the Cu-In-Ga-Se layer occurs, sulphur diffusion into the ZnO layer is found, too.

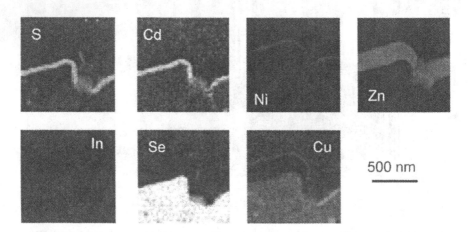

**FIGURE 66.** Elemental distribution maps of elements in a Cu-In-Ga-Se thin-film solar cell acquired with EDX.

## 6. NEW DEVELOPMENTS IN ELECTRON MICROSCOPY

A new generation of transmission electron microscopes is currently developed and installed at several research sites. Some of these instruments have been in use for a few years now and incorporate complex electron-optical systems to correct for lens aberrations, to improve the resolution of electron energy analyzers, and to obtain a highly monochromatic electron beam. These developments are based on new concepts of theoretical electron optics, but also on advancements in the production and stability control of the electron optical components.

Obtaining data on interatomic bonding is one of the goals of analytical electron microscopy. This requires an energy resolution better than 0.2 eV. The energy spread of electrons in current field-emission transmission electron microscopes is about 0.6 eV. Therefore, an electrostatic monochromator behind the electron gun was developed that significantly reduces the energy distribution (Kahl, 1999; Batson *et al.*, 2000). This monochromator allows only electrons through its aperture with energies up to 0.1 eV above and below the most probable energy in the electron energy distribution of the source. Therefore, a monochromator eliminates most of the electrons from the source. Two benefits are obtained when using a monochromator for the incident electrons.

The first is related to electron spectroscopic work, where a higher energy resolution yields more accurate data on electronic bonding states and on the energy levels of the unoccupied states. These spectroscopic data can be used to refine models for the electronic structures of materials. This will become especially important for nanomaterials where surface and interface effects may influence the electronic structure. However, for a high-energy resolution of spectroscopic data improved electron energy spectrometers have to be available that are capable of measuring the energy loss spectrum with a resolution better than 0.05 eV (Brink *et al.*, 2003). With an energy width of 0.18 eV of the monochromated beam Erni and Browning (2005) have studied the band structure of a high-temperature superconductor using valence electron energy-loss spectroscopy at a spatial resolution better than 1 nm.

The second advantage of a monochromator directly affects the information limit of modern microscopes. The chromatic aberration is one of the parameters influencing the information limit and the damping (Fig. 36(c)) of the contrast transfer function (Eq. (17)). With a highly monochromatic beam the contrast transfer can be extended to higher spatial frequencies, as $\Delta E$ is reduced. Simultaneously, efforts aim at improving the stability of the high voltage and the lens currents to reduce the effects of chromatic aberration (Kisielowski *et al.*, 2001). The improved monochromatic beam positively affects both conventional high-resolution micrographs as well as high-resolution STEM.

Another development addresses the correction of the spherical aberration $C_s$ of a transmission electron microscope (Haider *et al.*, 1998; Rose, 2003). A $C_s$ corrector is built of several hexapoles and round-lens transfer doublets. The second-order aberrations of the first hexapole have no rotational symmetry and are compensated by a second hexapole element (Rose, 1990). These two hexapoles induce a rotationally symmetrical third-order residual aberration, that is proportional to the square of the hexapole strength. This corrector system has a negative spherical aberration that can be used to compensate the spherical aberration of the objective lens. With such a system, the spherical aberration can be eliminated or adjusted to specific values (Kabius *et al.*, 2002). With the correction of the spherical aberration the point resolution of a microscope can be significantly improved. This is important for research on many materials where individual atomic columns along specific crystallographic directions cannot be resolved conventionally. The tunability of the spherical aberration also helps to improve the contrast for specific atomic distances.

Using a combination of $C_s$ and $C_c$ correction HRTEM images are directly interpretable up to a point resolution of 1 Å in a 200 kV microscope (Freitag *et al.*, 2005). For special defocus values image delocalization is reduced as the contrast transfer function does not show strong modulations. The correction of both spherical and chromatic aberration is also useful for scanning transmission electron microscopes to form a sub-angstrom electron probe (Krivanek *et al.* 2003).

With the advancement of the new filter and corrector systems spatial and energy resolution can be improved for medium-voltage microscopes. Also, systems with lower acceleration voltages can be used without sacrificing resolution. For lower voltages radiation damage is reduced. Therefore, more radiation-sensitive materials become available for high-resolution studies in a transmission electron microscope. The introduction of a $C_s$ corrector for low-resolution objective lenses with a wide pole piece gap will also be interesting for applications requiring high sample tilt angles in the goniometer. Currently, the tilt angles for high-resolution pole pieces are limited to 20°-40°. These higher tilt angles are especially desirable for 3D structural studies.

## 7. ACKNOWLEDGMENTS

The author wants to thank S. Senapati, H. Nukala, and A. Halbe from the University of Central Florida for their help in the preparation of this manuscript.

## REFERENCES

1. Al-Kassab, T., Wollenberger, H., Schmitz, G., Kirchheim, R., 2003, Tomography by atom probe field ion microscopy, in: *High-Resolution Imaging and Spectroscopy of Materials*, Eds. F. Ernst, M. Rühle, Springer Series in Materials Science, Vol. **50**, pp. 271-320, Springer-Verlag, Newy York.
2. Andersen, R., Klepeis, S. J., 1997, A new tripod polisher method for preparing TEM specimens of particles and fibers, Eds. R.M. Andersen, S.D. Walck, *Mater. Res. Soc. Proc.* **480**, Pittsburgh, 187-192.
3. Aroyo, M. I., Perez-Mato, J. M., Capillas, C., Kroumova, E., Ivantchev, S., Madariaga, G., Kirov, A., Wondratschek, H., 2006, Bilbao Crystallographic Server I: Databases and crystallographic computing programs, *Z. F. Kristallographie* **221**: 15-27. http://www.cryst.ehu.es/.

4. Barna, A., Pecz, B., Menyhard, M., 1998, Amorphization and surface morphology development at low-energy ion milling, *Ultramicroscopy* **70**: 161-171.
5. Batson P. E., Mook, H. W., Kruit, P., 2000, High brightness monochromator for STEM, in: *International Union of Microbeam Analysis 2000, Institute of Physics*, Bristol, U.K., Vol. 165, pp. 213-214.
6. Beeli, C., Doudin, B., Ansermet, J.-Ph., Stadelmann, P. A., 1997, Measurement of the remanent magnetization of single Co/Cu and Ni nanowires by off-axis TEM electron holography, *Ultramicroscopy* **67**: 143-151.
7. Bera, D., Kuiry, S. C., McCutchen, M., Seal, S., Heinrich, H., Slane, G. C., 2004, In situ synthesis of carbon nanotubes decorated with palladium nanoparticles using arc-discharge in solution method, *J. Appl. Phys.* **96**: 5152-5157.
8. Bera, D., Johnston, G., Heinrich, H., Seal, S., 2006, A parametric study on the synthesis of carbon nanotubes through arc-discharge in water, *Nanotechnology* **17**: 1722-1730.
9. Brink, H. A., Barfels, M. M. G., Burgner, R. P., Edwards, B. N., 2003, A sub-50 meV spectrometer and energy filter for use in combination with 200 kV monochromated (S)TEMs, *Ultramicroscopy* **96**: 367-384.
10. Cliff, G., Lorimer, G. W., 1975, The quantitative analysis of thin specimens, *J. Micr.* **103**: 203.
11. Coene, W. M. J., Thust, A., Op de Beeck, M., Van Dyck, D., 1996, Maximum-likelihood method for focus-variation image reconstruction in high-resolution transmission electron microscopy, *Ultramicroscopy* **64**: 109-135.
12. Crystal Lattice Structures, http://cst-www.nrl.navy.mil/lattice/index.html, U.S. Naval Research Laboratories.
13. De Graef, M., 2003, *Introduction to Conventional Transmission Electron Microscopy*, University Press, Cambridge.
14. Disko, M. M., 1992, Transmission electron energy-loss sprectrometry in materials science, Eds.: Disko, M.M., Ahn, C.C., Fultz, B., in: *Transmission Electron Energy Loss Sprectroscopy in Materials Science*, Minerals, Metals & Materials Society, Warrendale, PA.
15. Egerton, R. F., 1996, *Electron Energy-Loss Spectroscopy in the Electron Microscope*, Plenum Press, New York.
16. Erni, R., 2003, Atomic-scale analysis of precipitates in Al-3 at.% Ag: transmission electron microscopy. Dissertation ETH Zürich, Switzerland, No. 14988.
17. Erni, R., Browning, N. D., 2005, Valence electron energy-loss spectroscopy in monochromated scanning transmission electron microscopy, *Ultramicroscopy* **104**: 176-192.
18. Erni, R., Heinrich, H., Kostorz, G., 2003a, On the internal structure of Guinier-Preston zones in Al-3 at.% Ag, *Phil. Mag. Lett.* **83**: 599-609.
19. Erni, R., Heinrich, H., Kostorz, G., 2003b, Quantitative characterisation of chemical inhomogeneities in Al-Ag using high-resolution Z-contrast STEM, *Ultramicroscopy* **94**: 125-133.
20. Fan, Y., Wang, Y., Lou, J., Xu, S., Zhang, L., Heinrich, H., An, L., 2006, Formation of silicon-doped boron nitride bamboo structures via pyrolysis of a polymeric precursor, *J. Am. Ceram. Soc.*, **89**: 740-742.
21. Formánek, P., Kittler, M., 2004, Electron holography on silicon microstructures and its comparison to other microscopic techniques, *J. Phys.: Condens. Matter* **16**: S193-S200.
22. Freitag, B., Kujawa, S., Mul, P. M., Ringnalda, J., Tiemeijer, P. C., 2005, Breaking the spherical and chromatic aberration barrier in transmission electron microscopy, *Ultramicroscopy* **102**: 209-214.
23. Fultz, B., Howe, J.M., 2002, *Transmission Electron Microscopy and Diffractometry of Materials*, Springer-Verlag, Berlin.
24. Gabor, D., 1948, A new microscopic principle, *Nature* **161**: 777-778.
25. Gai, P. L., Boyes, E. D., 2003, *Electron Microscopy in Heterogeneous Catalysis*, Institute of Physics, London.

26. Goodhew, P. J., 1985, *Thin foil Preparation for Electron Microscopy*, Elsevier, Amsterdam.

27. Guinier, A., 1949, Precipitation dans les alliages, *Physica* **15**: 148-160.

28. Haider, M., Rose, H., Uhlemann, S., Kabius, B., Urban, K., 1998, Towards 0.1 nm resolution with the first spherically corrected transmission electron microscope, *J. Electr. Micr.* **47**: 395-405.

29. Hattenhauer, R., Schmitz, G., Wilbrandt, P. J., Haasen, P., 1993, Z-contrast TEM on precipitates in AlAg, *Phys. Stat. Sol. A* **137**: 429-434.

30. Heinrich, H., Kostorz, G., 2000, Bloch waves and weak-beam imaging of crystals, *J. Electron Micros.* **49**: 61-65.

31. Heinrich, H., Senapati, S., Kulkarni, S. R., Halbe, A. R., Rudmann, D., Tiwari, A. N., 2005, Defects and interfaces in Cu(In,Ga)Se2-based thin-film solar cells with and without Na diffusion barrier, *Mater. Res. Soc. Symp. Proc.* **865**: 137-142.

32. Henry, N. F., Lonsdale, K., Eds., 1969, *International Tables for X-Ray Crystallography*, Vol. 1, Kynoch Press, Birmingham.

33. Hillyard, S., Silcox, J., 1995, Detector geometry, thermal diffuse scattering and strain effects in ADF STEM imaging, *Ultramicroscopy* **58**: 6-17.

34. Hofmann, D., Ernst, F., 1994, Quantitative high-resolution transmission electron microscopy of the incoherent $\Sigma 3$ (211) boundary in Cu, *Ultramicroscopy* **53**: 205-221.

35. Howe, J. M., Dahmen, U., Gronski, R., 1987, Atomic mechanisms of precipitate plate growth, *Phil. Mag. A* **56**: 31-61.

36. Hÿtch, M.J., Snoeck, E., Kilaas, R., 1998, Quantitative measurements of displacement and strain fields from HREM micrographs, *Ultramicroscopy* **74**: 131-146.

37. Iijima, S., 1991, Helical microtubules of graphitic carbon, *Nature* **354**: 56-58.

38. Ishizuka, K., 2002, A practical approach for STEM image simulation based on the FFT multislice method, *Ultramicroscopy* **90**: 71-83.

39. Jouneau, P.-H., Stadelmann, P., 1998, *EMS On Line*, http://cimesg1.epfl.ch/CIOL/ems.html.

40. Kabius, B., Haider, M., Uhlemann, S., Schwan, E., Urban, K., Rose, H., 2002, First application of a spherical-aberration corrected transmission electron microscope in material science, *J. Elec. Micro.* **51**: 51-58.

41. Kahl, F., 1999, "Design eines Monochromators für Elektronenquellen." Ph.D. Thesis, Darmstadt University of Technology, Germany.

42. Kempshall, B. W., Sohn, Y. H., Jha, S. K., Laxman, S., Vanfleet, R. R., Kimmel, J., 2004, A microstructural observation of near-failure thermal barrier coating: a study by photostimulated luminescence spectroscopy and transmission electron microscopy, *Thin Solid Films* **466**: 128-136.

43. Kersker, M. M., 2001, The modern microscope today, Eds.: Zhang, X.-F., Zhang, Z., *Progress in Transmission Electron Microscopy 1*, Springer Series in Surface Sciences, Vol. **38**, Springer-Verlag, Berlin, pp. 1-34.

44. Keyse, R. E., Garratt-Reed, A. J., Goodhew, P. J., Lorimer, G. W., 1998, *Introduction to Scanning Transmission Electron Microscopy*, Microscopy Handbooks Vol. **39**, Springer, New York.

45. Kirkland, E. J., 1998, *Advanced Computing in Electron Microscopy*, Plenum Press, New York.

46. Kisielowski, C., Hetherington, C. J. D., Wang, Y. C., Kilaas, R., O'Keefe, M.A., Thust, A., 2001, Imaging columns of the light elements C, N, and O with sub-Angstrom resolution, *Ultramicroscopy* **89**: 243-263.

47. Kittel, C., 1995, *Introduction to Solid-State Physics*, Wiley, New York.

48. Kohler-Redlich, P., Mayer, J., 2003, Quantitative analytical transmission electron microscopy, Eds.: Ernst, F., Rühle, M., *High-Resolution Imaging and Spectrometry of Materials*, Springer, Berlin, pp. 119-188.

49. Konno, T.J., Okunishi, E., Ohsuna, T., Hiraga, K., 2004, HAADF-STEM study on the early stage of precipitation in aged Al-Ag alloys, *J. Electron. Microscopy* **53**: 611-616.

50. Kothleitner, G., Hofer, F., 2003, Elemental occurrence maps: a starting point for quantitative EELS spectrum image processing, *Ultramicroscopy* **96**: 491-508.

51. Krivanek, O.L., Nellist, P.D., Dellby, N., Murfitt, M.F., Szilagyi, Z., 2003, Towards sub-0.5 Å electron beams, *Ultramicroscopy* **96**: 229-237.

52. Krumeich, F., Muhr, H.-J., Niederberger, M., Bieri, F., Nesper, R., 2000, The cross-sectional structure of vanadium oxide nanotubes studied by transmission electron microscopy and electron spectroscopic imaging, *Z. Anorg. Allg. Chem.* **626**: 2208-2216.
53. Lehmann, M., Lichte, H., 2002, Tutorial on off-axis electron holography, *Microsc. Microanal.* **8**: 447-466.
54. Lehmann, M., Lichte, H., Geiger, D., Lang, G., Schweda, E., 1999, Electron holography at atomic dimensions: present state, *Materials Characterization* **42**: 249-263.
55. Lichte, H., 1997, Electron holography methods, Eds.: Amelinckx, S, van Dyck, D., van Landuyt, J., van Tendeloo, G., *Handbook of Microscopy: Applications in Materials Science, Solid-State Physics and Chemistry, Methods I*, VCH Weinheim, Germany, pp. 515-536.
56. Lichte, H., Geiger, D., Harscher, A., Heindl, E., Lehmann, M., Malamidis, D., Orchiwski, A., Rau, W.-D., 1996, Artefacts in electron holography, *Ultramicroscopy* **64**: 67-77.
57. Lichte, H., Lehmann, M., 2002, Electron holography: a powerful tool for the analysis of nanos-tructures, *Adv. Imaging and Elec. Phy.* **123**: 225–255.
58. Litynska, L., Dutkiewicz, J., Heinrich, H., Kostorz, G., 2004, Structure of precipitates in Al-Mg-Si-Sc and Al-Mg-Si-Sc-Zr alloys, *Acta Metall. Slovaca* **10**: 514-519.
59. Liu, J., Byeon, J. W., Sohn, Y.H., 2006, Effects of phase constituents/microstructure of thermally grown oxide on the failure of EB-PVD thermal barrier coating with NiCoCrAlY bond coat, *Surface & Coatings Technology* **200**: 5869-5876.
60. Malik, A., Schönfeld, B., Kostorz, G., Pedersen, J.S., 1996, Microstructure of Guinier-Preston zones in Al-Ag, *Acta Mater.* **39**: 4845-4852.
61. Malis, T., Cheng, S.C., Egerton, R,F., 1988, EELS log-ratio technique for specimen-thickness measurement in the TEM, *J. Elect. Microsc. Tech.* **8**: 193-200.
62. Möbus, G., Rühle, M., 1994, Structure determination of metal-ceramic interfaces by numerical contrast evaluation of HRTEM micrographs, *Ultramicroscopy* **56**: 54-70.
63. Müller, E., Kruse, P., Gerthsen, D., Schowalter, M., Rosenauer, A., Lamoen, D., Kling, R., Waag, A., 2005, Measurement of the mean inner potential of ZnO nanorods by transmission electron holography, *Appl. Phys. Lett.* **86**: 154108, 1-3.
64. Neddermeyer, H., Hanbüchen, M., 2003, Scanning tunneling microscopy (STM) and spectroscopy (STS), atomic force microscopy (AFM), in: *High-Resolution Imaging and Spectroscopy of Materials*, Eds. F. Ernst, M. Rühle, Springer Series in Materials Science Vol. 50, pp. 271-320, Springer-Verlag, New York.
65. Neumann, W., Kirmse, H., Häusler, I., Otto, R., Hähnert, I., 2004, Quantitative TEM analysis of quantum structures, *Journal of Alloys and Compounds* **382**: 2-9.
66. Op de Beeck, M., Van Dyck, D., Coene, W., 1996, Wave function reconstruction in HRTEM: the parabola method, *Ultramicroscopy* **64**: 167-183.
67. Pennycook, S.J., Jesson, D.E., Browning, N.D., 1995, Atomic-resolution electron energy loss spectroscopy in crystalline solids, *Nuclear Instruments and Methods in Physics Research Section B: Beam Interactions with Materials and Atoms* **96**: 575-582.
68. Pennycook, S.J., Jesson, D.E., Nellist, P.D., Chisholm, M.F., Browning, N.D., 1997, Scan-ning transmission electron microscopy: Z contrast, Eds.: Amelinckx, S, van Dyck, D., van Landuyt, J., van Tendeloo, G., *Handbook of Microscopy, Applications in Materials Science, Solid-State Physics and Chemistry, Methods II*, VCH Weinheim, Germany, pp. 595-620.
69. Portmann, M. J., Erni, R., Heinrich, H., Kostorz, G., 2004, Bulk interfaces in a Ni-rich Ni-Au alloy investigated by high-resolution Z-contrast imaging, *Micron* **35**: 695-700.
70. Qin, L.-C., 2001, Determining the helicity of carbon nanotubes by electron diffraction, Eds.: Zhang, X.-F., Zhang, Z., *Progress in Transmission Electron Microscopy*, Vol. 2, Springer-Verlag, Berlin, pp. 73-104.
71. Rau, W.D., Lichte, H., 1998, High-resolution off-axis electron holography, in: *Introduction to Electron Holography*, Eds. E. Völkl, L. F. Allard, and D. C. Joy, Kluwer Academic, New York, pp. 201-229.

72. Rau, W.D., Schwander, P., Baumann, F.H., Höppner, W., Ourmazd, A., 1999, Two-dimensional mapping of the electrostatic potential in transistors by electron holography, *Phys. Rew. Lett.* **82**: 2614-2617.

73. Reimer, L., 1989, Transmission Electron Microscopy, Springer-Verlag, Berlin.

74. Roberts, S., McCaffrey, J., Giannuzzi, L., Stevie, F., Zaluzec, N., 2001, Advanced techniques in TEM specimen preparation, Eds.: Zhang, X.-F., Zhang, Z., *Progress in Transmission Electron Microscopy 1*, Springer Series in Surface Sciences 38, Springer-Verlag, Berlin, pp. 301-361.

75. Rose, H., 1990, Outline of a spherically corrected semiaplanatic medium-voltage transmission electron microscope, *Optik* **85**: 19-24.

76. Rose, H., 2004, Advances in electron optics, Eds.: Ernst, F., Rühle, M., *High-Resolution Imaging and Spectrometry of Materials*, Springer-Verlag, Berlin, pp. 189-270.

77. Rosenauer, A., 2003, *Transmission Electron Microscopy of Semiconductor Nanostructures*, Ed. G. Höhler, Springer Tracts in Modern Physics 182, Springer-Verlag, Berlin.

78. Scherzer, O., 1936, Über einige Fehler von Elektronenlinsen, *Z. Phys.* **101**: 593-603.

79. Schwander, P., Kisielowski, C., Seibt, M., Baumann, F.H., Kim, Y., Ourmazd, A., 1993, Mapping projected potential, interfacial roughness, and composition in general crystalline solids by quantitative transmission electron microscopy, *Phys. Rev. Lett.* **71**: 4150-4153.

80. Senapati, S., Kabes, B., Heinrich, H., 2006, $Ag_2Al$ plates in Al-Ag alloys, *Zeitschrift für Metallkunde* **97**: 325-328.

81. Shindo, D., Oikawa, T., 2002, *Analytical Electron Microscopy for Materials Science*, Springer-Verlag, Tokyo.

82. Shiojiri, M., 2004, HAADF-STEM imaging and microscopy observations of heterostructures in electronic devices, *Electron Technology—Internet Journal* **36**, 3: 1-8.

83. Signoretti, S., Beeli, C., Liou, S.-H., 2004, Electron holography quantitative measurements on magnetic force microscopy probes, *J. Magn. Magn. Mater.* **272-276**: 2167-2168.

84. Signoretti, S., Del Bianco, L., Pasquini, L., Matteucci, G., Beeli, C., Bonetti, E., 2003, Electron holography of gas-phase condensed Fe nanoparticles, *J. Magn. Magn. Mater.* **262**: 142-145.

85. Tanaka, M., Terauchi, M., *Convergent-Beam Electron Diffraction I-III*, 1985, JEOL Ltd., Tokyo.

86. Tanaka, M., Terauchi, M., Tsuda, K., Saitoh, K., *Convergent-Beam Electron Diffraction IV*, 2002, JEOL-Marunzen, Tokyo.

87. Terheggen, M., 2003, "Microstructural Changes in CdS/CdTe Thin Film Solar Cells During Annealing with Chlorine," Dissertation ETH Zürich, Switzerland, No. 15214.

88. Twitchett, A.C., Dunin-Borkowski, R.E., Hallifax, R.J., Broom, R.F., Midgley, P.A., 2004, Off-axis electron holography of electrostatic potentials in unbiased and reverse biased focused ion beam milled semiconductor devices, *J. Microsc.* **214**: 287-296.

89. Velázquez-Salazar, J.J., Muñoz-Sandoval, E., Romo-Herrera, J.M., Lupo, F., Rühle, M., Terrones, H., Terrones, M., 2005, Synthesis and state of art characterization of BN bamboo-like nanotubes: Evidence of a root growth mechanism catalyzed by Fe, *Chem. Phys. Lett.* **416**: 342-348.

90. Villars, P., Calvert, L.D., Eds., 1991, *Pearson's Handbook of Crystallographic Data for Intermetallic Phases*, 2nd Edition, ASM International, Materials Park, OH.

91. Voelkl, E., Allard, L.F., Frost, B., 1997, Electron holography: recent developments, *Scanning Microscopy* **11**: 407-416.

92. von Heimendahl, M., 1980, *Electron Microscopy of Materials*, Academic Press, London.

93. Wang, S.Q., Wang, Y.M., Ye, H.Q., 2001, Quantitative analysis of high-resolution atomic images, Eds.: Zhang, X.-F., Zhang, Z., *Progress in Transmission Electron Microscopy 1*, Springer Series in Surface Sciences 38, Springer-Verlag, Berlin, pp. 162-190.

94. Wang, Z.L., 2001, Inelastic scattering in electron microscopy: effects, spectrometry and imaging, Eds.: Zhang, X.-F., Zhang, Z., *Progress in Transmission Electron Microscopy 1*, Springer, Series in Surface Sciences 38, Springer-Verlag, Berlin, pp. 113-159.

95. Williams, D.B., Carter, C. B., 1996, *Transmission Electron Microscopy*, Plenum Press, New York.

96. Yamazaki, T., Watanabe, K., Rečnik, A., Čeh, M., Kawasaki, M., Shiojiri, M., 2000, Simulation of atomic-scale high-angle annular dark field scanning transmission electron microscopy images, *J. Electr. Microsc.* **49**: 753-759.

97. Zandbergen, H. W., Træholt, C., 1997, Small particles, Eds.: Amelinckx, S., van Dyck, D., van Landuyt, J., van Tendeloo, G., *Handbook of Microscopy: Applications in Materials Science, Solid-State Physics and Chemistry: Applications*, VCH Weinheim, Germany, pp. 691-738.

98. Zhou, D., 2001, HREM study of carbon nanoclusters grown from carbon arc-discharge, Eds.: Zhang, X.-F., Zhang, Z., *Progress in Transmission Electron Microscopy 2*, Springer-Verlag, Berlin, pp. 25-71.

## QUESTIONS

1. Aluminum has a face-centered cubic structure with lattice parameter 0.405 nm. Determine the distance of neighboring atoms. What would be the magnification of an image, if the nearest neighbour distance of the atoms shown in the micrograph is 2 mm? How large would be the whole magnified image of an aluminum sample with a diameter of 3 mm?

2. The camera length of a TEM is a quantity describing the magnification of the diffraction pattern. With the small scattering angles in TEM, we can approximate the sine function in the Bragg equation by its argument, and we can assume a distance (the camera length $L$) of the imaged diffraction pattern from the sample (in reality there are several lenses between sample and viewing screen which magnify the diffraction pattern). The distance $R$ of a Bragg reflection from the undiffracted beam in a diffraction pattern depends on the camera length, the lattice spacing $d_{hkl}$, the electron wavelength, and the diffraction angle $2\Theta$. The camera constant is defined by $\lambda L = R d_{hkl}$.

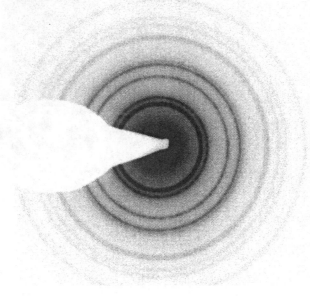

The micrograph above shows a Debye-Scherrer pattern from Au nanoparticles, the undiffracted beam is blocked by a beam stop. The lattice parameter of this fcc material is 0.40704 nm. Index the diffraction rings! Determine camera length and camera constant for 300 kV electrons!

3. Which sample orientations can be used for a bcc crystal structure with $a = 0.28$ nm to obtain high-resolution images of individual atomic columns so that a two-

dimensional lattice is resolved? Consider the point-resolution as the limiting factor to distinguish neighboring atomic columns ($C_s = 1.2\,\text{mm}$, $U = 300\,\text{kV}$).

4. The micrograph below shows a contrast simulation for TiAl. Identify and describe the defects. Determine the lattice parameter assuming that the $L1_0$ structure of TiAl does not significantly deviate from the fcc structure!

1 nm

5. Determine which superstructure reflections of $Ni_4Mo$ precipitates coherently embedded in a Ni-rich fcc matrix material appear in Fig. 27(b), when using the structural model below. Two consecutive (002) planes of the $Ni_4Mo$ structure are shown. Outlined symbols for the top layer, open circles for the bottom layer.

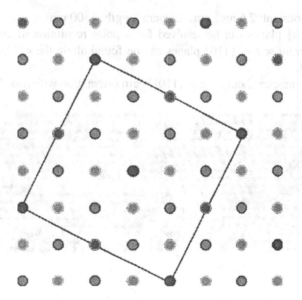

6. Determine the equations for the lengths of a real space vector $[xyz]$ and a reciprocal lattice vector $[hkl]$ for a monoclinic unit cell with $b = a/2, c = 2a, \gamma = 60°$!
   Determine the angles between
   (a) the [110] vector and the [100] vector in reciprocal space!
   (b) the [110] vector and the [100] vector in real space!
   (c) (110) planes and (100) planes in real space!
   (d) (110) planes and (100) planes in reciprocal space!
   (e) (110) planes and the [100] direction in reciprocal space!
   (f) (110) planes and the [100] direction in real space!

7. Use Fig. 50 to estimate the exponent $\alpha$ which determines the contrast parameters $C_\gamma$ and $C_{Al}$ of the two phases: $C = K(c_A Z_A^\alpha + c_B Z_B^\alpha)$. Here, $c_A$ and $c_B$ are the concentrations of the two components $A$ and $B$ in each phase, $K$ is a constant, and $Z_A$ and $Z_B$ are their corresponding atomic numbers. The intensity $I$ of a HAADF-STEM micrograph is given by the equations $I_\gamma = I_{bg} + I_c + t_\gamma C_\gamma$ and $I_{Al} = I_{bg} + I_c + t_{Al} C_{Al}$, where $t$ is the sample thickness, $I_{bg}$ is the background signal and $I_c$ is the signal from the contamination layer. The concentration of the $\gamma$-plate is 60 at.% Ag and 40 at.% Al, the Al matrix contains no silver.

## SOLUTIONS TO QUESTIONS

1. Magnification: 4.9 million times, whole sample magnified gives an image of 15 km.
2. Indices from inside to outside: 111, 200, 220, 311, 222, 400, 331, 420, 511 + 333,
   Camera constant: 2.6 nm* mm, Camera length: 1300 mm.
3. Only {110} planes can be resolved for a point resolution of about 0.2 nm. Two different sets of {110} planes can be found along the <111> and <100> directions.
4. Lattice parameter about 0.4 nm. [110] beam orientation with two {111} planes.

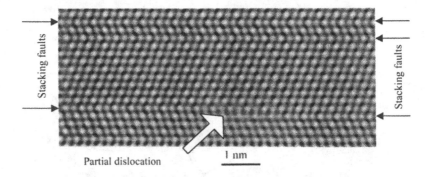

Partial dislocation                                    1 nm

5. Reflections from the intersections of the lines below. Indices of type {2/5,4/5,0).

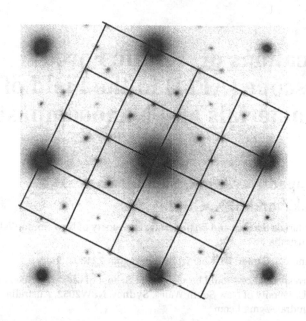

6. a: 90°, b: 19.1°, c: 90°, d: 19.1°, e: 40.9°, f: 30°.
7. $C_\gamma = 100$ counts/nm, $C_{Al} = 22$ counts/nm, $\alpha = 1.53$.

# Applications of Atomic Force Microscope (AFM) in the Field of Nanomaterials and Nanocomposites

## S. Bandyopadhyay,[1] S. K. Samudrala,[1,*] A. K. Bhowmick,[2] and S. K. Gupta[1]

[1] School of Materials Science and Engineering, University of New South Wales, Sydney, NSW 2052, Australia

[2] Rubber Technology Center, IIT Kharagpur, Kharagpur 721302, India

*To whom correspondence should be addressed. School of Materials Science and Engineering, University of New South Wales, Sydney, NSW2052, Australia. E-mail: sksamudrala@gmail.com

## 1. INTRODUCTION

Nanotechnology implies the capability to build up tailored nanostructures and devices for given functions by control at the atomic and molecular levels. Development of novel nanofabrication methods with effective control of structure, morphology and patterning at the nanometer-scale level is of principal importance for nanoscience and nanotechnology, advanced materials research, as well as for design of new functional nanostructures with predetermined and unique properties.[1] The study of such nanostructures can bridge the gap of knowledge between individual atoms and molecules where quantum mechanics laws are applied and vast volume phase in which most properties result from collective behavior of billion atoms. In the vast field of nanoscience and nanotechnology, this chapter focuses on an important aspect i.e. application of atomic force microscopy (AFM) in the field of nanoparticles/nanomaterials, and nanocomposites.

## 1.1. Nanomaterials

Nanomaterials are best defined as those comprising particles or molecules with at least one dimension in the nanometer range (<100 nm). They can be classified as:

- nanophase or nanoparticle materials;
- nanostructured materials.

While nanostructured materials are condensed bulk materials with grains of size in the nanometer range, nanoparticles are individual dispersed particles. These particles exist in different physical forms (vapor/gas, liquid, and solid phase, or all three phases may be present and interacting through vapor-liquid, solid-liquid, vapor-solid interfaces) and in different dimensions (one, two, or three-dimensional).[2] A wide range of morphologies can be expected for the nanoparticles, such as, spheres, flakes, platelets, dendritic structures, tubes, rods, coils, springs, and brushes.[3] In addition, nanoparticles can also exhibit diverse structures for the same material. For example, by controlling the growth kinetics, local growth temperature, and chemical composition of the source materials in a solid-state thermal sublimation process, a wide range of nanostructures of ZnO are synthesized (Fig. 1) and are reported by Zhong.[4]

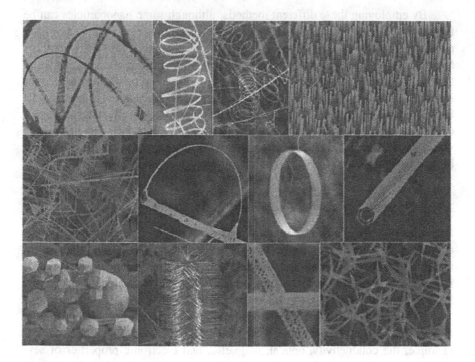

**FIGURE 1.** A collection of nanostructures of ZnO synthesized under controlled conditions by thermal evaporation of solid powders (Reprinted from Materials Today, 7(6), Z.L. Wang, Nanostructures of zinc oxide, 26-33, (2004), with permission from Elsevier.)

### 1.1.1. Routes to Synthesize Nanoparticles

In general there are four routes to synthesize nanoparticles, which include wet chemical, form-in-place, mechanical, and gas-phase synthesis. Among these:

1. *Wet chemical processes* include colloidal chemistry, hydrothermal methods, sol-gels and other precipitation processes. A large variety of inorganic, organic and metallic nanoparticles can be fabricated through the wet chemical methods.
2. *Mechanical processes* include grinding, milling, and mechanical alloying. Inorganic and metal nanoparticles are commonly synthesized by this method.
3. *Form-in-place processes* are generally lithography, physical and chemical vapor deposition, and spray coatings. These are mainly aimed at the production of nanostructure layers and coatings. However, they can be used to fabricate nanoparticles by scraping the deposits from the collector.
4. *Gas-phase processes* include flame pyrolysis, electro-explosion, laser ablation, high-temperature evaporation, and plasma synthesis techniques. Gas-phase processes are used to fabricate simple materials such as carbon black and fumed silica, and also many compounds.[3]

By employing these different methods, although same nanoparticles can be fabricated in some instances, the characteristics of the materials produced by each process are not always the same and possess different properties. There are three basic issues that must be considered in the design and synthesis of a new compound that will act as a nanomaterial: (a) properties of individual molecules (or even of individual functional groups within multifunctional molecules); (b) interactions between these molecules (or functional groups); (c) how they affect the properties of discrete ensembles of the molecules; and, finally, the interface with the outside world must be designed, that is, how the input of the energy or charge is going to be achieved, and how an output is going to be received or detected.[5]

### 1.1.2. Importance of Nanomaterials/Nanoparticles

Nanomaterials cover a diverse range of materials: polymers, metals, biomaterials and ceramics—hence their applications range from catalysis to food science; innovative electronic materials and devices to bio-detection and medicine; catalysts for pollution control, chemical processing and fuel cells to chemical sensing; and field emission panel displays to material reinforcement.[5-26]

The importance of these nanomaterials and/or nanoparticles is apparent if their properties are compared to their corresponding bulk material properties. For example, the conductivity, optical, magnetic, and electronic properties of several inorganic nanoparticles significantly change as their size is reduced from macroscale to micro- and nano-levels. In some cases, it has been reported that nanoparticles are more active than their traditional counterparts since they have higher percentages of atoms on their surfaces. Aymonier *et al.* have reported that silver nanoparticles exhibit biological (antimicrobial) activity; but on the contrary, silver metal does not.[27] Also, compared to bulk ferromagnetic materials where multiple magnetic domains exist, several small ferromagnetic particles consist of only a single magnetic domain resulting in superparamagnetism. Furthermore, nanoparticles possess exceptional properties such as:

- Surface properties (oxidation, adhesion, friction, and wear);
- Transport properties (diffusion and thermal conductivity);
- Mechanical properties (toughness, strength, elastic modulus, creep, etc);
- Rheological properties.

Therefore, nanoparticles are increasingly used as reinforcing fillers in different polymeric and ceramic materials. In comparison to conventional micrometric reinforced composites, these nanometric composites exhibit enhanced mechanical and physical properties. This is attributed to the confinement effects, large surface-to-volume ratio, interactions at length scales, and the possibility of generating new atomic and macromolecular structures. Table 1 lists some examples of such nanomaterials and their applicability.

**TABLE 1.** Typical nanostructures, nanomaterials and their applications in science and technology. (From Ref. 48, p. 218).

| Nanostructures | Size | Materials | Applications |
|---|---|---|---|
| Clusters, nanocrystals, quantum dots | Radius: 1-10 nm | Insulators, semiconductors, metals, magnetic materials | Catalysis, sensors |
| Nanowires | Diameter: 1-100 nm | Metals, semiconductors, oxides, sulfides, nitrides | Magnetic, electronic, optical, electro-optical devices or switches |
| Nanotubes | Diameter: 1-100 nm | Carbon, layered chalcogenides | Separation technologies |
| Surfaces, layers and ultrathin films, interfaces | Thickness: 1-1000 nm | Insulators, semiconductors, metals, ceramics | Electronics, medicine (drug delivery systems), energy (batteries, fuel cells, solar cells), optics, paints and dyes, cosmetics |
| Nanobiomaterials, nanobiorods | Radius: 5-10 nm; Diameter: 5 nm | Membrane protein DANN | Medicine (drug delivery systems), biology |
| 2D arrays of nanoparticles, 3D superlattices of nanoparticles | Area: several $nm^2$-$\mu m^2$; Radius: several nm | Metals, semiconductors, magnetic materials | Magnetic, electronic |

## 1.2. Nanocomposites

Nanocomposites are generally composed of different distinctly dissimilar materials (mostly organic and inorganic) having synergistic properties to obtain tailored chemical, physical, and mechanical properties depending on the applicability; for example, high modulus and strength, superior fracture toughness, good barrier and flame-retardant properties, enhanced electrical and thermal conductivity, magnetic properties, and so forth. They can be one-dimensional, two-dimensional, three-dimensional, crystalline or amorphous.[28] The properties of nanocomposite materials depend not only on the properties of their individual parents but also on their morphology and interfacial characteristics. The inorganic components can be:

- three-dimensional framework systems such as zeolites;
- two-dimensional layered materials such as clays, metal oxides, metal phosphates, chalcogenides;
- one-dimensional and zero-dimensional materials such as $(Mo_3Se_3\text{-})_n$ chains and clusters.

Nanocomposites promise new applications in many fields such as mechanically reinforced lightweight components, nonlinear optics, battery cathodes and ionics, nanowires, sensors and other systems.[29]

As mentioned above, the main advantage of nano-additives is their size, which brings about greater surface area-to-volume ratio when they are reinforced with different organic/inorganic matrices. Also, as they have higher percentages of atoms on their surfaces, they are expected to be more active.[30] This is important for bonding with the matrix and in improving the interface between them. Studies on polymer-based nanocomposites showed a significant improvement in properties, which are not displayed in the dual phases by their macro- and micro-composite counterparts. For example, fully exfoliated nylon 6/clay nanocomposites were fabricated where the silicate layers were well separated to 1 nm thick individual layers with other dimensions in 100-500 nm range. This nanocomposite exhibited enhanced improvements in tensile strength, modulus, and heat distortion temperature. Also, the nanocomposite exhibited lower water sensitivity, permeability to gases and thermal expansion coefficient.

Despite these large improvements, a vast range of potential engineering applications of polymeric silicate nanocomposites are limited due to their reduced fracture toughness. In general, three major characteristics define and form the basis of performance of polymer nanocomposites: nanoscopically confined matrix polymer chains, nanoscale inorganic constituents, and nanoscale arrangement of these constituents. It has been suggested based on the research on polymer nanocomposites for the past few decades that the full exploitation of these fundamental characteristics of nano-reinforcements in polymers should facilitate the achievement of enhanced properties in these nanocomposites.[29,31,32]

## 1.3. Characterization Techniques

Accurate characterization of the nanomaterials requires a precise understanding and measurement of the surface and interfacial phenomena along with the other factors such as size of the nanoparticles, crystallinity, and so forth. In general, characterization can be divided into two broad categories:

- Structural analysis;
- Properties measurement.

A number of ex situ analytical tools with high vertical, spatial and time resolutions have been used for this purpose and include:

- High-resolution, transmission electron microscope (HR-TEM);
- Field ion microscope (FIM);
- Scanning probe microscope (SPM);
- X-ray absorption near-edge spectroscopy (XANES);
- Time-of-flight secondary ion mass spectroscopy (ToF-SIMS);
- X-ray photoelectron spectroscopy (XPS);
- Auger electron spectroscopy (AES) and so forth.

An illustration of the resolution of surface analytical tools is given in Fig. 2 as viewed from time, size, and resolving power.[33] Although these powerful tools have increased the understanding of the structure of nanoparticles, the exact property analysis is still questionable. As the properties of nanostructures and nanocomposites depend strongly on the size and shape of nanoparticles, the structural characterization of nanoparticles and nanomaterials play a significant role.

The methods to characterize nanoparticles structure are generally categorized depending on the use of real or reciprocal space data as direct space methods and reciprocal space methods. Visualization of the atomic arrangements is possible using direct space methods, which include the usage of HRTEM, SPMs, and so on.[33] On the other hand, the interference and diffraction effects of the lattice planes, electrons or photons are utilized in reciprocal space methods for structural characterization of nanoparticles. The most common example is the characterization of extent of exfoliation of clay-platelets in a polymer nanocomposite using wide X-ray diffraction. While SAXS can provide information about the external form of nanoparticles or macromolecules from the electron density variations; i.e., it is used to estimate the radius of gyration of particles. Other reciprocal methods include measuring the motion of the particle in response to some force, such as gravity, centrifugal force, viscous drag, Brownian motion, or electrostatic force.

While reciprocal methods provide a means to characterize the nanoparticles, direct imaging is only possible using direct space methods. As nanoparticles are substantially nonspherical (e.g., rods, fibers, or cubes), microscope pictures are

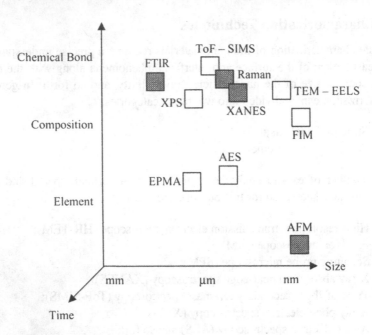

**FIGURE 2.** Schematic illustration of the resolution of surface analytical tools as viewed from time, size, and resolving power (open squares require high vacuum, while solid squares can be used in air). (Adopted from Ref. 33).

needed to clarify their size, size distribution, shape, and dispersion to correlate with their properties. Additionally, the dispersion of nanoparticles in a polymer or metal is a very complex process. The usage of reciprocal methods cannot fully identify the dispersion processes and can lead to false conclusions.

For example, in a recent study, Dasari et al.[29] have indicated that X-ray diffraction is *not* a reliable tool for analyzing the complex dispersion of clay layers in ternary nanocomposites, particularly, polymer-rubber-clay systems as no characteristic basal diffraction peak of clay was found in the range of 1-10° for all the nanocomposites, pointing to a complete exfoliation of clay. But the TEM results are completely different and did not support the XRD results. Also, most of the TEMs and SEMs are fitted with energy dispersive X-ray spectrometers (EDS) to identify the chemical composition of materials or particles. This technique relies on the energy of the X-rays that are released when the electron beam excites a certain spot on the sample and used to identify the atoms that are in the spot. This technique can also be used to produce a map of the concentrations of different species in a sample. An example of this is given in Fig. 3, which shows the Si mapping in polyamide 66/silica nanocomposites with varying content of silica.[34] The mapping clearly reveals the differences in dispersion behavior of the particles also.

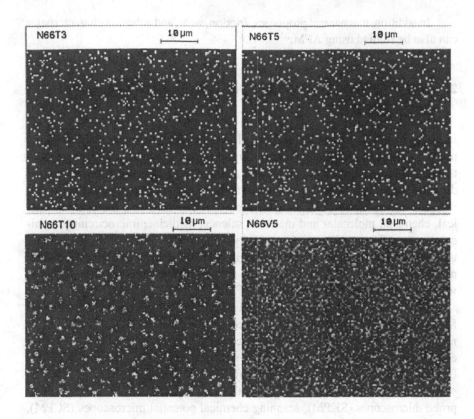

**FIGURE 3.** Si mapping of the unannealed N66T3, N66T5, N66T10 and N66V5 films. (Reprinted from Polymer 46(10), R. Sengupta, A. Bandyopadhyay, S. Sabharwal, T.K. Chaki and A.K. Bhowmick, Polyamide-6,6/in situ silica hybrid nanocomposites by sol-gel technique: synthesis, characterization and properties, 3343-3354, (2005), with permission from Elsevier).

Although TEMs, STEMs or SEMs provide a size distribution of individual particles or dispersed particles in a matrix, they cannot provide the three-dimensional structure of a particle. Also, the samples for SEM should be conductive; while for imaging dispersed nanoparticles in a matrix, the sample preparation for TEM is very cumbersome. Further, when using TEMs, questions arise on the stability of fine features to be imaged because of the high-energy electron beam usage. AFM offers a better alternative in some cases and as a supportive tool in other cases of characterizing nanoparticles or nanostructures.

The advantages of AFM include:

- Three-dimensional qualitative and quantitative;
- Can be operated in most of the media;
- Can image liquid dispersants;
- Has excellent resolution (see below in section 2 for more information).

In addition, mechanical properties, friction, wear, and adhesion characteristics can also be studied using AFM.[35]

## 2. ATOMIC FORCE MICROSCOPE INSTRUMENTATION AND SETUP

In the early 1980s, after being introduced into the world of microscopy, scanning probe microscopes (SPM) have emerged as powerful techniques capable of characterizing surface morphological features at ambient conditions and generating three-dimensional images of the surface topography with nanometer resolution.[36,37] So, they are extensively used for imaging in most of the disciplines including biological, chemical, molecular and materials science, medicine, microcircuitry, semiconductor industry, information storage systems, and so forth. SPMs are not only used for surface morphological characterization, but also for proximity measurements of magnetic, electrical, chemical, optical, thermal, spectroscopy, friction, wear, and other mechanical properties.[1,28,29,31,36-85]

The family of SPMs include scanning tunneling microscopes (STM), atomic force microscopes (AFM), friction force microscopes (FFM), scanning magnetic microscopes (SMM) (or magnetic force microscopes, MFM), scanning electrostatic force microscopes (SEFM), scanning near-field optical microscopes (SNOM), scanning thermal microscopes (SThM), scanning chemical force microscopes (SCFM), scanning electrochemical microscopes (SEcM), scanning Kelvin probe microscopes (SKPM), scanning chemical potential microscopes (SCPM), scanning ion conductance microscopes (SICM), and scanning capacitance microscopes (SCM).[86] The family of instruments that measures forces (e.g., AFM, FFM, SMM, and SEFM) is also referred to as scanning force microscopes (SFM).

The most important members of this family of SPMs are atomic force microscopes (AFM) and scanning tunneling microscopes (STM). Atomic force microscope has an added advantage for the high-resolution profiling of nonconducting surfaces. Also, AFM is a nondestructive technique and it does not require any specific sample preparations. Furthermore, the resolution capabilities of the AFM are near or equal to those of electron microscopes (an example illustrating the capability of AFM imaging is it can image as small as a carbon atom (0.25 nm in diameter) and as large as the cross-section of a human hair—80 $\mu$m in diameter)[87]; nonetheless, AFM differs from electron microscopes in that it does not have a lens, does not require coating or staining, and can be operated at atmospheric pressure, in fluids, under vacuum, low temperatures, and high temperatures.

The vertical resolution of AFM images is generally dictated by the interaction between the tip and the surface and the lateral resolution is determined by the size of the tip. Imaging in liquid allows the study of live biological samples, and it also eliminates water capillary forces present in ambient air present at the tip-sample interface.[86] Low-temperature (liquid helium temperatures) imaging is useful for the study of biological and organic materials and the study of low-temperature phenomena such as superconductivity or charge density waves.[86] Low-temperature operation is also advantageous for high-sensitivity force mapping due to the reduction in thermal vibration. Thus, AFM operation is relatively simple, artifacts are reduced and materials can be examined in their native state.

## 2.1. Principle of Operation

In simple terms, the underlying operational principle of AFM imaging is based on the interatomic force-distance concept (schematically illustrated in Fig. 4). As the AFM tip atom approaches the atoms on the surface being studied, a variety of forces are sensed depending on the interatomic distance, which include van der Waals, electrostatic, magnetic, capillary, or ionic repulsion forces. These forces cause a deflection of the lever on which the tip is mounted following even a minor change of topography, which is utilized to produce images of topography. More descriptive information on the operation of AFM is given below.

In a normal imaging mode of AFM, a sharp probe at the end of a cantilever bends in response to the force between the tip and the sample as it moves over the surface of a sample in a raster scan. The tip would ideally consist of only one

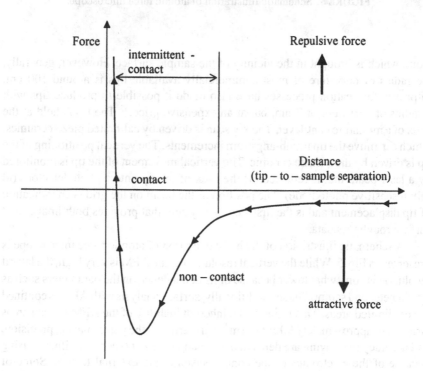

**FIGURE 4.** Schematic illustration of interatomic force-distance approach used in AFM.

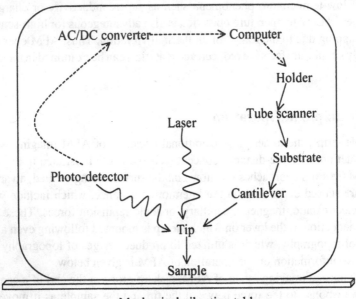

**FIGURE 5.** Schematic illustration of atomic force microscope.

atom, which is brought in the vicinity of the sample surface. However, generally, the radius of curvature of most commercially available tips is around 100 nm. Improved fabrication processes have also made it possible to produce tips with a radius of curvature of 2 nm, but at an expensive price.[87] The tip is held at the apex of a miniature cantilever. The x/y scan is driven by calibrated piezo ceramics, which can move the tip in sub-angstrom increments. The vertical positioning of the tip is driven by the z-piezo ceramic. The vertical movement of the tip is monitored by a laser beam that is reflected off the back of the tip onto a photodetector grid (photosensitive photodiode). The position of the beam on the grid is an indication of tip displacement and is the first step in a system that provides both images and surface roughness data.

A schematic illustration of the basic operation of atomic force microscope is presented in Fig. 5. While the vertical resolution of the AFM is very high, the lateral resolution is somewhat lower because of its dependence on the parameters such as tip diameter and sample shape. Additionally, surface analysis with AFM is confined to very limited areas. This is due to the inherent inability of the AFM to scan areas larger than approximately $90 \times 90 \ \mu m^2$. Furthermore, the performance, precision, and accuracy in imaging are dependent on many inherent (such as artifacts arising because of the electronics of the control system) and external factors. Some of which are briefly mentioned below.

## 2.2. Factors that Influence the Precision and Accuracy of AFM Imaging

### 2.2.1. Piezoelectric Ceramic Transducer

Piezoelectric ceramics are a class of materials that expand or contract in the presence of a voltage gradient or, conversely, create a voltage gradient when forced to expand or contract.[88] They are responsible for creating three-dimensional positioning devices of arbitrarily high precision. The piezoelectric scanners that are utilized in AFMs control either the motion of the cantilever probe with respect to a stationary sample or the motion of the sample with respect to a stationary probe. Most scanned-probe microscopes use tube-shaped piezoceramics because they combine a simple one-piece construction with high stability and large scan range. Four electrodes cover the outer surface of the tube, while a single electrode covers the inner surface.

Application of voltage to one or more of the electrodes causes the tube to bend or stretch, moving the sample in three dimensions. The amount of motion and direction of motion depends on the type of piezoelectric material, the shape of the material, and the field strength. A typical piezoelectric material will expand by about 1 nm per applied volt.[87] Thus, to get larger motions it is common to make piezoelectric transducers with hundreds of layers of piezoelectric materials. Besides these advantages of piezoelectric ceramics that are critical to the performance of AFMs, they also have exhibited drawbacks (nonlinearities). Examples include hysteresis and creep, which introduce uncertainties in measurements.[54,89,90] Sometimes, it may also overshoot during rapid movements.

### 2.2.2. Optical Lever Detection Systems

These systems are reported to have noise levels on the subnanometer level,[91] which allows for the excellent depth resolution of AFMs employing these. Despite this, the nonlinearities associated with the photodiode used in the optical lever detection system can significantly influence accurate measurements. For example, if the cantilever is deflected too far, the central portion of the spot can no longer cross the split in the photodiode. When the edge of the spot is crossing the split, a given spot movement produces less power, and therefore a lower sensitivity, than when the center of the spot crosses the split.[90]

### 2.2.3. Resolution and Probe-Related Image Distortions

As can be seen in the interatomic force-distance curve (Fig. 4), with increase in distance, the tip-surface interaction forces drop less steeply. This is very important for high-resolution imaging (imaging individual atoms). An AFM probe responds to the average force of interaction for a number of tip atoms, depending on the precision and sharpness of the tip, and so an AFM image does not show individual

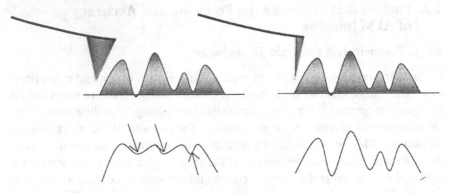

**FIGURE 6.** Comparison of AFM imaging of a surface structure with higher surface relief: (a) blunt probe and (b) sharp probe. Arrows indicate the artifacts.

atoms but rather an averaged surface. In addition to this, the sharpness of the tip also plays a dominant role in determining the lateral resolution of AFM when imaging surfaces with higher relief.[92,93]

If the surface features to be imaged are deeper than the probe length, it can result in artifacts (Fig. 6). A common artifact is the resulting image, which will be a combination of the actual sample surface and the shape of the cantilever. On the contrary, the vertical resolution in an AFM is established by relative vibrations of the probe above the surface. Sources for vibrations are acoustic noise, floor vibrations, and thermal vibrations. To obtain the maximum vertical resolution, it is necessary to minimize the vibrations of the instrument.

### 2.2.4. Tip-Surface Interactions

Mechanical forces that occur when the atoms of the probe physically interact with the atoms on the sample surface are the strongest forces between the probe and the surface. However, other forces between the probe and surface can have an impact on an AFM image.[94,95] These include:

- *Surface contamination*: In ambient air, all surfaces are covered with a very thin layer, <50 nm, of contamination. This contamination comprises water and hydrocarbons and depends on the environment the microscope is located in. When the AFM probe comes into contact with the surface contamination, capillary forces can pull the probe towards the surface, ultimately leading to a distorted image.
- *Electrostatic forces*: It is well known that insulating surfaces can store charges on their surfaces. When these charges interact with charges on the AFM probe or cantilever, the forces can be so strong that they even bend the cantilever when scanning a surface.
- *Surface material properties*: When imaging heterogeneous surfaces, which generally have regions of different hardness and friction, the interaction of

the probe with the surface can change when moving from one region to another. Such changes in forces on one hand are advantageous giving a contrast that is useful for differentiating between materials on a heterogeneous surface. On the other hand, they also can be disadvantageous if the changes in hardness, adhesion or frictional properties of the surface are drastic.

## 2.3. Different Modes of Imaging in AFM

Depending on the AFM tip and sample surface interaction, three conventional scanning modes have been used for imaging: contact mode, noncontact mode, and tapping mode.[36,37,96-101]

### 2.3.1. Contact Mode

In contact mode AFM, the tip and sample remain in close contact as the scanning proceeds (repulsive region of the interatomic force curve) (Fig. 7(a)), creating large forces on the sample. This mode is generally used for high-resolution imaging. Since the maximum vertical force is also controlled, the compression of the sample can be limited. In this mode, although the lateral forces as the tip moves over the surface can be a problem, in some situations they can actually be an advantage. For example, the lateral deflection can give information about the friction between the tip and the sample, and show areas that may have the same height but different chemical properties. Also, it is important to consider here that the set point value is the deflection of the cantilever, so a lower value of the set point gives a lower imaging force.

### 2.3.2. Noncontact Mode

In the noncontact operation, the probe is held at a small distance away from the surface and the cantilever is oscillated above the surface of the sample (Fig. 7(b)). However, due to the attractive forces there is a possibility of the tip coming into contact with the surface. So, this method is not widely used. The capillary force makes this even difficult to control in ambient conditions. Very stiff cantilevers are needed so that the attraction does not overcome the spring constant of the cantilever. Despite these disadvantages, the lack of contact with the sample makes this mode of imaging beneficial in that it causes a minimum damage to the surface.

### 2.3.3. Intermittent Contact or Tapping Mode

To obtain quality images, in general, the tip should not damage the surface being scanned but that it contacts the surface to obtain high-resolution topographic imaging of the sample surfaces. This is possible in tapping mode and so is the most commonly used mode of imaging in AFM. In this mode, the cantilever oscillates

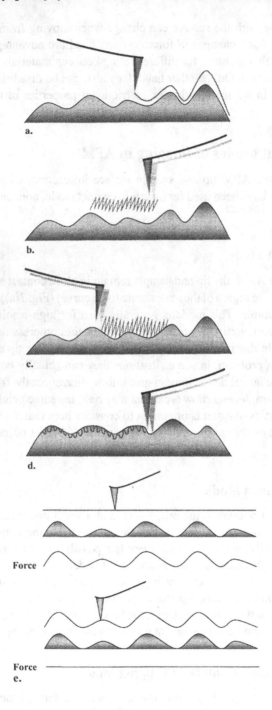

**FIGURE 7.** Schematic illustration of (a) contact mode, (b) noncontact mode, (c) intermittent contact mode, (d) force modulation mode of operation in AFM, and (e) constant height-force criterion.

and the tip makes repulsive contact with the surface of the sample at the lowest point of the oscillation (Fig. 7(c)). In other words, the cantilever is oscillated at its resonant frequency and positioned above the surface of the sample, so that it only taps the surface.

This method has the advantage that the tip does not damage the surface of the sample being scanned, and there is also a large reduction in the lateral forces, since the proportion of the time where the tip and sample are in contact is quite low. Also, the phase of the cantilever oscillation can give information about the sample properties, such as stiffness and mechanical information or adhesion. The resonant frequency of the cantilever depends on its mass and spring constant; normally, stiffer cantilevers have higher resonant frequencies. In this mode, the set point value is the amplitude of the oscillation, so a higher set point value means less damping by the sample and hence lower imaging forces.

### 2.3.4. Other Modes of Operating AFM

1. For micro/nano-tribology purposes, a different mode of AFM is used in which the AFM must have a heavy load mechanism for the tip and operated in a force mode (Fig. 7(d)).[102] This mode also resembles a dynamic form of contact mode as the tip does not leave the surface at all during the oscillation cycle. A force curve is produced, which is a plot of tip deflection as a function of the vertical motion of the scanner. Also, in these cases, the selection of spring constant of the cantilever probe also plays an important role for identifying the small differences in response.

2. AFM can also be operated as a chemical force microscope (CFM) to probe the chemical properties of surfaces.[103-107] In these cases, a specifically coated tip of known composition is brought into contact with a surface (generally, in a fluid environment), and then separated from the surface. During tip-sample separation, the pull-off force is recorded and related to the work of adhesion.[106]

3. In another mode called the vibrating material sensing mode, AFM cantilever can be vibrated to measure the force between a probe and surface during an AFM scan. The magnitude of amplitude damping and the amount of phase change of the cantilever depends on the surface chemical composition and the physical properties of the surface. Thus, on a heterogeneous sample, contrast can be observed between regions of varying mechanical or chemical composition. Typically, in the vibrating material sensing mode, if the amplitude is fixed by the feedback unit, then the contrast of the material is observed by measuring phase changes. This technique has many names including phase mode and phase detection microscopy.

4. In contact mode AFM it is also possible to monitor the torsion motions of the cantilever as it is scanned across a surface.[87] The amount of torsion of the cantilever is controlled by changes in topography as well as changes in surface chemical properties. If a surface is perfectly flat but has an interface

between two different materials, it is often possible to image the change in material properties on a surface. This technique is similar to lateral force microscopy (LFM).

5. Dynamic modes of operation are also possible in AFM where the cantilever vibrates, and this oscillation of the cantilever is measured rather than the static deflection of the tip. There are different ways to excite the oscillations—the cantilever substrate can be shaken directly, or a magnetic field can be used to drive the cantilever itself if it is coated with a ferromagnetic layer. In aqueous conditions, the most common technique is to drive the cantilever acoustically through the liquid. In all these cases, however, the measurement of the cantilever oscillation and control systems are the same and the cantilever is usually driven close to resonance.

## 2.4. Constant Force and Constant Height Criterion

The different modes of operation in AFM can be accomplished by utilization of either the constant force mode or the constant height mode criterion (Fig. 7(e)). In the constant force mode, the feedback loop is switched on (the positioning piezo responds to any change in the force by altering the tip-sample separation and adjusting the force to a predetermined value). In the constant height mode, the feedback loop is switched off (tip rasters the surface at a constant height from the surface of the sample). This mode is used for imaging relatively flat samples.

## 2.5. AFM in Nanotechnological Applications

Apart from simply imaging, as discussed above, AFM cantilevers can be used in many other modes of interaction with the surface. For example, by pushing the probe against a surface it is possible to measure how hard the surface is. Also, the ease by which the probe glides across a surface is a measure of the surface friction. Recently, the tip of the AFM has also been used to pattern the surface, move, and manipulate molecules or parts of the sample (the objects may be pushed, rolled around, or even picked up by the probe), or even to dissect the sample on a nanometer level, which leads to nanolithography, nanopatterning or nanofabrication.[108-110] Conventionally, there are many techniques available for fabrication of nanomaterials, ranging from milling techniques to nontraditional photolithographic and chemical methods. However, the major disadvantage of these methods lies in the difficulty in controlling the final morphology of the produced nanostructures. But mechanical scratching using AFM has provided several advantages over others—better control over the applied normal load, scan size and scanning speed, dry fabrication, and absence of chemical etching or electrical field. Nevertheless, the main disadvantage of using AFM in these cases is the formation of debris during scratching.[111] Table 2 lists different modes of AFM that can be used in nanotechnological applications.

**TABLE 2.** Main classifications of atomic force microscope (AFM.) (From Refs. 37, 86, 97.)

| Microscopy | Lithography |
| --- | --- |
| Tapping mode atomic force microscopy (TMAFM) | Dip-pen lithography (DPN) |
| Chemical force microscopy (CFM) | Mechanical lithography (indenting, ploughing, scribing) |
| Magnetic force microscopy (MFM) | Tip-induced oxidation |
| Electrical force microscopy (EFM) | |
| Current sensing atomic force microscopy (CS-AFM) | |
| Atomic force acoustic microscopy (AFAM) | |
| Lateral force microscopy (LFM) | |
| Friction force microscopy (FFM) | |
| Force spectroscopy | |

# 3. CONTRIBUTIONS OF AFM TO THE FIELD OF NANOTECHNOLOGY

## 3.1. Characterization of Nanoparticles/Nanomaterials

AFM ideally suits for characterizing nanoparticles and Table 3 lists some key attributes of AFM for characterization of nanoparticles. Characterization of the nano- and microstructure dispersion of particles is necessary to optimize the

**TABLE 3.** Features of AFM for the characterization of nanoparticles. (From Refs. 87, 99.)

Key attributes of AFM for characterization of nanoparticles

Qualitative analysis:
  3D visualization
  Material sensing
Quantitative analysis:
  Size
  Morphology
  Surface texture/roughness parameters like statistical
    information, particle counting, size distribution,
    surface area distribution, volume and mass
    distributions, spatial distribution
Medium:
  Gas ambient-air controlled environments
  Liquid dispersions
  Solid dispersions
Range:
  Particle size: 1 nm to 8 μm
  Scan range: up to 80 μm

**TABLE 4.** Classification of nanoparticles and their industrial applications. (From Refs. 37, 87.)

| Engineered nanoparticles (industries) | Non-engineered nanoparticles (industries) |
|---|---|
| Pharmaceuticals | |
| Food products | Environmental detection |
| Ceramics | Environmental monitoring |
| Quantum dots | Controlled environments |
| Chemical mechanical polishing | |
| Bio-detection and labeling | |
| Performance chemicals | |

structure-property relationship of different materials. Nanoparticles can generally be classified as one of two types: engineered or nonengineered (Table 4).[5,7,11,12,24,87] Engineered nanoparticles are intentionally designed and created with physical properties tailored to meet the needs of specific applications. They can be end-products in and of themselves, as in the case of quantum dots or pharmaceutical drugs, or they can be components later incorporated into separate end products, such as carbon black in rubber products.[11,24] Hence the particles' physical properties are extremely important to their performance and the performance of any product into which they are ultimately incorporated.

On the other hand, nonengineered nanoparticles are unintentionally generated nanoparticles, such as atmospheric nanoparticles created during combustion.[12] With nonengineered nanoparticles, physical properties also play an important role as they determine whether or not ill effects will occur as a result of the presence of these particles.[7,87] Some studies that have dealt with the usage of AFM in characterizing nanoparticles are discussed below pointing to their positive and negative aspects.

Nanostructured $TiO_2$ films obtained using sol-gel synthesis (hydrolysis (sol-product) and condensation (gel-product)) and $TiO_2$/acrylic-urethane nanocomposites were studied by Sung et al.[112] using AFM. Topographic and phase images were obtained simultaneously using a resonance frequency of approximately 300 kHz for the probe oscillation and a free-oscillation amplitude of 62 nm ± 2 nm. It has been reported that although AFM identifies the dispersion of nanoparticles, it is still necessary to subsidize the structure and dispersion of nanoparticles with other characterization techniques like SANS and neutron scattering.

An AFM investigation of size and surface properties of nanocrystalline ceria (synthesized by the micro-emulsion method) is reported by Gupta et al.[85] The investigation confirmed a relationship between the size and roughness of nanoceria as a function of the water-to-surfactant ratio. With increasing dilution of the surfactant, size distribution became narrow such that average particle size decreased linearly as the ratio increased without affecting lower size threshold of the particles (~10 nm). On the other hand, the surface roughness was found to increase with increasing water-to-surfactant ratio, implying diluted surfactant would provide

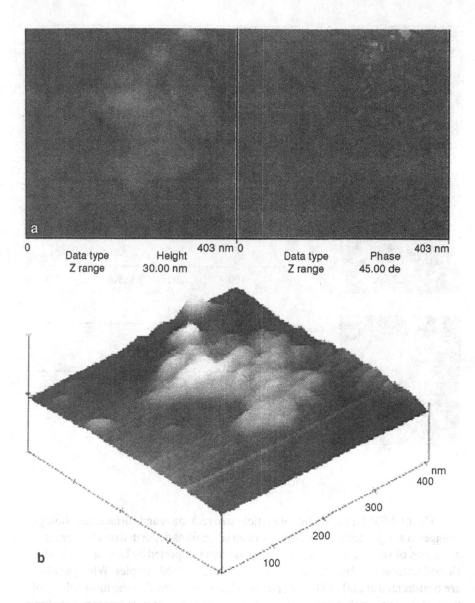

Data type      Height      403 nm    0     Data type      Phase     403 nm

Z range      30.00 nm         Z range      45.00 de

**FIGURE 8.** TM-AFM height and phase images of ceria particles; (ii) Illustration of measurement of cross-sectional profile of ceria particles: $Y$-axis represents the variation of vertical dimension in the selected regions of particles. (Reprinted with permission from Ref. 85. Copyright 2005 American Scientific Publishers)

rougher surface area of ceria nanoparticles (Fig. 8 (i)). The authors have reported that the information can be used to tailor the adhesion properties of nanoceria by optimizing the size distribution as well as surface roughness as a function of water to surfactant ratio. Figure 8 (ii) shows representative plots of section analysis of particles to monitor height variation with size.

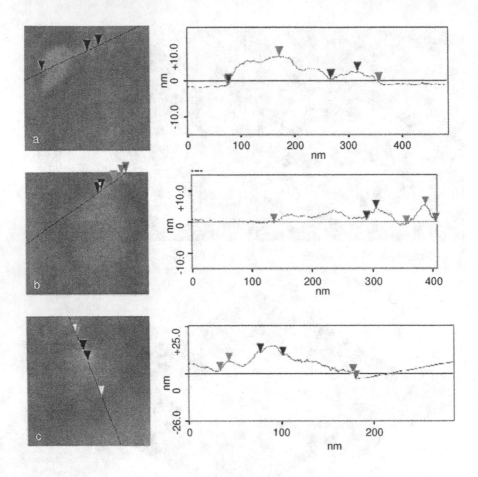

**FIGURE 8.** (*Continued*)

Use of AFM for the studies of particle size reduction and surface morphology changes in sol-gel derived nano-silver doped silica films deposited on glass slides as a function of varying heat treatment temperatures is reported by Li *et al*.[113] Needle-shaped features are observed on the surface in case of dried samples. When the films are heat-treated at 200, 300, 400 and 500°C, it is reported that the morphology of the surface particle changes to spherical, then a new feature with linear orientation, developing to a dendrite structure, and finally the flattening of the dendritic feature. Also it is reported that the major features on the surface flatten with an increase in temperature. A size reduction of >10 nm to 5 nm is also observed.

In another study of gold nanoparticles synthesized by sol-gel technique, AFM analysis reported the presence of ≤12-14 nm and 40-60 nm sized particles for $H_2S$/not heated and $H_2S$/heated samples respectively.[114] Extensive AFM characterization of surface morphology of Tin oxide (SnO$x$) semiconductor thin films coated on Pyrex glass (silica) substrates using the sol-gel dip-coating technique is also reported by Shukla *et al.*[115] Densely packed nanoparticles of near spherical shape with uniform particle size distribution have been observed. Average nanoparticle size is estimated to be $15 \pm 5$ nm, comparable with the resolution limit (tip radius) of AFM. Further HRTEM studies confirmed the size of the nano-crystallites to be 6-8 nm.

Parameters such as surface texture and surface roughness of nanoparticles play a determining role in some cases in altering the optical properties of materials, controlling adhesive properties of polymers, affecting the yield of processed silicon wafers, and also in controlling the density of stored magnetic materials.[116] There are numerous analytical methods for establishing the surface roughness as well as the visualization of surface texture. In addition to these techniques, AFM is ideal for quantitatively measuring the nanometer scale surface roughness and for visualizing the surface nano-texture on many types of material surfaces. Table 5 demonstrates the comparison of AFM with other techniques (optical and E beam).

**TABLE 5.** Comparison of AFM with other techniques for the measurement of nano surface texture-roughness. (From Refs. 86, 87, 185.)

| AFM | Optical techniques | Electron/ion beam methods |
|---|---|---|
| Nondestructive; Very high 3D spatial resolution; Line roughness measurements; Area roughness measurements | Horizontal resolution of all methods is limited to typically greater than $^1/_2$ μ | Capable of visualizing surface texture with horizontal resolutions of less than 1 nm. |
| | Require an optically opaque sample | Do not give quantitative 3D surface topograms; |
| | Microscopes: Excellent for visualization of surface texture Do not allow direct measurement of quantitative surface roughness parameters | Do not give contrast on flat homogeneous materials; Cross-sectioning of the sample can give accurate surface roughness values; |
| | Optical profilers: Ideal for rapidly measuring surface roughness parameters Horizontal resolution is greater than $^1/_2$ μ Large areas can be analyzed with optical profilers | Cross-sectioning can be difficult and the value may be changed during the process |
| | Scatterometry: Gives rapid surface roughness parameters Horizontal resolution is greater than 1 μ Not a direct measure of surface topography Does not allow visualization of surface texture | |

AFMs are well suited for visualization of surface texture, especially when the surface feature sizes are far below one micron. It is possible to measure the 1D surface roughness on a line (line roughness) in the horizontal or vertical direction in the AFM image. Optimal characterization of surface texture is often expressed with area roughness calculations that are made on the entire surface. Surface roughness calculations are similar to line roughness calculations but they include data in the $x$ and $y$ planes of the surface. It is also important to note here that there are also two primary potential sources of error when using an AFM for measuring surface textures.[117] The first is the probe geometry and the second is the length scale of the measurement.

1. *Probe geometry:* The surface texture that is measured with an AFM depends on the geometry of the probe tip. If the probe tip is larger than the features causing the surface texture, then the surface roughness measurements will appear smaller than they should be. This possible source of error is avoided by using the sharpest possible probe.
2. *Length scale:* Within the image used for the surface/area roughness calculation there must be an adequate sampling of the features giving rise to the surface texture. As a result, it is possible to get a different surface texture when the scan size is changed. This problem is avoided by using the same size scan range when surface roughness on several samples is being compared.

Some times just measuring the roughness of a sample is not enough to fully understand the affect of microstructure. This is particularly important for thin film optical coatings where a precise understanding of the relationship between the structure and optical scattering ability is needed as the amount of optical scattering depends not only on the roughness height of a structure but also on its lateral distribution. For this purpose power spectral density (PSD) function is used in AFM where a randomly rough surface is considered as a Fourier series of sinusoidal waves with different amplitudes, periods, and phases.[118] The grating equation shows that a single grating with spacing $d$ causes scatter into the angle, $s$, according to: $\sin \sigma = l/d$ where l is the wavelength of light; $d$ can be considered as one spatial wavelength present on the surface, or accordingly, $f = 1/d$ as one spatial frequency. At a randomly rough surface, many different spatial frequencies are present. This is quantitatively expressed by the PSD, giving the relative strength of each roughness component of a surface microstructure as a function of spatial frequency.[96]

## 3.2. Characterization of Nanocomposites

Polymer nanocomposites exhibit improved mechanical and physical properties because of their multicomponent and phase-separated morphology at the nanoscale.[21] A number of nanofillers (e.g., clays, graphite, carbon nanotubes, nanofibers, calcium carbonate, silica, alumina, titanate, etc.) are utilized for these purposes depending on the type of application, like, structural, electronics, thermal, barrier, and so forth.[32,119-128] Despite the advantages of using various nano-reinforcements in polymers, it is also important to note that with decreasing size of particles, their specific surface area becomes larger, and the probability of particles to agglomerate increases, leading to a number of loosened clusters of particles when added to polymers. Owing to this inhomogeneous distribution and dispersion of particles, under different loading or stressed conditions, this can result in poor performance of these nanocomposites. The ideal condition for enhanced improvement of mechanical, physical, and other properties of polymer nanocomposites can be recognized if all the nano-sized particles are dispersed uniformly in the matrix and with good bonding (either physical or chemical) to the matrix. In order to achieve homogeneous dispersion of nanoparticles, many different approaches have been used, including both physical and chemical methods like, surface treatment of nanoparticles, using coupling agents and compatibilizers, grafting agents, and so on.

For example, three general approaches have been adopted to modify the surface of carbon nanotubes to promote interfacial interaction between the matrix and CNTs: chemical, electrochemical, and plasma treatment. Velasco-Santos et al.[129] placed different organo-functional groups on multiwalled carbon nanotubes (MWCNTs) using an oxidation and silanization process. While Bubert et al.[130] modified the surface of CNTs using low-pressure oxygen plasma treatment and with the help of X-ray photoelectron spectroscopy (XPS), detected hydroxide, carbonyl, and carboxyl functionality on the surface layers of the CNTs. Also, silicate layers are generally organically modified to make them compatible and disperse in polymers.[131]

Amongst the vast nano-reinforcements available, clays have received special attention over the past two decades. Clay refers to a class of materials made up of layered silicates (similar to mica) for which the in-plane dimensions of the individual layers are on the order of a micron, and the thickness of a single clay nano-platelet is on the order of a nanometer (due to which they are the two-dimensional analogue of zero-dimensional quantum dots, nanoclusters and 1-D nanotube, nanowires materials).[132] These layered silicates are being used extensively to reinforce different thermoplastics and thermosets.

This is driven by the fact that once all the silicate layers are exfoliated in a polymer matrix, then with only a small percentage of filler loading, a range of properties improvement is observed, such as stiffness, strength, barrier, flame retardancy, and so on, which are important for lightweight automotive parts, packaging applications and, due to the limited oxygen and gas permeabilities, to enhance fire retardancy properties.[40,120,123,133,134] These property enhancements of nanocomposites, as mentioned in Section 1 are attributed to the unique properties of the nanoparticles, such as large interfacial area and particle/filler interaction that affects the glass transition temperature and polymer morphology characteristics.[119] The impact of nanoparticles on the crystallization process can result in changes in lamellar size, crystalline phases, supramolecular structure, degree of crystallinity, and rate of crystallization.[124,125] Recently, AFM has been used extensively to characterize the morphology of the PNCs, internal organization of the spherulites,

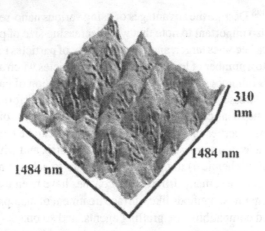

**FIGURE 9.** AFM image of PVC/clay nanocomposite showing the dispersion of fine clay layers [Reprinted with permission from L. Kovarova, A. Kalendova, J.F. Gerard, J. Malac, J. Simonik and Z. Weiss, Structure analysis of PVC nanocomposites, Macromolecular Symposia 221 (1), 105-114 (2005). Copyright (2005) WILEY – VCH).

the size of the nanoparticle aggregates, and the location of the nanoparticle aggregates.[119]

An AFM study of blend morphology and organoclay dispersion in an incompatible blend of polyamide-6(PA6) (as a matrix)/polypropylene(PP) (dispersed)/organoclay nanocomposite has been reported by Chow et al.[121] In a similar study, Kovarova et al.[32] have reported AFM studies in conjunction with the studies of TEM, SEM, and XRD of poly(vinyl chloride) (PVC)/clay nanocomposites prepared by melt intercalation process. Intercalation, exfoliation, nanophase dispersion and orientation of clay layers have been investigated.

Figure 9 is an AFM image of PVC/clay nanocomposite cut on ultramicrotome and plasma etched showing the dispersion of clay layers in the PVC matrix. Liu et al.[120] have also reported the morphology studies of hybrid epoxy nanocomposites modified with carboxyl-terminated butadiene acrylonitrile (CTBN) rubber and organoclay with AFM supported by XRD and SEM. A study of the morphology of rubber-based clay nanocomposites using AFM has been reported by Sadhu and Bhowmick.[17] Studies of qualitative and quantitative analysis of dispersion of nanoclay in nitrile rubber (NBR) and styrene-butadiene rubber (SBR) nanocomposites and the influence of nature of clay, copolymer composition, and polarity of the rubber on the intercalation and exfoliation (Fig. 10) processes have been reported.[134]

Dietsche et al.[123] used AFM to study the morphology of acrylic nanocomposites consisting of methyl methacrylate (MMA) / n-dodecyl methacrylate (LMA) copolymers and intercalated layered silicates. They have reported from both AFM and TEM studies that the resulting structures were very soft with respect to the polymer unlike the assumption of glasslike rigidity of the layered silicate in the literature.

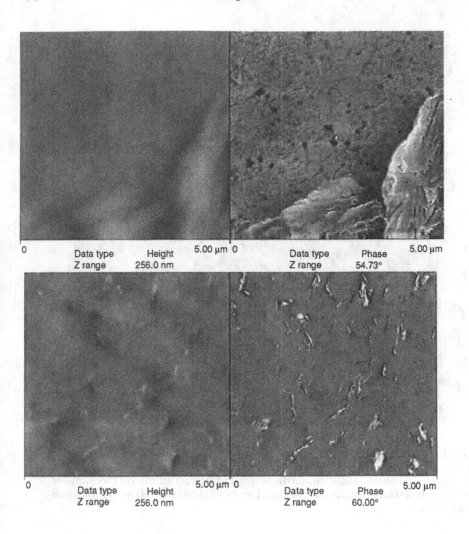

**FIGURE 10.** AFM images of SBR clay nanocomposite showing the exfoliated clay particles. (From S. Sadhu and A. Bhowmick, Morphology study of rubber based nanocomposites by transmission electron microscopy and atomic force microscopy, Journal of Materials Science 40(7), 1633-1642 (2005), with kind permission of Springer Science and Business Media).

While, Yalcin et al.[133] have reported the AFM studies of the individual montmorillonite platelets dispersed in PVC matrix (Fig. 11) and it has been observed that the edges of the particles were straight in some cases forming hexagonal angles and in other cases irregular. In another similar kind of study, AFM (tapping, contact, LFM and FFM modes) was used to study the clay nanoplatelets and their impurities. AFM, LFM and AFM, FFM images of clay deposited on mica and highly oriented pyrolytic graphite (HOPG) are shown in Fig. 12. It was concluded

**FIGURE 11.** AFM phase image (hard tapping) of plasticized PVC nanocomposite (10-wt% clay loading). (Reprinted from Polymer 45(19), B. Yalcin and A. Cakmak, The role of plasticizer on the exfoliation and dispersion and fracture behavior of clay particles in PVC matrix: a comprehensive morphological study, 6623-6638, (2004), with permission from Elsevier.)

that the clay sheets have substantial quantities of mobile impurities that cannot be separated from the clay nanoplatelets. The impurities are found to be water soluble (sodium or its counter ions), and the other is silicate based water-insoluble (found to be mobile on an atomically smooth surface).[40]

Amongst other reinforcements, carbon nanotubes, though expensive, have been used in the recent past extensively as they possess extraordinary mechanical properties as: tensile strength of 11-63 GPa, elastic bending modulus of ~200 MPa, Young's modulus values in the order of TPa (~1.7 TPa), and in addition, a high aspect ratio (100-1000) which enables the formation of a network structure in a matrix material at relatively low contents. Many researchers have also reported the intrinsic superconductivity,[135] field emission behavior,[136] potential as molecular quantum wires,[137] ability to store hydrogen,[138] unusually high

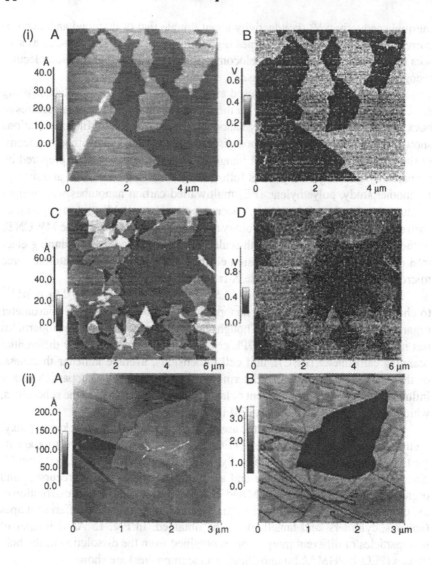

FIGURE 12. (i) AFM and LFM images of clay nanoplatelet samples on mica directly after deposition (top) and 5 days after deposition (bottom). (A) and (C) are contact mode images of the topology. (B) and (D) are LFM images, where lighter shades indicate higher friction. (ii) Force modulated microscopy of individual clay nanoplatelets on HOPG. (A) is the topology image, and (B) is the FMM image. In the FMM image, a darker shade indicates lower vibration amplitude of the AFM tip, hence greater damping of the FMM signal. (Reprinted with permission from R.D. Piner, T.T. Xu, F.T. Fisher, Y. Qiao and R.S. Ruoff, Atomic force microscopy study of clay nanoplatelets and their impurities, Langmuir 19(19), 7995-8001 (2003). Copyright (2003) American Chemical Society).

thermal conductivity[139] (conductivity about double that of diamond and electric-current-carrying capacity 1000 times higher than copper wire.[140]), usage as sensors for gas detection,[141] and the biocompatibility and potential for biomolecular recognition[142] of carbon nanotubes.[131]

AFM has been successfully used even to characterize the nanotubes. Wang et al.[143] have used AFM and SEM for the characterization of the nanostructures of buckypaper and buckypaper/epoxy nanocomposites. AFM and SEM observations showed that the SWNTs have a good dispersion in the buckypaper and nanocomposites. Figure 13 shows an AFM image of the buckypaper surface prepared by prepared by grinding BuckyPearls followed by multistep sonication and drying. In another study, polyethylene (PE) multiwalled carbon nanotubes with weight fractions ranging from 0.1 to 10 wt.% were prepared by melt blending using a mini-twin screw extruder.[131] The morphology and degree of dispersion of the MWCNTs in the PE matrix at different length scales was investigated using scanning electron microscopy (SEM), transmission electron microscopy (TEM), atomic force microscopy (AFM) and wide-angle X-ray diffraction (WAXD).

Small-angle light scattering (SALS) and AFM were used by Ma et al.[119] to characterize the effect of $TiO_2$ nanoparticles on the nanometer to micrometer organization of LDPE crystals. Though the presence of the $TiO_2$ nanoparticles has no effect on the degree of LDPE crystallinity (as determined by differential scanning calorimeter (DSC)), unit cell dimensions, average lamellar thickness, or the average spherulite size (determined by WXRD), the nanoparticles have influenced the internal arrangement of lamellae (or bundles) within the spherulites, which can be clearly seen in the AFM images (Fig. 14).

Renker et al. have described the use of poly(ethylene oxide)-block-poly(hexyl methacrylate) diblock copolymers (PEO-b-PHMA) as structure-directing agents for the synthesis of nanostructured polymer-inorganic hybrid materials. (from (3-glycidylpropyl) trimethoxysilane and aluminum sec-butoxide as precursors and organic, volatile solvents).[126] AFM and TEM studies showed that the dissolution of the composites rich in poly(hexyl methacrylate) nanoparticles of different shapes (spheres, cylinders, and lamellae) can be obtained. In Fig. 15 AFM images of nanoparticles of different morphologies obtained from the dissolution of the bulk phases (PEO-b-PHMA/aluminosilicate nanocomposites) are shown.

AFM study of the morphological investigation of organic-inorganic hybrid nanocomposites comprising polyamide-6,6 (PA66) and silica ($SiO_2$) synthesized through sol-gel technique (at ambient temperature) is reported by Sengupta et al.[34] Tapping mode AFM images of the nanocomposites revealed the dispersion of $SiO_2$ particles with dimensions of <100 nm in the form of network as well as linear structure. The lengths of the linear structures were in the micron range while the width varied between 50 and 70 nm. Discrete nanosized silica particles (<50 nm) are also dispersed throughout the matrix. Matsumura et al.[128] have also reported a study of cellulosic (heterogeneously partially hexanoylated cellulose samples embedded in epoxy resin) nanocomposites using AFM. The studies revealed biphasic morphology of the thermoplastic composites and the phase images indicated distinct

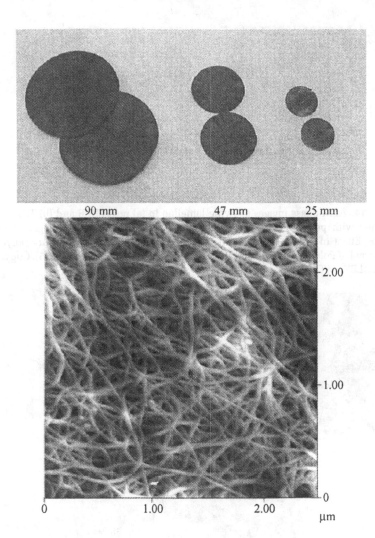

**FIGURE 13.** (a) Single-walled carbon nanotube buckypaper (thin (10-50 mm) membranes of nanotube networks produced by multiple steps of tube dispersion and suspension filtration); (b) AFM image of buckypaper surface (average diameter of the ropes is about 30-60 nm) (Reprinted from Composites Part a-Applied Science and Manufacturing 35(10), Z. Wang, Z.Y. Liang, B. Wang, C. Zhang and L. Kramer, Processing and property investigation of single-walled carbon nanotube (SWNT) buckypaper/epoxy resin matrix nanocomposites, 1225-1232, (2004), with permission from Elsevier).

periodicity on the scale of several tens of nanometers. AFM has also been used to characterize the topography of epoxy-silica nanocompoites that showed good transparency and miscibility as observed with AFM, SEM, and TEM.[144]

In addition to the morphological (phase and topography) and dispersion quality characterization of polymer nanocomposites, AFM has also been used

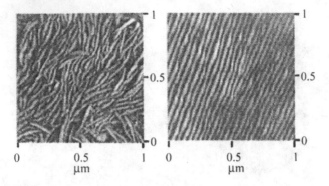

**FIGURE 14.** AFM phase images showing lamellae in (a) neat LDPE and (b) $TiO_2$/LDPE (Reprinted with permission from D.L. Ma, Y.A. Akpalu, Y. Li, R.W. Siegel and L.S. Schadler, Effect of Titania nanoparticles on the morphology of low density polyethylene, Journal of Polymer Science Part B-Polymer Physics 43(5), 488-497 (2005). Copyright (2005) WILEY – VCH).

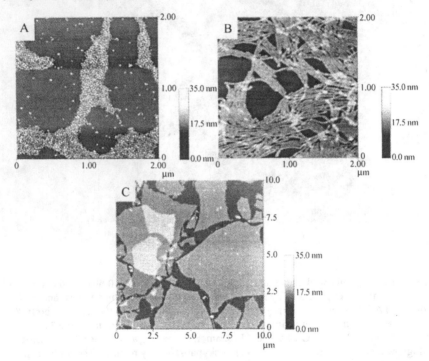

**FIGURE 15.** AFM images of calcined nano-objects (A) spheres (21 wt.% metal alkoxides), (B) cylinders (34 wt.% metal alkoxides) and (C) plates (53 wt.% metal alkoxides) (Reprinted with permission from S. Renker, S. Mahajan, D.T. Babski, I. Schnell, A. Jain, J. Gutmann, Y.M. Zhang, S.M. Gruner, H.W. Spiess and U. Wiesner, Nanostructure and shape control in polymer-ceramic hybrids from poly(ethylene oxide)-block-poly(hexyl methacrylate) and aluminosilicates derived from them, Macromolecular Chemistry and Physics 205(8), 1021-1030 (2004). Copyright (2004) WILEY – VCH).

to investigate the viscoelastic behavior, mechanical, and also tribological properties of these materials.[132] As most polymers exhibit viscoelastic behavior, the strain response lags the stress by a phase angle that is characteristic of the material. So, many studies have reported that the phase lag measured during tapping mode can be related to the attractive and repulsive, or adhesive, forces at the interface and the viscoelastic properties of the sample.[145-150] Scott and Bhushan have used the phase contrast in AFM to detect and quantify changes in composition across polymer nanocomposites (polyethylene terephthalate (PET) films with embedded ceramic particles, metal particle (MP) magnetic tape, and Si(100) with a non uniform Z-15 lubricant film) and molecularly thick lubricated surfaces by taking the advantage of contrast in viscoelastic properties of the different materials across the surface.[132]

Figure 16 shows the schematic of the AFM setting used by them. It is reported that very little correlation is found between phase angle images and friction force images for PET films with embedded ceramic particles and MP tape. Also, a numerical vibration model was developed and verified the viscoelastic properties and that low phase angle corresponds to low viscoelastic properties. More detailed discussion on the mechanical property and tribological characterization of a nanomaterial/nanocomposite with AFM is given in Section 3.4.

Also, as AFM provides easy access to observations under native and near-native conditions in contrast to SEM and TEM that are applicable only to biologically inactive, dehydrated samples and generally require extensive sample preparation, such as sectioning, staining or metal coating, AFM has often been applied to the study of cellulose fibers and crystals, cellulose derivatives, pulp and paper products, and wood under both dry and wet conditions.[128] Gray et al. used AFM to study the local surface variations in pulp, such as the orientation of micro fibrils and the aspect of fibrillation by beating and such parameters as lamellation for wood. Baker et al. studied the surface topography of Valonia cellulose I micro crystals under propanol and water and obtained images that revealed clear structural details consistent with the 0.54 nm repeat unit (glucose) along the cellulose chains.[128] An intermolecular spacing of about 0.6 nm was also described. AFM was also often used to evaluate the phase dimensions of cellulosic blends and composites.

Metal-insulator nanocomposites have useful properties such as electric, magnetic and optical properties. Hence the identification of nano-sized metal particles embedded in an insulating matrix with high resolution is very desirable, especially in the field of materials science.[151] AFM has been widely used in profiling the local electronic structure and morphology of various surfaces with high spatial resolution. However, normal AFM is not able to identify nano-sized metal particles in metal-insulator composites, (neither STM as the insulator prevents the use of STM) because it is chemically insensitive.[151] When the concentration of metal phase exceeds the percolation threshold, electron transport occurs in these materials via connected conducting networks (CNs). Along with the mapping of the CNs, electron transport (down to the scale of individual conducting paths)

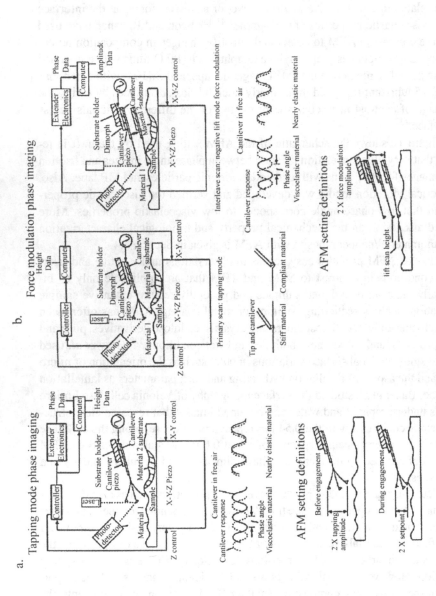

**FIGURE 16.** Schematic of (a) tapping mode used to obtain height and phase data and (b) force modulation mode used to obtain amplitude (stiffness) and phase data (Reprinted from Ultramicroscopy 97(1-4), W.W. Scott and B. Bhushan, Use of phase imaging in atomic force microscopy for measurement of viscoelastic contrast in polymer nanocomposites and molecularly thick lubricant films, 151-169, (2003), with permission from Elsevier).

can also be studied using the $c$-AFM. When the AFM tip contacts metal particle (during the surface scan), with an applied bias electrons can flow from the sample to the tip through which the CNs can be mapped. On the other hand when the AFM tip contacts the insulating phase there is no current. From the simultaneous observation of current and topographic images, metal particles from the insulating matrix can be identified.

Luo et al.[151] have presented a study on mapping CNs and identifying the metallic phase in percolating metal insulator nanocomposites $Ni_x(SiO2)_{1-x}$ and $Fe_x(SiO2)_{1-x}$ by $c$-AFM. Commercial $Si_3N_4$ cantilevers coated with a Cr (10 nm) layer and an Au layer of about 100 nm were used. Figure 17 shows the current and surface morphology images of $Ni_x(SiO2)_{1-x}$ sample. Tip-induced anodization

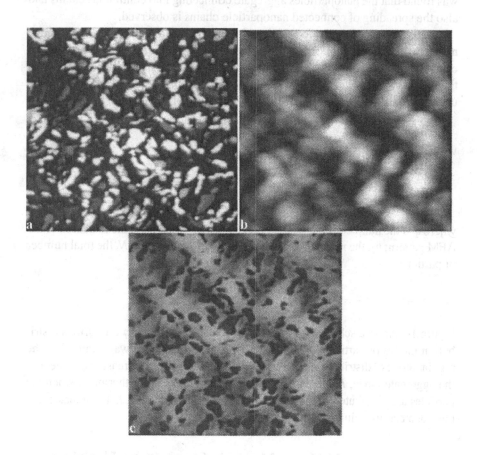

**FIGURE 17.** (a) Current images by $c$-AFM at a bias of 0.45 V on an $Ni_x(SiO2)_{1-x}$; (b) surface morphology image corresponds to $a$; (c) the superimposed image from $a$ and $b$. (From E.Z. Luo, J.B. Xu, W. Wu, I.H. Wilson, B. Zhao and X. Yan, Identifying conducting phase from the insulating matrix in percolating metal-insulator nanocomposites by conducting atomic force microscopy, Applied Physics a-Materials Science & Processing 66, S1171-S1174 (1998), with kind permission of Springer Science and Business Media).

(c-AFM operated in air) has been observed (increase in the FeO layer thickness) with the repeated scanning of the same area. They have also reported that the size of the metal particles observed in AFM images (30-40 nm) is far too higher than their actual size of 3-6 nm (obtained from HRTEM images) which can be explained by the tunneling effect and the tip size.

AFM studies of nanostructured metal (Pd, Sn, Cu)-polymer (poly-para-xylylene) and metal-oxide-polymer composites reveal the metal nanoparticles to have a size of 7-10 nm.[152] Three different types of surface morphology are distinguished based on the metal concentration in the composites. It has been reported that the surface of the Pd nanocomposites is sufficiently uniform and the spherical polymeric globules of a size up to 200 nm were well distinguishable. The size of the inorganic particles ranged from 7-10 nm. In the case of Sn nanocomposites it was found that the nanoparticles aggregate connecting into continuous chains and also the spreading of connected nanoparticle chains is observed.

AFM can also be used to calculate the specific surface area and edge-to-edge nearest neighbor length correlation distance through the size, shape, position of each feature in the images.[133] Quantitative analysis of the nano-filler dispersion in the matrix can be performed by the statistical processing of the AFM photographs using the quadrate method and Morishita's $I_\delta$ value. The $I_\delta$ index, is given by

$$I_\delta = q\delta$$

with

$$\delta \frac{\sum_{i=1}^{q} n_i (n_i - 1)}{N (N - 1)}$$

where $q$ is the number of elemental parts equally divided from the total area of the AFM pattern; $n_i$, the number of particles in the $i^{th}$ section; and $N$, the total number of particles:

$$N = \sum_{i=1}^{q} n_i$$

Figure 18 shows a schematic of the dependence of the $I_\delta$ on $q$ for various distribution modes of particles. For Poisson's distribution, $I_\delta$ is always unity. For the regular mode of distribution, $I_\delta$ gradually decreases with an increasing $q$ value. For the aggregate mode, $I_\delta$ increases as the $q$ value increases. Furthermore, when the particles are distributed in the regular mode in each aggregate, $I_\delta$ has a maximum peak at a certain value of $q$.[133]

### 3.3. Conductive AFM as a Means to Characterize Electrical Properties

Imaging of different nanomaterials/nanocomposites is possible by conventional atomic force microscopes, as shown in previous sections. In addition, AFM can

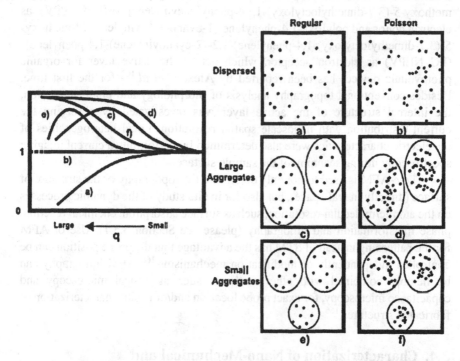

**FIGURE 18.** The dependence of Morishita's index $I_\delta$ on q for various distribution modes of particles (Reprinted from Polymer 45(19), B. Yalcin and A. Cakmak, The role of plasticizer on the exfoliation and dispersion and fracture behavior of clay particles in PVC matrix: a comprehensive morphological study, 6623-6638 (2004), with permission from Elsevier).

also be used to investigate force interactions and conductivity through organic films and polymers. The AFM topographic images not always reveal the real surface morphology. Harder surface areas might be less deformed by the tip during scanning and might appear higher in morphology images.[153] If the cantilever and tip are made from of electrically conducting material, besides topographical information of the sample, its electrical properties can be studied at the same time. On the other hand, phase mode of imaging is generally used for supporting topographical imaging of nonconducting heterogeneous surfaces.

Measuring the surface in contact mode with a voltage applied between tip and sample allows for obtaining the topography and the current distribution simultaneously. This method is called conductive or current sensing AFM (c-AFM). The resolution of c-AFM is as small as the tip-sample contact area, which can be less than 20 nm. This method can widely be used for the characterization of inorganic semiconductors[154-161] and its potential for the study of electrical properties of organic materials has been demonstrated.[162-164]

It has also been reported that the measurement of the I-V characteristics of the samples can help in determining several electrical parameters such as grain resistivity and tip-sample barrier height.[165] c-AFM study of the spatial distribution of electrical properties of semiconducting polymer blends (poly[2-methoxy-5-(3,7-dimethyloctyloxy)-1, 4-phenylenevinylene] (MDMO-PPV) as electron donor and poly[oxa-1,4-phenylene-(1-cyano-1, 2-vinylene)-(2-methoxy-5-(3,7-dimethyloctyloxy)-1,4-phenylene)-1,2-(2-cyanovinylene)-1,4-phenylene] (PC NEPV) as electron acceptor, which act as the active layer for organic photovoltaic devices has been reported by Alexeev et al.[162] for the first time. Besides conventional topography analysis of morphology and phase separation, the internal structure of the active layer was investigated by observing the current distribution with nanoscale spatial resolution. Local heterogeneities of the electric characteristics were also determined by performing a current imaging spectroscopy during scanning of the sample surface.

The c-AFM is used to not only determine the topography or the structure of surface (directly in real-space), but also for in situ study of the dynamic processes on the atomic/molecular resolution, such as surface adsorption, chemical reaction, phase transformation and lithography (please see Section 3.5.1). Use of AFM for the patterning instead of STM has the advantage that the probe position can be controlled independently of modification mechanisms.[166] AFM lithography can be flexibly combined with other methods, such as optical microscopy and capacitance microscopy, for exact probe location and/or in situ characterization of fabricated structures.[166]

## 3.4. Characterization of Nano-Mechanical and Nano-Tribological Properties

Different materials used in the nanotechnology applications like the nanopatterning, nanofabrication or for nanolithography requires the understanding of their mechanical properties at the very first surface layers as they are shown to govern their performance. Additionally, these properties are very important in determining the micro- and nano-tribological response of materials as in many advanced technological industries of semiconductor and data storage to help optimize polishing processes and lubrication of data storage substrates.[86,111,167-172] In traditional industries, such as, automotive and aerospace, tribological studies help increase the lifespan of mechanical components. Also, development of lubricants in the automobile industry depends on the adhesion of nanometer layers (mono layers) to a material surface.[170] Assembly of components can again depend critically on the adhesion of materials at the nanometer length scale.

For understanding the nanomechanical properties and nano-tribological response of different materials (bulk and thin films) at the surface and/or near-surface, depth-sensing instruments (DSI) are being widely used. These type of instruments have been developed in a number of research laboratories including that of Nishibori and Kinosita,[173] Pethica,[174] Tsukamoto et al.,[175] Oliver,[176] Weiler,[177] and CSIRO Division of Applied Physics.[178] The mechanical properties that are determined conveniently using these techniques are hardness, elastic modulus, yield stress, and fracture toughness. Despite these advantages of depth sensing instruments, the load resolution of these instruments is not good (not better than ±100 nN) and also most of them are unable to detect initial contact loads of less than 1 μN.[90] Additionally, for an accurate analysis of hardness of thin films using DSI, the film thickness should be at least five times the depth of penetration. Due to the miniaturization of components in electronics and computer industries, hardness measurements of ultrathin films (10 nm or less) is necessary, which demands the load and depth resolutions to be accurate. Furthermore, conventional nanoindenters does not allow precise positioning of indentations and thus, aligning the indenter tip with the feature of interest is difficult.[90,179] For these purposes AFMs are used which showed promising results even when the indentation depth is as low as 1 nm.[172,180-184] Also, for nano-hardness measurements, it is easier to calibrate the measurements with AFM than with any nanoindenter.

However, for these, the tip and the AFM set-up are different from the normal mode AFM. Generally, a very sharp diamond tip is used and the AFM must have a heavy load mechanism for the tip and operated in a force mode. A force curve is produced, which is a plot of tip deflection as a function of the vertical motion of the scanner. This curve is analyzed to produce the local mechanical response. Additionally, in these cases, the selection of spring constant of the cantilever probe also plays an important role for identifying the small differences in response.[90] Major advantages of the AFM for studying tribological behavior is that it can be routinely used on all types of materials and in all environments including vacuum and liquids. Materials commonly studied include: ceramics, metals, polymers, semiconductors, magnetic, optical, and biomaterials. Some examples of the application of AFM for sensing nano-tribological behavior are listed below:

- As a tool to scratch the surfaces at the nanoscale;
- Direct three-dimensional visualization of wear tracks, or scars on a surface;
- Measurement of the thickness of solid and liquid lubricants having nanometer or even monolayer thickness;
- Measurement of frictional forces at the nanometer scale;
- Surface characterization of morphology, texture, and roughness;
- Evaluation of mechanical properties such as hardness and elasticity, and plastic deformation at the nanometer scale;
- Other applications include, evaluating nanometric deformations, strain rate in nanostructures, strain relaxation and determination, dislocation behaviour, peeling analysis, surface displacements, cellular strain distribution, contrast and elasticity analysis, mapping of subsurface defects, thermal effect studies, topographic and spectroscopic imaging, and so forth.

Furthermore, using pulsed force mode in AFM, the stiffness of a sample at a matrix of locations is measurable. From this data it is possible to create a stiffness mapping of a surface.[87] Nonetheless, stiffness maps can only be made on samples where the stiffness of the surface is lower than the stiffness of the cantilever. Adding a fixture to the stage of the AFM makes the study of material behavior such as plastic deformation and fracture possible. The fixture permits creating forces on a sample while AFM images are being taken. Although various technical issues are associated with the AFM probing such as nonaxial loading, jump-into contact, nanoscale contact area, high local pressure, and topographical contributions, a number of successful applications of the AFM probing technique have been demonstrated to date.[179-181]

A variety of materials have been studied, such as polymer hydrogels, thin polymer films, fiber-reinforced composites, organic lubricants, self-assembled monolayers, polymer blends, block-copolymers, polymer brushes, individual macromolecules, biological materials,[180,182,183] and also including nanostructured materials such as nanocrystals, nanocomposites, nanograins, nanotubes, nano-ceramics, and nano-powders. Absolute values of the elastic modulus were measured in the range from 0.001 to 30 GPa and in a wide range of temperatures and frequencies, for organic films with thickness down to 2 nm, and with vertical and lateral resolution as low as 1-2 and 5-10 nm, respectively. Elastic modulus (loss and storage), surface glass transition temperatures, and relaxation times all have been obtained with reasonable confidence by applying direct force-distance measurements and/or cantilever modulated (vertically as well as laterally) modes.[180]

### 3.4.1. Evaluating Nano-Mechanical Properties

The conventional way of hardness determining from the indentation techniques depends on the computation of the residual indented area of contact, which was generally measured using the imaging techniques.[185] This type of technique, if used for measuring the contact area of nano-indentations, the errors in measurement will be large. Additionally, this technique can provide only the plastic response of the material tested and is not useful for obtaining elastic and viscoelastic/plastic properties of the materials. Therefore, compliance methodology (developed by Oliver and Pharr[176]) was introduced, which utilizes the force-displacement curve (Fig. 19) during both loading and unloading in determining the mechanical properties.

$$\text{Hardness, } H = P_{max}/A$$

The slope of the tangent drawn from the first part of the unloading curve represents the contact stiffness, $S$, evaluated at the maximum displacement (i.e., $S = (\partial P/\partial h)h_{max}$). Elastic modulus calculated using:

$$\frac{1}{E_r} = \frac{1 - v_{s^2}}{E_s} + \frac{1 - v_{i^2}}{E_i}$$

$P_{max}$ applied maximum normal load; $A$ is the projected contact area; $E_r$ is the reduced elastic modulus; $E_i$ and $E_s$ are the Young's modulus of indenter and sample; $v_s$ and $v_i$ are the Poisson's ratio of sample and indenter.

However, the compliance method to derive these properties relies on the analysis of the unloading segment of the load-displacement response by assuming it to be elastic, even if the contact is elastic-plastic. For most materials such as metals or ceramics, the estimation of the hardness and elastic modulus using this

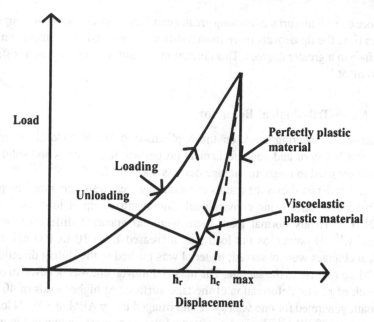

**FIGURE 19.** Schematic diagram of indentation load-displacement data for a viscoelastic-plastic material.

approach results in accurate data, but for polymeric materials, because of their time- and rate-dependent behavior, under load leads to an inadequate estimation of the properties.[185] Additionally, the hardness determined from the nano-indentation measurements using the Oliver and Pharr methodology tends to be overestimated due to the indentation size effect (ISE).[186,187] Further, phenomena like pile-up and sink-in observed during nano-indentation are still under constant debate.

In addition to the above procedure based on Oliver and Pharr[176] for the calculation of mechanical properties of different materials using force-displacement curves during nano-indentation with AFM, other methodologies are also used. In some cases, the surface was imaged before and immediately after the indentation and nano-hardness was calculated by dividing the indentation load by the projected residual area.[172,188] Recently, interfacial force microscope (IFM), a modified version of AFM was used to probe the indentation response of the polymeric materials.[87] The rigid displacement control of this modified version allowed the control of tip penetration into the sample with sub-nanometer resolution. This helps in ensuring much lower applied forces in IFM when compared to conventional DSI, giving nanoscale spatial resolution. Penetration of the tip into the sample is much smaller than the probe radius (70-500 nm) and so Hertzian analysis is used to evaluate the indentation response.

Negative lift force modulation technique is also applied for the stiffness measurement. The AFM cantilever/tip assembly is moved up and down by a bimorph with some applied load. Stiffer sample surfaces are deformed to a lesser extent than more compliant surfaces causing greater cantilever deflections than compliant surfaces (i.e., the tip deflects more from a stiffer surface and sinks into a compliant surface to a greater degree). This deflection measurement serves as a stiffness measurement.[132]

### 3.4.2. Nano-Tribological Behavior

Bhushan did pioneering work on the application of AFM to evaluate nano-tribological behavior and nanomechanical properties of thin films and solid surfaces that are used in magnetic storage devices and MEMS.[86,111,167–172,188] As an example, Fig. 20 (i) shows the results of nanoscale wear tests performed on polymeric magnetic tapes using conventional silicon nitride tip at loads of 10 and 100 nN. For 10 nN normal load, there were no apparent differences in the topography.[189] However, as the load was increased from 10 to 100 nN, topographical changes were observed; material was pushed in the sliding direction of the AFM tip relative to the sample. The material movement was believed to occur as a result of plastic deformation of the tape surface. At higher loads of 40 $\mu$N, wear mark generated for one scan cycle and imaged using AFM at 300 nN load is shown in Fig. 20 (ii).[190] The inverted map of the wear mark shown in Figure 20 (ii) (b) indicates the uniform material removal at the bottom of the wear mark.

Bhushan and Koinkar[172] used AFM with a specially prepared diamond tip (single-crystal natural diamond with a shape of three-sided pyramidal) having a radius of curvature of about 100 nm to measure the hardness of Si(111) wafer. The measurements were made in the normal load range of 10-150 $\mu$N and with a cantilever stiffness of 45 N/m. The nano-hardness and normal load as a function of indentation depth for the as-received Si (111) are shown in Fig. 21. As can be seen, the hardness measurements at even 1 nm depth are given. Also, the higher hardness values observed in the low-load indentation is attributed to the pressure-induced phase transformation during the nanoindentation. In another study, Kulkarni and Bhushan studied the nano indentation study of single-crystal aluminum (100) using a Berkovich indenter in conjunction with an AFM at loads ranging from 15 to 1000 $\mu$N.[191] Study of the load displacement behavior, hysteresis, hardness, Young's modulus of elasticity, creep and strain-rate sensitivity of single-crystal aluminum have been reported. It was also found that indentation hardness of aluminum decreases with an increase in indentation depth; the reason being the sampling volume effect. Deformation of aluminum on a nanoscale is found to be sensitive to strain rates.

The effect of reduced size on the elastic properties measured on silver and lead nanowires and on polypyrrole nanotubes with an outer diameter ranging between 30 and 250 nm was recently reported by Cuenot et al.[45] Resonant contact AFM is used to measure their apparent elastic modulus. The measured modulus of the nanomaterials with smaller diameters is significantly higher than that of the larger ones which was attributed to surface tension effects. It has also been reported that the surface tension of the probed material can be experimentally

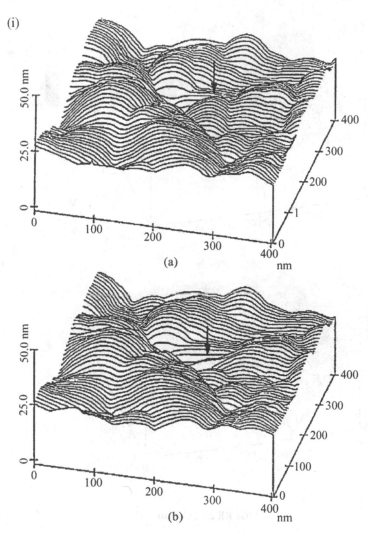

**FIGURE 20.** (i) Surface roughness maps of a polymeric magnetic tape at the applied normal load of (a) 10 nN and (b) 100 nN. Location of the change in surface topography as a result of nanowear is indicated by arrows. (ii) (a) Typical grayscale and (b) inverted AFM images of wear mark created using a diamond tip at a normal load of 40 N and one scan cycle on Si (100) surface (Reprinted from Wear 251, B. Bhushan, Nano- to microscale wear and mechanical characterization using scanning probe microscopy, 1105-1123 (2001), with permission from Elsevier).

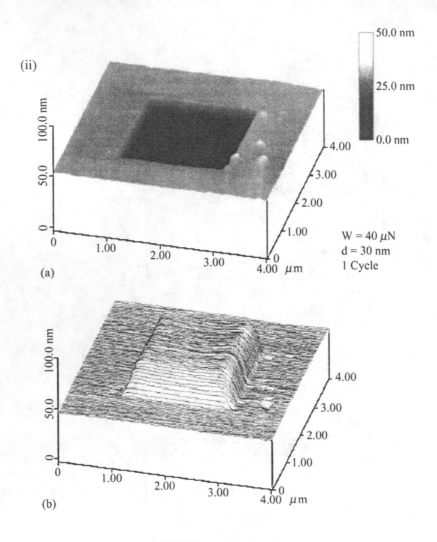

**FIGURE 20.** (*Continued*)

determined from the AFM measurements.[45] Investigation of the surface structure and elastic properties of calcium silicate hydrates (C-S-H), a main constituent of the cement phase at the nanoscale using AFM is reported for the first time by Plassard *et al.*[192] Because of it's nanocrystalline character,[193] C-S-H presents only a short range crystalline order[194,195] in cement paste.

A few techniques, such as magic angle spinning nuclear magnetic resonance (NMR) and X-ray absorption spectroscopy, are adapted to the study of crystalline structure of C-S-H; but these techniques were successful only to partly elucidate the crystalline nature. A contact mode AFM has been applied to study these C-S-H surfaces along with the structural properties. In the topographic mode, structural

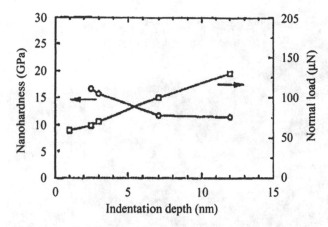

**FIGURE 21.** Nano-hardness and normal load as a function of indentation depth for the as-received Si (111) sample (Reprinted with permission from B. Bhushan and V.N. Koinkar, Nanoindentation Hardness Measurements Using Atomic-Force Microscopy, Applied Physics Letters 64(13), 1653-1655 (1994). Copyright [1994], American Institute of Physics).

information concerning the plan of the layers of C-S-H is demonstrated, while with nanoindentation, information in the direction perpendicular to the layers was obtained. Most of the atomic resolutions confirmed the ordered character of C-S-H and thus made it possible to determine the cell surface parameters of C-S-H.

Many researchers have used coated AFM tips with biological macromolecules to study their interactions with surfaces.[196-198] Among these colloid probe technique is well known. This technique involves the use of a sphere of known dimensions and material to construct a colloid probe, because the geometry of AFM tips is often unknown. Figure 22 shows an example of such a tip.[184] Ducker et al. first reported the use of colloid probes to study the influence of electrolyte on the interaction of silica surfaces. AFM instrument as a nanoindenter has been used by Kinney et al.[199] to study the elastic properties of human dentine. They determined the Young's moduli of peri- and intertubular dentine and measured only slight variation in the axial and transverse shear moduli with position in the tooth. An AFM tip is also used as a nano manipulation instrument acting as a nano-scalpel or nanotranslator tool. Xu and Ikai used the adhesive force between a silicon nitride tip and plasmid DNA strands bound to a mica surface to pick up a single molecule of DNA.[184]

As a force sensor and nanoindenter, AFM can directly measure properties such as the Young's modulus of surfaces or the binding forces of cells. AFM as a stress-strain gauge can study the stretching of single molecules or fibers and as a nano-manipulator it can dissect biological particles such as viruses or DNA strands. The adoption of an analysis of AFM probing of ultrathin (1-100 nm thick) polymer films and polymer films with a multilayered structure (molecularly thick hyperbranched polymer monolayers (<3 nm thick) on a solid substrate, and

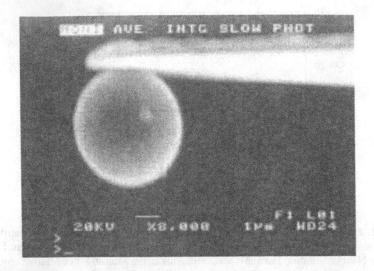

**FIGURE 22.** Scanning electron micrograph of a silica sphere immobilized at the apex of an AFM tipless cantilever to construct a colloid probe (From W.R. Bowen, R.W. Lovitt and C.J. Wright, Application of atomic force microscopy to the study of micromechanical properties of biological materials, Biotechnology Letters 22(11), 893-903 (2000), with kind permission of Springer Science and Business Media).

trilayered polymer films (20-40 nm thick) tethered to a solid substrate) is reported by Shulha et al.[180]

Further, a study on the usage of AFM to evaluate the nanomechanical and nano-tribological properties where the AFM tip slid against the SWNT reinforced epoxy composites was reported by Li et al.[200] It was shown that the SWNT reinforced polymer composites have shown only a moderate strength enhancement when compared with other hybrid materials.[201-203] The reason for this was attributed to the poor dispersion of SWNTs and poor bonding between the SWNTs and the polymer matrix. Also, the hardness, elastic modulus, and scratch resistance of SWNT reinforced epoxy composites increase with increasing weight % of the nanotubes.

In another study, to further enhance the luminescence and electroluminescence properties of poly(p-phenylene vinylene) (PPV) and its derivative poly(2-methoxy-5-(2′-ethylhexyloxy)-p-phenylene vinylene) (MEHPPV), they have been integrated with porous silicon.[204] To identify the deformation resistance of these polymers, nano-scratch tests were performed over the surfaces using an AFM with a diamond indenter (radius of around 50 nm).[204] Different normal loads ranging from 20 to 41 μN were used to scratch the samples. Three-dimensional AFM images of the nano-scratches on the four different samples together with the corresponding cross-sectional height profiles are shown in Fig. 23 and the nano-scratch penetration depths are listed in Table 6.

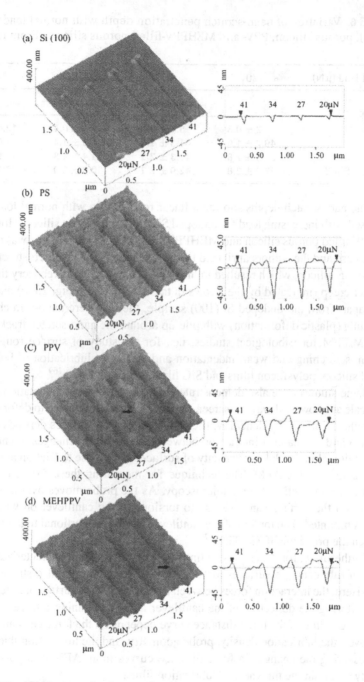

**FIGURE 23.** AFM topographs along with cross-sectional height profiles of nano-scratches made at four different normal loads on (a) undoped Si(100), (b) porous silicon, (c) PPV filled porous silicon, and (d) MEHPPV filled porous silicon (Reprinted with permission from H. Ni, X.D. Li, H.S. Gao and T.P. Nguyen, Nanoscale structural and. mechanical characterization of bamboo-like polymer/silicon nanocomposite films, Nanotechnology 16(9), 1746-1753 (2005). Copyright (2005) Institute of Physics Publishing).

**TABLE 6.** Variation of nano-scratch penetration depth with normal loads for Si(100), porous silicon, PPV- and MEHPPV-filled porous silicon. (From Ref. 188.)

| Normal load (μN) | 20 | 27 | 34 | 41 |
|---|---|---|---|---|
| Sample | | Nano-scratch depth (nm) | | |
| Si(100) | 3.2 ± 0.3 | 5.5 ± 0.3 | 9.1 ± 0.4 | 12.0 ± 0.3 |
| PS | 49.7 ± 1.6 | 61.3 ± 1.1 | 64.6 ± 1.4 | 74.0 ± 1.2 |
| PPV-filled | 35.5 ± 2.8 | 42.3 ± 2.1 | 44.4 ± 4.3 | 47.9 ± 3.8 |
| MEHPPV-filled | 37.9 ± 2.8 | 42.9 ± 2.1 | 50.0 ± 3.0 | 56.5 ± 2.6 |

The nanoscratch depths showed a linear relationship with normal load, i.e., increased with increasing load for undoped Si (100) and porous silicon. However, for, PPV-filled porous silicon and MEHPPV-filled porous silicon, this was clearly not the case and the authors attributed this to the heterogeneous microstructure of nanoporous silicon, which resulted in uneven local deformation recovery through polymer creep (indicated by the arrows in the 3D images in Fig. 23 (c) and (d)). Also, apart from the undoped Si (100) sample, the rest were nano-scratched by ploughing (plastic deformation) with pile-up around the nano-scratch tracks. Use of AFM/FFM for tribological studies, i.e., for the study of surface roughness, friction, scratching and wear, indentation and boundary lubrication of bulk and treated silicon, polysilicon films and SiC films is also reported.[169]

Some studies have also demonstrated that scratching and indentation on the nanoscale are powerful ways to screen for adhesion and resistance to deformation of ultrathin coatings.[205] Generally, friction between two surfaces depends on the chemical and mechanical interaction between the surfaces. Changes in chemical composition along with heterogeneity of surfaces giving rise to friction are easily measurable with the AFM. The technique for measuring these forces is called lateral force, or frictional force microscopy. As the probe moves over a surface, changes on the surface can give rise to torsions of the cantilever on which the probe is mounted. The torsion of the cantilever is then proportional to the friction between the probe and the surface.[87]

Furthermore, for developing optimized lubricating films, characterization of layers of lubricants on surfaces that are less than 100 nm are also studied with AFM. Here, the interaction forces between the probe and the surface are measured by monitoring the deflection of the cantilever and by obtaining a force-distance curve. The nature of the force-distance curve depends on the force constant of the cantilever, the lubrication density, probe geometry, and the lubrication thickness. By measuring the changes in force-distance curves in an AFM it is possible to directly ascertain the thickness of lubrication films.

## 3.5. Nanofabrication/Nanolithography

In addition to the above-mentioned applications of AFM, other important and novel applications of AFM are demonstrated for nanofabrication/nanomachining or nanolithography purposes.[188,206] Nanolithography, in broad, is the art and science of etching, writing, or printing at the microscopic level, where the dimensions of characters are on the order of nanometers. There are many conventional techniques available for fabrication of nanostructures, such as projection lithography, scanning beam (or maskless) lithography, milling techniques, photolithographic, X-ray lithography, and chemical methods. Amongst these, the most common method of nanolithography, used particularly in the creation of microchips, is photolithography. In this technique, the entire surface is drawn on in a single moment. However, the diffraction of the light used to expose a photoactivated film of resist limits the dimensions of the features produced by this technique.[87] Other common technique is scanning beam lithography, including electron beam and focused ion beam lithography. Here, a pattern is carved out line-by-line, by scanning a high-energy (>10 kV) beam of electrons or ions over a resist material.

The observation of conductance quantization, controlled transfer of single electrons, and the realization of almost perfect two-dimensional atoms are some examples of experiments involving nanostructures. However, the accuracy and resolution of these conventional optical and electronbeam lithographic techniques are restricted due to the constraint on the physical limitations.[207]

Probe microscopes proved to be a best alternative for fabricating nanostructures with high precision as well as for direct manipulation of atoms and molecules. They can be used to etch, write, or print on a surface in single-atom dimensions and also allow surface viewing in fine detail without necessarily modifying it. Additionally, there are different ways to fabricate nanostructures using probe microscopes, depending on the mode of application, like c-AFM and normal AFM used in force mode, and so on. STM can also be used in a field-emission mode for noncontact lithography, and form nano-sized features by localized heating or by inducing chemical reactions under the STM tip. It is generally operated in ultrahigh vacuum (UHV) to maintain a stable emission current. However operation in UHV can be cumbersome and time consuming. STM patterning also suffers from poor alignment capabilities since imaging may expose the resist.[110] Other disadvantages are that the tip sample distance, tunneling current, and tip voltage cannot be chosen independently and it is very difficult to use STM on partially insulating substrates such as semiconductor devices making it difficult to use it for nanofabrication/nanolithographic applications.[109]

### 3.5.1. Surface Modification Using Conductive AFM

When using a conducting tip AFM (c-AFM) for tip-induced oxidation, it has several advantages.[208,209] For example, with AFM, since the imaging is done by applied force as measured by the deflection of the cantilever supporting the tip, the imaging mechanism for AFM (contact force) is independent of the oxidation mechanism (tip bias). This implies that with c-AFM an oxidized pattern can be imaged with the same tip used to do the oxidation without the fear of further oxidation. Also, as only a top few monolayers of material have been exposed, and as extremely selective etches are available, both high resolution and effective pattern transfer are easily accomplished with this technique.

In the anodic oxidation process, an electrically conducting AFM tip that is operated in air is electrically biased with a negative polarity relative to a sample surface. The ambient humidity serves as an electrolyte such that the biased tip anodically oxidizes a small region of the surface. In other words (cathodic view), AFM tip acts as a cathode and the water meniscus formed between tip and surface is the electrolyte. This meniscus provides the oxy-anions ($OH^-$, $O^-$) needed to grow the oxide. Hence the strong localization of the electrical field lines near the tip apex gives rise to a nanometer-size oxide dot. The advantages of the anodic oxidation technique is that it provides a simple, reliable process for making a highly local chemical modification to a surface and also the oxidation process is fairly general and can be applied to most materials that can be anodized. The thin oxides fabricated by this technique can be used as a mask for pattern transfer by selective etching.[210]

Several AFM modes, such as contact, tapping mode, or noncontact have been used to perform local oxidation lithography.[211] Among these contact AFM mode of operation is the most extended mode for local oxidation. The meniscus is formed spontaneously in contact operation due to the condensation of water vapor in the nanometer-size cavities of the tip-surface interface.[210] However, the main drawback of this mode is the wear of the tip caused by the contact forces and also there is only a little control on the meniscus size (lateral dimensions).

To extend the tip's lifetime during the oxidation process and improve the reproducibility of the lithography, an AFM operated in noncontact mode has been proposed. In noncontact AFM oxidation, the liquid bridge is field induced by the application of an external voltage. Once the liquid bridge has been formed, its lateral dimensions can be decreased by increasing the average tip-sample separation.[147] It was also suggested that due to the control and smallness of the liquid meniscus, non-contact AFM oxidation can produce inherently smaller features (in the lateral dimension).[147,211] Nevertheless, the lateral dimensions of a local oxide depend on several other factors, such as the voltage and pulse duration, the relative humidity, hydrophobicity and dielectric constant of the material to be oxidized, doping of the substrate, chemical composition of the atmosphere, and the tip size and geometry.[211]

Tello et al.[211] compared contact and noncontact modes of AFM oxidation of p-type Si (100), oxidation performed at a constant force of 1 nN. Noncontact AFM oxidations were performed with doped $n^+$-type silicon cantilevers, whereas $Si_3N_4$ cantilevers are used for contact AFM experiments.

Figure 24 shows a sequence of oxide dots obtained by contact and non contact AFM operation. Noncontact AFM oxidation produces higher aspect ratios. This observation is reported to be a consequence of two effects. First, noncontact-AFM allows controlling the lateral size of the liquid meniscus which in turn, controls the lateral size of the oxide dot and second the vertical growth rate is smaller in contact AFM oxidation. In a similar study AFM operating in air is used to pattern

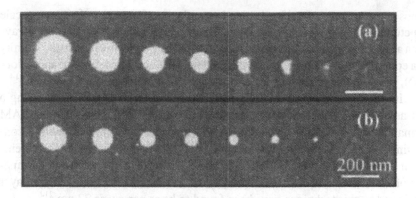

**FIGURE 24.** AFM image of a sequence of oxidation experiments performed for different pulse duration (3, 1, 0.3, 0.1, 0.03, 0.01, 0.003, and 0.001 s, from left to right) and constant voltage (20 V). (a) Contact AFM oxidation; (b) noncontact-AFM oxidation; (Reprinted with permission from M. Tello and R. Garcia, Nano-oxidation of silicon surfaces: Comparison of noncontact and contact atomic-force microscopy methods, Applied Physics Letters 79(3), 424-426 (2001). Copyright [2001], American Institute of Physics).

narrow features in a negative polymer resist in noncontact lithography mode.[110] A micromachined AFM cantilever with an integrated silicon probe tip acted as a source of electrons and the field emission current from the tip, sensitive to the tip-to-sample spacing is used as the feedback signal to control this spacing.

In another study, Campbell and Snow have reported the use of local electric field of a conducting tip atomic force microscope to write surface oxide patterns by local anodic oxidation, which further can be used as masks for selective etching to transfer the pattern into substrate.[209] They used this technique to fabricate side-gated Si field effect transistors with critical features as small as 30 nm. As an alternate to this approach, the oxide pattern is directly used as an active element of the device structure by Sugimura et al.[212] They demonstrated tip-induced anodic oxidation of thin Ti films. Matsumoto et al. used a conducting tip AFM to oxidize completely through thin Ti films and produce lateral metal-oxide-metal junctions and used these lateral junctions to fabricate a single-electron tunneling device that shows Coulomb blockade effects at room temperature.[213]

The use of conductive probe (a custom-made microfabricated silicon nitride probe coated with a 20 nm-thick NiCr layer, force constant 50.4 N/m) AFM for nanolithography on the basis of tip-induced electrochemistry (in a 60% relative humidity air atmosphere) has been reported by Sugimura et al.[166] The resist used for the studies was an organosilane monolayer composed of trimethylsilyl (TMS) groups. Since the TMS monolayer contacts both the substrate and probe electrodes, which are covered with surface oxide there is also a possibility to induce anodic oxidation of the monolayer with either the sample or the probe acting as anode. Although a number of techniques such as the electron beam, ultraviolet light, scanning probe lithography, and microcontact printing have been employed to form nanoscale structures of self-assembled monolayers, formation of the nanoscale structures of the SAMs (self-assembled monolayers) on selected areas has always been a challenge. The selective oxidation induced by the electric field in the vicinity of a conductive AFM probe, which is called field-induced oxidation (or anodation-FIO), is a promising method for fabricating such nanometer-scale structures.[214]

In a similar study, a nanoscale-patterning method on silicon oxide using a self-assembled monolayer was reported by Inoue et al.[214] Area-selective SAM formation was performed on Si oxide surfaces patterned using FIO with a conducting AFM. Contact angle measurement ($H_2O$), and simultaneous measurements of surface topography and friction force images using an AFM confirm the formation of SAM. Line structures of octadecyltrichlorosilane self-assembled monolayer fabricated by this technique have been found to be as narrow as 22 nm.[214]

### 3.5.2. Surface Modification Using Force Mode AFM

Some researchers also demonstrated mechanical scratching using AFM as a means to fabricate nanostructures with precise features. Commercial AFMs are generally modified to perform scratching experiments (e.g., lithography module in Nanoscope III, Digital Instruments). The major advantages in this mode are controlled movement of the tip with respect to the sample, better control over the applied normal load, scan size and speed, dry fabrication, and absence of chemical etching or electrical field.[188]

Figure 25(a) shows an example of nanofabrication.[205] The word "OHIO" was written on a (100) single-crystal silicon wafer by scratching the sample surface with a diamond tip at specified locations and scratching angles.[188,215] The normal load used for scratching (writing) was 50 $\mu$N and the writing speed was 0.2 $\mu$m/s. Each line was scribed manually; a few lines are not connected to each other due to the PZT drift and hysteresis. Also, if sufficient time is provided for the thermal stabilization of the PZT scanner, the hysteresis effects are minimized during nanofabrication. Nonetheless, the main disadvantage of using this type of mechanical scratching for nanofabrication purposes is the formation of debris during scratching, as evident in the Figure 25 (a). More complex patterns were also generated at a normal load of 15 $\mu$N and a writing speed of 0.5 $\mu$m/s (Figure 25(b)).[188] Such patterns are useful for resistor trimming (to increase path resistance) on a small scale. The separation between lines is about 50 nm.

### 3.5.3. Other Miscellaneous Modes of AFM for Surface Modification

In other modes of lithography using AFM, Mamin et al.[216] reported the use of a modulated laser beam to heat the AFM tip and deform a polymer substrate at high speed. The tip contact region on the PMMA surface softens with the heat from the tip and meanwhile the local tip pressure creates a pit. The size of the pits (several hundred angstroms to 1 $\mu$m) depends on the size of the laser pulse and the loading force on the tip. While in dip-pen nanolithographic technique (DPN),

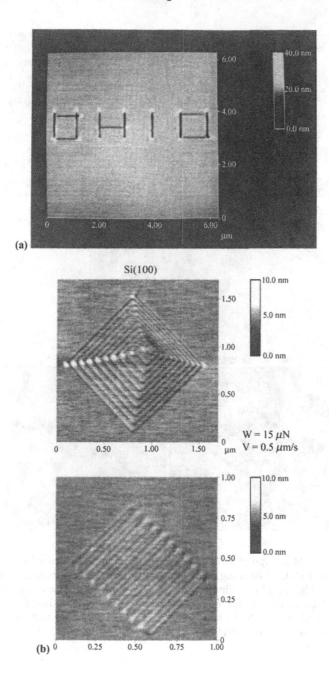

**FIGURE 25.** (a) The letters "OHIO" were generated by scratching a Si (111) surface using a diamond tip at a normal load of 50 μN and writing speed of 0.2 m/s. (b) Trim (top) and spiral (bottom) patterns generated by scratching a Si (100) surface using a diamond tip at a normal load of 15 μN and writing speed of 0.5 m/s (Reprinted from Wear 251, B. Bhushan, Nano- to microscale wear and mechanical characterization using scanning probe microscopy, 1105-1123 (2001), with permission from Elsevier).

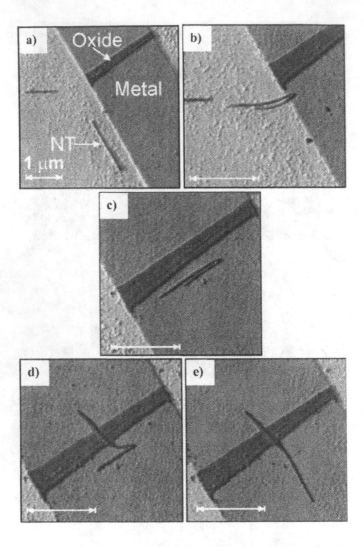

**FIGURE 26.** AFM manipulation of a single multiwall nanotube. Initially, the nanotube (NT) is located on the insulating (SiO$_2$) part of the sample. In a stepwise it is dragged up the 80 A° high metal thin film wire and finally is stretched across the oxide barrier (Reprinted from Applied Surface Science 141(3-4), P. Avouris, T. Hertel, R. Martel, T. Schmidt, H.R. Shea and R.E. Walkup, Carbon nanotubes: nanomechanics, manipulation, and electronic devices, 201-209 (1999), with permission from Elsevier).

an AFM tip is used to deliver molecules to a surface via a solvent meniscus, which naturally forms in the ambient atmosphere. This direct-write technique offers high-resolution patterning capabilities for a number of molecular and biomolecular inks on a variety of substrate types such as metals, semiconductors, and monolayer functionalized surfaces.

Wendel et al.[109] used AFM for integrated fabrication of a high electron mobility transistor (HEMT) device where a standard Hall bar device is prepared after which AFM is employed to pattern a hole array in a thin photoresist above a chosen area of the device. It has been reported that successful pattern transfer to the electron system is achieved both by wet etching and by ion beam irradiation. Preparation of hole arrays with a period down to 35 nm and deep lines of 50 nm width in various materials such as photoresist, PMMA, or gold is also reported. While fabrication of a single-electron transistor (SET) aligned with previously deposited contacts is reported by Bouchiat et al.[217] A PMMA resist trilayer process has been used that provided a rigid mask allowing large free-standing areas. AFM tip engraves a pattern of narrow furrows in the top soft polyimide layer. AFM-based trilayer lift-off technique proved to be a general-purpose nano-fabrication technique with alignment capability as well as an alternate to electron beam lithography technique for specific cases.

AFM was also used to manipulate CNTs position at room temperature by applying lateral forces of the appropriate magnitude with the tip. Contact mode AFM has been employed for the manipulation of a single CNT by Avouris et al.[108] The nanotube has been moved from an insulating silica substrate onto tungsten thin film wire and finally stretched across an insulating $WO_x$ barrier. Figure 26 shows a few steps of this process.[108]

# 4. CONCLUDING REMARKS

Though considerable progress of research has been made on the applications of AFM to characterize nanomaterials or dispersion of nanoparticles, it is very important to note that majority of the studies utilized other characterization techniques such as HRTEM, SEM, etc., support AFM observations in understanding the nature and structures of nanomaterials/nanocomposites. This is because of the inherent trade-offs when imaging with AFM like the dependency on the tip shape and size, thermal and mechanical drifts, vibrations, and so on. Hence, it is necessary to ensure that it is not what is being achieved, but it is the further development and understanding that must continue. For only this will enable us to utilize this fascinating technique properly and reliably in the vast field of nanotechnology.

# ACKNOWLEDGEMENTS

S.K. Samudrala acknowledges the University of New South Wales for the Visiting Research Associate (Academic) position with Dr. S. Bandyopadhyay in the School of Materials Science and Engineering. The authors thank Dr. R.S. Rajeev for his contribution towards Q4 in the "Questions" section of the chapter.

## REFERENCES

1. G.B. Khomutov, V.V. Kislov, R.V. Gainutdinov, S.P. Gubin, A.Y. Obydenov, S.A. Pavlov, A.N. Sergeev-Cherenkov, E.S. Soldatov, A.L. Tolstikhina and A.S. Trifonov, The design, fabrication and characterization of controlled-morphology nanomaterials and functional planar molecular nanocluster-based nanostructures, *Surf. Sci.* **532**, 287-293 (2003).
2. T. Cagin, J. Che, Y. Qi, Y. Zhou, E. Demiralp, G. Gao and W.A. Goddard, Computational materials chemistry at the nanoscale, *J. Nanopart. Res.* **1**(1), 51-69 (1999).
3. M.J. Pitkethly, Nanomaterials: the driving force, *Nanotoday* **Dec**, 20-29 (2004).
4. Z.L. Wang, Nanostructures of zinc oxide, *Mater. Today* **7**(6), 26-33 (2004).
5. A.C. Grimsdale and K. Mullen, The chemistry of organic nanomaterials, *Angew. Chem. Int. Ed.* **44**(35), 5592-5629 (2005).
6. H. Chander, Development of nanophosphors: A review, *Mater. Sci. Eng. R.* **49**(5), 113-155 (2005).
7. N. Savage and M.S. Diallo, Nanomaterials and water purification: Opportunities and challenges, *J. Nanopart. Res.* **7**(4-5), 331-342 (2005).
8. G.Q. Lu, Preface: nanomaterials in catalysis. Specially invited papers and selected papers from the workshop on nanostructured materials and catalysis from the Chemeca 2000 Conference, Perth, Australia, 9-12 July 2000, *Catal. Today* **68**(1-3), 1-1 (2001).
9. P.J. Sebastian, Nanomaterials for solar energy conversion, *Sol. Energy Mater. Sol. Cells* **70**(3), 243-244 (2001).
10. Y. Ito and E. Fukusaki, DNA as a "nanomaterial," *J. Mol. Cataly. B. Enzym.* **28**(4-6), 155-166 (2004).
11. R.A. Freitas, Current status of nanomedicine and medical nanorobotics, *J. Comput. Theor. Nanosci.* **2**(1), 1-25 (2005).
12. E.S. Kawasaki and A. Player, Nanotechnology, nanomedicine, and the development of new, effective therapies for cancer, *Nanomedi. Nanotechn., Biol. Med.* **1**(2), 101-109 (2005).
13. V. Balzani, Nanoscience and nanotechnotogy: A personal view of a chemist, *Small* **1**(3), 278-283 (2005).
14. G.M. Whitesides, Nanoscience, nanotechnology, and chemistry, *Small* **1**(2), 172-179 (2005).
15. M.F. Hochella, There's plenty of room at the bottom: nanoscience in geochemistry, *Geochimi. Cosmochimi. Acta* **66**(5), 735-743 (2002).
16. L.A. Bauer, N.S. Birenbaum and G.J. Meyer, Biological applications of high aspect ratio nanoparticles, *J. Mater. Chem.* **14**(4), 517-526 (2004).
17. D.R. Walt, Nanomaterials—Top-to-bottom functional design, *Nat. Mater.* **1**(1), 17-18 (2002).
18. N.S. Xu, S.Z. Deng and J. Chen, Nanomaterials for field electron emission: preparation, characterization and application, *Ultramicroscopy* **95**(1-4), 19-28 (2003).
19. E.F. Sheka and V.D. Khavryutchenko, Nanomaterials: reality and computational modelling, *Nanostruct. Mater.* **6**(5-8), 803-806 (1995).
20. C.N.R. Rao, New developments in nanomaterials, *J. Mater. Chem.* **14**(4), E4-E4 (2004).
21. J. Mackerle, Nanomaterials, nanomechanics and finite elements: a bibliography (1994-2004), *Model. Sim. Mater. Sci. Eng.* **13**(1), 123-158 (2005).
22. A.K. Cheetham and P.S.H. Grubstein, Nanomaterials and venture capital, *Nanotoday* **Dec**, 16-19 (2003).
23. H. Natter and R. Hempelmann, Tailor-made nanomaterials designed by electrochemical methods, *Electrochimi. Acta* **49**(1), 51-61 (2003).
24. V.L. Colvin, The potential environmental impact of engineered nanomaterials, *Nat. Biotech.* **21**(10), 1166-1170 (2003).
25. Z. Varga, G. Filipcsei, A. Szilagyi and M. Zrinyi, Electric and magnetic field-structured smart composites, *Macromole. Symp.* **227**, 123-133 (2005).
26. K. Schiffmann, Microwear experiments on metal-containing amorphous hydrocarbon hard coatings by AFM: Wear mechanisms and models for the load and time dependence, *Wear* **216**(1), 27-34 (1998).

27. C. Aymonier, U. Schlotterbeck, L. Antonietti, P. Zacharias, R. Thomann, J.C. Tiller and S. Mecking, Hybrids of silver nanoparticles with amphiphilic hyperbranched macromolecules exhibiting antimicrobial properties, *Chem. Commun.* (24), 3018-3019 (2002).

28. R.S. Rajeev, S.K. De, A.K. Bhowmick, G.J.P. Kao and S. Bandyopadhyay, Atomic force microscopy studies of short melamine fiber reinforced EPDM rubber, *J. Mater. Sci.* **36**(11), 2621-2632 (2001).

29. A. Dasari, Z.Z. Yu and Y.W. Mai, Effect of blending sequence on microstructure of ternary nanocomposites, *Polymer* **46**(16), 5986-5991 (2005).

30. X.C. Chen, B. You, S.X. Zhou and L.M. Wu, Surface and interface characterization of polyester-based polyurethane/nano-silica composites, *Surf. Interface Anal.* **35**(4), 369-374 (2003).

31. J. Jakubowicz, Application of AFM to study functional nanomaterials prepared by mechanical alloying, *J. Mater. Sci.* **39**(16-17), 5379-5383 (2004).

32. L. Kovarova, A. Kalendova, J.F. Gerard, J. Malac, J. Simonik and Z. Weiss, Structure analysis of PVC nanocomposites, *Macromol. Symp.* **221** (1), 105-114 (2005).

33. N. Ohmae, J.M. Martin and S. Mori, *Micro and Nanotribology* (ASME Press, New York, 2005).

34. R. Sengupta, A. Bandyopadhyay, S. Sabharwal, T.K. Chaki and A.K. Bhowmick, Polyamide-6,6/in situ silica hybrid nanocomposites by sol-gel technique: synthesis, characterization and properties, *Polymer* **46**(10), 3343-3354 (2005).

35. R.S. Rajeev, S.K. De, A.K. Bhowmick, B. Gong and S. Bandyopadhyay, in: *Atomic Force Microscopy in Adhesion Studies*, edited by J. Drelich and K.L. Mittal (VSP, Boston, 2005), pp. 657-678.

36. G. Binnig and H. Rohrer, Scanning tunneling microscopy, *Helv. Phys. Acta* **55**(6), 726-735 (1982).

37. G. Binnig, C.F. Quate and C. Gerber, Atomic force microscope, *Phys. Rev. Lett.* **56**(9), 930-933 (1986).

38. M. Rief, M. Gautel, F. Oesterhelt, J.M. Fernandez and H.E. Gaub, Reversible unfolding of individual titin immunoglobulin domains by AFM, *Science* **276**(5315), 1109-1112 (1997).

39. S.S. Wong, E. Joselevich, A.T. Woolley, C.L. Cheung and C.M. Lieber, Covalently functionalized nanotubes as nanometre-sized probes in chemistry and biology, *Nature* **394**(6688), 52-55 (1998).

40. R.D. Piner, T.T. Xu, F.T. Fisher, Y. Qiao and R.S. Ruoff, Atomic Force microscopy study of clay nanoplatelets and their impurities, *Langmuir* **19**(19), 7995-8001 (2003).

41. J.L. Hutter and J. Bechhoefer, Calibration of atomic-force microscope tips, *Rev. Sci. Instrum.* **64**(7), 1868-1873 (1993).

42. G.U. Lee, D.A. Kidwell and R.J. Colton, Sensing discrete streptavidin biotin interactions with atomic-force microscopy, *Langmuir* **10**(2), 354-357 (1994).

43. M.E. Greene, C.R. Kinser, D.E. Kramer, L.S.C. Pingree and M.C. Hersam, Application of scanning probe microscopy to the characterization and fabrication of hybrid nanomaterials, *Microsc. Res. Tech.* **64**(5-6), 415-434 (2004).

44. M.M. Miranda and M. Innocenti, AFM and micro-Raman investigation on filters coated with silver colloidal nanoparticles, *Appl. Surf. Sci.* **226**, 125-130 (2004).

45. S. Cuenot, C. Fretigny, S. Demoustier-Champagne and B. Nysten, Surface tension effect on the mechanical properties of nanomaterials measured by atomic force microscopy, *Phy. Rev. B* **69**(16) (2004).

46. H.G. Hansma, J. Vesenka, C. Siegerist, G. Kelderman, H. Morrett, R.L. Sinsheimer, V. Elings, C. Bustamante and P.K. Hansma, Reproducible imaging and dissection of plasmid DNA under liquid with the atomic force microscope, *Science* **256**(5060), 1180-1184 (1992).

47. Z. Wang, AFM study of gold nanowire array electrodeposited within anodic aluminum oxide template, *Appl. Phys. A* **74**, 563-565 (2002).

48. P. Staszczuk, World of nanostructures: Nanotechnology, surface properties of chosen nanomaterials. Determined by adsorption, Q-TG, AFM and SEM methods, *J. Therm. Analy. Calori.* **79**(3), 545-554 (2005).

49. M.Q. Li, Scanning probe microscopy (STM/AFM) and applications in biology, *Appl. Phys. Mater. Sci. Proc.* **68**(2), 255-258 (1999).

50. K.H. Chung, Y.H. Lee and D.E. Kim, Characteristics of fracture during the approach process and wear mechanism of a silicon AFM tip, *Ultramicroscopy* 102(2), 161-171 (2005).

51. P.K. Chu, R.G. Brigham and S.M. Baumann, Atomic-force microscopy (Afm) statistical process: control for microelectronics applications, *Mater. Chem. Phys.* 41(1), 61-65 (1995).

52. S. Kasas, N.H. Thomson, B.L. Smith, P.K. Hansma, J. Miklossy and H.G. Hansma, Biological applications of the AFM: from single molecules to organs, *Intl. J. Imag. Syst. Tech.* 8(2), 151-161 (1997).

53. L. Kordylewski, D. Saner and R. Lal, Atomic-force microscopy of freeze-fracture replicas of rat atrial tissue, *J. Micros. Oxford* 173, 173-181 (1994).

54. G. Meyer and N.M. Amer, Simultaneous measurement of lateral and normal forces with an optical-beam-deflection atomic force microscope, *Appl. Phys. Lett.* 57(20), 2089-2091 (1990).

55. S. Burnside, S. Winkel, K. Brooks, V. Shklover, M. Gratzel, A. Hinsch, R. Kinderman, C. Bradbury, A. Hagfeldt and H. Pettersson, Deposition and characterization of screen-printed porous multi-layer thick film structures from semiconducting and conducting nanomaterials for use in photovoltaic devices, *J. Mater. Sci. Mater. Electron.* 11(4), 355-362 (2000).

56. A. Diaspro and M. Aguilar, Proposal for a New Optical-Device to Sense Afm Forces, *Ultramicroscopy* 42, 1668-1670 (1992).

57. G.L. Caer and P. Delcroix, Characterization of nanostructured materials by Mossbauer spectrometry, *Nanostruct. Mater.* 7(1-2), 127-135 (1996).

58. H. Hahn, Unique features and properties of nanostructured materials, *Adv. Eng. Mater.* 5(5), 277-284 (2003).

59. S. Jakobs, A. Duparre and H. Truckenbrodt, AFM and light-scattering measurements of optical thin films for applications in the UV spectral region, *Int. J. Machine Tools Manufact.* 38(5-6), 733-739 (1998).

60. H.W. Liu and B. Bhushan, Nano-tribological characterization of molecularly thick lubricant films for applications to MEMS/NEMS by AFM, *Ultramicroscopy* 97(1-4), 321-340 (2003).

61. M. Villarroya, F. Perez-Murano, C. Martin, Z. Davis, A. Boisen, J. Esteve, E. Figueras, J. Montserrat and N. Barniol, AFM lithography for the definition of nanometer-scale gaps: application to the fabrication of a cantilever-based sensor with electrochemical current detection, *Nanotechnology* 15(7), 771-776 (2004).

62. J.I. Paredes, A. Martinez-Alonso and J.M.D. Tascon, Detecting surface oxygen groups on carbon nanofibers by phase contrast imaging in tapping mode AFM, *Langmuir* 19(18), 7665-7668 (2003).

63. T. Ono, C.C. Fan and M. Esashi, Micro-instrumentation for characterizing thermoelectric properties of nanomaterials, *J. Micromech. Microeng.* 15(1), 1-5 (2005).

64. R. Mahlberg, Application of AFM on the adhesion studies of oxygen-plasma-terated polypropylene and lignocellulosics, *Langmuir* 15, 2985-2992 (1999).

65. A. Razpet, G. Possnert, A. Johansson, A. Hallen and K. Hjort, Ion transmission and characterization of ordered nanoporous alumina, *Nuclear Instruments & Methods in Physics Research Section B-Beam Interactions with Materials and Atoms* 222(3-4), 593-600 (2004).

66. X.C. Sun, Microstructure characterization and magnetic properties of nanomaterials, *Mole. Phys.* 100(19), 3059-3063 (2002).

67. D. Roy and J. Fendler, Reflection and absorption techniques for optical characterization of chemically assembled nanomaterials, *Adv. Mater.* 16(6), 479-508 (2004).

68. B.T. Su, X.H. Liu, X.X. Peng, T. Xiao and Z.X. Su, Preparation and characterization of the TiO2/polymer complex nanomaterial, *Mater. Sci. Engin. Struct. Mater. Prop. Microstruct. Proc.* 349(1-2), 59-62 (2003).

69. Z.L. Wang, New developments in transmission electron microscopy for nanotechnology, *Adv. Mater.* 15(18), 1497-1514 (2003).

70. M. Tosa, Measurement of atomic forces between surface atoms using an ultra-clean AFM incorporated in an XHV integrated system, *Adv. Coll. Interf. Sci.* 71-2, 233-241 (1997).

71. D.P. Yu, Y.J. Xing, M. Tence, H.Y. Pan and Y. Leprince-Wang, Microstructural and compositional characterization of a new silicon carbide nanocables using scanning transmission electron microscopy, *Phys.-Low-Dimens. Syst. Nanostruct.* 15(1), 1-5 (2002).

72. G.D. Yuan, Environmental nanomaterials: Occurrence, syntheses, characterization, health effect, and potential applications, *J. Envi. Sci. Health Part a-Toxic/Hazardous Substances & Environmental Engineering* **39**(10), 2545-2548 (2004).

73. D. Vogel, A. Gollhardt and B. Michel, Micro- and nanomaterials characterization by image correlation methods, *Sens. Actuators A-Phys.* **99**(1-2), 165-171 (2002).

74. F. Lopour, R. Kalousek, D. Skoda, J. Spousta, F. Matejka and T. Sikola, Application of AFM in microscopy and fabrication of micro/nanostructures, *Surf. Interface Anal.* **34**(1), 352-355 (2002).

75. S.E. Harvey, P.G. Marsh and W.W. Gerberich, Atomic-force microscopy and modeling of fatigue-crack initiation in metals, *Acta Metall. Mater.* **42**(10), 3493-3502 (1994).

76. P. Boughton, S. Bandyopadhyay, C. Gregory, S. Patrick and H. Sinclair, Recent developments in water-based coatings for the transport industry, *Surf. Coatings Aus.* **37**(9), 20-25 (2000).

77. S. Ghosh, D. Khastagir, A.K. Bhowmick, S. Bandyopadhyay, G.J.P. Kao and L. Kok, Atomic force microscopy studies of molded thin films of segmented polyamides, *J. Mater. Sci. Lett.* **19**(23), 2161-2165 (2000).

78. R.S. Rajeev, S.K. De, A.K. Bhowmick, B. Gong and S. Bandyopadhyay, Atomic force microscopy, X-ray diffraction, X-ray photoelectron spectroscopy and thermal studies of the new melamine fiber, *J. Adhe. Sci. Technol.* **16**(14), 1957-1978 (2002).

79. A. Ghosh, R.A.K. Bhattacharya, A.K. Bhowmick, S.K. De, B. Wolpensinger and S. Bandyopadhyay, Atomic force microscopic studies on microheterogeneity of blends of silicone rubber and tetrafluoroethylene/propylene/vinylidene fluoride terpolymer, *Rubber Chem. Technol.* **76**(1), 220-238 (2003).

80. A. Ghosh, R.S. Rajeev, S.K. De, W. Sharp and S. Bandyopadhyay, Atomic force microscopic studies on the silicone rubber/fluororubber blend containing ground rubber vulcanizate powder, *J. Elast. Plast.* **38**(4), 119-132 (2006).

81. R.S. Rajeev, A.K. Bhowmick, S.K. De and S. Bandyopadhyay, Atomic force and scanning electron microscopic studies of effect of crosslinking system on properties of maleated ethylene-propylene-diene monomer rubber composites filled with short melamine fibers, *J. Appl. Polymer Sci.* **89**(5), 1211-1229 (2003).

82. S. Anandhan, P.P. De, S.K. De, S. Bandyopadhyay and A.K. Bhowmick, Mapping of thermoplastic elastomeric nitrile rubber/poly(styrene-co-acrylonitrile) blends using tapping mode atomic force microscopy and transmission electron microscopy, *J. Mater. Sci.* **38**(13), 2793-2801 (2003).

83. A.M. Shanmugharaj, S. Ray, S. Bandyopadhyay and A.K. Bhowmick, Surface morphology of styrene-butadiene rubber vulcanizate filled with novel electron beam modified dual phase filler by atomic force microscopy, *J. Adh. Sci. Tech.* **17**(9), 1167-1186 (2003).

84. S. Ray, A.K. Bhowmick and S. Bandyopadhyay, Atomic force microscopy studies on morphology and distribution of surface modified silica and clay fillers in an ethylene-octene copolymer rubber, *Rubber Chem. Technol.* **76**(5), 1091-1105 (2003).

85. S. Gupta, P. Brouwer, S. Bandyopadhyay, S. Patil, R. Briggs, J. Jain and S. Seal, TEM/AFM investigation of size and surface properties of nanocrystalline ceria, *J. Nanosci. Nanotech.* **5**(7), 1101-1107 (2005).

86. B. Bhushan, *Handbook of Micro/nanotribology* (CRC Press, Boca Raton, 1999).

87. www.pacificnano.com.

88. J.A. Gallegojuarez, Piezoelectric ceramics and ultrasonic transducers, *J. Phys. Sci. Instrum.* **22**(10), 804-816 (1989).

89. R. Howland and L. Benatar, A practical guide to scanning probe microscopy, Park Sceintific Instruments, Sunnyvale, 1996.

90. M.R. VanLandingham, J.S. Villarrubia, W.F. Guthrie and G.F. Meyers, Nanoindentation of polymers: An overview, *Macromol. Symp.* **167**, 15-43 (2001).

91. S.M. Hues, R.J. Colton, E. Meyer and H.J. Guntherodt, Scanning probe microscopy of thin films, *MRS Bull.* **18**(1), 41-49 (1993).

92. D.J. Keller and C.C. Chou, Imaging steep, high structures by scanning force microscopy with electron-beam deposited tips, *Surf. Sci.* **268**(1-3), 333-339 (1992).

93. J.H. Hoh and P.K. Hansma, Atomic force microscopy for high-resolution imaging in cell biology, *Trends Cell Biol.* **2**, 208-213 (1992).

94. A.L. Weisenhorn, P.K. Hansma, T.R. Albrecht and C.F. Quate, Forces in atomic force microscopy in air and water, *Appl. Phys. Lett.* **54**(26), 2651-2653 (1989).

95. N. Burnham and A.A. Kulik, in: *Handbook of Micro/nanotribology*, edited by B. Bhushan (CRC Press, Boca Raton, 1999), pp. 247-272.

96. www.veeco.com.

97. G. Binnig and H. Rohrer, Scanning tunneling microscopy: from birth to adolescence, *Rev. Mod. Phys.* **59**(3), 615-625 (1987).

98. R. Young, J. Ward and F. Scire, Topografiner: instrument for measuring surface microtopography, *Rev. Sci. Instrum.* **43**(7), 999-& (1972).

99. G. Binning, H. Rohrer, C. Gerber and E. Weibel, Surface studies by scanning tunneling microscopy, *Phys. Rev. Lett.* **49**(1), 57-61 (1982).

100. Y. Martin, C.C. Williams and H.K. Wickramasinghe, Atomic force microscope force mapping and profiling on a sub 100-a Scale, *J. Appl. Phys.* **61**(10), 4723-4729 (1987).

101. G. Meyer and N.M. Amer, Novel optical approach to atomic force microscopy, *Appl. Phys. Lett.* **53**(12), 1045-1047 (1988).

102. O. Marti, in: *Handbook of Micro/Nanotribology*, edited by B. Bhushan (CRC Press, Boca Raton, 1999), pp. 81-144.

103. R.W. Carpick and M. Salmeron, Scratching the surface: fundamental investigations of tribology with atomic force microscopy, *Chem. Rev.* **97**(4), 1163-1194 (1997).

104. E. Gnecco, R. Bennewitz, T. Gyalog and E. Meyer, Friction experiments on the nanometer scale, *J. Phys. Condens. Mat.* **13**(31), R619-R642 (2001).

105. G.J. Leggett, Friction force microscopy of self-assembled monolayers: probing molecular organization at the nanometer scale, *Anal. Chim. Acta* **479**(1), 17-38 (2003).

106. A. Noy, D.V. Vezenov and C.M. Lieber, Chemical force microscopy, *Ann. Rev. Mater. Sci.* **27**, 381-421 (1997).

107. H. Takano, J.R. Kenseth, S.S. Wong, J.C. O'Brien and M.D. Porter, Chemical and biochemical analysis using scanning force microscopy, *Chem. Rev.* **99**(10), 2845-+ (1999).

108. P. Avouris, T. Hertel, R. Martel, T. Schmidt, H.R. Shea and R.E. Walkup, Carbon nanotubes: nanomechanics, manipulation, and electronic devices, *Appl. Surf. Sci.* **141**(3-4), 201-209 (1999).

109. M. Wendel, S. Kuhn, H. Lorenz, J.P. Kotthaus and M. Holland, Nanolithography with an atomic-force microscope for integrated fabrication of quantum electronic devices, *Appl. Phys. Lett.* **65**(14), 1775-1777 (1994).

110. K. Wilder, C.F. Quate, D. Adderton, R. Bernstein and V. Elings, Noncontact nanolithography using the atomic force microscope, *Appl. Phys. Lett.* **73**(17), 2527-2529 (1998).

111. B. Bhushan, Nanoscale tribophysics and tribomechanics, *Wear* **225-229**(Part 1), 465-492 (1999).

112. L.P. Sung, S. Scierka, M. Baghai-Anaraki and D.L. Ho, Characterization of metal-oxide nanoparticles: synthesis and dispersion in polymeric coatings, in: *MRS Symposium I: Nanomaterialsfor Structural Applications*, edited by C. Berndt, T.E. Fischer, I. Ovid'ko, G. Skandan and T. Tsakalakos, Materials Research Society **740**, pp. 15.4.1-15.4.6 (2003).

113. W.Y. Li, S. Seal, E. Megan, J. Ramsdell, K. Scammon, G. Lelong, L. Lachal and K.A. Richardson, Physical and optical properties of sol-gel nano-silver doped silica film on glass substrate as a function of heat-treatment temperature, *J. Appl. Phys.* **93**(12), 9553-9561 (2003).

114. S. Shukla and S. Seal, Cluster size effect observed for gold nanoparticles synthesized by sol-gel technique as studied by X-ray photoelectron spectroscopy, *Nanostruct. Mater.* **11**(8), 1181-1193 (1999).

115. S. Shukla, S. Patil, S.C. Kuiry, Z. Rahman, T. Du, L. Ludwig, C. Parish and S. Seal, Synthesis and characterization of sol-gel derived nanocrystalline tin oxide thin film as hydrogen sensor, *Sens. Actuators B-Chem.* **96**(1-2), 343-353 (2003).

116. L. Boras and P. Gatenholm, Surface composition and morphology of CTMP fibers, *Holzforschung* **53**(2), 188-194 (1999).

117. M.R. Mucalo, C.R. Bullen, M. Manley-Harris and T.M. McIntire, Arabinogalactan from the Western larch tree: A new, purified and highly water-soluble polysaccharide-based protecting agent for maintaining precious metal nanoparticles in colloidal suspension, *J. Mater. Sci.* **37**(3), 493-504 (2002).

118. J.M. Bennett and L. Mattsson, Introduction to surface roughness and scattering, *Opt. Soc. America*, Washington (1993).

119. D.L. Ma, Y.A. Akpalu, Y. Li, R.W. Siegel and L.S. Schadler, Effect of titania nanoparticles on the morphology of low-density polyethylene, *J. Polym. Sci. Part B-Polym. Phys.* **43**(5), 488-497 (2005).

120. W.P. Liu, S.V. Hoa and M. Pugh, Morphology and performance of epoxy nanocomposites modified with organoclay and rubber, *Polym. Eng. Sci.* **44**(6), 1178-1186 (2004).

121. W.S. Chow, Z.A.M. Ishak and J. Karger-Kocsis, Atomic force microscopy study on blend morphology and clay dispersion in polyamide-6/polypropylene/organoclay systems, *J. Poly. Sci. Part B-Poly. Phys.* **43**(10), 1198-1204 (2005).

122. B.K. Satapathy, R. Weidisch, P. Potschke and A. Janke, Crack toughness behaviour of multiwalled carbon nanotube (MWNT)/polycarbonate nanocomposites, *Macromol. Rapid Commun.* **26**(15), 1246-1252 (2005).

123. F. Dietsche, Y. Thomann, R. Thomann and R. Mulhaupt, Translucent acrylic nanocomposites containing anisotropic laminated nanoparticles derived from intercalated layered silicates, *J. Appl. Polym. Sci.* **75**(3), 396-405 (2000).

124. A. Bafna, G. Beaucage, F. Mirabella and S. Mehta, 3D Hierarchical orientation in polymer-clay nanocomposite films, *Polymer* **44**(4), 1103-1115 (2003).

125. C. Saujanya and S. Radhakrishnan, Structure development and crystallization behaviour of PP/nanoparticulate composite, *Polymer* **42**(16), 6723-6731 (2001).

126. S. Renker, S. Mahajan, D.T. Babski, I. Schnell, A. Jain, J. Gutmann, Y.M. Zhang, S.M. Gruner, H.W. Spiess and U. Wiesner, Nanostructure and shape control in polymer-ceramic hybrids from poly(ethylene oxide)-block-poly(hexyl methacrylate) and aluminosilicates derived from them, *Macromol. Chem. Phys.* **205**(8), 1021-1030 (2004).

127. M. Templin, A. Franck, A. DuChesne, H. Leist, Y.M. Zhang, R. Ulrich, V. Schadler and U. Wiesner, Organically modified aluminosilicate mesostructures from block copolymer phases, *Science* **278**(5344), 1795-1798 (1997).

128. H. Matsumura and W.G. Glasser, Cellulosic nanocomposites. II. studies by atomic force microscopy, *J. Appl. Polym. Sci.* **78**(13), 2254-2261 (2000).

129. C. Velasco-Santos, A.L. Martinez-Hernandez, M. Lozada-Cassou, A. Alvarez-Castillo and V.M. Castano, Chemical functionalization of carbon nanotubes through an organosilane, *Nanotech.* **13**(4), 495-498 (2002).

130. H. Bubert, S. Haiber, W. Brandl, G. Marginean, M. Heintze and V. Bruser, Characterization of the uppermost layer of plasma-treated carbon nanotubes, *Diamond Relat. Mater.* **12**(3-7), 811-815 (2003).

131. T. McNally, P. Potschke, P. Halley, M. Murphy, D. Martin, S.E.J. Bell, G.P. Brennan, D. Bein, P. Lemoine and J.P. Quinn, Polyethylene multiwalled carbon nanotube composites, *Polymer* **46**(19), 8222-8232 (2005).

132. W.W. Scott and B. Bhushan, Use of phase imaging in atomic force microscopy for measurement of viscoelastic contrast in polymer nanocomposites and molecularly thick lubricant films, *Ultramicroscopy* **97**(1-4), 151-169 (2003).

133. B. Yalcin and A. Cakmak, The role of plasticizer on the exfoliation and dispersion and fracture behavior of clay particles in PVC matrix: a comprehensive morphological study, *Polymer* **45**(19), 6623-6638 (2004).

134. S. Sadhu and A. Bhowmick, Morphology study of rubber based nanocomposites by transmission electron microscopy and atomic force microscopy, *J. Mater. Sci.* **40**(7), 1633-1642 (2005).

135. M. Kociak, A.Y. Kasumov, S. Gueron, B. Reulet, Khodos, II, Y.B. Gorbatov, V.T. Volkov, L. Vaccarini and H. Bouchiat, Superconductivity in ropes of single-walled carbon nanotubes, *Phys. Rev. Lett.* **86**(11), 2416-2419 (2001).

136. A.G. Rinzler, J.H. Hafner, P. Nikolaev, L. Lou, S.G. Kim, D. Tomanek, P. Nordlander, D.T. Colbert and R.E. Smalley, Unraveling nanotubes: field-emission from an atomic wire, *Science* **269**(5230), 1550-1553 (1995).

137. S.J. Tans, M.H. Devoret, H.J. Dai, A. Thess, R.E. Smalley, L.J. Geerligs and C. Dekker, Individual single-wall carbon nanotubes as quantum wires, *Nature* **386**(6624), 474-477 (1997).

138. A.C. Dillon, K.M. Jones, T.A. Bekkedahl, C.H. Kiang, D.S. Bethune and M.J. Heben, Storage of hydrogen in single-walled carbon nanotubes, *Nature* **386**(6623), 377-379 (1997).

139. S. Berber, Y.K. Kwon and D. Tomanek, Unusually high thermal conductivity of carbon nanotubes, *Phys. Rev. Lett.* **84**(20), 4613-4616 (2000).

140. O. Lourie, D.M. Cox and H.D. Wagner, Buckling and collapse of embedded carbon nanotubes, *Phys. Rev. Lett.* **81**(8), 1638-1641 (1998).

141. J. Li, Y.J. Lu, Q. Ye, M. Cinke, J. Han and M. Meyyappan, Carbon nanotube sensors for gas and organic vapor detection, *Nano Lett.* **3**(7), 929-933 (2003).

142. M. Shim, N.W.S. Kam, R.J. Chen, Y.M. Li and H.J. Dai, Functionalization of carbon nanotubes for biocompatibility and biomolecular recognition, *Nano Lett.* **2**(4), 285-288 (2002).

143. Z. Wang, Z.Y. Liang, B. Wang, C. Zhang and L. Kramer, Processing and property investigation of single-walled carbon nanotube (SWNT) buckypaper/epoxy resin matrix nanocomposites, *Composites Part A* **35**(10), 1225-1232 (2004).

144. Y.L. Liu, C.Y. Hsu, W.L. Wei and R.J. Jeng, Preparation and thermal properties of epoxy-silica nanocomposites from nanoscale colloidal silica, *Polymer* **44**(18), 5159-5167 (2003).

145. B. Anczykowski, D. Kruger, K.L. Babcock and H. Fuchs, Basic properties of dynamic force spectroscopy with the scanning force microscope in experiment and simulation, *Ultramicroscopy* **66**(3-4), 251-259 (1996).

146. J. Tamayo and R. Garcia, Deformation, contact time, and phase contrast in tapping mode scanning force microscopy, *Langmuir* **12**(18), 4430-4435 (1996).

147. R. Garcia, J. Tamayo, M. Calleja and F. Garcia, Phase contrast in tapping-mode scanning force microscopy, *Appl. Phys. A* **66**, S309-S312 (1998).

148. J. Tamayo and R. Garcia, Effects of elastic and inelastic interactions on phase contrast images in tapping-mode scanning force microscopy, *Appl. Phys. Lett.* **71**(16), 2394-2396 (1997).

149. P.J. James, M. Antognozzi, J. Tamayo, T.J. McMaster, J.M. Newton and M.J. Miles, Interpretation of contrast in tapping mode AFM and shear force microscopy. A study of nafion, *Langmuir* **17**(2), 349-360 (2001).

150. B. Basnar, G. Friedbacher, H. Brunner, T. Vallant, U. Mayer and H. Hoffmann, Analytical evaluation of tapping mode atomic force microscopy for chemical imaging of surfaces, *Appl. Surf. Sci.* **171**(3-4), 213-225 (2001).

151. E.Z. Luo, J.X. Ma, J.B. Xu, I.H. Wilson, A.B. Pakhomov and X. Yan, Probing the conducting paths in a metal-insulator composite by conducting atomic force microscopy, *J. Phys. D-Appl. Phys.* **29**(12), 3169-3172 (1996).

152. S.A. Zavyalov, A.N. Pivkina and J. Schoonman, Formation and characterization of metal-polymer nanostructured composites, *Solid State Ionics* **147**(3-4), 415-419 (2002).

153. M. Knite, V. Teteris, B. Polyakov and D. Erts, Electric and elastic properties of conductive polymeric nanocomposites on macro- and nanoscales, *Mater. Sci. Eng. C* **19**(1-2), 15-19 (2002).

154. R.P. Lu, K.L. Kavanagh, S.J. Dixon-Warren, A. Kuhl, A.J.S. Thorpe, E. Griswold, G. Hillier, I. Calder, R. Ares and R. Streater, Calibrated scanning spreading resistance microscopy profiling of carriers in III-V structures, *J. Vac. Sci. Tech. B* **19**(4), 1662-1670 (2001).

155. P. De Wolf, T. Clarysse, W. Vandervorst, L. Hellemans, P. Niedermann and W. Hanni, Cross-sectional nano-spreading resistance profiling, *J. Vac. Sci. Technol. B* **16**(1), 355-361 (1998).

156. P. Dewolf, J. Snauwaert, T. Clarysse, W. Vandervorst and L. Hellemans, Characterization of a point-contact on silicon using force microscopy-supported resistance measurements, *Appl. Phys. Lett.* **66**(12), 1530-1532 (1995).

157. C. Shafai, D.J. Thomson, M. Simardnormandin, G. Mattiussi and P.J. Scanlon, Delineation of semiconductor doping by scanning resistance microscopy, *Appl. Phys. Lett.* **64**(3), 342-344 (1994).

158. R.P. Lu, K.L. Kavanagh, S.J. Dixon-Warren, A.J. SpringThorpe, R. Streater and I. Calder, Scanning spreading resistance microscopy current transport studies on doped III-V semiconductors, *J. Vac. Sci. Technol. B* **20**(4), 1682-1689 (2002).

159. P. Eyben, M. Xu, N. Duhayon, T. Clarysse, S. Callewaert and W. Vandervorst, Scanning spreading resistance microscopy and spectroscopy for routine and quantitative two-dimensional carrier profiling, *J. Vac. Sci. Technol. B* **20**(1), 471-478 (2002).

160. D. Ban, E.H. Sargent, S.J. Dixon-Warren, T. Grevatt, G. Knight, G. Pakulski, A.J. SpringThorpe, R. Streater and J.K. White, Two-dimensional profiling of carriers in a buried heterostructure multi-quantum-well laser: calibrated scanning spreading resistance microscopy and scanning capacitance microscopy, *J. Vac. Sci. Technol. B* **20**(5), 2126-2132 (2002).

161. D. Ban, E.H. Sargent, S. Dixon-Warren, I. Calder, A.J. SpringThorpe, R. Dworschak, G. Este and J.K. White, Direct imaging of the depletion region of an InP p-n junction under bias using scanning voltage microscopy, *Appl. Phys. Lett.* **81**(26), 5057-5059 (2002).

162. A. Alexeev, J. Loos and M.M. Koetse, Nanoscale electrical characterization of semiconducting polymer blends by conductive atomic force microscopy (C-AFM), *Ultarmicroscopy* **106**(3), 191-199 (2006).

163. G. Leatherman, E.N. Durantini, D. Gust, T.A. Moore, A.L. Moore, S. Stone, Z. Zhou, P. Rez, Y.Z. Liu and S.M. Lindsay, Carotene as a molecular wire: Conducting atomic force microscopy, *J. Phys. Chem. B* **103**(20), 4006-4010 (1999).

164. X.D. Cui, A. Primak, X. Zarate, J. Tomfohr, O.F. Sankey, A.L. Moore, T.A. Moore, D. Gust, L.A. Nagahara and S.M. Lindsay, Changes in the electronic properties of a molecule when it is wired into a circuit, *J. Phys. Chem. B* **106**(34), 8609-8614 (2002).

165. T.W. Kelley and C.D. Frisbie, Point contact current-voltage measurements on individual organic semiconductor grains by conducting probe atomic force microscopy, *J. Vac. Sci. Technol. B* **18**(2), 632-635 (2000).

166. H. Sugimura, K. Okiguchi, N. Nakagiri and M. Miyashita, Nanoscale patterning of an organosilane monolayer on the basis of tip-induced electrochemistry in atomic force microscopy, *J. Vac. Sci. Technol. B* **14**(6), 4140-4143 (1996).

167. B. Bhushan, Micro/nanotribology using atomic force microscopy/friction force microscopy: state of the art, *Proceedings of the Institution of Mechanical Engineers Part J: Journal of Engineering Tribology* **212**(J1), 1-18 (1998).

168. B. Bhushan, Nanoscale tribophysics and tribomechanics, *Wear* **229**, 465-492 (1999).

169. B. Bhushan, Tribology on the macroscale to nanoscale of microelectromechanical system materials: a review, *Proc. Inst. Mech. Engrs. Part J. J. Engg. Tribol.* **215**(J1), 1-18 (2001).

170. B. Bhushan, J.N. Israelachvili and U. Landman, Nanotribology: friction, wear and lubrication at the atomic scale, *Nature* **374**(6523), 607-616 (1995).

171. B. Bhushan and T. Kasai, A surface topography-independent friction measurement technique using torsional resonance mode in an AFM, *Nanotechnol.* **15**(8), 923-935 (2004).

172. B. Bhushan and V.N. Koinkar, Nanoindentation hardness measurements using atomic-force microscopy, *Appl. Phys. Lett.* **64**(13), 1653-1655 (1994).

173. M. Nishibori and K. Kinosita, Ultra-microhardness of vacuum-deposited films. 1. Ultra-microhardness tester, *Thin Solid Films* **48**(3), 325-331 (1978).

174. J.B. Pethica, in: *Ion Implantation in Metals*, edited by V. Ashworth, W. Grant and R. Proctor (Pergamon Press, Oxford, 1982), p. 147.

175. Y. Tsukamoto, H. Yamaguchi and M. Yanagisawa, Mechanical properties of thin films: measurements of ultramicroindentation hardness, young modulus and internal stress, *Thin Solid Films* **154**(1-2), 171-181 (1987).

176. W.C. Oliver, C.J. McHargue and S.J. Zinkle, Thin film characterization using a mechanical properties microprobe, *Thin Solid Films* **153**, 185-196 (1987).

177. W. Weiler, Hardness testing: a new method for economical and physically meaningful micro-hardness testing, *Brit. J. NonDestruct. Test.* **31**(5), 253-258 (1989).

178. A.C. Fischer-Cripps, Study of analysis methods of depth-sensing indentation test data for spherical indenters, *J. Mater. Res.* **16**(6), 1579-1584 (2001).

179. M.R. Vanlandingham, S.H. McKnight, G.R. Palmese, J.R. Elings, X. Huang, T.A. Bogetti, R.F. Eduljee and J.W. Gillespie, Nanoscale indentation of polymer systems using the atomic force microscope, *J. Adh.* **64**(1-4), 31-59 (1997).

180. H. Shulha, A. Kovalev, N. Myshkin and V.V. Tsukruk, Some aspects of AFM nanomechanical probing of surface polymer films, *Eur. Polym. J.* **40**(5), 949-956 (2004).

181. R.M. Overney, Nanotribological Studies on Polymers, *Trends Polym. Sci.* **3**(11), 359-364 (1995).

182. T. Kurokawa, J.P. Gong and Y. Osada, Substrate effect on topographical, elastic, and frictional properties of hydrogels, *Macromolecules* **35**(21), 8161-8166 (2002).

183. P. Eaton, F.F. Estarlich, R.J. Ewen, T.G. Nevell, J.R. Smith and J. Tsibouklis, Combined nanoin-dentation and adhesion force mapping using the atomic force microscope: Investigations of a filled polysiloxane coating, *Langmuir* **18**(25), 10011-10015 (2002).

184. W.R. Bowen, R.W. Lovitt and C.J. Wright, Application of atomic force microscopy to the study of micromechanical properties of biological materials, *Biotechnol. Lett.* **22**(11), 893-903 (2000).

185. B.J. Briscoe, L. Fiori and E. Pelillo, Nano-indentation of polymeric surfaces, *J. Phys. D* **31**(19), 2395-2405 (1998).

186. R. Rodriguez and I. Gutierrez, Correlation between nanoindentation and tensile properties Influ-ence of the indentation size effect, *Mater. Sci. Eng. A* **361**(1-2), 377-384 (2003).

187. N.I. Tymiak, D.E. Kramer, D.F. Bahr, T.J. Wyrobek and W.W. Gerberich, Plastic strain and strain gradients at very small indentation depths, *Acta Mater.* **49**(6), 1021-1034 (2001).

188. B. Bhushan, Nano- to microscale wear and mechanical characterization using scanning probe microscopy, *Wear* **251**, 1105-1123 (2001).

189. B. Bhushan and J.A. Ruan, Atomic-scale friction measurements using friction force microscopy 2: application to magnetic media, *J. Tribology Trans. ASME* **116**(2), 389-396 (1994).

190. X. Zhao and B. Bhushan, Material removal mechanism of single-crystal silicon on nanoscale and at ultralow loads, *Wear* **223**, 66-78 (1999).

191. A.V. Kulkarni and B. Bhushan, Nano/picoindentation measurements on single-crystal aluminum using modified atomic force microscopy, *Mater. Lett.* **29**(4-6), 221-227 (1996).

192. C. Plassard, E. Lesniewska, I. Pochard and A. Nonat, Investigation of the surface structure and elastic properties of calcium silicate hydrates at the nanoscale, *Ultramicroscopy* **100**(3-4), 331-338 (2004).

193. S. Gauffinet, E. Finot, R. Lesniewska and A. Nonat, Direct observation of the growth of calcium silicate hydrate on alite and silica surfaces by atomic force microscopy, *Comptes Rendus De L Academie Des Sciences Serie Ii Fascicule a-Sciences De La Terre Et Des Planetes* **327**(4), 231-236 (1998).

194. D. Viehland, J.F. Li, L.J. Yuan and Z.K. Xu, Mesostructure of calcium silicate hydrate (C-S-H) gels in Portland cement paste: Short-range ordering, nanocrystallinity, and local compositional order, *J. American Cera. Soc.* **79**(7), 1731-& (1996).

195. Z. Xu and D. Viehland, Observation of a mesostructure in calcium silicate hydrate gels of portland cement, *Phys. Rev. Lett.* **77**(5), 952-955 (1996).

196. E.L. Florin, V.T. Moy and H.E. Gaub, Adhesion forces between individual ligand-receptor pairs, *Science* **264**(5157), 415-417 (1994).

197. U. Dammer, O. Popescu, P. Wagner, D. Anselmetti, H.J. Guntherodt and G.N. Misevic, Binding Strength between cell-adhesion proteoglycans measured by atomic-force microscopy, *Science* **267**(5201), 1173-1175 (1995).

198. H. Mueller, H.J. Butt and E. Bamberg, Force measurements on myelin basic protein adsorbed to mica and lipid bilayer surfaces done with the atomic force microscope, *Biophysic. J.* **76**(2), 1072-1079 (1999).

199. J.H. Kinney, M. Balooch, G.M. Marshall and S.J. Marshall, A micromechanics model of the elastic properties of human dentine, *Arch. Oral Biol.* **44**(10), 813-822 (1999).

200. X. Li, H. Gao, W.A. Scrivens, D. Fei, X. Xu, M.A. Sutton, A.P. Reynolds and M.L. Myrick, Nanomechanical characterization of single-walled carbon nanotube reinforced epoxy composites, *Nanotechnology* **15**(11), 1416-1423 (2004).

201. P.C.P. Watts, W.K. Hsu, G.Z. Chen, D.J. Fray, H.W. Kroto and D.R.M. Walton, A low resistance boron-doped carbon nanotube-polystyrene composite, *J. Mater. Chem.* **11**(10), 2482-2488 (2001).

202. A.A. Mamedov, N.A. Kotov, M. Prato, D.M. Guldi, J.P. Wicksted and A. Hirsch, Molecular design of strong single-wall carbon nanotube/polyelectrolyte multilayer composites, *Nature Mater.* **1**(3), 190-194 (2002).

203. R. Haggenmueller, H.H. Gommans, A.G. Rinzler, J.E. Fischer and K.I. Winey, Aligned single-wall carbon nanotubes in composites by melt processing methods, *Chem. Phys. Lett.* **330**(3-4), 219-225 (2000).

204. H. Ni, X.D. Li, H.S. Gao and T.P. Nguyen, Nanoscale structural and. mechanical characterization of bamboo-like polymer/silicon nanocomposite films, *Nanotechnology* **16**(9), 1746-1753 (2005).

205. B. Bhushan, Micro/Nanotribology and Its Applications to Magnetic Storage Devices and Mems, *Tribology Intr.* **28**(2), 85-96 (1995).

206. L.L. Sohn and R.L. Willett, Fabrication of Nanostructures Using Atomic-Force-Microscope-Based Lithography, *Appl. Phys. Lett.* **67**(11), 1552-1554 (1995).

207. T.H. Fang, W.J. Chang and S.L. Tsai, Nanomechanical characterization of polymer using atomic force microscopy and nanoindentation, *Microelectronics J.* **36**(1), 55-59 (2005).

208. P.M. Campbell and E.S. Snow, Proximal probe-based fabrication of nanostructures, *Semiconductor Sci. Technol.* **11**(11), 1558-1562 (1996).

209. P.M. Campbell and E.S. Snow, Proximal probe-based fabrication of nanometer-scale devices, *Mater. Sci. Eng. B* **51**(1-3), 173-177 (1998).

210. J. Israelachvili, *Intermolecular and Surface Forces* (Academic, London, 1992).

211. M. Tello and R. Garcia, Nano-oxidation of silicon surfaces: comparison of noncontact and contact atomic-force microscopy methods, *Appl. Phys. Lett.* **79**(3), 424-426 (2001).

212. H. Sugimura, T. Uchida, N. Kitamura and H. Masuhara, Nanofabrication of titanium surface by tip-induced anodization in scanning tunneling microscopy, *Jap. J. Appl. Phys. Part 2-Lett.* **32**(4A), L553-L555 (1993).

213. K. Matsumoto, M. Ishii, K. Segawa, Y. Oka, B.J. Vartanian and J.S. Harris, Room temperature operation of a single electron transistor made by the scanning tunneling microscope nanooxidation process for the TiOx/Ti system, *Appl. Phys. Lett.* **68**(1), 34-36 (1996).

214. A. Inoue, T. Ishida, N. Choi, W. Mizutani and H. Tokumoto, Nanometer-scale patterning of self-assembled monolayer films on native silicon oxide, *Appl. Phys. Lett.* **73**(14), 1976-1978 (1998).

215. M.S. Bobji and B. Bhushan, In situ microscopic surface characterization studies of polymeric thin films during tensile deformation using atomic force microscopy, *J. Mater. Res.* **16**(3), 844-855 (2001).

216. H.J. Mamin and D. Rugar, Thermomechanical writing with an atomic force microscope Tip, *Appl. Phys. Lett.* **61**(8), 1003-1005 (1992).

217. V. Bouchiat and D. Esteve, Lift-off lithography using an atomic force microscope, *Appl. Phy. Lett.* **69**(20), 3098-3100 (1996).

218. J.B. Pethica and W.C. Oliver, Tip surface interactions in STM and AFM, *Phys. Scripta* **T19A**, 61-66 (1987).

219. E.Z. Luo, J.B. Xu, W. Wu, I.H. Wilson, B. Zhao and X. Yan, Identifying conducting phase from the insulating matrix in percolating metal-insulator nanocomposites by conducting atomic force microscopy, *Appl. Phys. A* **66**, S1171-S1174 (1998).

## QUESTIONS

1. How to eliminate probe artefacts when imaging nanoparticles/nanocomposites?
2. Describe how to characterize surface roughness of nano composite surfaces accurately with AFM?
3. List the different types of operational forces between the AFM probe and the sample surface as the probe approaches the surface
4. Describe how to identify the fiber-rubber interphase

# Index